Advances in Algal Research

Kaustubha Mohanty • S. Venkata Mohan
Himabindu Vurimindi
Satya Sundar Mohanty
Editors

Advances in Algal Research

Springer

Editors
Kaustubha Mohanty
Department of Chemical Engineering
Indian Institute of Technology Guwahati
Guwahati, Assam, India

S. Venkata Mohan
CSIR- National Environmental Engineering Research Institute
Nagpur, Maharashtra, India

Himabindu Vurimindi
Centre for Environment
Jawaharlal Nehru Technological University
Hyderabad, Telangana, India

Satya Sundar Mohanty
Division of Biotechnology
Karunya Institute of Technology & Sciences
Coimbatore, Tamil Nadu, India

ISBN 978-981-95-7233-5 ISBN 978-981-95-7234-2 (eBook)
https://doi.org/10.1007/978-981-95-7234-2

This Springer imprint is published by the registered company Springer Nature Singapore Pte Ltd.
The registered company address is: 152 Beach Road, #21-01/04 Gateway East, Singapore 189721, Singapore

Preface

Algae have increasingly come to the forefront of global scientific attention as the world seeks sustainable, resilient, and nature-based solutions to environmental and economic challenges. Over the past decade, algal biotechnology has evolved from a niche discipline into a multidisciplinary field intersecting biology, engineering, materials science, agriculture, and environmental restoration. This rapid expansion of research activity and technological innovation created a clear need for a comprehensive reference that captures the latest advancements while offering an integrated perspective on current trends, emerging applications, and future opportunities in algal science. *Advances in Algal Research* was conceived to fulfill this need.

The primary purpose of this volume is to bring together high-quality, contemporary work from researchers who are contributing to the cutting edge of algal biotechnology. The chapters included here reflect the breadth and maturity of the field—from sustainable agriculture and functional foods to engineered biofactories, circular bioeconomy strategies, environmental remediation, and next-generation green technologies. By compiling these diverse yet interconnected themes, this book serves as both a foundational resource and a forward-looking guide for understanding the transformative potential of algae in global sustainability initiatives.

This book is intended for a broad community of readers. Researchers, faculty, and graduate students will find detailed scientific insights and advanced methodologies relevant to ongoing investigations in algal biology, bioprocessing, and environmental applications. Industry professionals working in biotechnology, nutraceuticals, biofuels, bioplastics, and environmental management will benefit from the translational perspectives and technological developments highlighted across chapters. Policymakers, sustainability strategists, and innovators seeking nature-based solutions will discover practical approaches aligned with circular economy and carbon-neutrality goals. By addressing these varied audiences, the volume aims to support informed decision-making, inspire future research, and strengthen cross-sector collaborations.

The book is organized into three thematic sections designed to provide a logical and coherent narrative. The first section explores the role of algae in sustainable agriculture, food security, and human health, emphasizing functional foods, prebiotics, nutraceuticals, and phytohormone-based interventions. The second section focuses on algal biotechnology and bioengineering, showcasing advances in metabolic optimization, pigment production, high-value by-products, diatom engineering, and synergies contributing to carbon neutrality. The third combined section integrates environmental remediation, bioenergy, and emerging green technologies. This merged section brings together research on elimination of organic contaminants, macroalgal and microalgal remediation strategies, synthetic biology tools, bioplastic and polyhydroxyalkanoate production, pyrolysis routes, and forward-looking innovations such as algae-powered bioelectronics. By consolidating environmental and energy-related topics, the section offers a holistic perspective on algae as engines of ecological restoration, renewable energy development, and sustainable technological advancement.

The contributors to this volume were selected for their expertise, scholarly reputation, and pioneering work in algal research. Many are recognized leaders from premier academic institutions and research laboratories, whose contributions represent a wide spectrum of disciplinary perspectives and regional contexts. Their collective efforts ensure that the book reflects not only current scientific knowledge but also emerging opportunities that will shape the future of algal biotechnology.

We gratefully acknowledge all authors for their dedication, timely contributions, and commitment to scientific rigor. We also extend our appreciation to the reviewers and editorial team for their thoughtful evaluations and guidance throughout the compilation process. Finally, we wish to thank Springer Nature for their support in bringing this volume to publication and for their continued commitment to advancing scientific knowledge.

We hope that *Advances in Algal Research* will serve as a valuable reference for researchers, practitioners, and innovators worldwide, and that it will inspire further exploration into the remarkable potential of algae as a cornerstone of sustainable development.

Guwahati, Assam, India — Kaustubha Mohanty
Nagpur, Maharashtra, India — S. Venkata Mohan
Hyderabad, Telangana, India — Himabindu Vurimindi
Coimbatore, Tamil Nadu, India — Satya Sundar Mohanty

Contents

Editors and Contributors

About the Editors

Kaustubha Mohanty is a Professor in the Department of Chemical Engineering at Indian Institute of Technology Guwahati. He has established a distinguished research career in the areas of biofuels and biomass pyrolysis, microalgae-based biorefinery, membrane-based separations, and heterogeneous catalysis for environmental and energy applications. He has authored more than 250 peer-reviewed journal papers, 5 books, and over 35 book chapters, and his work has been widely cited, with an h-index of 63. Scholar GPS has ranked him at 36 (Lifetime All) and 1 (Prior 5 Years) under "Biomass." Also, 9 (Prior 5 Years) under "Pyrolysis" and 56 (Prior 5 Years) in "Chemical Engineering."

S. Venkata Mohan, a prominent scientist in environmental bioengineering and sustainable technologies, is currently serving as the Director of CSIR-National Environmental Engineering Research Institute (NEERI), Nagpur. Prior to it, he was Chief Scientist and INAE-SERB Abdul Kalam Technology Innovation National Fellow at the CSIR-Indian Institute of Chemical Technology (CSIR-IICT), Hyderabad. His illustrious career spans over two decades, marked by groundbreaking contributions to environmental sustainability, carbon capture, bio-based chemicals and fuels, advanced wastewater treatment, and the circular bioeconomy. His achievements include the prestigious Vigyan Shri award, Shanti Swarup Bhatnagar Prize for Science and Technology, INAE Abdul Kalam Fellowship, and numerous awards such as the Tata Innovation Fellowship and VASVIK Industrial Research Award. His prolific research output comprises 452 refereed journal articles, 15 patents, and a book, showcasing an impressive h-index of 102 with over 35,500 citations.

Himabindu Vurimindi is working as Professor, Centre for Environment, and Coordinator, Centre for Alternative Energy Options, CEN, UCEST, JNTU Hyderabad, India. She has done MSc in Environmental Chemistry and MTech in Environment Management and earned PhD in Chemistry from JNTU Hyderabad. Her research focuses on monitoring of air, water, and soil pollutants and their control technologies, biofuels production, energy materials, sequestration of CO_2 gases from industrial air emissions and hydrogen energy, and sustainable technologies for environmental remediation. She has a teaching experience of total 23 years. She has authored 149 publications in reputed national and international journals. She has successfully executed 23 R&D projects funded by state and central government. She holds a patent on biohydrogen production by dark fermentation from distillery spent wash.

Satya Sundar Mohanty is currently working as an Assistant Professor at the Division of Biotechnology in Karunya Institute of Technology and Sciences (Deemed-to-Be University), Coimbatore, Tamil Nadu, India. He holds a master's degree in Biotechnology and PhD in Chemical Engineering from the National Institute of Technology Rourkela (NIT Rourkela), where his doctoral research focused on sustainable bioprocesses, bioremediation, and optimization of engineered biological systems. Before joining Karunya, he was working as a Post-doctoral Fellow at Indian Institute of Technology Guwahati, Assam, India, prior to which he has a teaching experience in National Institute of Technology Raipur, Chhattisgarh. His research focuses on algal biotechnology, wastewater treatment, bioremediation of emerging contaminants, bioprocess engineering, and waste-to-value bioconversion. His work particularly emphasizes the use of microalgae, microbial systems, and engineered bioprocesses for nutrient removal, degradation of antibiotics and pesticides, and valorization of industrial and municipal waste streams.

Contributors

Makaranga Abdalah Omics of Algae Group, Industrial Biotechnology, International Centre for Genetic Engineering and Biotechnology (ICGEB), New Delhi, India

Department of Agriculture, Earth and Environmental Sciences (DAES), Faculty of Science, Mwenge Catholic University (MWECAU), Moshi, Kilimanjaro, Tanzania

Ruchika Agarwal Department of Environmental Science, School of Earth Sciences, Central University of Rajasthan, Kishangarh, Ajmer, India

Gopika Ajit Department of Biotechnology & Biochemical Engineering, Sree Buddha College of Engineering, Alappuzha, Kerala, India

Athulya Babu Department of Biotechnology, Sahrdaya College of Engineering and Technology, Kodakara, Thrissur, Kerala, India

Vanavil Balakrishnan Department of Biotechnology, Kalasalingam Academy of Research and Education, Krishnankoil, Tamil Nadu, India

Paramasivan Balasubramanian BioEnergy and Environmental (BEE) Lab, Department of Biotechnology and Medical Engineering, National Institute of Technology Rourkela, Rourkela, Odisha, India

Prathyusha Bogguru Department of Biotechnology, Kalasalingam Academy of Research and Education, Krishnankoil, Tamil Nadu, India

Rayanee Chaudhuri BioEnergy and Environmental (BEE) Lab, Department of Biotechnology and Medical Engineering, National Institute of Technology Rourkela, Rourkela, Odisha, India

Anjani Devi Chintagunta Department of Biotechnology, Vignan's Foundation for Science, Technology and Research, Vadlamudi, Andhra Pradesh, India

Gayatri Dave Department of Biological Sciences, P D Patel Institute of Applied Sciences, CHARUSAT University, Changa, Gujarat, India

Shailesh Dave Loyola Centre for Research and Development, Xavier Research Foundation, Navrangpura, Ahmedabad, Gujarat, India

S. Dhanush Division of Biotechnology, Karunya Institute of Technology and Sciences, Coimbatore, Tamil Nadu, India

Chinchu Elezebeth Department of Biotechnology & Biochemical Engineering, Sree Buddha College of Engineering, Alappuzha, Kerala, India

Shirin Fathima M Department of Biotechnology, Sahrdaya College of Engineering and Technology, Kodakara, Thrissur, Kerala, India

Vitasta Jad Biological Oceanography Division, CSIR- National Institute of Oceanography, Dona Paula, Goa, India

Academy of Scientific and Innovative Research (AcSIR), Ghaziabad, Uttar Pradesh, India

Brigita Jain Loyola Centre for Research and Development, Xavier Research Foundation, Navrangpura, Ahmedabad, Gujarat, India

Department of Life Science, University School of Science, Gujarat University, Ahmedabad, Gujarat, India

Adeline Juanita Division of Biotechnology, Karunya Institute of Technology and Sciences, Coimbatore, Tamil Nadu, India

Pannaga Pavan Jutur Omics of Algae Group, Industrial Biotechnology, International Centre for Genetic Engineering and Biotechnology (ICGEB), New Delhi, India

Ajay Kumar Kataria Department of Biological Sciences and Engineering, Netaji Subhas University of Technology, New Delhi, India

Jayshri Khadilkar Department of Biological Sciences and Biotechnology, Institute of Chemical Technology, Mumbai, Maharashtra, India

Boda Ravi Kiran Department of Biosciences and Bioengineering, Indian Institute of Technology (IIT) Roorkee, Roorkee, Uttarakhand, India

Poonam Kumari Bioengineering and Environmental Sciences Lab, Department of Energy and Environmental Engineering, CSIR-Indian Institute of Chemical Technology (CSIR-IICT), Hyderabad, India

Academy of Scientific and Innovative Research (AcSIR), Ghaziabad, India

Sushma Lavudya Centre for Environment, UCEST, JNTUH, Hyderabad, Telangana, India

Cathrine Sumathi Manohar Biological Oceanography Division, CSIR- National Institute of Oceanography, Dona Paula, Goa, India

Academy of Scientific and Innovative Research (AcSIR), Ghaziabad, Uttar Pradesh, India

Simran Maratha Department of Environmental Science, School of Earth Sciences, Central University of Rajasthan, Kishangarh, Ajmer, India

Sanjeev Mishra Biochemical Conversion Division, Sardar Swaran Singh National Institute of Bio-Energy, Kapurthala, India

S. Venkata Mohan Bioengineering and Environmental Sciences Lab, Department of Energy and Environmental Engineering, CSIR-Indian Institute of Chemical Technology (CSIR-IICT), Hyderabad, India

Academy of Scientific and Innovative Research (AcSIR), Ghaziabad, India

CSIR-National Environmental Engineering Research Institute (CSIR-NEERI), Nagpur, Maharashtra, India

Kaustubha Mohanty Department of Chemical Engineering, Indian Institute of Technology Guwahati, Guwahati, Assam, India

Satya Sundar Mohanty Division of Biotechnology, Karunya Institute of Technology and Sciences, Coimbatore, Tamil Nadu, India

Madhumita Mondal Department of Biotechnology, Visva Bharati, Santiniketan, West Bengal, India

Nikhil Sai Nachu Department of Biotechnology, Vignan's Foundation for Science, Technology and Research, Vadlamudi, Andhra Pradesh, India

Senthil Nagappan Biochemical Conversion Division, Sardar Swaran Singh National Institute of Bio-Energy, Kapurthala, India

Asha Arumugam Nesamma Omics of Algae Group, Industrial Biotechnology, International Centre for Genetic Engineering and Biotechnology (ICGEB), New Delhi, India

Vaishnavi Newaskar Department of Biological Sciences, P D Patel Institute of Applied Sciences, CHARUSAT University, Changa, Gujarat, India

Saravanan Palanivishwanath Biochemical Conversion Division, Sardar Swaran Singh National Institute of Bio-Energy, Kapurthala, India

Department of Chemical Engineering, Indian Institute of Technology Roorkee, Roorkee, Uttarakhand, India

Sana Parveen Department of Environmental Science, School of Earth Sciences, Central University of Rajasthan, Kishangarh, Ajmer, India

Shailesh Kumar Patidar Department of Environmental Science, School of Earth Sciences, Central University of Rajasthan, Kishangarh, Ajmer, India

Krishna Mohan Poluri Department of Biosciences and Bioengineering, Indian Institute of Technology (IIT) Roorkee, Roorkee, Uttarakhand, India

Centre for Transportation Systems, Indian Institute of Technology (IIT) Roorkee, Roorkee, Uttarakhand, India

Gunjan Prakash Department of Biological Sciences and Biotechnology, Institute of Chemical Technology, Mumbai, Maharashtra, India

Praveena P Department of Biotechnology, Sahrdaya College of Engineering and Technology, Kodakara, Thrissur, Kerala, India

Associate Professor, Department of Biotechnology, Sahrdaya College of Engineering and Technology, Thrissur, Kerala, India

Aanandhalakshmi Raghuraman Department of Biotechnology, Kalasalingam Academy of Research and Education, Krishnankoil, Tamil Nadu, India

Nikitha Ramalal Department of Biotechnology, Sahrdaya College of Engineering and Technology, Kodakara, Thrissur, Kerala, India

B Razia Sultana Department of Microbiology and Botany, School of Sciences, JAIN (Deemed-to-be University), Bengaluru, Karnataka, India

Ankita Roy Department of Biosciences and Bioengineering, Indian Institute of Technology (IIT) Roorkee, Roorkee, Uttarakhand, India

Chandrima Roy Centre for Environment, UCEST, JNTUH, Hyderabad, Telangana, India

Sudha Sahay Loyola Centre for Research and Development, Xavier Research Foundation, Navrangpura, Ahmedabad, Gujarat, India

Department of Microbiology and Biotechnology, University School of Science, Gujarat University, Ahmedabad, Gujarat, India

N. S. Sampath Kumar Department of Biotechnology, Vignan's Foundation for Science, Technology and Research, Vadlamudi, Andhra Pradesh, India

Meera Bai Santhakumari Department of Biotechnology, Kalasalingam Academy of Research and Education, Krishnankoil, Tamil Nadu, India

Department of Biotechnology & Biochemical Engineering, Sree Buddha College of Engineering, Alappuzha, Kerala, India

Soumya Sasmal Department of Biotechnology, Visva Bharati, Santiniketan, West Bengal, India

Pramita Sen Department of Chemical Engineering, Heritage Institute of Technology, Kolkata, West Bengal, India

Krushna Prasad Shadangi Department of Chemical Engineering, Veer Surendra Sai University of Technology Burla, Sambalpur, Odisha, India

Sk. Shaheda Department of Biotechnology, Vignan's Foundation for Science, Technology and Research, Vadlamudi, Andhra Pradesh, India

Kumkum Sharma Department of Environmental Science, School of Earth Sciences, Central University of Rajasthan, Kishangarh, Ajmer, India

Srividya Shivakumar School of Allied Healthcare and Sciences (SAHS), JAIN (Deemed-to-be University), Bengaluru, Karnataka, India

Aryan Singh Department of Environmental Science, School of Earth Sciences, Central University of Rajasthan, Kishangarh, Ajmer, India

Bharmjeet Singh Omics of Algae Group, Industrial Biotechnology, International Centre for Genetic Engineering and Biotechnology (ICGEB), New Delhi, India

Pooja Singh Department of Chemical Engineering, Indian Institute of Technology Guwahati, Guwahati, Assam, India

B. R. Sreelekhmy Division of Environmental Biotechnology, Department of biosciences, Rajagiri College of Social Sciences (Autonomous), Kalamassery, Kochi, India

Amrutha V. Sunil Department of Biotechnology, Sahrdaya College of Engineering and Technology, Kodakara, Thrissur, Kerala, India

Antony Suresh Loyola Centre for Research and Development, Xavier Research Foundation, Navrangpura, Ahmedabad, Gujarat, India

St. Xavier's College, Navrangpura, Ahmedabad, Gujarat, India

Jibu Thomas Division of Biotechnology, Karunya Institute of Technology and Sciences, Coimbatore, Tamil Nadu, India

Komal Tripathi Department of Chemical Engineering, Indian Institute of Technology Roorkee, Roorkee, Uttarakhand, India

Devanshi Trivedi Loyola Centre for Research and Development, Xavier Research Foundation, Navrangpura, Ahmedabad, Gujarat, India

Sabeela Beevi Ummalyma Division of Environmental Biotechnology, Department of biosciences, Rajagiri College of Social Sciences (Autonomous), Kalamassery, Kochi, India

N. Vinuba Division of Biotechnology, Karunya Institute of Technology and Sciences, Coimbatore, Tamil Nadu, India

Himabindu Vurimindi Centre for Environment, UCEST, JNTUH, Hyderabad, Telangana, India

Part I

Algae for Food, Health and Sustainable Agriculture

1 Algae-Based Solutions for Sustainable Agriculture and Food Security

Sabeela Beevi Ummalyma and B. R. Sreelekhmy

Abstract

Algae are photosynthetic micro and macroscopic plants that can fix CO_2 and produce biomass faster than their terrestrial counterparts. Algal biomass is enriched with a plethora of biochemicals such as phytohormones, vitamins, minerals, carbohydrates, proteins, and lipids. These molecules act as potent growth- and crop-productivity-increasing molecules and also stimulate soil health by improving microbial community, physicochemical properties, and soil texture. These improvements to crops and soils can aid climate-resilient agricultural practices, playing a critical role in ensuring security in food. Biostimulants, biofertilizers, and soil conditioners from algae can encourage sustainable organic agricultural activities by reducing the application of chemical fertilizers and pesticides in the field. This approach attracts a more resilient agricultural system that is better adapted to cope with the challenges of water and soil pollution by chemicals, pesticides, and herbicide inputs. This chapter's aim is algae-based solutions for sustainable agriculture to meet food security. It highlights the algae's role in crop improvements, productivity, yield, soil health in terms of fertility, and microbiome improvements.

Keywords

Biofertilizers · Biostimulant · Sustainable agriculture · Food security · Soil fertility · Crop improvement

S. B. Ummalyma (✉) · B. R. Sreelekhmy
Division of Environmental Biotechnology, Department of biosciences, Rajagiri College of Social Sciences (Autonomous), Kalamassery, Kochi, India
e-mail: sabeela@rajagiri.edu

K. Mohanty et al. (eds.), *Advances in Algal Research*,
https://doi.org/10.1007/978-981-95-7234-2_1

1.1 Introduction

It has been reported that the increase in global population is expected to escalate up to 10 billion by 2050. However, high pressure in agriculture to shift to more sustainable production of crops yields to increased production to feed the increasing populations (Minaoui et al. 2024). The overdosing of chemical fertilizers and pesticides to compensate for the nitrogen, potassium, phosphorus, and pest control affected the soil fertility and human health due to environmental pollution. In order to overcome these issues, farming is rooted towards more sustainable practices by applying soil conditioning by using plant- and algae-based sources. During chemical fertilizer applications, plants emit CO_2, along with H_2, NH_2, and CO, all of which contribute to global warming. Over use of chemical fertilizers can lead to negative consequences, including contamination of groundwater sources and the environment. This is primarily due to the xenobiotics accumulation and changes in the indigenous microbiome of the soil. In addition to this, it leads to acidification and soil hardening, and the loss of beneficial microorganisms' biodiversity leads to a deterioration in soil fertility, which finally affects the root strength and respiration (Sadvakasova et al. 2023; Kumar et al. 2022; Bisht and Chauhan 2020). Adapting a climate-resilient organic farming system using biofertilizers or biostimulants derived from natural photoautotrophic microorganisms is the solution to this problem. Biostimulants are natural organic molecules that enhance and promote plant growth and resilience. It has the capacity to address the bottleneck posed by soil degradation, food security, and climate change (Rouphael and Colla 2020). Reports indicate that phototrophic microorganisms such as microalgae significantly contribute to soil fertility, acting as natural biostimulants/biofertilizers. Biofertilizer applications utilize microalgae like *Chlorella* sp., *Nostoc* sp., and *Scenedesmus* sp. These algae have potent for producing a variety of plant growth-promoting hormones and amino acids. These phytochemicals from algae enhance crop productivity and yield and provide resistance to pathogens (Sadvakasova et al. 2023). Microalgae, including cyanobacteria, play a vital role in maintaining soil fertility by acting as natural biofertilizers. Microalgae mainly enhance phosphate accessibility in the soil by releasing organic acids, produce phytohormones such as auxins, gibberellic acids, cytokinins, abscisic acid, and ethylene, improving water-holding capacity and soil porosity due to their adhesive and gelatinous constituents, which reduce soil salinity, and controlling weeds (Sadvakasova et al. 2023). A recent study highlighted that microalgae and its extract have a positive impact on stimulating germination, length of shoot, and root biomass of crops such as peppers, tomatoes, amaranth, lettuce, and wheat (Minaoui et al. 2024; Munaro et al. 2024; Barone et al. 2018).

Seaweeds are macroalgae distributed in the coastline areas from tropics to polar regions. They have a vital role in marine ecosystems and act as a primary producer in the ecosystems. Seaweeds are also alternative resources for application as soil conditioners used since Roman times, currently used by countries like France, China, Spain, Britain, and Japan (Gul et al. 2024). Previously algae, including seaweeds, were used as a source of food and biofuels. Recently, due to overcoming the climatic change, algae-based biostimulants are given more preference for climate-resilient agriculture practices. Algae supply plenty of macro- and micronutrients into the soil, which slowly release nutrients and continuously for long durations, which enhance microbial diversity and nutrient acquisitions (Gul et al. 2024). Microalgae and seaweeds provide plant growth-promoting hormones, proteins, essential amino acids, vitamins, fatty acids, and carotenoids. This improves plants' productivity, crop yield, and pest resistance and acts as a protection against phytopathogens. In this chapter, we are providing the insights of different microalgal metabolites' role as biofertilizers, their role in plant growth promotions, and their application strategies. This chapter finally concludes with challenges and future directions.

1.2 Algal Metabolites as Biostimulant

Different algal metabolites can be extracted from microalgae and seaweeds that have a significant role as biostimulants. Major algal metabolites that act as biostimulants are phytohormones (auxins, cytokines, abscisic acids (ABA), gibberellins, ethylene, and jasmonic acids), vitamins, exopolysaccharides and polysaccharides such as heteropolysaccharides composed of glucose, galactose, and xylose, protein hydrolysate, humic acid- similar substances and amino acids (Sánchez-Quintero et al. 2023). Each metabolite that has an important role in crop yield and productivity is discussed in the following section.

1.3 Polysaccharides

Algal polysaccharides are important molecules that have wide applications in microbiology, food and cosmetics, dairy, and medical products. Due to their structural complexity, such as glycosidic linkages, molecular weight, degree of sulfation pattern, and branching configurations, their functional properties are influenced. For example, carrageenan has the potential to stabilize emulsions and has been extensively used in dairy, cosmetic, and medical products. Alginate has the capacity to form a gel, which has wide application in industries such as pharmaceutical, biotechnological, food, and beverages (Tariq et al. 2025). Agar is widely used in microbiological experiments as a vital solidifying agent for culture media. Recently algal polysaccharides have been explored as biostimulants due to challenges associated with agricultural sectors due to climate change and global warming issues. Algal polysaccharides have a multifaceted impact on soil stress tolerance, health, crop yield, and productivity. Alginate and ulvan have a potential role in improving retention of soil moisture, which is crucial for crops cultivated in arid and semi-arid regions which are prone to drought. This functional property can couple with their potential to enhance the uptake of water and nutrients, which supports plant productivity under below optimal conditions (Shukla et al. 2019). It has been reported that the antioxidant and antistressor potential of polysaccharides plays an important role in mitigating oxidative damages due to stress such as phytopathogen attacks, extreme temperatures, and salinity (Tariq et al. 2025). These qualities of algal polysaccharides make them a supreme biostimulant option for sustainable agricultural practices by optimizing crop yield and minimizing applications of synthetic fertilizers. There are several challenges associated with these algal metabolites in addition to their potential effect on soil fertility and crop yield. Polysaccharide composition varies based on algal species and geographical locations, and seasonal variations also limit the development of consistent product formulations (Tariq et al. 2025). Moreover, the high cost in the processing and extraction of polysaccharides remains a key bottleneck for their widespread application and adoption in agriculture practices. The sulfated polysaccharides (extracellular) obtained from *Chlamydomonas rheinense* and *Rhodella reticulata* proved inhibitory effects on various Gram-positive and negative bacteria such as *Streptococcus species, Bacillus subtilis, Neisseria mucosa, Bacillus cereus, Salmonella typhimurium, Escherichia coli, Streptococcus pyogenes,* and *Staphylococcus aureus* (Zhou et al. 2022; Roy et al. 2025). The report shows that it can act as an elicitor functional property that can trigger defense responses in plants. Oligosaccharides from this polysaccharide have the potential to function as elicitors by inducing the expression of certain genes involved in defense systems and enhancing antibacterial properties and protecting the crops from infections (Côté et al. 1998).

1.4 Peptides and Proteins

Peptides serve as building blocks for the synthesis of proteins. It has been reported that certain proteins are conjugated with phytohormones, which play a synchronized role in shaping plant growth response to adverse environmental conditions. The immune system of crop consists of several amino acid metabolic pathways in addition to the stress hormone response. Algal biostimulant contain several amino acids and proteins that can positively affect the defense mechanisms of crops and are widely used in sustainable agriculture (Roy et al. 2025; Bulgari et al. 2014). Protein hydrolysate of algal extract contains polypeptides and oligopeptides, free amino acids conjugated with fats, phytohormones, micronutrient elements, and polysaccharides (Colla et al. 2017). Industries including Siapton, Macro-sorb Foliar, Alfalfa, and Aminoplant from Spain, the United States, Italy, China, and India are produced similar substitute from animal processing waste or biomass either by alkali or acid, thermal, or the enzymatic hydrolysis processes (Calvo et al. 2014; Kapoore et al. 2021). Proteins and peptides have a direct effect on carbon and nitrogen metabolism modulations in plants by regulating enzymes involved in the tricarboxylic acid cycle (isocitrate dehydrogenase, malate dehydrogenase, and citrate synthase). It can also influence nutrient availability and nutrient uptake in plants through promoting soil respiration and growth of plant-growth-promoting bacteria in the plant microbiome (Bulgari et al. 2014). Amino acids in algal extract enhance synthesis of phytohormones, proteins, and pigments responsible for crop growth (Kapoore et al. 2021; Guedes et al. 2018). Some microalgae are reported to contain more than 40% of proteins and amino acids, such as 46.8% in *A. platensis*, 44.9% in

A. maxima, Chlorella sp. rich in 44.3%, *Neochloris* sp. (42%), and *C. saccharophila* (42.4%) respectively (Ummalyma et al. 2023; Hempel et al. 2012). This high protein content of algae makes it the ideal resource for biostimulant development. The lyophilized algal suspension from *A. platensis* to red beet through foliar spray methods enhances protein, sugars, free amino acids, and chlorophyll with improved hypocotyl growth (Mógor et al. 2018). The effect could be due to the presence of essential amino acids (arginine, aspartic acid, threonine, glutamic acid, isoleucine, and alanine, leucine) released by the algae. Foliar application of protein hydrolysate derived from *Arthrospira* sp., and *Scenedesmus* biomass (10 g/L) to the *Petunia x hybrida* showed an high number of flowers, the fresh and dry matter, and the dry weight of the root (Plaza et al. 2018).

1.5 Phytohormones

Phytohormones are naturally occurring small molecules acts as role in signaling pathways. Hormones such as auxins, cytokinins, and gibberellins play a critical role in regulating different developmental processes and signaling pathways associated with plant responses to abiotic and biotic stress. Phytohormones act as chemical messengers within crops and regulate responses to environmental stimuli. The presence of these hormones is well documented in algal species. Microalgae and seaweeds can store and secrete these hormones. It has been reported that these hormones have similar functions in their cells as those in their terrestrial counterpart (higher plant) (Kapoore et al. 2021). Algal extract contains plant hormones or hormone-like molecules that can modulate shoot and root growth and crop development. In addition to this, it helps plants respond against phytopathogens and abiotic stress. *Scenedesmus* biomass contains high concentrations of cytokinins and auxins (7–21 mg/g Dried Weight (DW)). Application of this algal extract as a biostimulant increases the shoot weight of lettuce by 13% (Álvarez-González et al. 2025). Bioactives extracted from cyanobacteria (*Nostoc commune* MBDU 101) characterization showed key plant hormones such as indole 3-butyric acid, Gibberellic acid, and auxin conjugates. This extract, applied in hydroponic growth of *Zeamays L.*, showed substantial enhancement in seedling length, leaf dimension, biomass, lateral root formations, and root colonization (Gayathri et al. 2025).

1.6 Humic Acids

Humic acids are soil organic matter, formed by the decomposition of biomass residues, or the metabolic activity of soil microorganisms on organic matter (Kapoore et al. 2021). Based on their solubility and molecular weight humic substances are categorized into fulvic acids, humins, and humic acids. Major humic substance currently used in agriculture is obtained from mineral deposit, peat, and coal. It has been used as a biostimulant and improved the growth of 16 tested plants (Calvo et al. 2014). Reports highlighted that humic substances are also produced in algal biomass. Brown filamentous algae such as *Pilayella littoralis* produced 10% humic acids in biomass (Ghabbour et al. 1994). Acid treatment of *C. reinhardtii* biomass resulted in brown solid precipitate of humic acids (Heilmann et al. 2011). *C. intermedia* isolated from Pony Lake of Antarctica were enriched with fulvic acids accumulated from the lake (Mao et al. 2007). The role of this humic substance in microalgae and plant growth needs to be explored more, and further investigation is needed. It has been reported that it has root growth stimulation and subsequently stimulates H^+ ATPase stimulated root growth in organically grown onion seedlings (Gemin et al. 2019).

1.7 Vitamins and Minerals

Microalgal biomass is enriched with vitamin B complexes and other essential vitamins. It has a potential role in the growth and productivity of crops along with the many metabolic processes of plants. It has been reported that the cyanobacterium, *Cylindrospermum muscicola* produced 1.2–1.5 μg g^{-1} of B12 vitamins (Venkataraman and Neelakantan 1967). Application of this cyanobacteria as biofertilizers showed increment in root growth of rice plants compared with control (Venkataraman and Neelakantan 1967). Certain marine algae produce commercially relevant vitamins such as vitamin E, nicotinic acids, biotin, and pantothenic acids. Spirulina sp. produces vitamin A and vitamin B12 in high parentage (Chakdar et al. 2012; Kapoore et al. 2021). However, this species has potential industrial value for the extraction of vitamins for biostimulant applications. In addition to this, marine algal biomass contains many essential minerals, which act as cofactors in several metabolisms of crop growth and productivity. Figure 1.1 represents the potential of microalgae metabolites as promising biostimulant for the growth and yield of crops.

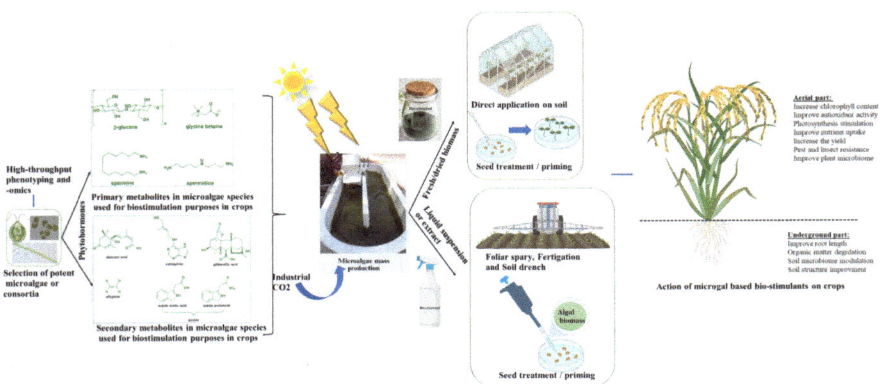

Fig. 1.1 Microalgae as a promising and eco-friendly biostimulant for the growth and yield of crops

1.8 Role of Microalgal Metabolites as in Plant Growth

Microalgae is a natural biostimulant/biofertilizer, which can improve the crop quality and yield. The production of a wide range bioactive molecules, such as N-fixing enzymes, soluble amino acids, polysaccharides, and phytohormones, by the microalgae plays a vital function in plant growth and soil health. Metabolites of microalgae function in improving soil fertility, resistance of plants against abiotic stress, nutrient uptake, and more (Berthon et al. 2021). Microalgae and its metabolites impart 3 different modes of growth-promoting properties, namely (1) biofertilizers—to enrich soil and plant rhizosphere, (2) biostimulants—to stimulate plant growth and development, and (3) biopesticides—to control the plant pathogens and to impart resistance to the plants (Lee and Ryu 2021).

1.9 Role of Microalgae as a Biofertilizer

1.9.1 Role in Soil Physiology

Cyanobacteria and microalgae act as natural biofertilizers in restoring soil physiology and their mobilization to crop plants. The primary role of microalgae, particularly cyanobacteria, is to fix atmospheric nitrogen, which in turn increases the soil's nitrogen content. Cyanobacteria acts as diazotrophs, which utilize the nitrogenase enzyme complex to produce ammonia from atmospheric nitrogens. Different microalgae impart different mechanisms for nitrogen fixation. Cyanobacteria adopted many strategies for nitrogen fixation, as the presence of light and oxygen inhibits the N fixation. In heterocystous cyanobacteria, like *Anabaena* sp., and *Nostoc* sp., N fixation is spatially separated from the photosynthesis process between heterocysts and vegetative cells (Maldener and Muro-Pastor 2010). *Lyngbya* sp., and *Cyanothece* sp., are non-heterocyst exhibit separation of photosynthesis temporarily and fix nitrogen in presence of light and dark phases, respectively (Misra 1999). *Gleothece* sp. adopt a combination of process (Compaoré and Stal 2010), and in *Trichodesmium* sp., reduction on genome and oxygenic photosynthesis lose are observed (Bergman et al. 2013). Among these, heterocyst-based N_2 fixation is more predominant. The major advantages are that (1) less chance of leaching compared to chemical-based N fertilizers (Mulbry et al. 2005) and (2) volatilization of NH_3 negligible with the application of dried algae biomass (de Siqueira Castro et al. 2017). Algae play a vital role in phosphate solubilization for plants. In soil, phosphate is available as bound to calcium or aluminum, depending on the pH. Cyanobacteria solubilize this unavailable phosphate either by releasing chelators that bind calcium ions or by organic acids release for solubilization (Alvarez et al. 2021). *Anabaena variabilis* and *Westelliopsis* sp. secrete phthalic acid, which facilitates solubilization of phosphate from tricalcium phosphate and phosphate rock (Yandigeri et al. 2011). Moreover, algae play a role in phosphorus mineralization by converting organic phosphorus into bioavailable forms. Microalgae produce phosphate hydrolyzing enzymes such as phosphodiesterases, alkaline phosphatases, phytases, 5′nucleotidases to mineralize phosphorous from organic phosphorus compounds, including phytates and phosphoesters (Markou et al. 2014). Algae such as *Tetraselmis suecia, Nannochloropsis gaditana, and Nanochloropsis oceania* showed high uptake mechanism for phosphate, where they accumulate it as polyphosphate granules intracellularly (Cañavate et al. 2017). This reservoir can be further potentially harnessed as a slow-release nutrient source for plant growth. In addition to macronutrients, microalgae and its metabolites enhance micronutrient availability in the soil. *Anabaena* sp. and *Synechococcus* sp. secrete siderophores, nitrogenous compounds that have a strong affinity towards Fe^{3+} ions. This helps in solubilization and mobilization of Fe into plants (Chakraborty et al. 2019; Årstøl and Hohmann-Marriott 2019). The extracellular polymeric substances secreted by certain microalgae play a crucial role in regulating soil water dynamics by absorption, stability, and retaining moisture (Potts 2001). This creates a microenvironment in the soil that helps to increase soil water holding capacity and reduce the trans-evaporation loss to the atmosphere and promote aggregation of soil particles (Mager and Thomas 2011; Watanabe 1984). The biological soil crust, a hybrid of algal species along with fungi and bacteria, plays a crucial role in improving soil fertility and crop productivity (Abinandan et al. 2025) in the harsh environment.

1.9.2 Soil Microbiome

Soil microbiomes play a vital role in plant growth and yield. However, prolonged application of chemical fertilizers disrupts the natural balance of the soil microbial community and in turn adversely affects the soil fertility and health. The loss of diversity in soil microbiome can disturb the ecosystem by altering the nutrient cycle and decreasing soil fertility (Hartmann and Six 2023). Thus, restoring the soil microbiome is crucial. Studies have shown that microalgal biomass contributes to the restoration by releasing polymeric macromolecules that improve the moisture retention, stimulate the proliferation of beneficial microbes, and support their colonization. In addition, phytohormones such as Indole—3 acetic acid (IAA) released by microalgae have been reported to augment the soil microbiome quality (Karthikeyan et al. 2009; Lee and Ryu 2021). Besides enriching the soil with essential nutrients, algae also foster microbial diversity and are themselves a natural component of the soil microbiome. Many plants engage in symbiotic interaction with algae; for example, Nostoc has been documented to colonize in the plants' roots and enhance nitrogen fixation (Lee and Ryu 2021).

1.10 Role of Microalgae as a Biostimulant

In addition to the nutrients such as nitrogen, phosphorous and potassium (NPK), and calcium (Ca), microalgae biomass is a reservoir of diverse plant growth-promoting compounds. These include primary metabolites such as betaines, amino acids, peptides, vitamins, polyamines, and polysaccharides, as well as secondary metabolites in the form of phytohormones such as ethylene, auxins, abscisic acid, and more (Stirk et al. 2013a, b; Tate et al. 2013). Collectively these biomolecules influence multiple physiological and developmental processes in plants, including regulation of cell division, stem and root elongation, seed germination, differentiation, and more (Tarakhovskaya et al. 2007).

1.10.1 Seed Germination

Gibberellic acid is a key phytohormone that plays an important role in the seed propagation process. It stimulates the production of hydrolytic enzymes, such as α-amylase and protease, which break stored food reserves in the seed. The result facilitates the mobilization of nutrients from the endosperm, providing energy and resources for the growth of the seedlings (Shah et al. 2023). Interestingly, microalgae is a reservoir of such phytohormones (Stirk et al. 2013a, b). Microalgae could produce exogenous (Prieto et al. 2011) and endogenous phytohormones (Stirk et al. 2002, 2014). Thus, either extracts of microalgae/microalgae biomass or spent medium after algal growth can be used for the formulation of biostimulants. The concentrations of phytohormones in these products largely depend on the algal strains used and the extraction techniques applied to recover active compounds. Navarro-López et al. reported that application of non-hydrolyzed algal biomass of *Scenedesmus obliquus has* the ideal concentration of amino acids and/or polyamines to stimulate the seed germination of watercress seeds. This results in achievement of a higher germination index of 139.1%, *which* is 40% higher than the control. While disrupted or hydrolyzed biomass could increase the concentration of amino acids and polyamine compounds, which leads to an inhibition of the seeds' growth (Navarro-López et al. 2020). Such concentration-dependent inhibitory responses have also been reported in *Triticum aestivum* var. Pusa Gold (Kumar and Sahoo 2011). In contrast, Van Do et al. reported that water extract of *Chlorella sorokiniana* contains a two-fold higher concentration of gibberellins. This accelerated seed germination of tomato and rice plants in 1 and 2 days, respectively (Do et al. 2020). The aqueous extract of *Acutodesmus* dimorphus culture has been used as a seed primer for Roma tomato to trigger the seed germination in 2 days earlier than the untreated one (Garcia-Gonzalez and Sommerfeld 2016). Alling et al. compared the biostimulant potential of two Nordic microalgae strains, namely, *Chlorella vulgaris* and *Scenedesmus obliquus*, on tomato and barley seeds. The results indicate that seeds treated with *C. vulgaris* especially with intact cells or culture supernatant, achieved up to a 25% higher germination percentage than seeds treated with *S. obliquus* (Alling et al. 2023). Ferreira et al. explored the biostimulant effects of *Tetradesmus obliquus*, *Chlorella vulgaris,* and *Synechocystis* sp. on the germination of cucumber, wheat, tomato, barley, watercress, and soybean seeds. The study showed that microalgae produce diverse bioactive compounds with species-specific effects on different crops. *C. vulgaris* was most effective for tomato, cucumber, wheat seeds, while *T. obliquus* had the greatest impact on barley and watercress seeds. For soybean, it stands out differently since it needs significantly high nitrogen demand for seed germination. *Synechocystis* sp. is potent for fixing atmospheric nitrogen and can work symbiotically for seed germination and improved the germination index by 65% (Ferreira et al. 2021). All these results evidence the fact that microalgae can synthesize a remarkable diversity of bioactive molecules; however, the mode of action of these compounds in seed germination is different for different seeds and still not well explored.

1.10.2 Root and Shoot Development

Beyond seed germination, the diverse bioactive compounds in microalgae support overall plant growth and resilience to environmental stress. They also stimulate key physiological processes such as cell division, elongation, fresh weight, and shoot and root development. Additionally, microalgae contribute to improving chlorophyll content, enhancing heat tolerance, and strengthening plant performance and stress adaptability. Microalgae such as *Chlorella* sp. (Tate et al. 2013), *Coenochloris* sp., *Scenedesmus* sp. (Navarro-López et al. 2020), *and Chlorococcum* sp. (Chen et al. 2024) have been reported to produce auxins, IAA and Indole-3-acetamide (Kapoore et al. 2021; Parmar et al. 2023). IAA derived from microalgae has the major role in root formation and elongation. Foliar spray of *Scenedesmus* sp. and *A. dimorphus* extract on petunia and tomato plants resulted in increased dry root weight, enhancing plant height, number of flowers, and plant branches (Supraja et al. 2020; Plaza et al. 2018; Garcia-Gonzalez and Sommerfeld 2016). Similarly, colonization of endophytic cyanobacteria, namely, *Nostoc* sp., associated with rice and wheat plants promotes the plant growth and yield due to the production of equal volume of cytokinins and IAA (Hussain et al. 2015). Here auxins act as a signaling molecule during cyanobacteria plant interactions, especially for colonization of root (Lee and Ryu 2021). In addition, these bioactive compounds trigger the accumulation of metabolites in primary and secondary biosynthetic pathways and thus enhance plant growth. This in turn facilitates the plant's ability to explore soil surface and access deep soil layers and thus improves the nutrients efficiency. In addition to phytohormones, microalgae also produce other bioactive compounds like amines, brassinosteroids, salicylic acid, and more, which facilitate stress response mechanisms in plants. For example, polysaccharides extracted from the microalgae play a major role in the enhancement of plant growth and also alleviate the effects of saline stress on plants, and brassinosteroids take part in enzymatic and non-enzymatic defense responses. The cytokinins from microalgae are also reported to play a role in abiotic stress tolerance in plants. For example, extracts of *Nannochloropsis* sp. highlighted alleviated water and nutrients stress in tomato plants due to its free radical scavenging functional properties (Oancea et al. 2013; Lu et al. 2014; Fahad et al. 2015). It is also highlighted that application of microalgae-based biostimulants not only improves the yield but also triggers the biochemical processes that improve the quality, such as improvement in the sugar and carotenoid content of the fruits (Coppens et al. 2016), essential oil content of leaves in peppermint (Shariatmadari et al. 2015), and more.

1.11 Role of Microalgae as Biopesticides

Many of the microalgae biomass and its extracts have been reported to have biopesticidal properties. It acts either by inhibiting or restricting the growth of the plant pathogens (Costa et al. 2019a, b). Many algal groups, including cyanobacteria produce antimicrobial compounds such as norharmane, diterpenoids, β-ionone, and more. These compounds inhibit the growth of the pathogens either by inhibiting key metabolic enzymes such as monoamine oxidase, indoleamine 2,3-dioxygenase, and nitric oxide synthase (Volk 2008) or by disrupting cellular processes like protein synthesis and cell division (Volk and Furkert 2006; Abdel-Hafez et al. 2015). Some strains produce hydrolytic enzymes such as β-1,4-glucanase, chitosanase, and β-1,3-glucanase that hydrolyze the cell wall of fungal pathogens (Senousy et al. 2022; Gupta et al. 2011; Natarajan et al. 2013). Sometimes, microalgal extracts trigger plant defense responses by enhancing the activity of hydrolytic enzymes, polyphenol oxidase, and peroxidase (Roberti et al. 2015; Rachidi et al. 2021; Ghazal et al. 2025), thereby strengthening systemic resistance against infection (Singh et al. 2011; Babu et al. 2015). Some antimicrobial compounds like siderophores restrict the pathogen proliferation by competitively binding to essential nutrients such as iron (Fe) and copper (Cu) and depriving the pathogens of the nutrients (Lee and Ryu 2021). Moreover, microalgae possess nematicidal properties by the secretion of neurotoxins such as anatoxin-A and nodularins (Mankiewicz et al. 2003). Some secondary metabolites produced by microalgae, such as methoxyphenyl, hexamethyl phenol, and flavonoids inhibit nematode growth by reducing gall formation in roots of *Cucumis sativus* plants (El-Eslamboly et al. 2019). Additionally, *Anabaena* sp. and *Scytonema* sp. produce peptide toxins that act as repellents to insects, and some microalgae extracts impart sterility in insects by suppressing oviposition. All these reports evidence the fact that the success of a biostimulant treatment depends on many factors, including the microalgal and cyanobacteria species used, the method of extraction and processing of algae biomass, the plant species or seeds on which it is applied, and, moreover, the concentration and method of application. Table 1.1 represents the biostimulant and biopesticide potential of algal biomass and its mode of action.

1.12 Microalgae Biostimulant Application Methods

Different application methods are adopted for biostimulant applications to improve crop productivity, yield, and soil fertility. Application methods include soil drench/drip fertigation, foliar spray, and seed treatments (Colla and Rouphael 2020). The seed treatment process consists of seed priming, seed dipping, and seed coating (Parmar et al. 2023). Soil drench methods are more efficient than foliar spray for application of *Chlorella vulgaris* extract as a biostimulant in tomato plants (Suchithra et al. 2022). Biostimulant prepared from algal consortium containing microalgae and cyanobacteria (*Chlorella* sp., *Scenedesmus* sp., *Spirulina* sp., *and Synechocystis* sp.) application in seed treatment is more effective than foliar spray in

Table 1.1 Biostimulant/biopesticides property of microalgae and its mode of action

Microalgae/ consortia	Host plant	Biostimulant activity	Mode of action	References
Nostoc spp.	*Solanum lycopersicum L.*	Fungicidal activity against Sclerotinia sclerotiorum	Increase nutrient uptake High adventitious root formation	Biondi et al. (2004)
Anabaena, Oscillatoria, Nostoc, Nodularia, Calothrix spp.	NR	Fungicidal activity against Alternaria alternate, Botrysis cinerea, Rhizopus stolonifer, and more	Reduce fungal colonization Enhance plant defense mechanisms	Kim et al. (2006)
Anabaena, Oscillatoria, Nostoc, Nodularia, Calothrix spp.	NR	Fungicidal activity by enhancing plant defense mechanism	Enhance plant defense responses by the production of hydrolytic enzymes such as peroxidase, chitinase, and more	Kim and Kim (2008)
Spirulina platensis and amphora cofeaeformis	*Cucumis sativus*	Nematicidal activity	Secondary metabolites of microalgae such as asetamide and phenolic compounds act on the nematicides	El-Eslamboly et al. (2019)
Spirulina platensis	*Moringa oleifera* LAM	Fungicidal activity against Fusarium oxysporum, F. Solani	Control damping off, root rot, and wilt diseases. Increase shoot weight and growth of plant	Imara et al. (2021)
Anabaena variabilis and Westelliopsis spp.	NR	Facilitates P solubilization and acts as biofertilizer	Secretes pthalic acid, which facilitates P solubilization from tricalcium phosphate and phosphate rock	Yandigeri et al. (2011)
Nanochloropsis oceania, Nannochloropsis gaditana, Tetraselmis suecia	NR	Luxury P uptake and act as biofertilizer	Accumulate P as polyphosphate granules intracellularly	Cañavate et al. (2017)
Cyanobacteria	NR	Convert organic phosphorus into bioavailable forms	P hydrolyzing enzymes such as alkaline phosphatases, phosphodiesteratses,	Markou et al. (2014)

(continued)

Table 1.1 (continued)

Microalgae/ consortia	Host plant	Biostimulant activity	Mode of action	References
			5′nucleotidases, and phytases to internalize P from phytates and phosphoesters	
Anabaena spp., *and Synechococcus* spp.	NR	Contribute to the solubilization and mobilization of Fe into plants	Secrete siderophores, nitrogenous compounds, which have strong affinity for Fe3+ ions	Chakraborty et al. (2019)
Scenedesmus obliquus	Watercress seeds	Accelerate seed germination	Non-hydrolyzed algal biomass have the optimal concentration of amino acids and/or polyamines and stimulate the seed germination	Navarro-López et al. (2020)
Chlorella sorokiniana	Tomato and rice	Accelerated seed germination	Water extract of *Chlorella sorokiniana* contains two-fold higher concentration of gibberellins	Van Do et al. (2020)
Acutodesmus dimorphus	Roma tomato	Trigger the seed germination 2 days earlier	Applied as seed primer	Garcia-Gonzalez and Sommerfeld (2016)
Nostoc spp.	Rice and wheat	Promote plant growth and yield	Produce phytohormones	Hussain et al. (2015)
Aphanothece sp.	Tomato	Increase the chlorophyll content and root growth	Production of secondary metabolites Cytokinin promote chlorophyll content and Giberelin brassinosteroids trigger root elongation	Mutale-Joan et al. (2020)
Dunaliella salina, Chlorella ellipsoidea, Aphanothece sp., and *Arthrospira maxima*	Tomato	Plant growth and tolerance under saline condition	Proline accumulation was significantly enhancing the plant growth and adjusting the osmosis and ion homeostasis	Mutale-Joan et al. (2021)

growth rate of tomato plants (Supraja et al. 2020). Another report showed that application of polysaccharides from algal biomass in soil presented significant influence on the increase in leaf numbers, nodes, and shoot weight of tomato plants (Rachidi et al. 2021). The study highlighted that foliar applications of extract from *Chlorella* sp. in fenugreek plants showed enhancement in plant heights, branch numbers, and leaves. It can be improved by mixing biostimulant with some wetting agent, such as ethoxylated isodecyl alcohol, which will improve the efficiency of foliar spray application to the leaves (Ördög et al. 2021).

Algae-based biostimulant products are available in the market either as a liquid or in powder form. Usually biostimulant/biofertilizers from algae are applied as spraying or irrigation methods. Algal extract can be mixed with other agricultural products such as fertilizers. Similarly, soil drench methods are also adopted for biofertilizer applications (Mazzafera 2025). Li et al. (2022) showed that application of foliar or irrigation in the soil in 180 field trials resulted in high productivity and yield of crops. This improved yield is due to the changes of microbiota in the soil. It could be due to symbiotic interactions, enhancing the nitrogen fixation by the bacterial community. Authors hypothesized that better yield is probably due to the border positioning of the products within the crops' phenological cycles, protecting the plants from stress such as high salinity, temperature, and light intensity. Another study showed the success of foliar spray depends on the time of application, preferably morning, when the ambient temperature is lower and stomatal opening is greater, which improves overall efficiency (Hong et al. 2021). The methods of application of biostimulants depend on the crop-specific requirement, such as micronutrient enrichment, disease control, and nutrient supplementation (Sánchez-Quintero et al. 2023).

1.13 Algae Based Approaches for Food Security

The global food demand is projected to rise between 2019 and 2050 by about 50 to 60% (Falcon et al. 2022). Traditional agricultural practices are no longer relied upon to tackle the growing challenges of food security (Mok et al. 2020). In addition, pressing issues such as land and soil degradation, water scarcity, biodiversity loss, and climate change are intensifying the need for sustainable alternatives to conventional food and feed sources. This method is essential to meet the increasing global demand while ensuring nutrition security and promoting healthier lives for future generations. Rather than the algal advancement in sustainable agricultural practices, algae ranging from unicellular microalgae to multicellular macroalgae also has the potential to be used as an alternate sustainable food and feed. For centuries, algae have been an integral part of human diets across diverse cultures, and the global interest is increasing due to their numerous functional and health benefits (Wells et al. 2017). *Spirulina* sp., a blue-green alga, and *Chlorella vulgaris* are among the most extensively studied algal species due to their exceptionally high protein content (40–60% and 40–55%, respectively, of dry biomass), similar to the protein level of meat and soybeans (Singh et al. 2023; Wang et al. 2024; Yaishana et al. 2025). In addition, it is a rich source of essential omega-3 fatty acids such as eicosapentaenoic acid (EPA), docosahexaenoic acid, and more (Costa et al. 2019a, b). *Nannochloropsis* sp. are particularly promising due to their high lipid content (up to 68%), primarily composed of polyunsaturated fatty acids (Georgianna and Mayfield 2012). *Dunaliella salina* is commercially exploited for β-carotene production, which can constitute around 10% of its dry biomass (Lamers et al. 2008). Similarly, *H. lacustris* is recognized as the richest natural source of astaxanthin, a red carotenoid pigment renowned for its exceptional antioxidant properties (Li et al. 2020). Most of the macroalgae are exceptionally high in mineral content, including essential and trace minerals such as iron, zinc, selenium, and more, which are essential for human nutrition in the range of 8–40% compared to terrestrial plants and animals' products (Puri et al. 2022). Microalgae in the diet can provide nearly all essential vitamins required by humans, frequently exceeding those found in terrestrial plants. Microalgae are particularly rich in vitamins D and K, which are typically scarce in terrestrial plants and fruits. Moreover, microalgae provide an abundant supply of several B vitamins, including vitamins B12, B9, and B6, along with other essential vitamins such as A, C, and E (Puri et al. 2022; Del Mondo et al. 2020). Microalgae are also rich in healthy dietary fibers such as non-starch polysaccharides such as pectin, cellulose, hemicellulose, and more. Particularly, *Euglena* contains B-1,3-glucan paramylon (Barros de Medeiros et al. 2022).

Conventional agriculture practices demand hectares of arable land, large quantities of fresh water, and favorable environmental conditions to sustain crop production and meet the needs of the growing population globally. However, climate change and land and soil degradation due to industrialization and natural calamities pose significant challenges to the dependence on domesticated plants as reliable food sources. In contrast, high biomass yield per unit area and capacity to grow on non-arable land using saline and even non-potable water make the algae a more suitable alternative source for food and feed (Torres-Tiji et al. 2020). Thus, algae cultivation does not directly compete with traditional agriculture for land and also stimulates economic activities in areas unsuitable for conventional farming (Trentacoste et al. 2015). Furthermore, algae cultivation can alleviate the strains on freshwater resources by utilizing low-quality water sources such as brackish and seawater that are unfit for agriculture and domestic use (Chen et al. 2015). All these attributes highlight the potential of algae to adequately meet human nutritional needs and contribute significantly to future food security.

All these algae-based approaches, including biopesticides, biostimulants, and food, contribute significantly to improving food availability, accessibility, and long-term food security through multiple channels. Algae as biofertilizers enhance the soil fertility by supplying essential nutrients and organic matter for plant growth and productivity. It also maintains and restores essential soil microbiomes, which directly or indirectly influence plant productivity and improve food availability. At the same time, algae as biostimulants provide bioactive compounds such as phytohormones, amino acids, and more, which stimulate plant growth and nutrient uptake. Moreover, it improves the plant's resilience to environmental stresses such as drought and salinity and ensures sustained yield. Compared to other chemical fertilizers and stimulants, algae-based product inputs long-term soil health and ecological balance, thereby ensuring the accessibility and sustainability of food systems. Moreover, algae can directly serve as a sustainable and alternative food resource to meet human nutritional needs and contribute significantly to food security.

1.14 Market Feasibility and Regulations

The demand for biostimulants and biofertilizers is expected to rise in the future due to the growing emphasis on sustainable and eco-friendly agricultural practices. It is reported that the market value of biofertilizer is increasing and is expected to reach USD 4.47 million by 2025 from USD 1106 million in 2016 (Soni et al. 2022; Behera et al. 2021). Microalgae-based biofertilizers/biostimulants are gaining attention. Recently, several companies, including Hindustan Bioenergy Limited and Soley Biotech, have begun exploring microalgal strains such as *Chlorella* sp., *Scenedesmus* sp., *and Nannochloropsis* sp. as a promising source of bioactive compounds for biostimulant applications (Behera et al. 2021). The formulation of biofertilizers/biostimulants using cyanobacteria, mostly *Anabaena* sp., *Nostoc* sp., and more, stands out and is generally used in countries such as the United States, the Philippines, Egypt, India, and China (Faheed and Abd-El Fattah 2008). Currently, companies in countries like Jordan, India, and the United States are producing and commercializing algae-based liquid biofertilizers with successful application reported in fruits, vegetables, and field crops (Allouzi et al. 2022). The global market forecast size for the biofertilizers industry for the year 2029 is expected to be USD 5377.8 million, with a CAGR rate of 15.2% from 2024 to 2029 (Biofertilizers Market—Innovative Markets Forecast (2029) Report). The Asia–Pacific region held the largest participation in the world market for biofertilizer industries in the year 2024 (USD 1323.8 million—56.1% share), followed by North America, with a share of 23.0% (USD 542.7 million), and in third place is the European continent, with USD 418.2 million (17.7% share) (Mordor Intelligence 2023). Additionally, countries like China and India are also emerging leaders in this industry. For example, in the Indian scenario, the biostimulants market was valued at USD 71.23 million in 2017 and is projected to grow at a CAGR of 16.49%, reaching approximately a total value of USD 180.95 million in 2024. On the other hand, the revenue generated by the global microalgae-based biofertilizer market in 2021 exceeded USD 11.33 million and is expected to generate profits worth USD 22.28 million in 2028. Some of the companies driving the global microalgae fertilizers market include Algaenergy (Spain), Algatech Ltd. (USA), Algenol Biotech (USA), Rejuve Biotech Co. Ltd. (China), Roquette Frères (France), AstaReal (Sweden), Australian Spirulina (Australia), and Archimede Ricerche SRL (Italy). Additionally, governments and policymakers around the world are promoting the development of biofertilizers/biostimulants. However, large-scale commercial production of biofertilizers based on microalgae continues to present challenges, with marketing limitations due to the short shelf life of bioformulated products, lack of customer knowledge, adequate storage, and poor consumer awareness, among others.

1.15 Future Directions and Conclusion

Major challenges in algae-based biostimulants are wide acceptance and affordable price. Focus should be needed on reducing the price to the production cost reduction in algal biomass generation and downstream processing. In this case, further research is needed in the identification and cultivation optimization of commercial algal strains or consortia; metabolomic analysis for the identification of novel bioactive molecules has biostimulant potential; the coupling of a circular bioeconomy approach using waste streams for algal biomass generation and a biorefinery approach; and low-cost extraction process identification for the bioactives extractions from algae for back-to-soil applications. Studies are required for the deep understanding of economic feasibility with the use of life cycle analysis and techno-economic assessment tools in real applications before commercialization. Validation of lab-to-field application of biostimulant is needed before commercialization. Biostimulant application methods are either foliar/seed/soil or carrier-based, and time duration and dosages require validation for each biostimulant development for different crops. More exploration and understanding of the mode of action of bioactive present in biostimulants on plant performance, synergistic interactions of biostimulants with crop variety and rhizosphere microbiome modulations, and colonization of the microbial community and improving soil fertility..

Algae-based biostimulants have immense potential and scope for sustainable, climate-resilient agriculture. Algal biostimulants are enriched with polysaccharides, amino acids, proteins, vitamins, minerals, and phytohormones. In addition, these microalgae offer ecosystem services such as CO_2 sequestration and bioremediation of waste streams. Data showed that algal biostimulants have improved soil microbiome, soil texture, and overall fertility, along with crop yield and productivity. For long-term commercial production of algal biostimulant is required to resolve field-based challenges. Currently few companies are selling algal-based biostimulants for organic farming, and in the near future, algae-based biostimulants will be affordable to all organic farmers for ensuring sustainable agriculture and food security.

Acknowledgments The authors gratefully acknowledge the Management of Rajagiri College of Social Sciences, Kochi, India, for providing infrastructure and administrative assistance. SBU acknowledges Rajagiri College of Social Sciences for the Major Research Project grant (RCSS/IQAC/BB-S60/2025/160) for financial support. We have used Grammarly and QuillBot tools for language corrections and fluency.

References

Abdel-Hafez SI, Abo-Elyousr KA, Abdel-Rahim IR (2015) Fungicidal activity of extracellular products of cyanobacteria against *alternaria porri*. Eur J Phycol 50(2):239–245. https://doi.org/10.1080/09670262.2015.1028105

Abinandan S, Kuppan P, Venkateswarlu K, Mukunthan K, Megharaj M (2025) Pyroligneous acid as a multifunctional biostimulant enhances microalgal growth and soil beneficial metabolites for sustainable agriculture. World J Microbiol Biotechnol 41(8):1–7. https://doi.org/10.1080/07388551.2019.1654972

Alling T, Funk C, Gentili FG (2023) Nordic microalgae produce biostimulant for the germination of tomato and barley seeds. Sci Rep 13(1):3509

Allouzi MMA, Allouzi SMA, Keng ZX, Supramaniam CV, Singh A, Chong S (2022) Liquid biofertilizers as a sustainable solution for agriculture. Heliyon 8(1):E12609

Alvarez AL, Weyers SL, Goemann HM, Peyton BM, Gardner RD (2021) Microalgae, soil and plants: a critical review of microalgae as renewable resources for agriculture. Algal Res 54:102200. https://doi.org/10.1016/j.algal.2021.102200

Álvarez-González A, Serrano L, Gorchs G, Uggetti E (2025) Exploring the biostimulant potential of Scenedesmus sp. grown in wastewater: impacts on plant growth and photosynthetic activity of lettuce. Chemosphere 382:144494. https://doi.org/10.1016/j.chemosphere.2025.144494

Årstøl E, Hohmann-Marriott MF (2019) Cyanobacterial siderophores—physiology, structure, biosynthesis, and applications. Mar Drugs 17(5):281. https://doi.org/10.3390/md17050281

Babu S, Bidyarani N, Chopra P, Monga D, Kumar R, Prasanna R, Kranthi S, Saxena AK (2015) Evaluating microbe-plant interactions and varietal differences for enhancing biocontrol efficacy in root rot disease challenged cotton crop. Eur J Plant Pathol 142(2):345–362. https://doi.org/10.1007/s10658-015-0619-6

Barone V, Baglieri A, Stevanato P, Broccanello C, Bertoldo G, Bertaggia M, Cagnin M, Pizzeghello D, Moliterni VM, Mandolino G, Fornasier F (2018) Root morphological and molecular responses induced by microalgae extracts in sugar beet (Beta vulgaris L.). J Appl Phycol 30(2):1061–1071

Barros de Medeiros VP, da Costa WK, da Silva RT, Pimentel TC, Magnani M (2022) Microalgae as source of functional ingredients in new-generation foods: challenges, technological effects, biological activity, and regulatory issues. Crit Rev Food Sci Nutr 62(18):4929–4950

Behera B, Supraja KV, Paramasivan B (2021) Integrated microalgal biorefinery for the production and application of biostimulants in circular bioeconomy. Bioresour Technol 339:125588

Bergman B, Sandh G, Lin S, Larsson J, Carpenter EJ (2013) Trichodesmium–a widespread marine cyanobacterium with unusual nitrogen fixation properties. FEMS Microbiol Rev 37(3):286–302. https://doi.org/10.1111/j.1574-6976.2012.00352.x

Berthon JY, Michel T, Wauquier A, Joly P, Gerbore J, Filaire E (2021) Seaweed and microalgae as major actors of blue biotechnology to achieve plant stimulation and pest and pathogen biocontrol–a review of the latest advances and future prospects. J Agric Sci 159(7–8):523–534. https://doi.org/10.1017/S0021859621000885

Biondi N, Piccardi R, Margheri MC, Rodolfi L, Smith GD, Tredici MR (2004) Evaluation of Nostoc strain ATCC 53789 as a potential source of natural pesticides. Appl Environ Microbiol 70(6):3313–3320. https://doi.org/10.1128/AEM.70.6.3313-3320.2004

Bisht N, Chauhan PS (2020) Excessive and disproportionate use of chemicals cause soil contamination and nutritional stress. In: Soil contamination. IntechOpen. https://doi.org/10.5772/intechopen.94593

Bulgari R, Cocetta G, Trivellini A, Vernieri P, Ferrante A (2014) Biostimulants and crop responses: a review. Biol Agric Hortic 17(31):1–17

Calvo P, Nelson L, Kloepper JW (2014) Agricultural uses of plant biostimulants. Plant Soil 383(1):3–41. https://doi.org/10.1007/s11104-014-2131-8

Cañavate JP, Armada I, Hachero-Cruzado I (2017) Interspecific variability in phosphorus-induced lipid remodelling among marine eukaryotic phytoplankton. New Phytol 213(2):700–713. https://doi.org/10.1111/nph.14179
Chakdar H, Jadhav SD, Dhar DW, Pabbi S (2012) Potential applications of blue green algae. J Sci Ind Res 71:13–20
Chakraborty S, Verma E, Singh SS (2019) Cyanobacterial siderophores: ecological and biotechnological significance. In: Cyanobacteria. Academic Press, pp 383–397
Chen G, Zhao L, Qi Y (2015) Enhancing the productivity of microalgae cultivated in wastewater toward biofuel production: a critical review. Appl Energy 137:282–291
Chen X, Ye X, Yu X, Zhao J, Song M, Yin D, Yu J (2024) Analysis of the regulatory mechanism of exogenous IAA-mediated tryptophan accumulation and synthesis of endogenous IAA in Chlorococcum humicola. Chemosphere 354:141633
Colla G, Rouphael Y (2020) Microalgae: new source of plant biostimulants. Agronomy 10(9):1240
Colla G, Hoagland L, Ruzzi M, Cardarelli M, Bonini P, Canaguier R, Rouphael Y (2017) Biostimulant action of protein hydrolysates: unraveling their effects on plant physiology and microbiome. Front Plant Sci 8:2202. https://doi.org/10.3389/fpls.2017.02202
Compaoré J, Stal LJ (2010) Oxygen and the light–dark cycle of nitrogenase activity in two unicellular cyanobacteria. Environ Microbiol 12(1):54–62. https://doi.org/10.1111/j.1462-2920.2009.02034.x
Coppens J, Grunert O, Van Den Hende S, Vanhoutte I, Boon N, Haesaert G, De Gelder L (2016) The use of microalgae as a high-value organic slow-release fertilizer results in tomatoes with increased carotenoid and sugar levels. J Appl Phycol 28(4):2367–2377
Costa JA, Freitas BC, Cruz CG, Silveira J, Morais MG (2019a) Potential of microalgae as biopesticides to contribute to sustainable agriculture and environmental development. J Environ Sci Health B 54(5):366–375. https://doi.org/10.1080/03601234.2019.1571366
Costa JA, Freitas BC, Rosa GM, Moraes L, Morais MG, Mitchell BG (2019b) Operational and economic aspects of Spirulina-based biorefinery. Bioresour Technol 292:121946
Côté F, Ham KS, Hahn MG, Bergmann CW (1998) Oligosaccharide elicitors in host-pathogen interactions: generation, perception, and signal transduction. Plant-Microbe Interact 29:385–432
de Siqueira CJ, Calijuri ML, Assemany PP, Cecon PR, de Assis IR, Ribeiro VJ (2017) Microalgae biofilm in soil: greenhouse gas emissions, ammonia volatilization and plant growth. Science Total Environ 574:1640–1648
Del Mondo A, Smerilli A, Sané E, Sansone C, Brunet C (2020) Challenging microalgal vitamins for human health. Microb Cell Factories 19(1):201
Do TC, Tran DT, Le TG, Nguyen QT (2020) Characterization of endogenous auxins and gibberellins produced by Chlorella sorokiniana TH01 under phototrophic and mixtrophic cultivation modes toward applications in microalgal biorefinery and crop research. J Chem 2020(1):4910621
El-Eslamboly AA, Abd El-Wanis MM, Amin AW (2019) Algal application as a biological control method of root-knot nematode Meloidogyne incognita on cucumber under protected culture conditions and its impact on yield and fruit quality. Egypt J Biol Pest Control 29(1):18. https://doi.org/10.1186/s41938-019-0122-z
Fahad S, Hussain S, Matloob A, Khan FA, Khaliq A, Saud S et al (2015) Phytohormones and plant responses to salinity stress: a review. Plant Growth Regul 75(2):391–404. https://doi.org/10.1007/s10725-014-0013-y
Faheed FA, Abd-El Fattah Z (2008) Effect of Chlorella vulgaris as bio-fertilizer on growth parameters and metabolic aspects of lettuce plant. J Agri Soc Sci 4:165–169
Falcon WP, Naylor RL, Shankar ND (2022) Rethinking global food demand for 2050. Popul Dev Rev 48(4):921–957
Ferreira A, Melkonyan L, Carapinha S, Ribeiro B, Figueiredo D, Avetisova G, Gouveia L (2021) Biostimulant and biopesticide potential of microalgae growing in piggery wastewater. Environ Adv 4:100062

Garcia-Gonzalez J, Sommerfeld M (2016) Biofertilizer and biostimulant properties of the microalga Acutodesmus dimorphus. J Appl Phycol 28(2):1051–1061. https://doi.org/10.1007/s10811-015-0625-2
Gayathri M, Shunmugam S, Muralitharan G (2025) Unveiling the potency of rice field cyanobacterial strains for plant growth promotion and root colonization by deciphering the molecular elicitors. Algal Res:104131. https://doi.org/10.1016/j.algal.2025.104131
Gemin LG, Mógor ÁF, Amatussi JD, Mógor G (2019) Microalgae associated to humic acid as a novel biostimulant improving onion growth and yield. Sci Hortic 256:108560. https://doi.org/10.1016/j.scienta.2019.108560
Georgianna DR, Mayfield SP (2012) Exploiting diversity and synthetic biology for the production of algal biofuels. Nature 488(7411):329–335
Ghabbour EA, Khairy AH, Cheney DP, Gross V, Davies G, Gilbert TR, Zhang X (1994) Isolation of humic acid from the brown alga Pilayella littoralis. J Appl Phycol 6(5):459–468. https://doi.org/10.1007/BF02182399
Ghazal MF, Hammad A, Hewait HM, Salem G (2025) Cyanobacteria for sustainable management of Lupine Root rot. Egypt J Soil Sci 65(1):203–218
Guedes WA, Araújo RHCR, Rocha JLA, de Lima JF, Dias GA, de Oliveira AMF, de Lima RF, Oliveira LM (2018) Production of papaya seedlings using Spirulina platensis as a biostimulant applied on leaf and root. J Exp Agric Int 28:1–9
Gul S, Nawaz MF, Yousaf MT, Rashid MH, Adnan MY, Tausif S, Javed A, Abideen Z, El-Keblawy A (2024) Brown macro-seaweeds derived agro-biostimulant for Zea mays farming in saline conditions: growth enhancement and optimum biochemical and ion feedback. Biocatal Agric Biotechnol 57:103105. https://doi.org/10.1016/j.bcab.2024.103105
Gupta V, Natarajan C, Kumar K, Prasanna R (2011) Identification and characterization of endoglucanases for fungicidal activity in Anabaena laxa (Cyanobacteria). J Appl Phycol 23(1):73–81. https://doi.org/10.1007/s10811-010-9539-1
Hartmann M, Six J (2023) Soil structure and microbiome functions in agroecosystems. Nat Rev Earth Environ 4(1):4–18
Heilmann SM, Jader LR, Harned LA, Sadowsky MJ, Schendel FJ, Lefebvre PA, Von Keitz MG, Valentas KJ (2011) Hydrothermal carbonization of microalgae II. Fatty acid, char, and algal nutrient products. Appl Energy 88(10):3286–3290. https://doi.org/10.1016/j.apenergy.2010.12.041
Hempel N, Petrick I, Behrendt F (2012) Biomass productivity and productivity of fatty acids and amino acids of microalgae strains as key characteristics of suitability for biodiesel production. J Appl Phycol 24:1407–1418
Hong J, Wang C, Wagner DC, Gardea-Torresdey JL, He F, Rico CM (2021) Foliar application of nanoparticles: mechanisms of absorption, transfer, and multiple impacts. Environ Sci Nano 8(5): 1196–1210. https://doi.org/10.1039/D0EN01129K
Hussain A, Shah ST, Rahman H, Irshad M, Iqbal A (2015) Effect of IAA on in vitro growth and colonization of Nostoc in plant roots. Front Plant Sci 6:46. https://doi.org/10.3389/fpls.2015.00046
Imara DA, Zaky WH, Ghebrial EW (2021) Performance of soil type, cyanobacterium spirulina platensis and biofertilizers on controlling damping-off, root rot and wilt diseases of moringa (Moringa oleifera lam.) in Egypt. Egyptian. J Phytopathol 49(2):10–28. https://doi.org/10.21608/ejp.2021.79498.1036
Kapoore RV, Wood EE, Llewellyn CA (2021) Algae biostimulants: a critical look at microalgal biostimulants for sustainable agricultural practices. Biotechnol Adv 49:107754. https://doi.org/10.1016/j.biotechadv.2021.107754
Karthikeyan N, Prasanna R, Sood A, Jaiswal P, Nayak S, Kaushik BD (2009) Physiological characterization and electron microscopic investigation of cyanobacteria associated with wheat rhizosphere. Folia Microbiol 54(1):43–51. https://doi.org/10.1007/s12223-009-0007-807754

Kim J, Kim JD (2008) Inhibitory effect of algal extracts on mycelial growth of the tomato-wilt pathogen, Fusarium oxysporum f sp lycopersici. Mycobiology 36(4):242–248. https://doi.org/10.4489/MYCO.2008.36.4.242

Kim KH, Tsao R, Yang R, Cui SW (2006) Phenolic acid profiles and antioxidant activities of wheat bran extracts and the effect of hydrolysis conditions. Food Chem 95(3):466–473. https://doi.org/10.1016/j.foodchem.2005.01.032

Kumar G, Sahoo D (2011) Effect of seaweed liquid extract on growth and yield of Triticum aestivum var. Pusa Gold. J Appl Phycol 23(2):251–255

Kumar S, Sindhu SS, Kumar R (2022) Biofertilizers: an ecofriendly technology for nutrient recycling and environmental sustainability. Curr Res Microb Sci 3:100094. https://doi.org/10.1016/j.crmicr.2021.100094

Lamers PP, Janssen M, De Vos RC, Bino RJ, Wijffels RH (2008) Exploring and exploiting carotenoid accumulation in Dunaliella salina for cell-factory applications. Trends Biotechnol 26(11):631–638

Lee SM, Ryu CM (2021) Algae as new kids in the beneficial plant microbiome. Front Plant Sci 12: 599742. https://doi.org/10.3389/fpls.2021.599742

Li X, Wang X, Duan C, Yi S, Gao Z, Xiao C, Agathos SN, Wang G, Li J (2020) Biotechnological production of astaxanthin from the microalga *Haematococcus pluvialis*. Biotechnol Adv 43: 107602

Li J, Van Gerrewey T, Geelen D (2022) A meta-analysis of biostimulant yield effectiveness in field trials. Front Plant Sci 13:836702. https://doi.org/10.3389/fpls.2022.836702

Lu Y, Tarkowská D, Turečková V, Luo T, Xin Y, Li J, Wang Q, Jiao N, Strnad M, Xu J (2014) Antagonistic roles of abscisic acid and cytokinin during response to nitrogen depletion in oleaginous microalga N annochloropsis oceanica expand the evolutionary breadth of phytohormone function. Plant J 80(1):52–68. https://doi.org/10.1111/tpj.12615

Mager DM, Thomas AD (2011) Extracellular polysaccharides from cyanobacterial soil crusts: a review of their role in dryland soil processes. J Arid Environ 75(2):91–97

Maldener I, Muro-Pastor AM (2010) Cyanobacterial heterocysts. eLS. https://doi.org/10.1002/9780470015902.a0000306.pub2

Mankiewicz J, Tarczyńska M, Walter Z, Zalewski M (2003) Natural toxins from cyanobacteria (blue-green algae). Acta Biol Cracoviensia 45:19–20

Mao J, Cory RM, McKnight DM, Schmidt-Rohr K (2007) Characterization of a nitrogen-rich fulvic acid and its precursor algae from solid state NMR. Org Geochem 38:1277–1292

Markou G, Vandamme D, Muylaert K (2014) Microalgal and cyanobacterial cultivation: the supply of nutrients. Water Res 65:186–202. https://doi.org/10.1016/j.watres.2014.07.025

Mazzafera P (2025) Microalgae as biostimulant of plant growth and yield. In: Microalgae and one health. Academic Press, pp 571–577. https://doi.org/10.1016/B978-0-443-22080-7.00007-7

Minaoui F, Hakkoum Z, Chabili A, Douma M, Mouhri K, Loudiki M (2024) Biostimulant effect of green soil microalgae Chlorella vulgaris suspensions on germination and growth of wheat (Triticum aestivum var. Achtar) and soil fertility. Algal Res 82:103655

Misra HS (1999) Oxygen implication in the diazotrophic growth of Plectonema boryanum in dark–light cycles. Plant Sci 143(2):135–142. https://doi.org/10.1016/S0168-9452(99)00047-3

Mógor ÁF, de Oliveira Amatussi J, Mógor G, de Lara GB (2018) Bioactivity of cyanobacterial biomass related to amino acids induces growth and metabolic changes on seedlings and yield gains of organic red beet. Am J Plant Sci 9(5):966–978. https://doi.org/10.4236/ajps.2018.95074

Mok WK, Tan YX, Chen WN (2020) Technology innovations for food security in Singapore: a case study of future food systems for an increasingly natural resource-scarce world. Trends Food Sci Technol 102:155–168

Mordor Intelligence (2023) E-Commerce Market in Oman—Trends, Growth, Covid-19 Impact, and Forecasts (2023-2028)

Mulbry W, Westhead EK, Pizarro C, Sikora L (2005) Recycling of manure nutrients: use of algal biomass from dairy manure treatment as a slow release fertilizer. Bioresour Technol 96(4): 451–458

Munaro D, Mazo CH, Bauer CM, da Silva Gomes L, Teodoro EB, Mazzarino L, da Rocha Veleirinho MB, e Silva SM, Maraschin M (2024) A novel biostimulant from chitosan nanoparticles and microalgae-based protein hydrolysate: improving crop performance in tomato. Sci Hortic 323:112491

Mutale-Joan C, Redouane B, Najib E, Yassine K, Lyamlouli K, Laila S, Zeroual Y, Hicham EA (2020) Screening of microalgae liquid extracts for their bio stimulant properties on plant growth, nutrient uptake and metabolite profile of Solanum lycopersicum L. Sci Rep 10(1):2820

Mutale-Joan C, Rachidi F, Mohamed HA, Mernissi NE, Aasfar A, Barakate M, Mohammed D, Sbabou L, Arroussi HE (2021) Microalgae-cyanobacteria–based biostimulant effect on salinity tolerance mechanisms, nutrient uptake, and tomato plant growth under salt stress. J Appl Phycol 33(6):3779–3795

Natarajan C, Gupta V, Kumar K, Prasanna R (2013) Molecular characterization of a fungicidal endoglucanase from the cyanobacterium Calothrix elenkinii. Biochem Genet 51(9):766–779. https://doi.org/10.1007/s10528-013-9605-x

Navarro-López E, Ruíz-Nieto A, Ferreira A, Acién FG, Gouveia L (2020) Biostimulant potential of Scenedesmus obliquus grown in brewery wastewater. Molecules 25(3):664

Oancea F, Velea S, Fătu V, Mincea C, Ilie L (2013) Micro-algae-based plant biostimulant and its effect on water stressed tomato plants. Rom J Plant Prot 6(2013):104–117. https://doi.org/10.1016/j.algal.2015.07.006

Ördög V, Stirk WA, Takács G, Pőthe P, Illés A, Bojtor C, Széles A, Tóth B, van Staden J, Nagy J (2021) Plant biostimulating effects of the cyanobacterium Nostoc piscinale on maize (Zea mays L.) in field experiments. South Afri J Botany 140:153–160

Parmar P, Kumar R, Neha Y, Srivatsan V (2023) Microalgae as next generation plant growth additives: functions, applications, challenges and circular bioeconomy based solutions. Front Plant Sci 14:1073546

Plaza BM, Gómez-Serrano C, Acién-Fernández FG, Jimenez-Becker S (2018) Effect of microalgae hydrolysate foliar application (Arthrospira platensis and Scenedesmus sp.) on Petunia x hybrida growth. J Appl Phycol 30(4):2359–2365. https://doi.org/10.1007/s10811-018-1427-0

Potts M (2001) Desiccation tolerance: a simple process? Trends Microbiol 9(11):553–559

Prieto CRE, Cordoba CNM, Montenegro JAM, González-Mariño GE (2011) Production of indole-3-acetic acid in the culture medium of microalga Scenedesmus obliquus (UTEX 393). J Braz Chem Soc 22:2355–2361

Puri M, Gupta A, McKinnon RA, Abraham RE (2022) Marine bioactives: from energy to nutrition. Trends Biotechnol 40(3):271–280

Rachidi F, Benhima R, Kasmi Y, Sbabou L, Arroussi HE (2021) Evaluation of microalgae polysaccharides as biostimulants of tomato plant defense using metabolomics and biochemical approaches. Sci Rep 11(1):930

Roberti R, Galletti S, Burzi PL, Righini H, Cetrullo S, Perez CJ (2015) Induction of defence responses in zucchini (Cucurbita pepo) by Anabaena sp. water extract. Biol Control 82:61–68. https://doi.org/10.1016/j.biocontrol.2014.12.006

Rouphael Y, Colla G (2020) Biostimulants in agriculture. Front Plant Sci 11:40. https://doi.org/10.3389/fpls.2020.00040

Roy A, Girish TR, Nori SS, Narayan SS, Bhattacharjee S, Hallan V (2025) Marine algae oligosaccharides: eco-friendly elicitors for sustainable agriculture. Algal Res:104093. https://doi.org/10.1016/j.algal.2025.104093

Sadvakasova AK, Kossalbayev BD, Bauenova MO, Balouch H, Leong YK, Zayadan BK, Huang Z, Alharby HF, Tomo T, Chang JS, Allakhverdiev SI (2023) Microalgae as a key tool in achieving carbon neutrality for bioproduct production. Algal Res 72:103096

Sánchez-Quintero Á, Fernandes SC, Beigbeder JB (2023) Overview of microalgae and cyanobacteria-based biostimulants produced from wastewater and CO_2 streams towards sustainable agriculture: a review. Microbiol Res 277:127505

Senousy HH, El-Sheekh MM, Saber AA, Khairy HM, Said HA, Alhoqail WA, Abu-Elsaoud AM (2022) Biochemical analyses of ten cyanobacterial and microalgal strains isolated from egyptian habitats, and screening for their potential against some selected phytopathogenic fungal strains. Agronomy 12(6):1340

Shah SH, Islam S, Mohammad F, Siddiqui MH (2023) Gibberellic acid: a versatile regulator of plant growth, development and stress responses. J Plant Growth Regul 42(12):7352–7373

Shariatmadari Z, Riahi H, Abdi M, Hashtroudi MS, Ghassempour AR (2015) Impact of cyanobacterial extracts on the growth and oil content of the medicinal plant Mentha piperita L. J Appl Phycol 27(6):2279–2287

Shukla PS, Mantin EG, Adil M, Bajpai S, Critchley AT, Prithiviraj B (2019) Ascophyllum nodosum-based biostimulants: sustainable applications in agriculture for the stimulation of plant growth, stress tolerance, and disease management. Front Plant Sci 10:655

Singh DP, Prabha R, Yandigeri MS, Arora DK (2011) Cyanobacteria-mediated phenylpropanoids and phytohormones in rice (Oryza sativa) enhance plant growth and stress tolerance. Antonie Van Leeuwenhoek 100(4):557–568

Singh SK, Shukla L, Yadav N, Singh PK, Singh SM, Yadav MK, Kaushalendra, Kumar A (2023) Spirulina: from ancient food to innovative super nutrition of the future and its market scenario as a source of nutraceutical. In: Cyanobacterial Biotechnology in the 21st century. Springer Nature Singapore, Singapore, pp 51–61

Soni G, Kumar S, Raj R, Mahto RV, Mangla SK, Mittal ML, Lim WM (2022) A decision-making framework for Industry 4.0 technology implementation: The case of FinTech and sustainable supply chain finance for SMEs. Technol Forecast Social Chang 180:121686

Stirk WA, Ördög V, Van Staden J, Jäger K (2002) Cytokinin-and auxin-like activity in Cyanophyta and microalgae. J Appl Phycol 14(3):215–221

Stirk WA, Bálint P, Tarkowská D, Novák O, Strnad M, Ördög V, Van Staden J (2013a) Hormone profiles in microalgae: gibberellins and brassinosteroids. Plant Physiol Biochem 70:348–353

Stirk WA, Ördög V, Novák O, Rolčík J, Strnad M, Bálint P, van Staden J (2013b) Auxin and cytokinin relationships in 24 microalgal strains1. J Phycol 49(3):459–467

Stirk WA, Bálint P, Tarkowská D, Novak O, Maróti G, Ljung K, Turečková V, Strnad M, Ördög V, Van Staden J (2014) Effect of light on growth and endogenous hormones in Chlorella minutissima (Trebouxiophyceae). Plant Physiol Biochem 79:66–76

Suchithra MR, Muniswami DM, Suba Sri M, Usha R, Ahamed Rasheeq A, Antrose Preethi B, Dineshkumar R (2022) Effectiveness of green microalgae as biostimulants and biofertilizer through foliar spray and soil drench method for tomato cultivation. South Afri J Botany 146: 740–750

Supraja KV, Behera B, Balasubramanian P (2020) Efficacy of microalgal extracts as biostimulants through seed treatment and foliar spray for tomato cultivation. Indust Crops Prod 151:112453

Tarakhovskaya ER, Maslov YI, Shishova MF (2007) Phytohormones in algae. Russ J Plant Physiol 54(2):163–170

Tate JJ, Gutierrez-Wing MT, Rusch KA, Benton MG (2013) The effects of plant growth substances and mixed cultures on growth and metabolite production of green algae Chlorella sp.: a review. J Plant Growth Regul 32(2):417–428

Tariq TB, Karishma UM, Mubeen-ur-Rehman (2025) The potential of seaweed-derived polysaccharides as sustainable biostimulants in agriculture. Int J Biol Macromol 298:140009

Torres-Tiji Y, Fields FJ, Mayfield SP (2020) Microalgae as a future food source. Biotechnol Adv 41:107536

Trentacoste EM, Martinez AM, Zenk T (2015) The place of algae in agriculture: policies for algal biomass production. Photosynth Res 123(3):305–315

Ummalyma SB, Chiang A, Herojit N, Arumugam M (2023) Sustainable microalgal cultivation in poultry slaughter house wastewater for biorefinery products and pollutant removal. Bioresour Technol 374:128790

Venkataraman GS, Neelakantan S (1967) Effect of the cellular constituents of the nitrogen-fixing blue-green alga, Cylindrospermum muscicola, on the root growth of rice plants. J Gen Appl Microbiol 13:53–61

Volk RB (2008) Screening of microalgae for species excreting norharmane, a manifold biologically active indole alkaloid. Microbiol Res 163(3):307–313. https://doi.org/10.1016/j.micres.2006.06.002

Volk RB, Furkert FH (2006) Antialgal, antibacterial and antifungal activity of two metabolites produced and excreted by cyanobacteria during growth. Microbiol Res 161(2):180–186. https://doi.org/10.1016/j.micres.2005.08.005

Wang N, Xiang Y, Ma Y, Zhang P, Zhou X, Zhu H, Zhang Z, Li Z, Xiao X, He M, Mehmood MA (2024) Fermented Spirulina platensis alleviated DSS-induced ulcerative colitis by regulating gut microbiota and MyD88/TLR4 signaling pathway. Asian J Agric Biol 2024(01)

Watanabe I (1984) 10. Nitrogen Fixation in Wetland Rice Field

Wells ML, Potin P, Craigie JS, Raven JA, Merchant SS, Helliwell KE, Smith AG, Camire ME, Brawley SH (2017) Algae as nutritional and functional food sources: revisiting our understanding. J Appl Phycol 29(2):949–982

Yaishana T, Bora S, Afzia N, Ghosh T (2025) A comprehensive review on spirulina composition, properties, extraction of protein using ultrasonication and its food application. Food Sci Biotechnol:1–2

Yandigeri MS, Yadav AK, Srinivasan R, Kashyap S, Pabbi S (2011) Studies on mineral phosphate solubilization by cyanobacteria Westiellopsis and Anabaena. Microbiology 80(4):558–565. https://doi.org/10.1134/S0026261711040229

Zhou Y, Chen X, Chen T, Chen X (2022) A review of the antibacterial activity and mechanisms of plant polysaccharides. Trends Food Sci Technol 123:264–280. https://doi.org/10.1016/j.tifs.2022.03.020

Cyanobacteria in Agriculture: Harnessing Phytohormones for Crop Growth and Soil Fertility

2

Nikhil Sai Nachu, N. S. Sampath Kumar, and Anjani Devi Chintagunta

Abstract

Cyanobacteria play a pivotal role in sustainable agriculture by enhancing soil health and promoting plant growth. These photosynthetic microorganisms contribute to key ecological processes, such as nitrogen fixation and soil stabilization, which are vital for agricultural productivity. This chapter explores the diverse functions of cyanobacteria in agricultural ecosystems, focusing on their ability to produce bioactive metabolites, particularly phytohormones, which influence plant development. The biosynthetic pathways of phytohormones in cyanobacteria and higher plants are discussed, highlighting the potential for manipulating these pathways to augment plant growth. Experimental evidence demonstrating the role of auxins in promoting seed germination and root development is presented, alongside the application of cyanobacterial consortia as biofertilizers for sustainable agriculture. This chapter concludes by discussing future directions for optimizing cyanobacterial applications, emphasizing the need for further research into their genetic manipulation and large-scale implementation in agricultural systems to enhance food security and environmental sustainability.

Keywords

Agriculture · Auxin · Cyanobacteria · Phytohormones · Climate-smart agriculture · Environmental sustainability

N. S. Nachu · N. S. Sampath Kumar · A. D. Chintagunta (✉)
Department of Biotechnology, Vignan's Foundation for Science, Technology and Research, Vadlamudi, Andhra Pradesh, India

K. Mohanty et al. (eds.), *Advances in Algal Research*,
https://doi.org/10.1007/978-981-95-7234-2_2

2.1 Introduction

Global food demand continues to rise while farmland is shrinking and climate change intensifies environmental pressures. To maintain productivity, farmers often depend on synthetic fertilizers, pesticides, and intensive tillage. Although these practices can temporarily boost yields, they also degrade soils, contaminate water, reduce biodiversity, and release greenhouse gases. At the same time, crops are increasingly vulnerable to drought, salinity, extreme temperatures, and pest attacks, all of which disrupt physiological processes and reduce yields (Shahwar et al. 2023). To address these challenges, sustainable agriculture is shifting toward microbial bio-inoculants that improve nutrient uptake, enhance water-use efficiency, and stress tolerance while minimizing ecological impact. Many microbes also act as natural biopesticides by suppressing pathogens through competitive exclusion, antimicrobial compounds, and activation of plant defense mechanisms.

Among these beneficial microbes, cyanobacteria are particularly promising. These ancient photosynthetic prokaryotes inhabit a wide range of ecosystems, from freshwater and marine environments to deserts, hot springs, and polar regions. Their structural and metabolic versatility enables them to fix atmospheric nitrogen, solubilize phosphorus, and secrete phytohormones, amino acids, and extracellular polysaccharides (EPS), thereby enhancing plant growth and improving stress tolerance (Robati 2024; Yang et al. 2025). Ecologically, they act as keystone organisms by performing oxygenic photosynthesis, contributing to nitrogen fixation, and in arid lands form biological soil crusts that stabilize soils, enhance fertility, and promote carbon sequestration. They also establish symbiotic relationships with plants, fungi, and heterotrophic bacteria, facilitating nutrient and metabolite exchange. For instance, in the *Azolla–Anabaena* association, the cyanobacterium provides nitrogen while the host offers protection, whereas in microbial mats, their excreted organic matter is recycled by heterotrophs into nutrients and vitamins, increasing ecosystem resilience through functional redundancy (Álvarez et al. 2023). Recent studies further emphasize the role of cyanobacteria as key agents in climate-smart agriculture, where their integration into farming systems enhances soil fertility, nutrient-use efficiency, and resilience to abiotic stresses under changing climatic conditions (Nawaz et al. 2025; Jing et al. 2025).

From an agricultural perspective, these traits can either substitute for or complement chemical inputs. Cyanobacteria contribute bioavailable nitrogen through heterocysts, release organic acids and phosphatases that mobilize phosphorus, and secrete EPS that enhance soil aggregation and facilitate root colonization. In addition, they synthesize phytohormones such as auxins, gibberellins, and cytokinins, along with amino acids (e.g., glutamic acid, aspartic acid, tryptophan) and siderophores that enhance nutrient uptake (Toribio et al. 2020). Beyond promoting plant growth, many strains contribute to bioremediation by binding or degrading heavy metals, hydrocarbons, pesticides, and dyes, thereby improving both fertility and ecological stability. Evidence shows that inoculation with cyanobacteria can stabilize degraded soils, increase water-holding capacity, and improve nutrient availability within weeks. Moreover, their EPS mitigate salinity stress by chelating sodium ions and facilitating biofilm formation; for example, *Anabaena* application in saline rice fields reduces fertilizer requirements while improving yields (Li et al. 2019).

2.2 Soil Health and Cyanobacterial Functions

Cyanobacteria play a vital role in nutrient cycling. Heterocystous genera such as *Anabaena*, *Nostoc*, and *Calothrix* fix nitrogen using the enzyme nitrogenase, which is protected from oxygen by either temporal separation (night fixation) or spatial compartmentalization within heterocysts (Kollmen and Strieth 2022). They also contribute to carbon sequestration by increasing organic matter inputs and stabilizing soil carbon pools (Elbasiouny et al. 2022). In terms of phosphorus mobilization, cyanobacteria release organic acids and chelators that solubilize insoluble minerals, stimulate phosphatase activity, and recycle phosphorus through biomass turnover (Robati 2024). Additionally, cyanobacteria aid in the detoxification of contaminated soils through EPS-mediated biosorption and enzyme-driven degradation, with laccases and peroxidases playing key roles in pollutant breakdown (Priya et al. 2022). Altogether, they represent multifunctional, eco-friendly agents for sustainable agriculture. By improving soil structure, cycling nutrients, enhancing plant resilience, and remediating pollutants, they offer a sustainable alternative to chemical-intensive farming and contribute to the development of climate-resilient food systems.

2.3 Phytohormone Biosynthetic Pathways in Plants and Cyanobacteria

2.3.1 Auxin Biosynthesis

Auxin, mainly indole-3-acetic acid (IAA), is a central regulator of growth and development in both higher plants and certain cyanobacteria. Although tryptophan-independent routes have been suggested, the majority of IAA production proceeds through tryptophan (Trp)-dependent mechanisms (Fig. 2.1). In higher plants, the indole-3-pyruvic acid (IPA/IPyA) pathway is the predominant and most conserved route, and its regulation is crucial for developmental processes such as organ patterning and meristem maintenance. Alternative pathways, including the indole-3-acetamide (IAM), tryptamine (TAM), and indole-3-acetaldoxime (IAOx) routes, function in specific taxa or under specialized conditions. Notably, the IAOx pathway is largely restricted to Brassicales, where it links auxin biosynthesis with secondary metabolite production, including glucosinolates and camalexin. The detection of IAM in diverse species despite the absence of *iaaM* (tryptophan monooxygenase) homologs indicates that plants may use alternative or yet unidentified enzymes for IAM formation. Such alternative pathways confer metabolic flexibility, linking auxin biosynthesis to environmental signals and physiological processes (Tang et al. 2023).

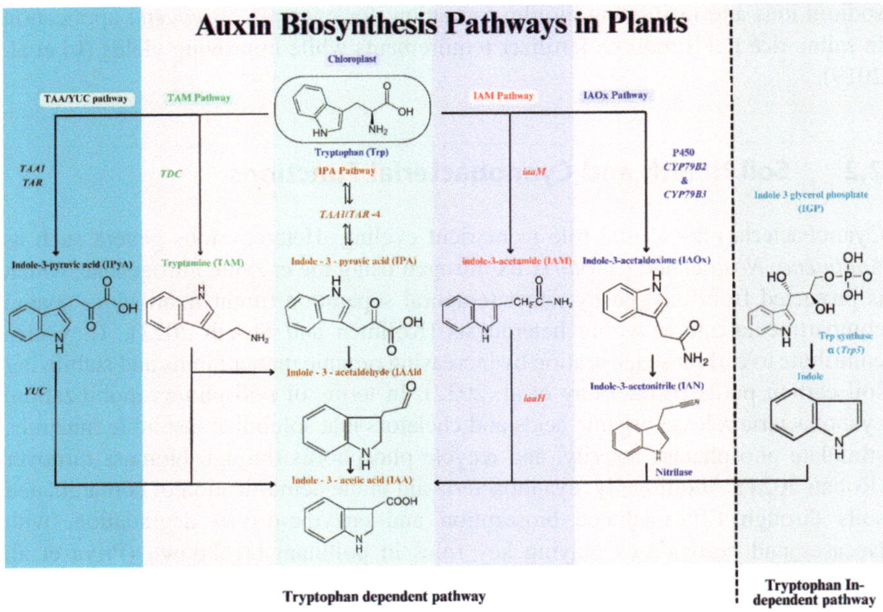

Fig. 2.1 Auxin biosynthesis in plants

Cyanobacterial auxin biosynthesis exhibits notable similarities to that of higher plants. The (IPA/IPyA) pathway is the most conserved, with homologs of plant TAA/TAR and YUC enzymes identified in genera such as *Anabaena*, *Nostoc*, and *Synechocystis* (Mano and Nemoto 2012). The IAM pathway occurs particularly in symbiotic or plant-associated strains, reflecting its bacterial origin. Metabolites of the TAM pathway are evident, yet the enzymatic machinery driving this route is not well understood. Unlike plants, cyanobacteria may also utilize tryptophan-independent mechanisms, as IAA production has been reported in the absence of exogenous Trp. This suggests a possible role for indole or indole-3-glycerol phosphate from the shikimate pathway, although genetic evidence is still lacking. Notably, the IAOx pathway has not been identified in cyanobacteria.

2.3.2 Gibberellin Biosynthesis

Gibberellins (GAs) are diterpenoid phytohormones that profoundly influence plant growth and development, including seed germination, stem elongation, leaf expansion, and reproductive transitions (Sponsel and Hedden 2004). Their biosynthesis is highly compartmentalized across plastids, the endoplasmic reticulum (ER), and the cytosol, a spatial separation that allows strict regulatory control (Fig. 2.2).

The regulation of GA metabolism is characterized by a fine-tuned balance between the production of bioactive forms and their subsequent deactivation.

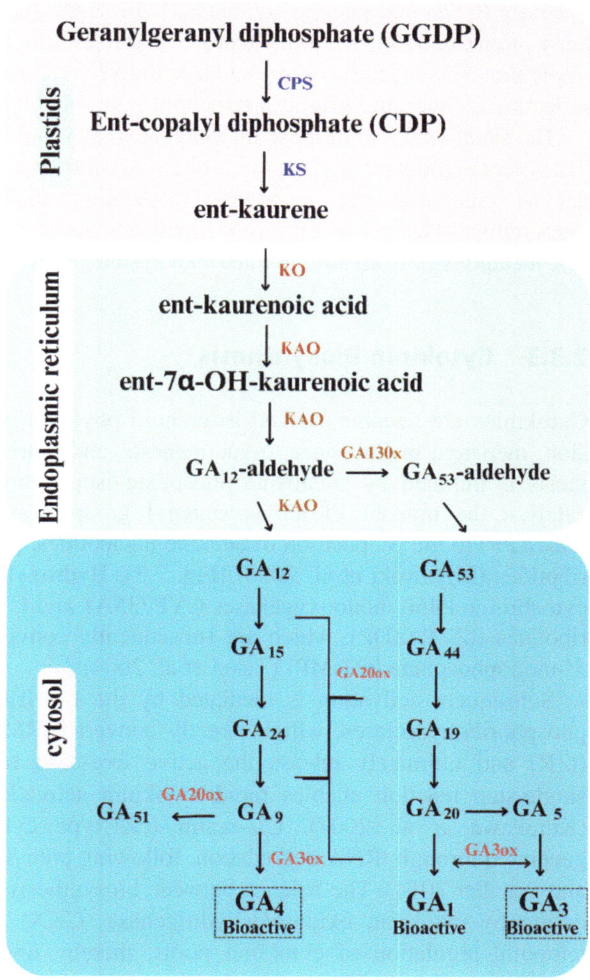

Fig. 2.2 Gibberellin biosynthesis in plants. (Reproduced from Ritonga et al. (2023))

Bioactive GAs (notably GA_1, GA_3, and GA_4) are synthesized through the action of dioxygenases, while GA2-oxidases deactivate excess molecules to avoid uncontrolled growth responses (Salazar-Cerezo et al. 2018). This dynamic turnover ensures developmental plasticity, allowing plants to fine-tune GA levels in response to environmental signals such as photoperiod, temperature, and nutrient availability. Additionally, GA biosynthesis integrates with other hormonal networks, particularly auxins and brassinosteroids, to coordinate complex developmental processes. This cross-talk underscores the role of GAs as central regulators, rather than acting in isolation, thereby enhancing their significance in adaptive growth responses (Hedden and Thomas 2012).

In cyanobacteria, direct biochemical evidence for a complete GA biosynthetic pathway remains limited. Nevertheless, comparative genomics has revealed homologs of early diterpenoid biosynthetic enzymes, such as copalyl diphosphate synthase (CPS) and kaurene synthase (KS), suggesting the potential to synthesize ent-kaurene or related intermediates (Salazar-Cerezo et al. 2018). These findings imply that cyanobacteria may retain a primitive version of the GA pathway, possibly reflecting an ancestral origin of terpenoid hormone biosynthesis.

The functional role of these intermediates in cyanobacteria is not yet unresolved. They may participate in signaling processes, contribute to ecological interactions, or act as precursors for secondary metabolites distinct from canonical plant gibberellins. This raises intriguing questions about the evolutionary trajectory of GA metabolism, from simple microbial systems to its elaboration in higher plants.

2.3.3 Cytokinin Biosynthesis

Cytokinins are adenine-derived isoprenoid phytohormones that regulate cell division, meristem maintenance, organogenesis, and nutrient allocation. Their biosynthesis is initiated by adenosine phosphate–isopentenyltransferases (IPTs), which catalyze the transfer of an isopentenyl group from dimethylallyl diphosphate (DMAPP) to the N^6 position of adenine nucleotides, producing isopentenyladenine ribotides (Miyawaki et al. 2006) (Fig. 2.3). Hydroxylation of these precursors by cytochrome P450 monooxygenases CYP735A1 and CYP735A2 yields trans-zeatin ribotides (tZTP/tZDP), which are subsequently converted to trans-zeatin riboside 5′-monophosphate (tZRMP) (Takei et al. 2004).

Subsequent activation is mediated by the LONELY GUY (LOG) family of phosphoribohydrolases, which directly convert tZRMP into trans-zeatin riboside (tZR) and ultimately release the active free-base form, tZ (trans-zeatin). This single-step reaction enables rapid cytokinin activation at specific cellular sites (Kurakawa et al. 2007). Cis-zeatin (cZ)–type cytokinins are predominantly generated through tRNA degradation, following prenylation by tRNA-IPTs (Kieber and Schaller 2014). The balance between biosynthesis, interconversion, and degradation by cytokinin oxidase/dehydrogenase (CKX) ensures precise spatial and temporal regulation of cytokinin pools, thereby fine-tuning developmental and environmental responses.

Cyanobacteria possess homologs of plant IPT genes, suggesting that cytokinin biosynthesis originated prior to the emergence of land plants. In *Nostoc* sp. PCC 7120, the IPT enzyme NoIPT1 (Nostoc isopentenyltransferase 1) preferentially utilizes DMAPP as a prenyl donor, a specificity that aligns with plant IPTs but contrasts with the broader substrate flexibility observed in many bacterial IPTs (Wu et al. 2021). This supports the notion of evolutionary conservation of cytokinin biosynthesis across photosynthetic lineages.

Although the downstream enzymatic machinery is not fully characterized in cyanobacteria, the production of isopentenyladenine ribotides (iP-type cytokinins) has been demonstrated, while trans-zeatin (tZ) derivatives are inferred from genomic and metabolomic analyses (Uniyal et al. 2022). Functionally, cytokinin production in cyanobacteria is proposed to contribute to cell differentiation, stress tolerance, and symbiotic competence, indicating that the signaling role of cytokinins predates the divergence of cyanobacteria and higher plants.

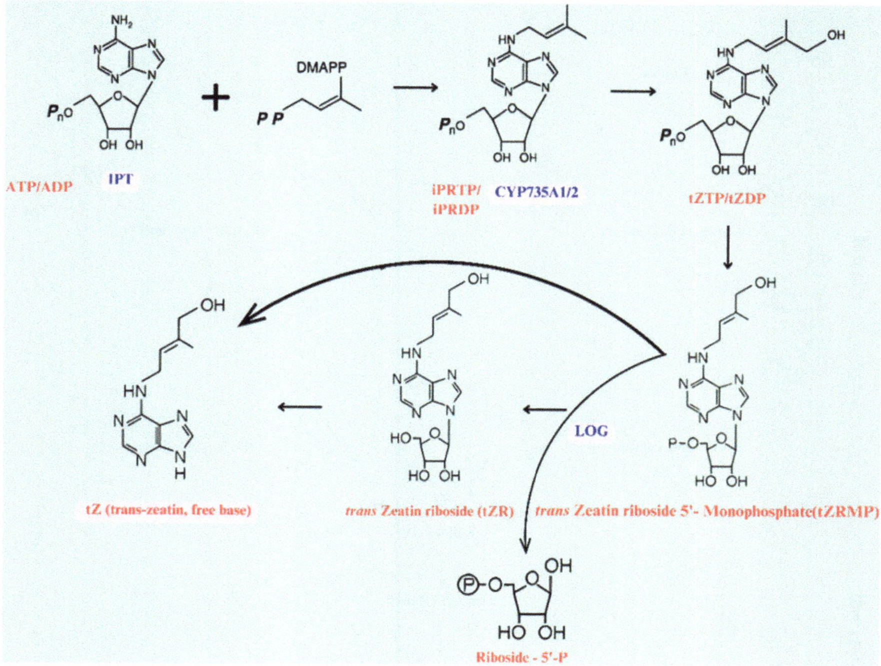

Fig. 2.3 Cytokinin biosynthesis in plants. (Reproduced from Kieber and Schaller (2014))

2.3.4 Abscisic Acid (ABA) Biosynthesis

ABA biosynthesis in higher plants is a plastid-to-cytosol metabolic process tightly linked to environmental stress adaptation. Rather than being a simple linear conversion, this pathway is controlled by multiple regulatory nodes, among which the activity of 9-cis-epoxycarotenoid dioxygenase (NCED) is the major rate-limiting step. Induction of NCED transcription under drought, osmotic stress, and salinity directly governs the extent of ABA accumulation, thereby coupling carotenoid metabolism to abiotic stress perception (Wu et al. 2023).

Another distinctive feature of the pathway is its compartmentalization (Fig. 2.4). The initial carotenoid-derived precursors are synthesized and cleaved within plastids, while the final oxidative reactions leading to ABA formation occur in the cytosol. This inter-organellar metabolic relay provides additional checkpoints for regulation, ensuring precise spatial and temporal control of ABA levels in response to developmental and stress cues. Beyond stress signaling, ABA plays essential roles in seed dormancy, stomatal regulation, and modulation of root architecture, underscoring its significance as a central integrator of growth–stress trade-offs in higher plants.

Cyanobacteria exhibit fundamental differences in ABA biosynthesis compared to higher plants. Genomic analyses reveal the absence of canonical NCED homologs, indicating that they do not rely on carotenoid cleavage as in plants (González-

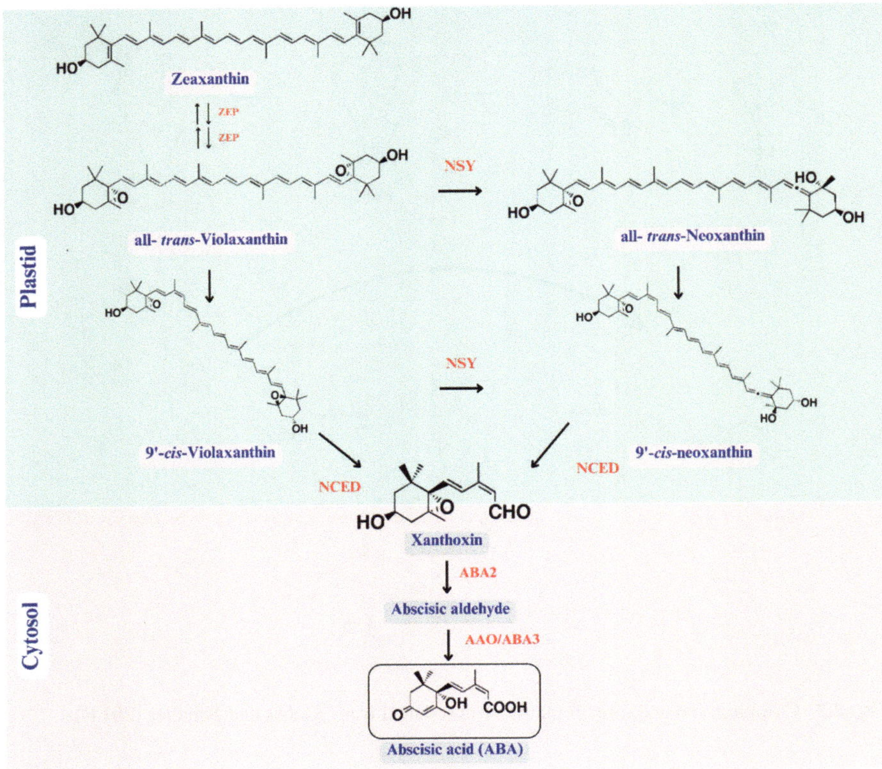

Fig. 2.4 Abscisic acid biosynthesis in plants. (Reproduced from Wu et al. (2023))

Grandío et al. 2017). Instead, accumulating evidence suggests that cyanobacteria may utilize an alternative metabolic route, possibly originating from farnesyl diphosphate (FDP) through the mevalonic acid (MVA) pathway (Inomata et al. 2005). Although ABA has been detected in multiple cyanobacterial lineages, the precise enzymatic machinery remains poorly resolved. This suggests the presence of a divergent biosynthetic strategy, which may reflect an evolutionarily ancient mechanism of isoprenoid-derived ABA formation. Elucidating this pathway will provide critical insights into the evolutionary origins of ABA signaling and its early role in stress acclimation prior to the colonization of terrestrial habitats.

2.3.5 Ethylene Biosynthesis

Ethylene production in plants is predominantly regulated at the level of 1-aminocyclopropane-1-carboxylic acid synthase (ACS), which is encoded by a multigene family with highly stress- and tissue-specific expression patterns (Fig. 2.5). ACS activity is considered the principal rate-limiting step and is tightly

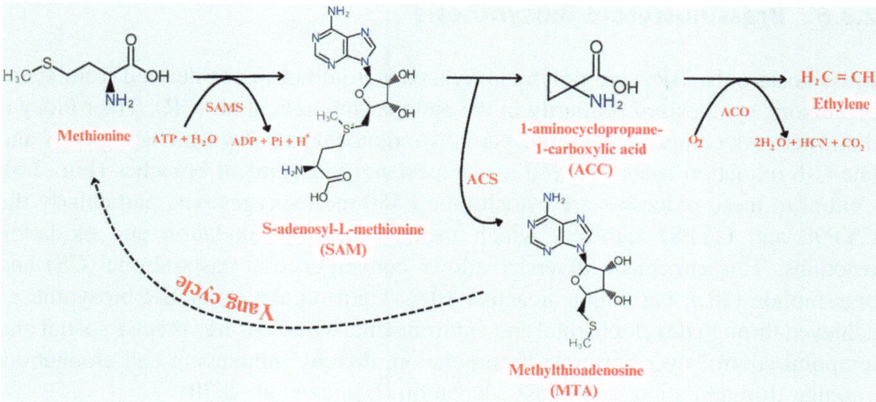

Fig. 2.5 Ethylene biosynthesis in higher plants. (Reproduced from Pattyn et al. (2021))

controlled by transcriptional induction, phosphorylation, and protein turnover, ensuring precise modulation of ethylene levels during growth, development, and stress responses (Ouyang et al. 2000). Another critical regulatory node lies in 1-aminocyclopropane-1-carboxylic acid (ACC) metabolism. Plants can reversibly divert ACC into inactive conjugates such as malonyl-ACC (MACC) and γ-glutamyl-ACC (GACC), or into jasmonyl-ACC (JA-ACC), which integrates ethylene with jasmonate signaling pathways. This conjugation–deconjugation mechanism acts as a buffering system to fine-tune the pool of bioactive ACC in response to fluctuating environmental and developmental signals.

Additionally, the by-product hydrogen cyanide (HCN), released during ethylene synthesis, necessitates detoxification by β-cyanoalanine synthase, linking ethylene biosynthesis to nitrogen and cyanide metabolism (Fatma et al. 2022). Such integration reflects the broader role of ethylene as a hub metabolite at the interface of primary metabolism and stress signaling.

In cyanobacteria, ethylene formation relies on an alternative route mediated by the ethylene-forming enzyme (EFE), which utilizes 2-oxoglutarate and L-arginine as substrates. This pathway directly connects ethylene output with central carbon flux (via the TCA cycle intermediate 2-oxoglutarate) and nitrogen metabolism (via arginine), suggesting a role in balancing metabolic homeostasis under stress conditions (Wu et al. 2021). Cyanobacteria, unlike higher plants, lack homologs of ACS and ACO, indicating an independent evolutionary origin of ethylene biosynthesis. Much of our understanding comes from heterologous expression of the *efe* gene in model cyanobacteria such as *Synechococcus elongatus*, where ethylene production responds to environmental cues including nitrogen limitation and oxidative stress (Veetil et al. 2017). This stress-linked regulation implies that ethylene may function as a signaling molecule in cyanobacteria, potentially mediating adaptive responses, inter-microbial communication, or even influencing plant–microbe interactions.

2.3.6 Brassinosteroid Biosynthesis

Brassinosteroids (BRs) are polyhydroxylated steroid hormones derived from sterol precursors, synthesized primarily in the endoplasmic reticulum (ER). Their biosynthetic network comprises multiple, partially redundant pathways, including early and late C-6 oxidation routes as well as campestanol-independent branches (Fig. 2.6). Central to these processes are cytochrome P450 monooxygenases, particularly the CYP90 and CYP85 families, which mediate key hydroxylation and oxidation reactions. This enzymatic diversity allows convergence at castasterone (CS) and brassinolide (BL), the major bioactive BRs. Tight regulation of BR biosynthesis, achieved through developmental and environmental cues, ensures precise spatial and temporal control over hormone accumulation, thereby influencing cell elongation, vascular differentiation, and stress adaptation (Bajguz et al. 2020).

Cyanobacteria lack canonical sterol-derived BR pathways but exhibit genomic evidence for alternative biosynthetic routes. Genes encoding cytochrome P450s homologous to plant CYP90/85 families, as well as short-chain dehydrogenases and nicotinamide adenine dinucleotide phosphate (NADPH)-dependent oxidoreductases, suggest a potential capacity for BR-like molecule biosynthesis. Unlike plants, which rely on campesterol as the sterol precursor, cyanobacteria utilize the 2-C-methyl-D-erythritol 4-phosphate (MEP) pathway to generate

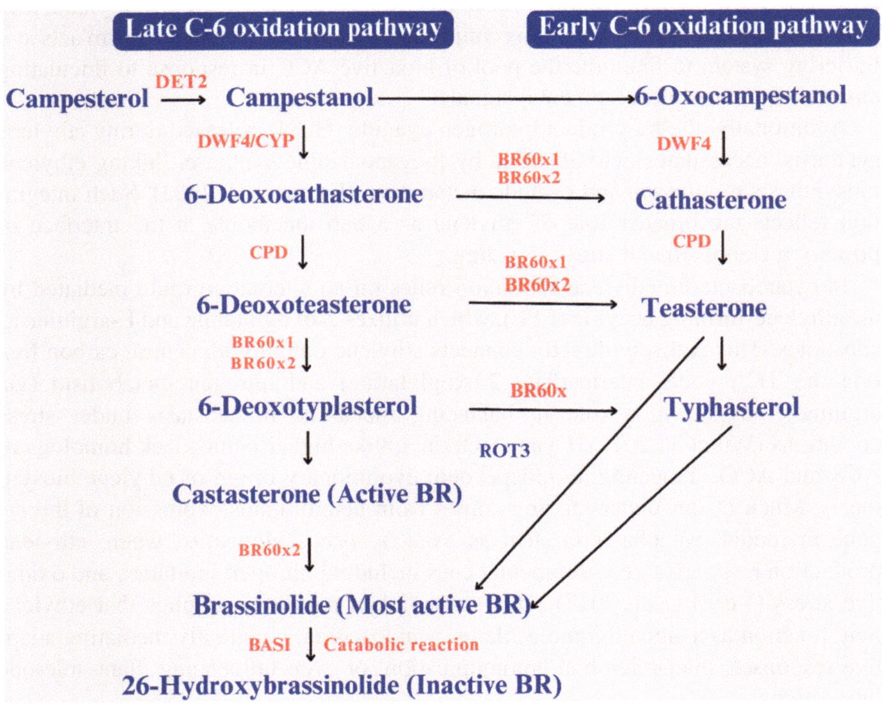

Fig. 2.6 Brassinosteroids biosynthesis in plants. (Reproduced from Bhanu (2019))

isoprenoid intermediates that could serve as precursors for triterpenoid derivatives. These putative BR analogs hypothesized to arise through iterative hydroxylation, oxidation, and structural rearrangements, although their biochemical identities are still unresolved. Further metabolomic and enzymatic analyses are anticipated to resolve the capacity of cyanobacteria can produce functional BR analogs and to define their roles in cellular regulation and interspecies signaling (Martin-Creuzburg and Merkel 2016).

2.4 Strategies to Augment Phytohormone Metabolism in Cyanobacteria

The elucidation of phytohormone biosynthetic pathways in cyanobacteria highlighted their potential as biofertilizers. Nonetheless, endogenous hormone production is often strain-specific and suboptimal under field conditions (Poveda 2021). To address these limitations, diverse strategies including genetic engineering, ecological manipulation, environmental modulation, and strain selection are being developed to enhance phytohormone metabolism and improve agricultural applications.

2.4.1 Genetic and Metabolic Engineering

Metabolic engineering allows precise modification of biosynthetic pathways to optimize hormone production. This involves overexpression of biosynthetic genes, repression of degradation enzymes, and heterologous integration of plant or microbial genes (Satta et al. 2023).

- *Auxins* (*IAA*): Overexpression of *ipdC* (indole-3-pyruvate decarboxylase) and *iaaM* enhances IAA synthesis. Expression of *iaaM* alone induces auxin overproduction in higher plants, suggesting feasibility in cyanobacteria (Tang et al. 2023).
- *Cytokinins* (*CKs*): The *ipt* gene catalyzes cytokinin nucleotide synthesis. Cyanobacterial *ipt* homologs regulate cell division and stress tolerance; overexpression enhances plant–microbe interactions (Uniyal et al. 2022).
- *Gibberellins* (*GAs*): Although early enzymes such as (*CPS*, *KS*) exist, complete GA biosynthesis is unconfirmed. Heterologous expression of *GA20ox* and *GA3ox* could enable bioactive GA production (Salazar-Cerezo et al. 2018).
- *Abscisic acid* (*ABA*): Cyanobacteria lack canonical NCED genes, suggesting an alternative FDP-derived pathway (Inomata et al. 2005). Engineering efforts are still in early development.
- *Ethylene*: Cyanobacteria lack native ethylene pathways; however, heterologous expression of *efe* from *Pseudomonas syringae* enables efficient ethylene synthesis under photoautotrophic conditions (Kuchmina et al. 2017).

- *Brassinosteroids* (*BRs*): No functional homologs are known, although sterol-like intermediates are reported. Heterologous expression of plant BR biosynthetic genes remains a subject for future research.

2.4.2 Co-culturing with Plants or Microbiota

Plant root exudates (flavonoids, sugars, organic acids) stimulate cyanobacterial hormone biosynthesis and improve colonization. Rhizospheric bacteria can also metabolize hormone precursors, enhancing overall phytohormone output. Co-cultivation has been shown to increase IAA and cytokinin production while facilitating nutrient exchange and root association (Kollmen and Strieth 2022).

2.4.3 Environmental Modulation

- Abiotic stresses regulate phytohormone metabolism by altering redox balance and reactive oxygen species (ROS) signaling (Odenthal et al. 2024). Cyanobacteria display unique adaptations to environmental fluctuations that can be leveraged to stimulate hormone synthesis:
- *Light*: Cyanobacteria adapt across broad PAR ranges (0.1–1500 μmol m^{-2} s^{-1}) through pigment remodeling and chromatic adaptation for spectral optimization. UV exposure further induces ROS and protective metabolites (e.g., scytonemin, MAAs), indirectly influencing hormone regulation (Yadav et al. 2022).
- *Temperature*: Cold stress activates desaturases (*DesA–D*) for membrane fluidity (Sinetova and Los 2016), whereas heat stress induces HSPs, influencing hormone-related enzymes (Hongsthong et al. 2009).
- *pH*: Cyanobacteria tolerate pH 5–10, but extremes disrupt hormone biosynthesis. Buffering enzymes (carbonic anhydrase, oxalate decarboxylase) help stabilize intracellular homeostasis (Rezayian et al. 2019).
- *Salinity*: Na^+/H^+ antiporters, osmolyte accumulation, and antioxidants contribute to salt tolerance. Salinity modulates phytohormone levels, supporting growth under saline soils (Jan et al. 2023).
- *Desiccation*: Poikilohydric cyanobacteria (e.g., *Nostoc commune*) survive prolonged dehydration by metabolic arrest and rapid recovery upon rehydration (Lan et al. 2014).
- *Heavy metals*: Cyanobacteria tolerate metals via transporters, sequestration, and metallothionein synthesis. Metal stress perturbs ROS balance, indirectly modulating hormone pathways (Miyazaki and Asanuma 2023).

These stress-responsive traits highlight non-GMO strategies to enhance phytohormone metabolism by controlled modulation of environmental factors.

2.4.4 Selection of High-Producing Strains

Cyanobacteria exhibit substantial inter and intra genus variability in hormone biosynthesis. Screening strategies include the following:

- Colorimetric assays (e.g., Salkowski reagent for IAA) (Ahmed et al. 2010).
- Chromatographic profiling (HPLC, LC-MS/MS analysis) (Almeida and Mithöfer 2019).
- Plant bioassays for growth promotion experiments (Toribio et al. 2020).
- Genomic mining for biosynthetic clusters identification via (antiSMASH, KEGG) (Zhang et al. 2023).
- Omics-based prediction via transcriptomics and metabolomics signatures guide pre-screening (Omari Alzahrani 2024).

Recent advances in transcriptomics and metabolomics have enabled detailed mapping of metabolic pathways and gene expression patterns linked to phytohormone biosynthesis. These molecular signatures allow predictive identification of high-producing strains prior to experimental validation. Moreover, environmental adaptability, especially responsiveness to plant-derived signaling molecules, is a critical determinant of strain performance under stress (Robati 2024). Integrating molecular profiling with ecological adaptation thus provides a holistic framework to accelerate the development of cyanobacterial strains optimized for plant growth promotion.

2.5 Experimental Evidence of Auxin-Mediated Plant Growth Promotion

Recent advances in plant–microbe interaction studies have clarified the pivotal role of auxins in regulating plant growth and development. Experimental evidence demonstrates that auxin-producing cyanobacteria enhance plant performance either through direct secretion of IAA or by influencing endogenous auxin pathways. Inoculation of roots or exogenous application of IAA-producing strains consistently results in improved seed germination, enhanced root elongation, lateral root proliferation, and overall seedling vigor across a range of crops. These effects have been documented across a wide range of crops, including rice, wheat, and chickpea, using cyanobacterial genera such as *Nostoc*, *Synechocystis*, *Leptolyngbya*, and *Fischerella* (Parmar et al. 2023). For instance, *Fischerella muscicola* isolates with high IAA production capacity were associated with accelerated flowering and enhanced shoot biomass in rice.

Co-inoculation studies highlight synergistic effects between cyanobacteria and plant growth-promoting rhizobacteria (PGPR). For example, combined inoculation of *Anabaena* spp. with *Azospirillum brasilense* increased wheat yield by 40%, attributed to improved root surface area and activation of auxin-responsive genes (Gavilanes et al. 2020). Similarly, consortia or biofilm-based inoculants have been shown to significantly improve root-to-shoot ratios, photosynthetic pigments, and dry biomass relative to uninoculated controls (Kholssi et al. 2021). Exogenous application of auxins and cytokinins also enhanced dry matter yield, improved plant stability, reduced disease incidence, and increased mineral content (K^+, Ca^{2+}) in plant tissues (Sosnowski et al. 2023).

Beyond growth regulation, auxins interact with other phytohormones, including abscisic acid, cytokinins, gibberellins, and ethylene, to mediate adaptive responses under stress conditions such as drought, heat, and heavy metal toxicity (Gao et al. 2024). Specifically, auxin signaling has been shown to mitigate heavy metal stress by promoting chelation and sequestration, reducing uptake, and alleviating oxidative damage, with responses varying by metal type and concentration (Yang and Wang 2023). Collectively, these findings establish that cyanobacteria-derived auxins function not only as growth enhancers but also as regulatory signals that shape hormonal cross-talk, nutrient mobilization, and stress resilience. This dual role underscores their potential as critical components in advanced biofertilizer formulations designed to support sustainable and climate-resilient agriculture. Field-level evaluations have also confirmed these benefits, with cyanobacterial inoculants improving crop yield stability and soil health under variable environmental conditions, thereby reinforcing their practical applicability beyond controlled laboratory and greenhouse studies (Hakkoum et al. 2025; Haraguchi and Shimizu 2025).

2.6 Cyanobacterial Consortia for Sustainable Agriculture

The application of cyanobacterial consortia offers a promising alternative to single-strain inoculants in sustainable farming systems (Toribio et al. 2020). Unlike monocultures, such consortia combine cyanobacteria with other beneficial microbes including PGPR, fungi (e.g., arbuscular mycorrhizal fungi), and algae providing functional diversity and complementarity (Chandran et al. 2021). Consortia design often integrates multiple plant-beneficial traits, such as nitrogen fixation, phosphate solubilization, and phytohormone production, resulting in synergistic effects on crop performance (Bargaz et al. 2018).

For instance, *Calothrix* contributes robust phytohormone synthesis, whereas *Nostoc* and *Trichormus* excel in nitrogen fixation, collectively supporting improved nutrient acquisition and stress tolerance (Osman et al. 2010). Similarly, co-inoculation with phosphate or potassium solubilizing bacteria enhances nutrient bioavailability, resulting in improved biomass, root length, and yield relative to single-strain inoculations (El Sabagh et al. 2022).

Beyond nutrient mobilization, cyanobacterial consortia also play a protective role against pathogens by producing bioactive compounds that interfere with microbial quorum sensing, thereby suppressing fungal and bacterial virulence. Pot and field experiments confirm the agronomic value of such combinations. Mixtures of *Anabaena*, *Nostoc*, and *Calothrix* improved rice yields and soil nitrogen status (Parmar et al. 2023), while co-inoculation with PGPR enhanced nodulation and biomass accumulation in legumes such as chickpea (Bidyarani et al. 2016). From a practical standpoint, formulation and carrier selection are critical for consortium stability and field applicability. Carriers such as peat, biochar, alginate beads, and vermiculite improve microbial survival, shelf life, and delivery efficiency. Compatibility screening and strain–strain interaction studies further optimize functional performance in target agroecosystems, ensuring consistent outcomes under field conditions.

2.7 Challenges and Future Scope

Although cyanobacteria show great potential as bio-stimulants, many applications still rely on a few well-known strains that often fail to perform effectively under diverse field conditions due to ecological mismatches and lack of local adaptation. Additional challenges include difficulties in maintaining culture viability, short shelf life of formulations, and limited compatibility with conventional farming practices. Regulatory frameworks, biosafety concerns, and low farmer awareness further restrict large-scale adoption.

Future efforts should focus on developing stress-resilient, high-performing strains, designing effective consortia, and using novel carriers to improve culture stability and shelf life. Omics-based approaches and genetic engineering can enhance phytohormone production, while long-term field trials and farmer-focused strategies will be critical for translating laboratory success into practical applications. Collectively, overcoming these challenges will enable cyanobacterial bio-stimulants to contribute to sustainable and climate-smart agriculture.

2.8 Conclusion

Cyanobacterial consortia represent a sustainable alternative to chemical inputs by enhancing nutrient availability, phytohormone production, soil health, and stress resilience. Their application holds great promise for climate-smart agriculture, with future efforts needed to optimize strain selection, field scalability, and integration into existing farming systems.

Acknowledgments The authors sincerely thank Vignan's Foundation for Science, Technology & Research (Deemed to be University) for providing a conducive environment and ANRF project "TAR/2023/000036" for supporting this work.

Declaration of Competing Interest The authors declare that they have no known competing financial interests or personal relationships that could have influenced the work reported in this paper.

References

Ahmed M, Stal L, Hasnain S (2010) Production of indole-3-acetic acid by the cyanobacterium *Arthrospira platensis* strain MMG-9. J Microbiol Biotechnol 20(9):1259–1265

Almeida M, Mithöfer A (2019) Quantification of phytohormones by HPLC-MS/MS including phytoplasma-infected plants: methods and protocols. Methods Mol Biol 1992:345–358

Álvarez C, Jiménez-Ríos L, Iniesta-Pallarés M, Jurado-Flores A, Molina-Heredia FP, Ng CK, Mariscal V (2023) Symbiosis between cyanobacteria and plants: from molecular studies to agronomic applications. J Exp Bot 74(19):6145–6157

Alzahrani O (2024) A comprehensive analysis of the 9-cis epoxy carotenoid dioxygenase gene family and their responses to salt stress in *Hordeum vulgare* L. Plants 13(23):3327

Bajguz A, Chmur M, Gruszka D (2020) Comprehensive overview of the brassinosteroid biosynthesis pathways: substrates, products, inhibitors, and connections. Front Plant Sci 11:1034

Bargaz A, Lyamlouli K, Chtouki M, Zeroual Y, Dhiba D (2018) Soil microbial resources for improving fertilizers efficiency in an integrated plant nutrient management system. Front Microbiol 9:1606

Bhanu AN (2019) Brassinosteroids: relevance in biological activities of plants and agriculture. J Plant Sci Res 35:1–5

Bidyarani N, Prasanna R, Babu S, Hossain F, Saxena AK (2016) Enhancement of plant growth and yields in chickpea (*Cicer arietinum* L.) through novel cyanobacterial and biofilmed inoculants. Microbiol Res 188–189:97–105

Chandran H, Meena M, Swapnil P (2021) Plant growth-promoting rhizobacteria as a green alternative for sustainable agriculture. Sustainability 13(19):10986

El Sabagh A, Islam MS, Hossain A, Iqbal MA, Mubeen M, Waleed M, Reginato M, Battaglia M, Ahmed S, Rehman A, Arif M, Athar HUR, Ratnasekera D, Danish S, Raza MA, Rajendran K, Mushtaq M, Skalicky M, Brestic M, Abdelhamid MT (2022) Phytohormones as growth regulators during abiotic stress tolerance in plants. Front Agron 4:765068

Elbasiouny H, El-Ramady H, Elbehiry F, Rajput VD, Minkina T, Mandzhieva S (2022) Plant nutrition under climate change and soil carbon sequestration. Sustainability 14(2):2

Fatma M, Asgher M, Iqbal N, Rasheed F, Sehar Z, Sofo A, Khan NA (2022) Ethylene signaling under stressful environments: analyzing collaborative knowledge. Plants 11(17):2211

Gao J, Zhuang S, Zhang W (2024) Advances in plant auxin biology: synthesis, metabolism, signaling, interaction with other hormones, and roles under abiotic stress. Plants 13(17):2523

Gavilanes FZ, Andrade DS, Zucareli C et al (2020) Co-inoculation of *Anabaena cylindrica* with *Azospirillum brasilense* increases grain yield of maize hybrids. Rhizosphere 15:100224

González-Grandío E, Pajoro A, Franco-Zorrilla JM, Tarancón C, Immink RG, Cubas P (2017) Abscisic acid signaling is controlled by a BRANCHED1/HD-ZIP I cascade in *Arabidopsis* axillary buds. Proc Natl Acad Sci USA 114(2):E245–E254

Hakkoum Z, Minaoui F, Tazart Z, Chabili A, Douma M, Mouhri K, Loudiki M (2025) Impact of a soil cyanobacteria consortium-based bioinoculant on tomato growth, yield, and fruit quality. Plants 14(13):2034

Haraguchi Y, Shimizu T (2025) Crop cultivation without nitrogen fertiliser using nitrogen-fixing cyanobacterial extracts for low environmental impact. Sci Rep 15(1):18365

Hedden P, Thomas SG (2012) Gibberellin biosynthesis and its regulation. Biochem J 444(1):11–25

Hongsthong A, Sirijuntarut M, Yutthanasirikul R, Senachak J, Kurdrid P, Cheevadhanarak S, Tanticharoen M (2009) Subcellular proteomic characterization of the high-temperature stress response of the cyanobacterium *Spirulina platensis*. Proteome Sci 7:33

Inomata M, Hirai N, Yoshida R, Ohigashi H (2005) Biosynthesis of abscisic acid by the direct pathway via ionylideneethane in a fungus, *Cercospora cruenta*. Biosci Biotechnol Biochem 68(12):2571–2580

Jan Z, Ali S, Rahim HU, Akbar WA, Sehrish AK, Taj A, Rahim T, Iqbal M (2023) Cyanobacterial strains reclaimed induced-salinity stress attributes and improved the physicochemical properties of induced-saline soil. Acta Ecol Sin 43(6):1074–1079

Jing H, Zhou N, Tantai Y, Zhao Y (2025) Artificial cyanobacteria crusts can improve soil fertility and plant growth in a semi-arid area, northern China. J Arid Land 17(6):808–822
Kholssi R, Marks E, Miñón J et al (2021) A consortium of cyanobacteria and plant growth promoting rhizobacteria for wheat growth improvement in a hydroponic system. S Afr J Bot 142:247–258
Kieber JJ, Schaller GE (2014) Cytokinins. Arabidopsis Book 12:e0168
Kollmen J, Strieth D (2022) The beneficial effects of cyanobacterial co-culture on plant growth. Life (Basel) 12(2):223
Kuchmina E, Klähn S, Jakob A, Bigott W, Enke H, Dühring U, Wilde A (2017) Ethylene production in *Synechocystis* sp. PCC 6803 promotes phototactic movement. Microbiology 163(12):1937–1945
Kurakawa T, Ueda N, Maekawa M, Kobayashi K, Kojima M, Nagato Y, Sakakibara H, Kyozuka J (2007) Direct control of shoot meristem activity by a cytokinin-activating enzyme. Nature 445(7128):652–655
Lan S, Wu L, Zhang D, Hu C (2014) Desiccation provides photosynthetic protection for crust cyanobacteria *Microcoleus vaginatus* from high temperature. Physiol Plant 152(2):345–354
Li H, Zhao Q, Huang H (2019) Current states and challenges of salt-affected soil remediation by cyanobacteria. Sci Total Environ 669:118–129
Mano Y, Nemoto K (2012) The pathway of auxin biosynthesis in plants. J Exp Bot 63(8): 2853–2872
Martin-Creuzburg D, Merkel P (2016) Sterols of freshwater microalgae: potential implications for zooplankton nutrition. J Plankton Res 38:1–11
Miyawaki K, Tarkowski P, Matsumoto-Kitano M, Kato T, Sato S, Tarkowska D, Tabata S, Sandberg G, Kakimoto T (2006) Roles of *Arabidopsis* ATP/ADP isopentenyltransferases and tRNA isopentenyltransferases in cytokinin biosynthesis. Proc Natl Acad Sci USA 103(44): 16598–16603
Miyazaki I, Asanuma M (2023) Multifunctional metallothioneins as a target for neuroprotection in Parkinson's disease. Antioxidants 12(4):894
Nawaz T, Fahad S, Gu L, Xu L, Zhou R (2025) Harnessing nitrogen-fixing cyanobacteria for sustainable agriculture: opportunities, challenges, and implications for food security. Nitrogen 6(1):16
Odenthal K, Nunes E, Nunes N, Fernandes T, Fernandes IA, de Carvalho MA (2024) Microalgae and cyanobacteria as natural sources of antioxidant enzymes and enzyme inhibitors for Alzheimer's and diabetes. Algal Res 82:103610
Osman MEH, El-Sheekh MM, El-Naggar AH, Gheda SF (2010) Effect of two species of cyanobacteria as biofertilizers on some metabolic activities, growth, and yield of pea plant. Biol Fertil Soils 46(8):861–875
Ouyang J, Shao X, Li J (2000) Indole-3-glycerol phosphate, a branchpoint of indole-3-acetic acid biosynthesis from the tryptophan biosynthetic pathway in *Arabidopsis thaliana*. Plant J 24(3): 327–333
Parmar P, Kumar R, Neha Y, Srivatsan V (2023) Microalgae as next generation plant growth additives: functions, applications, challenges and circular bioeconomy based solutions. Front Plant Sci 14:1073546
Pattyn J, Vaughan-Hirsch J, Van de Poel B (2021) The regulation of ethylene biosynthesis: a complex multilevel control circuitry. New Phytol 229(2):770–782
Poveda J (2021) Cyanobacteria in plant health: biological strategy against abiotic and biotic stresses. Crop Prot 141:105450
Priya AK, Gnanasekaran L, Dutta K, Rajendran S, Balakrishnan D, Soto-Moscoso M (2022) Biosorption of heavy metals by microorganisms: evaluation of different underlying mechanisms. Chemosphere 307:135957
Rezayian M, Niknam V, Ebrahimzadeh H (2019) Stress response in cyanobacteria. J Appl Phycol 31(5):2773–2787

Ritonga FN, Zhou D, Zhang Y, Song R, Li C, Li J, Gao J (2023) The roles of gibberellins in regulating leaf development. Plants 12(6):1243

Robati SMS (2024) Cyanobacteria: a promising future for sustainable agriculture. In: Severo IA, Martínez-Burgos WJ, Ordonez J (eds) Insights into algae: fundamentals, culture techniques and biotechnological uses of microalgae and cyanobacteria. IntechOpen, London, pp 1–19. https://doi.org/10.5772/intechopen.1005021

Salazar-Cerezo S, Martínez-Montiel N, García-Sánchez J, Pérez-y-Terrón R, Martínez-Contreras RD (2018) Gibberellin biosynthesis and metabolism: a convergent route for plants, fungi and bacteria. Microbiol Res 208:85–98

Satta A, Esquirol L, Ebert B (2023) Current metabolic engineering strategies for photosynthetic bioproduction in cyanobacteria. Microorganisms 11(2):455

Shahwar D, Mushtaq Z, Mushtaq H, Alqarawi AA, Park Y, Alshahrani TS, Faizan S (2023) Role of microbial inoculants as biofertilizers for improving crop productivity: a review. Heliyon 9(6):e16736

Sinetova MA, Los DA (2016) New insights in cyanobacterial cold stress responses: genes, sensors, and molecular triggers. Biochim Biophys Acta 1860(11 Pt A):2391–2403

Sosnowski J, Truba M, Vasileva V (2023) The impact of auxin and cytokinin on the growth and development of selected crops. Agriculture 13(3):724

Sponsel VM, Hedden P (2004) Gibberellin biosynthesis and inactivation. In: Davies PJ (ed) Plant hormones: biosynthesis, signal transduction, action! Springer, Dordrecht, pp 63–94

Takei K, Ueda N, Aoki K, Kuromori T, Hirayama T, Shinozaki K et al (2004) AtIPT3 is a key determinant of nitrate-dependent cytokinin biosynthesis in *Arabidopsis*. Plant Cell Physiol 45(8):1053–1062

Tang J, Li Y, Zhang L, Mu J, Jiang Y, Fu H et al (2023) Biosynthetic pathways and functions of indole-3-acetic acid in microorganisms. Microorganisms 11(8):2077

Toribio AJ, Suárez-Estrella F, Jurado MM, López MJ, López-González JA, Moreno J (2020) Prospection of cyanobacteria producing bioactive substances and their application as potential phytostimulating agents. Biotechnol Rep 26:e00449

Uniyal S, Bhandari M, Singh P, Singh RK, Tiwari SP (2022) Cytokinin biosynthesis in cyanobacteria: insights for crop improvement. Front Genet 13:933226

Veetil TP, Angermayr SA, Hellingwerf KJ (2017) Ethylene production with engineered *Synechocystis* sp. PCC 6803 strains. Microb Cell Factories 16(1):34

Wu W, Du K, Kang X, Wei H (2021) The diverse roles of cytokinins in regulating leaf development. Hortic Res 8(1):118

Wu W, Cao S, Shi L, Chen W, Yin X, Yang Z (2023) Abscisic acid biosynthesis, metabolism and signaling in ripening fruit. Front Plant Sci 14:1279031

Yadav P, Singh R, Rana S, Joshi D, Kumar D, Bhardwaj N et al (2022) Mechanisms of stress tolerance in cyanobacteria under extreme conditions. Stress 2(4):531–549

Yang S, Wang J (2023) Auxin homeostasis in plant responses to heavy metal stress. Plant Physiol Biochem 205:108210

Yang Y, Deng Y, Cai C, Wang F, Fang Y, Duan X (2025) Biocrusts' impact on hydrological processes and erosion dynamics: a review. Geoderma 459:117383

Zhang JW, Wang R, Liang X, Han P, Zheng YL, Li XF, Gao DZ, Liu M, Hou LJ, Dong HP (2023) Novel gene clusters for natural product synthesis are abundant in the mangrove swamp microbiome. Appl Environ Microbiol 89(6):e00102–e00123

Algae-Derived Functional Foods, Advances in Bioactive Compounds, Health Benefits, and Applications

3

Madhumita Mondal, Ajay Kumar Kataria, Krushna Prasad Shadangi, and Soumya Sasmal

Abstract

Algae-derived functional foods are a rapidly growing field in global nutrition, healthcare, and sustainable food production. Functional foods can effectively leverage the biochemical repertoire of algae (both micro and macro), which includes high-quality protein, bioactive peptides, complex polysaccharides such as fucoidan, alginate, carrageenan, and beta-glucans. Algae are also rich in lipids enriched with essential ω-3 polyunsaturated fatty acids (notably EPA and DHA) as well as antioxidant fat-soluble pigments, carotenoids (like astaxanthin, fucoxanthin) and water-soluble pigment phycobilin, and various phenolic compounds. These components exhibit a broad spectrum of health-promoting effects, including antioxidant, anti-inflammatory, immunomodulatory, prebiotic, cardioprotective, and neuroprotective activities. This chapter provides a comprehensive overview of recent advances in algae-derived functional foods, highlighting the bioactive compounds, health effects, and innovative applications in the food industry. However, production variability, lack of quality control, high production costs, limitations due to taste and odor, low bioavailability, and a strict regulatory framework are hindering widespread application in this field. Recent

M. Mondal · S. Sasmal (✉)
Department of Biotechnology, Visva Bharati, Santiniketan, West Bengal, India
e-mail: soumya.sasmal@visva-bharati.ac.in

A. K. Kataria
Department of Biological Sciences and Engineering, Netaji Subhas University of Technology, New Delhi, India

K. P. Shadangi
Department of Chemical Engineering, Veer Surendra Sai University of Technology Burla, Sambalpur, Odisha, India

K. Mohanty et al. (eds.), *Advances in Algal Research*,
https://doi.org/10.1007/978-981-95-7234-2_3

studies highlighted the potential of advanced extraction technologies (enzyme-assisted ultrasound, supercritical fluid extraction), nano-encapsulation, genetic optimization, hybrid cultivation methods, and biosensor-based quality control systems. In the future, large-scale clinical validation, standardized protocols, consumer-acceptable formulations, and the inclusion of blue economy frameworks will play a significant role in making this industry viable and sustainable. The aim of this chapter is to highlight the role of algae as a sustainable solution for future nutrition and health, and to outline the challenges and opportunities that will shape the next generation of functional food products.

Keywords

Algae · Functional food · Bioactive · Nutraceuticals · Sustainable Development Goals

Abbreviations

ABTS	2,2′-Azino-bis (3-ethybenzothiazoline-6-sulfonic acid)
COX-2	Cyclooxygenase-2
DHA	Docosahexaenoic acid
DNA	Deoxyribonucleic acid
DPPH	2,2-Diphenyl-1-picrylhydrazyl
EPA	Eicosapentaenoic acid
FDA	Food and Drug Administration
HDL	High density lipoprotein
IL-6	Interleukin-6
iNOS	Inducible nitric oxide synthase
LDL	Low density lipoprotein
MAAs	Mycosporine-like amino acids
MAPKs	Mitogen-activated protein kinases
NF-κB	Nuclear factor kappa-light-chain-enhancer of activated B cells
NK	Natural killer cells
PUFAs	Poly unsaturated fatty acids
ROS	Reactive oxygen species
SCFA	Short-chain fatty acid
SDG	Sustainable Development Goals
TALENs	Transcription activator-like effector nucleases
TG	Triglycerides
TNF-alpha	Tumor necrosis factor-alpha
UV	Ultraviolet

3.1 Introduction

The current global food system is not just about satisfying hunger. It wants to play a significant role in improving the overall health of people and preventing diseases. Rapid urbanization, changes in lifestyle and eating habits, and an increase in fast food and processed foods are continuously increasing the incidence of diabetes, obesity, and other lifestyle-related diseases. These diseases cannot be cured by medicine alone, but balanced nutrition and healthy eating habits have become essential. The main challenges of the current global food system are overpopulation, climate change, and ensuring food security. It is essential to ensure not only the quantity of food but also the supply of healthy food and its diverse nutritional value. In this context, there is a growing interest in functional foods that not only provide health benefits but also help in increasing immunity.

Functional foods provide essential nutrition while actively supporting key bodily functions, such as disease prevention, inflammation reduction, cellular protection, immune system enhancement, and long-term health maintenance. Notably, algae are gaining increasing recognition for their role in these benefits (Temple 2022).

Algae have been ranked as an important functional food due to their bioactive components (such as polyphenols, carotenoids, ω-3 fatty acids, and polysaccharides). Algae species are rich in nutrients like protein, fiber, vitamins (especially B12), and minerals (such as iodine and iron). Antioxidant-rich compounds are present in algae that increase immunity, control cholesterol, and exhibit anti-inflammatory and anti-cancer properties in the human body (Mirza et al. 2025). Especially microalgae such as *Spirulina* sp. and *Chlorella* sp. can play a role in addressing food and nutritional deficiencies, which can be an important solution for developing countries. Algae are an alternative source for vegan diets. Algae have recently attracted significant interest because of their wide-ranging potential applications in biofertilizers, food products, nutraceuticals, pharmaceuticals, animal feed, and biofuel production (Singh et al. 2023).

Microalgae, rich in essential nutrients and bio-compounds for life, have become an effective food ingredient to help achieve the United Nations Sustainable Development Goals (SDGs). In addition to their nutritional and therapeutic potential, algae also represent a pivotal renewable resource for sustainable energy production (Kataria et al. 2025a). The Sustainable Development Goals, adopted by the United Nations in 2015, are a global call to end poverty, protect the planet, and ensure prosperity for all by 2030. These 17 interconnected goals provide a blueprint for addressing complex economic, social, and environmental challenges (Fig. 3.1). Achieving these goals requires creative, multifaceted, and nature-based solutions. In this context, from tiny microalgae to large-scale seaweed, they are increasingly being recognized as a versatile and sustainable resource. In addition, the inclusion of microalgae as functional foods, food supplements, and medicines makes it possible to contribute to public health and socio-economic development. Although the potential of algae is immense, some challenges still exist, such as limitations in large-scale production, the cost of technology, the policy and legal framework to ensure product acceptance in the market, and the safety of the product. Overcoming

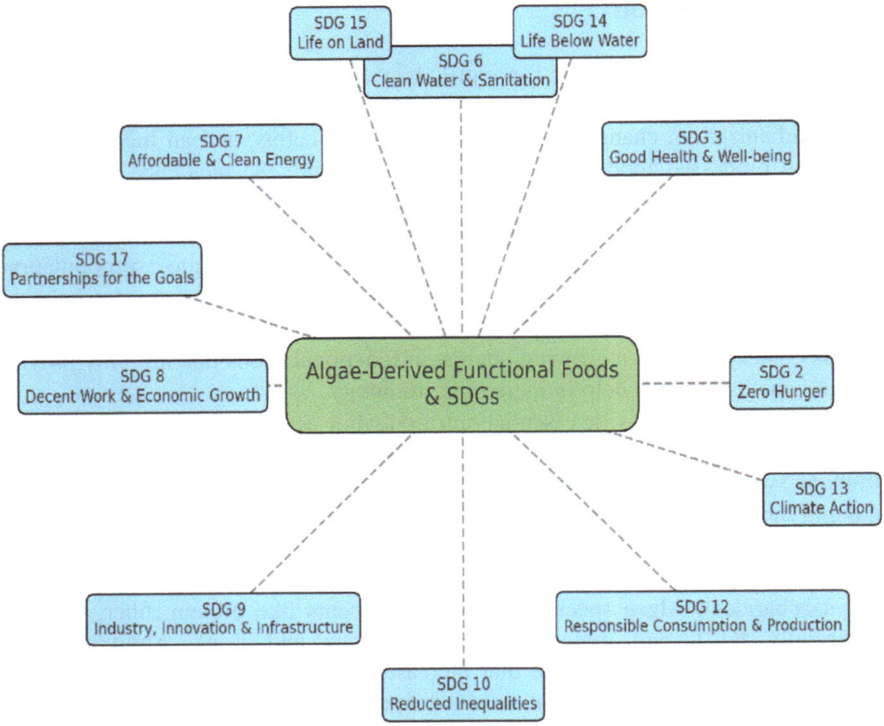

Fig. 3.1 Linkages between algae-derived functional foods and the United Nations Sustainable Development Goals (SDGs). Algae contribute to nutrition and health (SDG 2, SDG 3), environmental protection (SDG 6, SDG 13, SDG 14, SDG 15), sustainable energy and industry (SDG 7, SDG 9, SDG 12), social equity and economic growth (SDG 8, SDG 10), and global collaboration (SDG 17), underscoring their role as a sustainable solution for future food systems

these obstacles requires a coordinated policy, investment in research and development, and the creation of a supportive market environment. The production capacity of microalgae is increasing with the help of biotechnology innovations (Hosny et al. 2025). Therefore, including algae in the global food system means not only meeting nutritional needs but also moving toward a sustainable and eco-friendly food and future.

The purpose of this chapter is to provide a comprehensive overview of algae-derived functional foods, with a focus on their bioactive compounds, nutritional significance, health benefits, and potential applications in the food and nutraceutical industries. It aims to highlight the role of microalgae and macroalgae as sustainable and versatile resources that contribute to human health, food security, and environmental sustainability. This chapter also examines recent technological advancements in extraction, processing, and product development, while addressing the current limitations of bioavailability, production costs, and regulatory challenges. By doing so, it underscores the potential of algae as a next-generation solution for achieving global nutrition, public health, and Sustainable Development Goals (SDGs) within the framework of the blue and circular bioeconomy.

In recent years, technological innovation has played a pivotal role in unlocking the full potential of algae-derived functional foods. Advances in extraction, processing, and formulation have enabled efficient recovery of bioactive compounds such as carotenoids, polysaccharides, and ω-3 fatty acids while preserving their nutritional integrity (Kataria et al. 2025b). Emerging green extraction methods—including enzyme-assisted, ultrasound, and supercritical CO_2 techniques—along with nanoencapsulation and biopolymer-based delivery systems, are enhancing the stability and bioavailability of algal ingredients. These developments, supported by biosensor-based quality control and evolving regulatory frameworks, are shaping a new generation of sustainable, consumer-ready functional food products derived from algae.

3.2 Types of Algae

Table 3.1 highlights commonly used microalgae and macroalgae in functional foods, detailing their key nutrients, bioactive compounds, and associated health benefits. Species such as *Spirulina* sp. and *Chlorella* sp. provide high-quality proteins and immunity-boosting compounds, while *Dunaliella* and *Haematococcus pluvialis* are rich in carotenoids with strong antioxidant properties. Macroalgae like *Nori*, *Chondrus crispus*, Kelp, and *Ulva* contribute essential minerals, polysaccharides, and fibers with roles in gut health, metabolic regulation, and disease prevention. Collectively, these algae demonstrate significant potential as sustainable ingredients in nutraceuticals and functional food applications.

Table 3.1 Types of algae used in functional foods: nutrients, health benefits

Algae	Key nutrients/bioactive	Health benefits/applications
Spirulina (microalgae)	Protein, vitamin B complex, iron, magnesium	Boosts immunity, controls cholesterol, increases energy, anti-inflammatory
Chlorella (microalgae)	Chlorophyll, protein, iron, vitamin C	Detoxification, immunity enhancement, digestion improvement
Dunaliella (microalgae)	Beta-carotene, protein, vitamin A precursor, ω-3 fatty acids	Antioxidant activity, eye health, immune support; used in food and cosmetics
Haematococcus pluvialis (microalgae)	Astaxanthin (potent antioxidant)	Strong antioxidant, neuroprotective, cardioprotective; popular in nutraceuticals

Ampofo and Abbey (2022)

3.3 Nutritional Composition of Algae

3.3.1 Protein and Amino Acid

Algae play an important role as a functional food in the food system due to their high-quality protein and amino acid content. Microalgae are particularly rich sources of protein and amino acids and have health benefits. Algae contain 9 essential amino acids out of 20 amino acids that are not synthesized in our body, and it is taken through diet. Essential amino acids are leucine, isoleucine, valine, lysine, methionine, threonine, tryptophan, phenylalanine, and histidine. The amino acid profile of algae protein is generally equal to or superior to conventional protein sources such as soybeans and eggs (Moniruzzaman et al. 2023). *Scenedesmus obliquus* species have high levels of essential amino acids such as phenylalanine and leucine, making this species a nutritious protein source (García-Encinas et al. 2025).

3.3.2 Carbohydrates

The carbohydrate content of algae is rich in a variety of bioactive compounds. These carbohydrates include simple sugars, polysaccharides, and fiber, which not only provide energy but also act as prebiotics, antioxidants, and immune modulators. Studies have shown that microalgae contain high levels of carbohydrates, such as cellulose, which are beneficial for both health and the production of biofuels. For example, *Tetraselmis* sp., *Dunaliella* sp., and *Chlorella* sp. are rich in carbohydrates. Spirulina sp. is rich in glycogen and rhamnose, which are easily digestible and suitable as prebiotics. *Porphyridium* sp. contains sulfated polysaccharides that are effective in preventing inflammation and viruses. Marine red and brown algae contain high-value polysaccharides such as fucoidan, alginate, and carrageenan, and their antiviral, anti-tumor, and anti-coagulant activity has been shown to be effective. Algal carbohydrates also help to grow beneficial bacteria in the gut and improve gut health. In addition, some microalgae polysaccharides can increase insulin sensitivity, which is helpful in diabetes management (Conde et al. 2021).

3.3.3 Fatty Acids and Lipids

Algae, especially microalgae, are considered as rich source of lipids and explored for an effective way to obtain their nutritional lipids and beneficial fatty acid (Kataria et al. 2023). The lipids present in algae include important PUFAs such as ω-3 and ω-6, which have great efficacy on disease prevention, brain development, and inflammation control (Chen et al. 2023).

Microalgae species like *Chlorella vulgaris* and *Nannochloropsis gaditana* can contain 20%–60% lipids in dry weight (Chua and Schenk 2017). It was previously considered that ω-3s were only found in marine fish or fish oil, but modern research has shown that the main source of these ω-3 fatty acids is algae consumed by fish in the food chain. The microalgae species *Crypthecodinium cohnii* and *Schizochytrium* sp. have shown to produce particularly high levels DHA and in some cases EPA. ω-3 made from algae is being widely used as a soft gel capsule, children's syrup, and pregnancy supplement (Kunj et al. 2025).

3.3.4 Vitamins and Minerals

Algae are an excellent source of various water- and fat-soluble vitamins. *Spirulina* sp. and *Chlorella* sp. are rich in provitamin A, B12, and vitamin C. Red and green algae are high in beta-carotene and vitamin E (Menaa et al. 2021). Algae contain large amounts of sodium, potassium, calcium, magnesium, phosphorus, iron, zinc, copper, and iodine. The mineral content of algae is generally higher than that of terrestrial plants and foods, especially iodine, iron, and calcium. Iodine is important for thyroid hormones and is easily obtained from algae. Algae are naturally nutrient-dense foods, rich in vitamins and minerals that can help meet the nutritional needs when included in the diet regularly (Cadar et al. 2025).

3.3.5 Dietary Fiber

Algae, especially microalgae, play an important role in the diet because they contain a large amount of dietary fiber. Studies have shown that microalgae can make up about 14%–17% of the dry weight of food, which helps in the body's digestive system, gut health, and cholesterol control. Dietary fiber from microalgae is mainly polysaccharides that act as prebiotics and help the growth of beneficial bacteria in the body, such as polysaccharides obtained from red microalgae *Porphyridium* sp., which help reduce blood cholesterol levels and improve intestinal structure. Dietary fiber from algae can play a role in regulating the body's metabolism, including antioxidant and anti-inflammatory properties. It contains natural soluble fiber, which can increase immunity and reduce inflammation. Recent studies have supported these important functions of microalgae in the food chain, and therefore, algae-derived foods have been found in the market as nutraceuticals and functional foods.

3.4 Advances in Bioactive Compounds of Algae as Functional Food

Naturally produces a variety of bio-compounds that have important nutritional, therapeutic, and nutraceutical properties. Recent studies have created new perspectives on the efficacy of these algae-derived compounds, which can be used to protect human health and prevent disease.

3.4.1 Polysaccharide

Research has found important biological activities of polysaccharide compounds in seaweed, for example, antiviral, antioxidant, and other effects. The sulfated polysaccharides, which are found in red algae (carrageenan) and brown algae (fucoidan), help prevent viral infections and boost the body's immune system. The sulfated polysaccharides protect the body by scavenging reactive oxygen species (ROS) and have beneficial effects on the body. The polysaccharides obtained from seaweed have a variety of high molecular weights and complex structures, which ensure their various medicinal and industrial uses. These are mainly antiviral drugs, immunomodulators, and have potential in the nutritional and cosmetic fields. Different types of polysaccharides are as follows:

3.4.1.1 Carrageenan

Carrageenan is a type of sulfated polysaccharide mainly collected from red algae like *Chondrus crispus* and *Kappaphycus alvarezii*. It is widely used in the food industry to thicken and increase stability of food, gelling agent. Carrageenan is essentially a repeating chain of galactose units with sulfate groups attached. There are three types of carrageenan. The first is a continuous chain and one sulfate group that forms a hard gel. The second type of carrageenan, bearing two sulfate groups, forms a soft, elastic gel. The third one does not form a gel but rather increases the density. Carrageenan exhibits numerous biological properties (Hossain et al. 2024). These biological properties are:

Antiviral: Carrageenan prevents viruses from entering cells and is effective in preventing viral infections due to its high sulfate content, because it interferes with viral interactions at the cell surface (Rupert et al. 2022). It is effective in preventing sexually transmitted infections and respiratory tract infections. Studies have shown that it can reduce the transmission and spread of influenza viruses when used as a nasal spray, which is effective in preventing infections.

Antioxidant: Carrageenan can neutralize free radicals such as superoxide, hydroxyl, and dioxygen radicals. Free radicals damage the DNA, proteins, and lipids of cells. Carrageenan reduces oxidative stress in cells by neutralizing free radicals. Oxidative stress increases the risk of heart disease, diabetes, cancer, and other degenerative diseases. Studies have shown that the rich in carrageenan extracts DPPH, ABTS, and other scavenging ash show a strong antioxidant function (Dong et al. 2021).

Anti-inflammatory: Carrageenan significantly reduces the production of inflammatory cytokines like TNF-alpha, IL-6, and IL-1 beta in immune cells. These cytokines are the root cause of inflammation. When these are reduced, the level of inflammation also reduces. Carrageenan blocks inflammatory pathways and puts down oxidative stress. Most of the research has been done on in vitro and animal models where carrageenan has been shown to be effective to reduce inflammations (Zamanabadi et al. 2025).

3.4.1.2 Alginate

Alginate is a natural polymer and dietary fiber usually extracted from brown seaweed. It is biocompatible and non-toxic for human consumption, reported by the USFDA as GRAS (Generally Recognized as Safe). It can be easily produced and used in various biomedical and functional food applications (Li et al. 2025). The uses of alginate are:

Gut health and probiotic effect: Alginate helps in the growth of beneficial bacteria in the gut and increases the production of short chain fatty acids which are helpful in preventing major diseases of the colon such as ulcerative colitis (Mollah et al. 2021).

Reduce cholesterol: Alginate binds to cholesterol and fatty acids in the intestines and prevents them from being absorbed, thereby reducing the levels of cholesterol and fat in the blood. Regular consumption of it improves the blood lipid profile and helps reduce the risk of heart disease (Li et al. 2025).

3.4.1.3 Beta Glucans

Beta glucans, especially algal-derived beta glucans, such as paramylons, have been identified in recent studies as a bioactive compound that is important for as functional foods and nutraceuticals. Paramylon is usually extracted from Euglenoid algae (*Euglena gracilis*) and is a beta one three glucan type polysaccharide that is water soluble and not easily digested. Studies have shown that beta-glucan acts as a soluble fiber in the intestinal environment, thereby helping to regulate bowel movements and nutrient absorption, and plays a significant role in controlling blood glucose and cholesterol. With regular consumption, it can reduce LDL cholesterol and reduce the risk of heart disease. It was to be effective in reducing inflammation and cancer by making health claims approved by the FDA in the United States. It also reduces the risk of obesity and metabolic syndrome and improves glycemic control (El Khoury et al. 2012). The unique structural and functional properties of paramylon make it highly suitable for use in food supplements and pharmaceutical products. Recent studies have clearly demonstrated the antioxidant, anti-inflammatory, and immunomodulatory properties of algal beta-glucan, establishing it as an important ingredient for modern functional foods (Singla et al. 2024).

3.4.1.4 Fucoidan

Fucoidan is a sulfated polysaccharide derived from a brown alga (*Phaeophyceae*). Its unique structure comprising L-Fucose, sulfate group, uronic acid, and other monosaccharides confer versatile bio-properties that are very important for use in functional foods and nutraceuticals (Yue et al. 2025). The activities shown by fucoidan are as follows:

Antioxidant activity: Fucoidan is a powerful free radical scavenger and antioxidant due to degree of sulfation. Recent studies have shown that Fucoidan from *Sargassum fusiforme* can protect liver cells from harmful oxidative stress (Yue et al. 2025).

Anticancer activity: Fucoidan helps in causing cell death of cancer cells, preventing cell division, and reducing metastasis. It works through proteins such as caspase-3 (Yue et al. 2025) and has shown to have strong anti-cancer efficacy in In-vitro and In-silico studies (Mualimin and Arum 2025).

Antimicrobial and antiviral properties: Fucoidan shows antibacterial, anti-fungal, and antiviral effects. Fucoidan can be used as a food preservative and natural therapy against *E. coli*, *Bacillus cereus*, *Candida albicans*, etc. (Trang Thuy and Men 2025).

Metabolic and cardiovascular benefits: Fucoidan helps in controlling metabolic syndrome, maintaining gut health, and exhibits anti-inflammatory properties, helping in preventing heart disease and metabolic problems (Anisha et al. 2022).

Food packaging and drug delivery: Fucoidan has film-forming ability and biocompatibility that make it suitable for food packaging, edible foods coating, and transporting pharmaceuticals (Anisha et al. 2022). Fucoidan is suitable as a promising vehicle for drug delivery (notably in nanodrug delivery system) due to its non-toxic, biodegradable properties (George and Shrivastav 2023).

3.4.2 Peptide

Algae has great potential as a functional food due to its high content of amino acids and bio-peptides (Baravia et al. 2025). Microalgae and seaweed contain a large amount of essential amino acids and short chain peptides, which play a role to prevent diseases and control metabolism. Short-chain peptide (2–20 amino acid) prepared from algae proteins through enzymatic hydrolysis show various bio-properties.

The digestibility of some algae proteins is relatively low. Large-scale clinical studies are also needed on the development of seven flavors and textures and their effectiveness in humans (Karabulut et al. 2024).

3.4.3 ω-3 Fatty Acid (EPA, DHA)

Algae, which is rich in ω-3 fatty acids, is a very important bioactive compound in functional foods (Kunj et al. 2025). ω-3 fatty acids are a type of polyunsaturated fat that is essential for the human body and cannot be produced by the body itself. Although they are usually found in fish and seafood, their main source is seaweed and phytoplankton, from which fish also obtain this element (Derbyshire et al. 2024).

3.4.4 Carotenoids

Carotenoids are natural pigments found in abundance in microalgae and seaweed. Among them Astaxanthin, fucoxanthin, beta-carotene, and lutein are significant. Recent studies have shown that these compounds are very important for the antioxidant, anti-inflammatory, and health properties in functional food. Carotenoid

Table 3.2 Comparative chart summarizing the different types of carotenoids derived from algae, their main algal sources, functional food applications, and key health benefits (Generalić Mekinić et al. 2023; Razzak 2024)

Carotenoids	Algal source(s)	Functional food applications	Health benefits
Astaxanthin	*Haematococcus pluvialis*	Supplements, beverages, dairy	Strong antioxidant, anti-inflammatory, neuroprotective, cardiovascular support
Fucoxanthin	Brown algae (*Undaria*, *Phaeodactylum*)	Functional snacks, bakery, dairy alternatives	Anti-obesity, anti-diabetic, antioxidant, anti-cancer
Beta carotene	*Dunaliella salina*, *Chlorella*, *Spirulina*	Dairy alternatives, food supplements	Provitamin A, antioxidant, immune support
Lutein	*Chlorella*, *Spirulina*	Beverages, supplements, functional foods	Eye health, antioxidant, anti-inflammatory
Zeaxanthin	*Spirulina*, *Chlorella*	Food colorant, supplements	Eye health, antioxidant

prevents cancer, aging, and other diseases by blocking free radicals. Algae derived carotenoids are revolutionizing the health and food industries, but more research is needed to standardize production technology and clinical validity (Razzak 2024; Huang et al. 2025). The different types of carotenoids, their algal sources, functional food applications, and key health benefits are shown in Table 3.2.

3.4.5 Polyphenols

Algae, especially seaweed, are considered an important source of bio-compounds such as polyphenols. These compounds play an important role in protecting health and are suitable for use in functional foods. Brown algae contain a special type of polyphenol called phlorotannins. In addition, marine algae also contain flavonoids and phenolic acids (Nagahawatta et al. 2024). Algae-derived polyphenols are currently being used in functional foods and supplements as natural ingredients for health protection and food preservation. They are especially suitable for vegetarians and health-conscious human beings. The number of polyphenols extracted from algae is still relatively low, but new technology and research are making it possible to overcome this problem.

3.4.6 Phycobiliproteins

Phycobiliproteins (absorbs light having wavelengths of 450–650 nm) are brightly colored water-soluble proteins found in aquatic algae and cyanobacteria that play an important role in photosynthesis, mainly through light absorption (Tiwari et al. 2020).

Types of phycobiliproteins:

(a) Phycocyanin: Blue absorbs light at a wavelength of 610–625 nm.
(b) Phycoerythrin: Red or pink absorbs light at a wavelength of 490–570 nm.
(c) Allophycocyanin: Located in the center of phycobilisome and absorbs light of about 650–660 nm.

One of the most important properties of phycobiliproteins are their antioxidant and anti-inflammatory properties. They protect cells by neutralizing harmful reactive oxygen species and reduce the chemical effects of inflammation. In addition, Phycobiliproteins have been found to have antimicrobial, antiviral, and anti-cancer effects in studies.

Various studies have developed advanced technologies for the extraction and purification of phycobiliproteins, which have resulted in the production of high purity phycocyanin and allophycocyanin. These proteins are used in food, medicine, and biotechnology as natural colors, as healthy antioxidants, and as ingredients in anti-inflammatory drugs. For large-scale production, cyanobacteria and red algae are cultivated using advanced technology where light and nutrient conditions are controlled.

3.5 Health Benefits of Algal Derived Functional Food

3.5.1 Metabolic Health

3.5.1.1 Glucose and Lipid Control

Recent clinical and animal studies have shown that regular consumption of algae and algae derived bioactive compounds can help control blood glucose, blood lipid levels and algae derived PUFAs, Polysaccharides, and peptides are effective in preventing diabetes and high cholesterol.

A study conducted with brown algae (Fucoxanthin) showed reduced insulin sensitivity and liver fat accumulation (Kim et al. 2023).

3.5.1.2 Glucose and Insulin Homeostasis

Fucoxanthin and Carotenoids Fucoxanthin from brown algae and carotenoids in species like *Phaeodactylum tricornutum* boosts glucose uptake by activating AMPK pathways in muscle, supporting increased fat burning and better glycemic control (Oliyaei et al. 2021).

Bioactive Peptides and Polysaccharides Few microalgae peptides (e.g., *Spirulina* sp.) stimulate incretin hormone release, enhancing insulin secretion. Polysaccharides from *Chlorella* sp. and *Porphyridium* sp. retard glucose absorption, reducing post-meal glucose spikes, antidiabetic and anti-obesity effects (García-Encinas et al. 2025).

Polyphenols (Flavonoids, Phlorotannin) Polyphenols from brown and red algae inhibit carbohydrate-digesting enzymes (α-amylase, α-glucosidase), helping control blood sugar after meals (Thitame and Aher 2025).

Fiber and Prebiotics Microalgal fibers act as prebiotics, improving the gut microbiome, enhancing metabolic function, and reducing systemic inflammation.

Antioxidant and Anti-inflammatory Compounds Carotenoids, lutein, vitamin E, and phycocyanin in microalgae fight oxidative stress and inflammation, improving insulin sensitivity and protecting liver health (Razzak 2024).

3.6 Algal Derived Functional Food Application

3.6.1 Dairy and Bakery Products

Microalgae (*Spirulina* sp., *Chlorella* sp.) are added to foods like yogurt, cheese, and bread to increase protein, vitamins, minerals, and antioxidants content that provide health benefits. Microalgae fortified foods not only add nutrients but also adding natural color to the food (Hernández et al. 2022; Andrade-Bustamante et al. 2025).

3.6.2 Drinks and Gel

Algae produced extract is being used in various health drinks and in gel because it contains antioxidants, anti-inflammatory and immune-boosting ingredients, and gelling agents. Phycocyanin from spirulina and astaxanthin from *Haematococcus* sp. are popular for their health-promoting effects (Martínez-Ruiz et al. 2025).

3.7 Technological Advances in Algal-Based Functional Food

Recent progress in extraction, processing, formulation, and quality control has transformed the utilization of algae from traditional food supplements into scientifically validated functional food ingredients. These technologies aim to enhance the yield, stability, sensory quality, and bioavailability of valuable bioactive while maintaining sustainability and economic viability.

3.7.1 Advance Extraction Techniques

Efficient recovery of bioactive compounds from algae requires disruption of the robust cell wall and selective extraction of intracellular metabolites. The recent advancements have replaced conventional solvent extraction methods by greener, energy-efficient techniques (Mishra et al. 2025). The currently adopted techniques include the following:

Enzyme-assisted extraction (EAE): Hydrolytic enzymes such as cellulase, pectinase, and protease break down the complex polysaccharide matrix, increasing yields of pigments, peptides, and polyphenols with minimal solvent use. EAE from *Chlorella* and *Porphyridium* has improved recovery of phycobiliproteins and antioxidant polysaccharides.

Ultrasound-assisted extraction (UAE): Cavitation enhances cell disruption, reduces extraction time, and lowers solvent consumption. UAE has been successfully applied for fucoxanthin, astaxanthin, and omega-3 fatty acids.

Microwave-assisted extraction (MAE): Rapid volumetric heating allows efficient recovery of thermolabile compounds such as β-carotene and phenolics from *Dunaliella salina* and *Sargassum* sp.

Supercritical CO_2 extraction (SCE): Recognized as a green, solvent-free method for lipid-soluble compounds. SCE enables isolation of EPA/DHA-rich oils and carotenoids from micro-algae with high purity and without residual solvents.

Aqueous two-phase and deep eutectic solvent systems: New eco-friendly media provide selective partitioning of phycocyanin and fucoidan while maintaining their functional integrity.

Integration of multiple techniques, such as enzyme-ultrasound or microwave-supercritical combinations, has further enhanced efficiency and reduced processing costs, aligning with green chemistry principles.

3.7.2 Processing and Stabilization of Algal Bioactive

Post-extraction stabilization is critical because algal bioactive are sensitive (mostly) to light, oxygen, and temperature. Innovative encapsulation and formulation strategies are emerging:

Nano- and micro-encapsulation: Bioactive are embedded within nano emulsions or biopolymeric matrices (alginate, chitosan, carrageenan), improving dispersibility, taste masking, and controlled release. Phycocyanin-loaded alginate nanoparticles and fucoxanthin nano emulsions have demonstrated enhanced antioxidant retention and intestinal absorption.

Spray drying and freeze drying: Widely used for producing stable algal powders for incorporation into bakery, dairy, and beverage systems. Controlled drying protects heat-labile pigments and fatty acids.

Emulsification and lipid-based carriers: Self-emulsifying systems improve solubility of hydro-phobic carotenoids and ω-3 oils, providing better sensory acceptance and extended shelf life.

Biopolymer blending: Combining algal hydrocolloids (alginate, carrageenan, β-glucan) with plant proteins produces novel textures suitable for vegan meat analogues and functional gels.

These technological advances bridge the gap between laboratory discovery and marketable food products.

3.7.3 Product Development and Industrial Integration

Modern functional food formulations now incorporate algae at different stages of processing:

Dairy and plant-based analogues: Microalgae biomass enhances protein, vitamin B12, and antioxidant levels in yogurts, cheeses, and fermented beverages while imparting natural color.

Bakery and cereal products: Addition of *Spirulina* or *Chlorella* powder improves amino-acid balance and antioxidant properties without compromising sensory attributes when optimized at low inclusion levels.

Beverages and nutraceuticals: Encapsulated astaxanthin, phycocyanin, and fucoxanthin are employed in fortified juices, smoothies, and health drinks with improved stability.

Edible films and packaging: Polysaccharide-based coatings derived from fucoidan and alginate act as biodegradable, antioxidant films, extending food shelf life and reducing plastic use.

Pilot-scale biorefineries have now adopted integrated valorization, using one algal biomass, to produce multiple products (proteins, lipids, pigments); thereby achieving circular bioeconomy goals.

3.7.4 Quality Control and Regulatory Considerations

Ensuring safety, consistency, and consumer trust remains a challenge. The related advances include the following:

Biosensor-based quality control: Optical and electrochemical biosensors detect oxidation, pigment degradation, and microbial contamination in real time, enabling traceable quality assurance.

Standardization and certification: The U.S. Food and Drug Administration (FDA) recognize several algal products (e.g., *Spirulina platensis*, *Schizochytrium* sp. oil) as Generally Recognized as Safe (GRAS). European Food Safety Authority (EFSA) and Food Safety and Standards Authority of India (FSSAI) in India are expanding novel-food approvals.

Safety assessment: Genomic and metabolomic profiling helps identify potential toxins and ensures non-GMO labeling.

Consumer perception and sensory optimization: Formulation scientists employ flavor-masking and color-balancing technologies to enhance acceptance, a crucial factor for large-scale commercialization.

3.7.5 Outlook

Technological convergence is reshaping the algae-based food sector. Green extraction technologies lower environmental impact, nanotechnology improves delivery efficiency, and smart biosensors guarantee safety and traceability. The next phase of research will integrate AI-driven process optimization and biotechnological strain improvement (CRISPR/TALEN-based engineering) to boost metabolite yield. Collectively, these innovations will reduce production costs, overcome bioavailability constraints, and promote algae-derived ingredients as mainstream components of the global functional food market.

3.8 Conclusion

This chapter underscores algae as a nutritionally rich, therapeutic, and industrially versatile bioresource that can transform the landscape of functional food innovation. Both microalgae (such as *Spirulina* sp., *Chlorella* sp., *Dunaliella* sp., and *Haematococcus* sp.) and macroalgae (such as kelp, *Ulva*, *Nori*, and *Chondrus crispus*) provide a unique biochemical matrix of high-value compounds—proteins, peptides, polysaccharides (fucoidan, carrageenan, alginate, β-glucans), ω-3 fatty acids (EPA, DHA), carotenoids (astaxanthin, fucoxanthin, β-carotene, lutein), polyphenols, and phycobiliproteins. These bioactive demonstrate multifaceted physiological benefits, including antioxidant, anti-inflammatory, immunomodulatory, cardiometabolic, neuroprotective, and gut-health–promoting effects, validating algae's role as a foundation for preventive nutrition and therapeutic formulations.

Recent advances in extraction and formulation science, particularly enzyme-assisted, ultrasound, microwave, and supercritical CO_2 extraction, are replacing traditional solvent-based methods, reducing environmental impact while improving yield, purity, and bioactive stability. Concurrently, encapsulation technologies (nanoencapsulation and microencapsulation) have revolutionized the delivery of algal compounds, enhancing solubility, bioavailability, and controlled release within the digestive tract. These innovations have facilitated the incorporation of algal ingredients into functional foods, beverages, nutraceuticals, and drug delivery platforms with improved sensory quality and consumer acceptance.

From a systems perspective, integrated algal biorefineries and circular bioeconomy frameworks now enable the co-production of proteins, pigments, polysaccharides, and lipids from a single biomass stream, markedly improving resource utilization and cost efficiency. Such sustainable models align with global priorities for climate resilience, waste minimization, and low-carbon production. The application of digital tools, including automation, photobioreactor optimization, and machine learning, further enhances scalability and industrial viability.

Overcoming sensory and regulatory challenges remains critical to reaching full market maturity. Techniques such as biopolymer-based encapsulation (alginate, chitosan, carrageenan) and natural flavor balancing are successfully addressing taste and odor limitations, while evolving international guidelines ensure safety, traceability, and product standardization.

Conclusively, technological, biotechnological, and industrial synergies are rapidly bridging the gap between laboratory discovery and commercial application. Algae-derived functional foods exemplify the convergence of nutrition science, biotechnology, and sustainability offering a realistic, scalable pathway toward global food security, public health advancement, and environmental preservation. Continued interdisciplinary research and innovation will be essential to fully harness algae's potential as a cornerstone of next-generation functional foods and to advance the United Nations Sustainable Development Goals (SDGs) through a blue and circular bioeconomy paradigm.

References

Ampofo J, Abbey L (2022) Microalgae: bioactive composition, health benefits, safety and prospects as potential high-value ingredients for the functional food industry. Foods 11(12):1744

Andrade-Bustamante G, Martínez-Ruiz FE, Ortega-García J, Renganathan P, Gaysina LA, Mahendhiran M et al (2025) Microalgae-based functional foods: a blue-green revolution in sustainable nutrition and health. Appl Microbiol 5(2):39

Anisha GS, Padmakumari S, Patel AK, Pandey A, Singhania RR (2022) Fucoidan from marine macroalgae: biological actions and applications in regenerative medicine, drug delivery systems and food industry. Bioengineering (Basel) 9(9):472

Baravia PA, Qhabibi FR, Hadinata E, Katuuk SHH, Alfaray RI, Rampengan DD et al (2025) Algal peptides for future food: key players and multifaceted agents in the fight against cardiometabolic syndrome. J Agric Food Res 20:101786

Cadar E, Popescu A, Dragan AML, Pesterau AM, Pascale C, Anuta V et al (2025) Bioactive compounds of marine algae and their potential health and nutraceutical applications: a review. Mar Drugs 23(4):152

Chen W, Li T, Du S, Chen H, Wang Q (2023) Microalgal polyunsaturated fatty acids: hotspots and production techniques. Front Bioeng Biotechnol 11:1146881

Chua ET, Schenk PM (2017) A biorefinery for Nannochloropsis: induction, harvesting, and extraction of EPA-rich oil and high-value protein. Bioresour Technol 244:1416–1424

Conde TA, Neves BF, Couto D, Melo T, Neves B, Costa M et al (2021) Microalgae as sustainable bio-factories of healthy lipids: evaluating fatty acid content and antioxidant activity. Mar Drugs 19(7):357

Derbyshire EJ, Birch CS, Bonwick GA, English A, Metcalfe P, Li W (2024) Optimal omegas–barriers and novel methods to narrow omega-3 gaps. A narrative review. Front Nutr 11:1325099

Dong Y, Wei Z, Xue C (2021) Recent advances in carrageenan-based delivery systems for bioactive ingredients: a review. Trends Food Sci Technol 112:348–361

El Khoury D, Cuda C, Luhovyy BL, Anderson GH (2012) Beta glucan: health benefits in obesity and metabolic syndrome. J Nutr Metab 2012:851362

García-Encinas JP, Ruiz-Cruz S, Juárez J, Ornelas-Paz JDJ, Del Toro-Sánchez CL, Márquez-Ríos E (2025) Proteins from microalgae: nutritional, functional and bioactive properties. Foods 14(6): 921

Generalić Mekinić I, Šimat V, Rathod NB, Hamed I, Čagalj M (2023) Algal carotenoids: chemistry, sources, and application. Foods 12(14):2768

George A, Shrivastav PS (2023) Fucoidan, a brown seaweed polysaccharide in nanodrug delivery. Drug Deliv Transl Res 13(10):2427–2446

Hernández H, Nunes MC, Prista C, Raymundo A (2022) Innovative and healthier dairy products through the addition of microalgae: a review. Foods 11(5):755

Hosny S, Elshobary ME, El-Sheekh MM (2025) Unleashing the power of microalgae: a pioneering path to sustainability and achieving the sustainable development goals. Environ Sci Pollut Res Int 32:1–31

Hossain MM, Sultana F, Khan S, Nayeema J, Mostafa M, Ferdus H et al (2024) Carrageenans as biostimulants and bio-elicitors: plant growth and defense responses. Stress Biol 4(1):3

Huang JJ, Xie Q, Lin S, Xu W, Cheung PCK (2025) Microalgae-derived astaxanthin: bioactivities, biotechnological approaches and industrial technologies for its production. Crit Rev Food Sci Nutr 65:1–35

Karabulut G, Purkiewicz A, Goksen G (2024) Recent developments and challenges in algal protein and peptide extraction strategies, functional and technological properties, bioaccessibility, and commercial applications. Compr Rev Food Sci Food Saf 23(3):e13372

Kataria AK, Mohanty K, Sasmal S (2023) Algal biofuel and its prospects. In: Bioenergy engineering. Woodhead Publishing, pp 299–314

Kataria AK, Dubey AK, Mohanty K, Sasmal S (2025a) Pretreatment strategy for enhanced lipid extraction from algal biomass in bio-oil production. Indian J Chem Technol 32(2):190–202

Kataria AK, Dubey AK, Mohanty K, Sasmal S (2025b) Pyrolysis-derived algal oil: a sustainable and high-performance green bio-lubricant. Indian J Chem Technol 32(5):584–592

Kim YR, Park MJ, Park SY, Kim JY (2023) Brown seaweed consumption as a promising strategy for blood glucose management: a comprehensive meta-analysis. Nutrients 15(23):4987

Kunj P, Sahu S, Singh G, Arya SK (2025) A review on algal oils as a sustainable source of essential omega-3 fatty acids. Bioresour Technol Rep 30:102154

Li Y, Liu J, Yao L (2025) Preparation and properties of drug-carrying microspheres of konjac glucomannan/sodium alginate/hydroxyapatite. Biochem Biophys Res Commun 761:151708

Martínez-Ruiz FE, Andrade-Bustamante G, Holguín-Peña RJ, Renganathan P, Gaysina LA, Sukhanova NV et al (2025) Microalgae as functional food ingredients: nutritional benefits, challenges, and regulatory considerations for safe consumption. Biomass 5(2):25

Menaa F, Wijesinghe U, Thiripuranathar G, Althobaiti NA, Albalawi AE, Khan BA et al (2021) Marine algae-derived bioactive compounds: a new wave of nanodrugs? Mar Drugs 19(9):484

Mirza SS, Akbar S, Ijaz MU, Morowvat MH, Ishaque A, Fatima K (2025) Nutritional health connection of algae and its pharmaceutical value as anticancer and antioxidant constituents of drugs. Recent Pat Biotechnol 19(1):19–34

Mishra A, Bhandari M, Pant T, Tiwari H, Kataria AK, Khandelwal V, Chand N, Prajapati SK (2025) Advancement in microalgal metabolites extraction techniques: multi-factorial considerations. J Microbiol Methods 238:107280

Mollah MZI, Zahid HM, Mahal Z, Faruque MRI, Khandaker MU (2021) The usages and potential uses of alginate for healthcare applications. Front Mol Biosci 8:719972

Moniruzzaman M, Bhuiyan MNH, Mondol MMH, Ahmed NU, Akter MM, AlBashera M et al (2023) Amino acids profile of six freshwater algae in the northern region of Bangladesh: a solution for the protein supplement of poultry feed. J Agric Food Res. 14:100828

Mualimin L, Arum MS (2025) In silico study of bioactive compounds fucoxanthin and fucoidan from *Sargassum echinocarpum* as anti-cancer agents targeting caspase-3 protein. Aquat Life Sci 1(2):45–52

Nagahawatta DP, Liyanage NM, Jayawardena TU, Jeon YJ (2024) Marine polyphenols in cardiovascular health: Unraveling structure–activity relationships, mechanisms, and therapeutic implications. Int J Mol Sci 25(15):8419

Oliyaei N, Moosavi-Nasab M, Tamaddon AM, Tanideh N (2021) Antidiabetic effect of fucoxanthin extracted from *Sargassum angustifolium* on streptozotocin-nicotinamide-induced type 2 diabetic mice. Food Sci Nutr 9(7):3521–3529

Razzak SA (2024) Comprehensive overview of microalgae-derived carotenoids and their applications in diverse industries. Algal Res 78:103422

Rupert R, Rodrigues KF, Thien VY, Yong WTL (2022) Carrageenan from *Kappaphycus alvarezii* (Rhodophyta, Solieriaceae): metabolism, structure, production, and application. Front Plant Sci 13:859635

Singh SK, Kumar Y, Sasmal S (2023) Perspectives of HMF and LA from microalgal biomass. Algal Res 72:103133

Singla A, Gupta OP, Sagwal V, Kumar A, Patwa N, Mohan N et al (2024) Beta-glucan as a soluble dietary fiber source: origins, biosynthesis, extraction, purification, structural characteristics, bioavailability, biofunctional attributes, industrial utilization, and global trade. Nutrients 16(6):900

Temple NJ (2022) A rational definition for functional foods: a perspective. Front Nutr 9:957516

Thitame SN, Aher AA (2025) Algal biomolecules in cardiovascular disease: a review of current evidence and emerging therapeutic avenues. J Pharm Bioallied Sci 17(3):S12–S15

Tiwari ON, Sasmal S, Kataria AK, Devi I (2020) Application of microbial extracellular carbohydrate polymeric substances in food and allied industries. 3 Biotech 10(5):221

Trang Thuy NN, Men TT (2025) Phytochemical and bioactive analysis of extracted brown macroalgae (*Dictyota implexa*) collected in Vietnam. Biochem Res Int 2025:9461117

Yue Q, Liu Y, Li F, Hong T, Guo S, Cai M et al (2025) Antioxidant and anticancer properties of fucoidan isolated from *Saccharina japonica* brown algae. Sci Rep 15(1):8962

Zamanabadi SN, Hemmati AA, Zalaghi R, Sohrabi, M, Samarin ME, Nejaddehbashi F, Kiashi F (2025) Assessment of the anti-inflammatory effects of several native macrolichens on carrageenan-induced paw edema in rats. Jundishapur J Nat Pharm Prod 20(20)

Microalgae as Functional Ingredients: Exploring Prebiotic Potential for Gut Health

4

Sk. Shaheda, Anjani Devi Chintagunta, and N. S. Sampath Kumar

Abstract

Microalgae are increasingly recognized as valuable sources of bioactive compounds with significant prebiotic potential. These photosynthetic microorganisms produce a variety of non-digestible polysaccharides, such as sulfated galactans and β-glucans, which selectively stimulate the growth and activity of beneficial gut microbiota. This chapter explores the structural diversity and functionality of prebiotic compounds derived from microalgae, their mechanisms of action in promoting gastrointestinal health, and the evidence supporting their prebiotic effects. It also highlights the emerging concept of microalgae-based synbiotics, where microalgal prebiotics are combined with specific probiotic strains for synergistic benefits. This chapter further examines the integration of microalgae in functional foods and nutraceuticals, outlines regulatory considerations, and discusses future prospects for their use in gut health and wellness. Overall, microalgae represent a sustainable, multifunctional platform with untapped potential in shaping the future of human nutrition and microbiome-targeted therapies.

Keywords

Microalgae · Prebiotics · Synbiotics · Gut Microbiota · Nutraceuticals

S. Shaheda · A. D. Chintagunta · N. S. Sampath Kumar (✉)
Department of Biotechnology, Vignan's Foundation for Science, Technology and Research, Vadlamudi, Andhra Pradesh, India

K. Mohanty et al. (eds.), *Advances in Algal Research*,
https://doi.org/10.1007/978-981-95-7234-2_4

4.1 Introduction

The critical role of the gut in maintaining overall health has been recognized since ancient times, as reflected in the Hippocratic perspective that many diseases begin in the gastrointestinal system (Afzaal et al. 2022). Contemporary research substantiates this perspective, demonstrating that a prolonged disruption of gut microbial balance leads to dysbiosis—a state associated with inflammatory bowel disease, obesity, diabetes, allergies, and neurodegenerative disorders (Origüela and Lopez-Zaplana 2025). The composition and stability of gut microbiota are influenced by multiple factors, such as diet, lifestyle, antibiotic use, stress, and host genetic, with diet and lifestyle exerting the most significant influence (Shen et al. 2025). For instance, diets rich in animal-derived foods favor bile-tolerant microorganisms such as *Alistipes* sp., *Bilophila* sp., *Bacteroides* sp. while plant-based diet enhances saccharolytic bacteria such as *Roseburia* sp., *Eubacterium rectale* (David et al. 2014).

A state of balanced gut microbiota, termed eubiosis, is essential for maintaining host homeostasis. Dietary fibers act as the primary substrates for microbial fermentation in the colon, generating key metabolites that sustain the host-microbe symbiosis. Among these, short-chain fatty acids (SCFAs), such as acetate, propionate, and butyrate play pivotal roles by regulating intestinal barrier integrity, energy homeostasis, appetite control, and lipid metabolism (Valdes et al. 2018). SCFAs also influence microbial gene expression related to spore formation, virulence, drug resistance, and immunomodulation. In addition, gut microbes produce neuroactive compounds—including gamma-aminobutyric acid (GABA), 5-hydroxytryptophan (5-HTP), dopamine, and catecholamines—that modulate gut motility, immune responses, cognition, and stress-related behaviors through the gut-brain axis (Bhardwaj et al. 2025). These diverse roles of the gut microbiota are summarized in Fig. 4.1 which highlights the health benefits associated with eubiosis.

Growing recognition of the gut microbiota's role in health and disease has accelerated research into dietary components that promote eubiosis. Among these, polysaccharides and their derivatives—including dietary fibers have emerged as particularly effective (De Jesus Raposo et al. 2016). Such non-digestible dietary fibers that resist digestion in the upper gastrointestinal tract and selectively stimulate beneficial microorganisms are collectively referred to as prebiotics. These compounds are derived from diverse sources such as terrestrial plants, seaweeds, fungi, and even synthetically engineered oligosaccharides.

Among these diverse sources, microalgae have recently emerged as a sustainable and nutrient-rich reservoir of prebiotic polysaccharides. Their inherent structural traits and digestive resistance make them ideal candidates for functional foods designed to support gut microbiota and holistic health. This chapter therefore focuses on exploring the prebiotic potential of microalgae, with particular attention to their bioactive components, synbiotic applications, supporting scientific evidence, industrial integration, regulatory considerations, and future prospects in gut health.

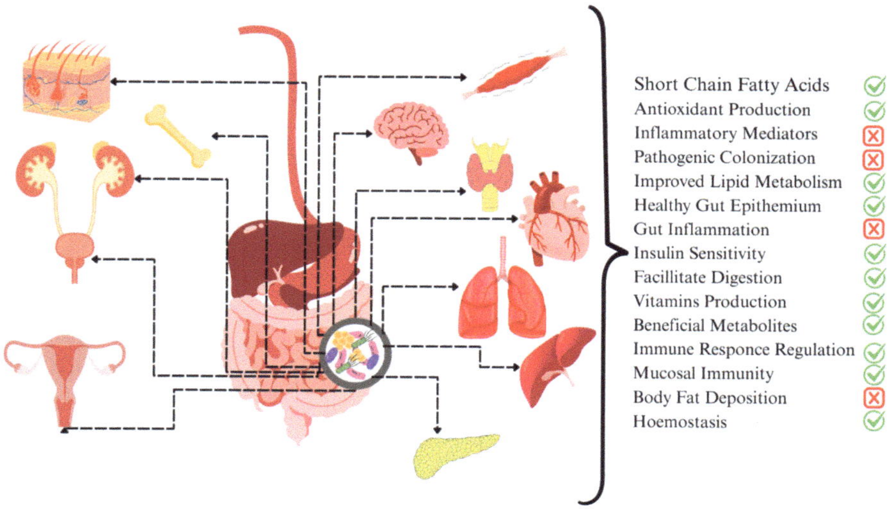

Fig. 4.1 Health benefits associated with gut microbial eubiosis. (Reproduced from Afzaal et al. (2022))

4.2 Microalgae Overview

Microalgae are single-celled, microscopic, photosynthetic organisms—both eukaryotic and prokaryotic—that have attracted significant attention due to their wide range of applications in pharmaceuticals, nutraceuticals, food, cosmetics, biofuels, chemicals from microalgae and aquaculture. They represent a taxonomically diverse group comprising major lineages such as Chlorophyta (green algae), Bacillariophyta (diatoms), and Cyanobacteria (blue-green algae), each categorized by unique structural and biochemical attributes that underpin their ecological and industrial significance (Narang and Yashpal 2024).

Commercial exploitation has focused on species such as *Arthrospira*, *Chlorella*, *Dunaliella*, *Nannochloris*, *Nitzschia*, *Crypthecodinium*, *Schizochytrium*, *Tetraselmis*, and *Skeletonema*, which are valued for their ability to produce proteins, essential fatty acids, antioxidants, pigments, and polysaccharides (Martínez-Ruiz et al. 2025). Their nutritional composition is summarized in Table 4.1, which illustrates interspecific variability in protein, lipid, and carbohydrate contents.

The table summarizes the protein, lipid, and carbohydrate contents (% dry weight) of representative microalgae commonly explored for food, nutraceutical, and biotechnological applications. Values highlight interspecific variability in macromolecular composition, reflecting species-specific metabolic traits and potential suitability for different industrial uses.

This biochemical diversity highlights their versatility across multiple industries. High lipidproducing species, for instance, are well suited for biofuel applications,

Table 4.1 Biochemical composition of microalgae

Microalgae species	Protein (% dry weight)	Lipid (% dry weight)	Carbohydrate (% dry weight)	Reference
Chlorella pyrenoidosa	57.47	2.63	26.55	Gorgônio et al. (2013)
Dunaliella tertiolecta	46.70	12.45	25.73	
Isochrysis galbana	30.56	19.07	23.25	
Tetraselmis gracilis	31.10	13.04	17.78	
Spirulina platensis	60–70	5–6	15–20	Barkia et al. (2019)
Nannochloropsis sp.	30–50	20–50	10–20	Kassim et al. (2020)
Scenedesmus sp.	40–55	12–14	10–17	Barkia et al. (2019)
Haematococcus pluvialis	20–40	20–30	15–20	Singh et al. (2011)
Chlamydomonas reinhardtii	48	21	17	Barkia et al. (2019)
Phaeodactylum tricornutum	~45	~30	~10	Parakh et al. (2023)
Botryococcus braunii	20–30	25–75	10–15	
Porphyridium cruentum	28–39	9–14	40–57	Barkia et al. (2019)
Ankistrodesmus falcatus	~50	15–20	10–20	
Chlorella vulgaris	40–55	10–20	20–30	Wong et al. (2022)
Skeletonema costatum	~30–40	15–20	10–15	Parakh et al. (2023)
Navicula sp.	~40	20–30	~10	
Thalassiosira pseudonana	~35	15–25	~10–15	
Neochloris oleoabundans	30–35	35–54	20–25	Kassim et al. (2020)

while polysaccharides and pigment-rich strains hold promise for functional foods and pharmaceuticals. Microalgae can also act as a renewable carbon source for producing platform chemicals such as HMF (hydroxymethyl furfural) and LA (levulinic acid) through catalytic conversion of biomass (Singh et al. 2023). Efficient utilization, however, depends on optimizing cultivation strategies, as biomass yield and metabolite productivity are strongly influenced by growth conditions.

Microalgae cultivation is broadly categorized into open systems and closed photobioreactors, each offering distinct advantages and limitations depending on species requirements, production scale, and target metabolites. The choice of

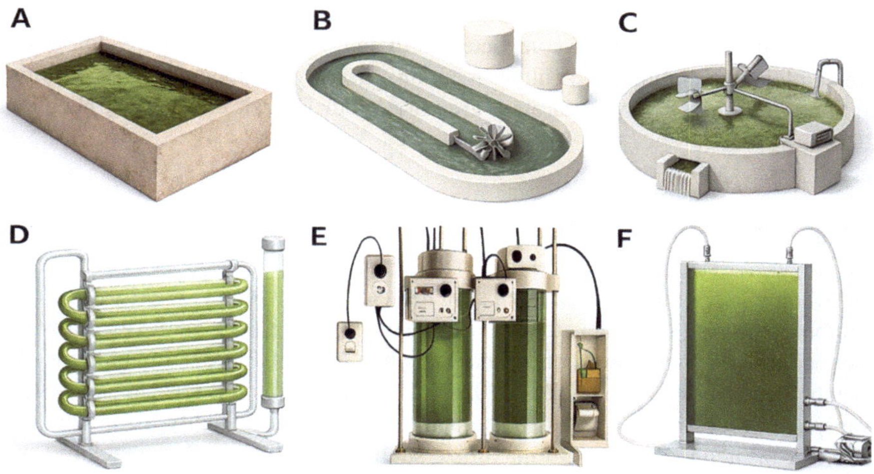

Fig. 4.2 Open systems—(**a**) unstirred pond, (**b**) raceway pond, (**c**) circular pond; closed systems—(**d**) horizontal tubular photo bioreactor, (**e**) bubble column vertical tubular photobioreactor, (**f**) flat-panel photobioreactor

cultivation system is therefore critical to ensuring economic feasibility and environmental sustainability. An overview of commonly employed cultivation designs is presented in Fig. 4.2, which distinguishes open systems (a–c) from closed systems (d–f).

4.2.1 Open Systems

Open pond systems represent one of the earliest and most widely adopted methods for microalgae cultivation (Costa et al. 2019). Their appeal lies in simple operation, low energy requirements, and ease of scalability, making them particularly attractive for bulk biomass production. These systems may utilize natural water bodies such as lakes and ponds, or artificially constructed units like unstirred ponds, circular ponds, and raceway ponds. In certain cases, containers-based approaches, including tanks, are employed for more controlled cultivation (Tan et al. 2020). Among these, unstirred open ponds (Fig. 4.2a) are shallow (≈50 cm depth) and typically lined with plastic film at the base. They represent the most cost-effective approach but are limited by low productivity and minimal control over environmental conditions (Abdur Razzak et al. 2024). Tank-based systems permit improved mixing through aeration or mechanical stirring, though they are mostly suited to small-scale operations.

Circular ponds (Fig. 4.2c), one of the earliest systems for large-scale cultivation, generally have a depth of 30–70 cm and are equipped with a centrally mounted rotating agitator that facilitates uniform mixing. Raceway ponds (Fig. 4.2b), now among the most widely used designs, consist of shallow channels (15–30 cm depth) constructed from concrete or compacted earth and lined with plastic. Paddlewheels are employed to circulate the culture medium, ensuring homogenous distribution of light and nutrients for optimal algal growth (Abdur Razzak et al. 2024; Hallmann 2015).

4.2.2 Closed Systems

To overcome the limitations of open systems, closed cultivation systems—primarily photobioreactors—were developed. Their main objective is to enable the growth of monocultures with minimal contamination, while allowing production of high-value metabolites (Abdur Razzak et al. 2024). The tubular photobioreactor consists of transparent tubes made of glass or polymer, arranged horizontally, vertically, helically, or in inclined formats to maximize sunlight absorption. Commercially, such systems are used for the production of *Haematococcus* (Germany) and *Chlorella* (Israel) (Tan et al. 2020).

The bubble column vertical tubular photobioreactor (Fig. 4.2e) functions as a basic airlift system, where transparent vertical tubes allow light penetration, while CO_2 is supplied via bubbling (Carvalho et al. 2006). Typically, these reactors are twice as tall as they are wide, offering a high surface-to-volume ratio that supports efficient mass and heat transfer (Abdur Razzak et al. 2024). Though commercial applications remain limited, experimental setups reactors have been trialed, such as a 40-L outdoor vertical column photobioreactor for *Chlorella zofingiensis* cultivation in Guangdong, China (Tan et al. 2020).

The flat-plate photobioreactor (Fig. 4.2f) employs a transparent, rectangular chamber (typically 1–5 cm light path, though industrial units up to ~10 cm are used) with a recirculating airlift for mixing (Tan et al. 2020). In a 500 L flat-plate glass unit (effective culture volume ~440 L; 10 cm light path) (Richmond and Cheng-Wu 2001), achieved volumetric biomass productivities up to 0.27 $g\ L^{-1}\ d^{-1}$ for *Nannochloropsis* under semi-continuous operation. Scaled versions (~1000–2000 L) are operational, and recent reviews note high areal productivities when panels are coupled with aeration and controlled bubbling (Abdur Razzak et al. 2024). Recent innovations, such as the electromagnetic photobioreactor (E-PBR), have further enhanced algal biomass productivity, especially by improving carbohydrate yields, making it a promising tool for optimizing both biomass accumulation and metabolite output (Singh et al. 2022).

Beyond their nutritional diversity and cultivation strategies, an important aspect of microalgae in the context of human health is their content of bioactive compounds. Among these, polysaccharides and derived oligosaccharides are notable for their ability to resist upper gastrointestinal digestion and act as selective substrates for beneficial gut microbiota. This functional property defines them as prebiotics, positioning microalgae as both a sustainable biomass source and a reservoir of gut-health-promoting compounds. The following section outlines the major classes of prebiotic molecules identified in microalgae and their potential role in modulating the gut microbiome.

4.3 Prebiotic Compounds in Microalgae

4.3.1 Types of Prebiotics

Prebiotics are defined as non-digestible dietary compounds that escape enzymatic digestion in the upper gastrointestinal tract, allowing them to reach the colon intact where they serve as fermentable substrates for beneficial gut microbes. Microalgae constitute a valuable source of such compounds, offering both native and modified polysaccharides with demonstrated prebiotic potential. These include β-glucans, galactans, arabinoxylans, xylo-oligosaccharides (XOS), galacto-oligosaccharides (GOS), agarose-derived oligosaccharides (AGAROS), alginate oligosaccharides (ALGOS), and neoagaro-oligosaccharides (NAOS). Beyond polysaccharides, polyphenolic compounds like phlorotannins and flavonoids in microalgae also exhibit selective modulatory effects on intestinal bacteria, thereby broadening their functional scope as prebiotics (Patel et al. 2021). Representative examples of prebiotic compounds and their microalgal sources are presented in Table 4.2.

The examples summarized above clearly indicate that dietary fibers derived from microalgae can serve as promising prebiotics, selectively modulating gut microbiota and thereby contributing to host health. Although microalgae can be consumed directly as whole biomass to confer prebiotic benefits, the targeted extraction and characterization of their polysaccharides enhances both the potency and precision of their functional activity. Optimized extraction methods not only improve yield but also help preserve the structural integrity of bioactive polysaccharides, which is essential for retaining their functional activity.

4.3.2 Extraction and Characterization

Polysaccharides extracted from microalgae, have gained increasing attention due to their diverse bioactive properties and applications in the pharmaceutical, food, and biomedical fields. The choice of extraction method is critical, as it directly influences not only the efficiency and yield but also the structural integrity and functional characteristics of the recovered polymers. Since bioactivity is often linked to fine structural features, careful selection and optimization of extraction strategies are essential for maximizing both scientific and industrial utility.

4.3.2.1 Types of Extraction Methods

Extraction methods are broadly classified into conventional and non-conventional (modern) techniques. Irrespective of the strategy employed, the underlying goal is to disrupt the cell wall matrix and solubilize the polysaccharide fraction while preserving bioactivity.

Conventional Methods

Hot water extraction remains one of the most widely practiced traditional method. In this approach, elevated temperatures disrupt the cell wall, releasing intracellular

Table 4.2 Different types of prebiotics and their microalgal source

S. no.	Type of prebiotic	Specific compound	Microalgal source	References
1	Polysaccharide	β-Glucans	*Euglena gracilis*, *Chlorella vulgaris*, *Phaeodactylum tricornutum*	Lihanová et al. (2023), Palol et al. (2024)
2	Polysaccharide	Sulfated galactans	*Porphyridium cruentum* (red microalga)	de Jesus Raposo et al. (2013)
3	Polysaccharide	Xyloglucans	*Chlamydomonas reinhardtii* (green alga)	Popper et al. (2011)
4	Polysaccharide	Ulvan (rhamnose-rich sulphated polysaccharide)	*Ulva lactuca* (green macroalgae; model for microalgal analogs)	O'Sullivan et al. (2010)
5	Polyphenols	Phlorotannins, flavonoids	*Spirulina platensis*, *Haematococcus pluvialis*, *chlorella* sp.	Pintea and Rugin (2025)
6	Oligosaccharide	Alginate oligosaccharides (AOS)	Derived from brown algae and modelled in microalgae	Andriamanantoanina and Rinaudo (2010)
7	Oligosaccharide	Xylo-oligosaccharides (XOS)	Derived from *chlorella*, *Chlorococcum* sp., *Scenedesmus obliquus*	Gouda et al. (2022)
8	Polysaccharide	Exopolysaccharides (EPS)	*Nannochloropsis*, *Arthrospira (spirulina)*	Laroche (2022)
9	Polysaccharide/ oligosaccharide	Laminarin (β-1,3-glucan)	Found in diatoms like *Phaeodactylum tricornutum*	Frick et al. (2023)

molecules into the solvent (Wu et al. 2014). Alkali extraction employs sodium hydroxide (NaOH), where hydroxide ions (OH^-) weaken hydrogen bonds within polysaccharides, enhancing their solubilization (Wu et al. 2014; Song et al. 2018). Similarly, acid extraction relies on hydrochloric acid (HCl); here, protons (H^+) destabilize hydrogen linkages, liberating polysaccharide into solution (Wu et al. 2014; Wassie et al. 2021).

Post-extraction, the crude material is typically concentrated and purified through a series of steps—ethanol precipitation, protein removal via the Sevag method, dialysis to eliminate low-molecular-weight impurities, and lyophilization to yield a stable polysaccharide powder (Wu et al. 2014). While simple and scalable, these conventional techniques often suffer from drawbacks such as low selectivity, extended processing times, and potential degradation of delicate structures.

Non-conventional Methods

To overcome the limitations of conventional approaches, several modern methods have been developed, offering higher efficiency, reduced solvent usage, and improved preservation of structural features (Fig. 4.3).

Enzyme-Assisted Extraction (EAE) This technique employs enzymes such as cellulase, hemicellulase, and protease to selectively hydrolyze cell walls components, thereby releasing polysaccharides with minimal structural alteration (Corrêa et al. 2021; Dobrinčić et al. 2020). Its mild operating conditions make EAE especially valuable when structural preservation is critical (Fig. 4.3b).

Microwave-Assisted Extraction (MAE) Considering one of the most effective extraction methods, MAE utilizes rapid volumetric heating induced by dipole rotation and ionic conduction. This localized heating disrupts cell walls, releasing intracellular polysaccharides directly into the solvent (Corrêa et al. 2021; Dobrinčić et al. 2020; Kapoore et al. 2018). MAE significantly reduces processing time while improving extraction efficiency, although optimization of parameters is essential to prevent degradation (Fig. 4.3e).

Ultrasound-Assisted Extraction (UAE) UAE applies high-frequency sound waves (15 and 20 kHz) to induce cavitation—the formation and collapse of microscopic bubbles within the liquid medium (Fig. 4.3c). The intense shear forces and localized energy release disrupt cell walls and enhance mass transfer, facilitating rapid polysaccharide recovery (Corrêa et al. 2021).

Subcritical Water Extraction (SWE) SWE also termed superheated or pressurized hot water extraction, SWE employs liquid water maintained at 100–374 °C under high pressure (up to 22.1 MPa). At these conditions, water's dielectric constant, viscosity, and surface tension decrease, while diffusivity increases, dramatically enhancing solvent penetration and compound solubility (Aminzai et al. 2025). This eco-friendly method allows efficient extraction of thermally stable polysaccharides with reduced use of organic solvents (Aminzai et al. 2025; Zhang et al. 2020)

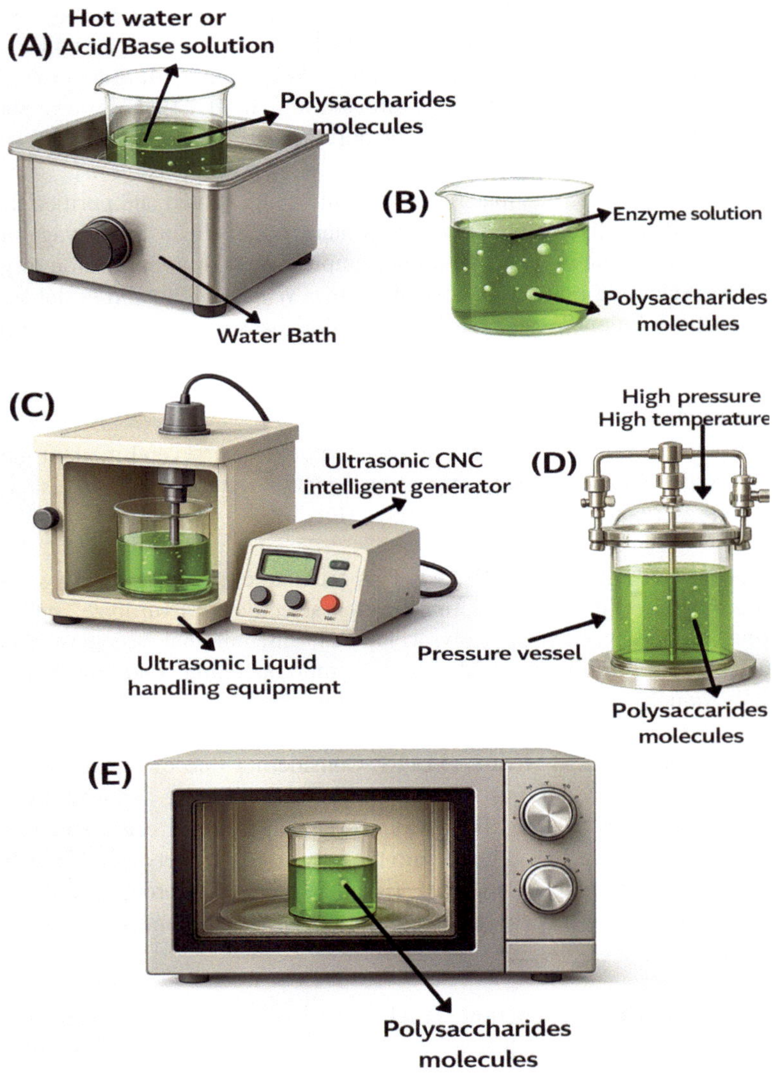

Fig. 4.3 Conventional methods—hot water or acid/base extraction (**a**); non-conventional methods—enzymatic extraction (**b**), subcritical extraction (**d**), ultrasonic assisted extraction (**c**), microwave assisted extraction (**e**). (Reproduced from Lu (2023))

4.3.3 Types of Characterization Methods

After extraction, characterization plays a crucial role in understanding the composition, molecular structure, and functional properties of polysaccharides, since these attributes directly influence their bioactivity and application potential.

Table 4.3 Major analytical methods for characterization of microalgal polysaccharides

Technique	Principle	Information obtained	References
FTIR (Fourier transform infrared spectroscopy)	Measures absorption of infrared light by molecular bonds	Identifies functional groups (−OH, −CH, glycosidic linkages)	Wan et al. (2020)
HPLC (high-performance liquid chromatography)	Separation based on interaction with stationary/mobile phases	Quantification of monosaccharide composition after hydrolysis	Lv et al. (2022)
GC–MS (gas chromatography–mass spectrometry)	Separation of volatile derivatives + fragmentation patterns	Detailed monosaccharide profiling and molecular weight	Chen et al. (2018)
NMR (nuclear magnetic resonance)	Nuclear spins interact with external magnetic field	Structural configuration, linkage types, anomeric carbons	Wan et al. (2020)
SEC/GPC (size exclusion/ gel permeation chromatography)	Separation based on hydrodynamic volume	Molecular weight distribution and polydispersity	Ciric et al. (2012)
AFM (atomic force microscopy)	Probe scans surface at nanoscale	3D morphology, surface topology	Wan et al. (2020)
SEM (scanning electron microscopy)	Electron beam imaging of surfaces	Surface texture, roughness, and morphology	Wang et al. (2018)
XRD (X-ray diffraction)	Diffraction of X-rays by crystalline structures	Degree of crystallinity vs. amorphous nature	Mishra et al. (2011)
TGA/DSC (thermogravimetric analysis/differential scanning calorimetry)	Measures weight change and heat flow under controlled heating	Thermal stability, decomposition temperature, glass transition	Mishra et al. (2011)
MALDI-TOF MS (matrix-assisted laser desorption ionization–time of flight mass spectrometry)	Laser ionization + TOF analyzer	Precise molecular weight of oligosaccharides, repeating units	Mishra et al. (2011)

Polysaccharides are complex macromolecules exhibiting diverse structures and biological functions. Their characterization requires various analytical techniques to analyze their chemical composition, molecular weight (MW), structural configuration, and surface characteristics. A summary of the most widely employed methods is presented in Table 4.3, which highlights their principles, applications, and key outcomes.

By combining these techniques, researchers can gain a holistic understanding of polysaccharide chemical identity, structural configuration, functional groups, and physical properties. While techniques like FTIR, HPLC, and NMR are indispensable for defining chemical and structural details, advanced tools such as MALDI-TOF MS, XRD, and DSC provide deeper insights into molecular architecture, crystallinity, and stability. Integrating multiple characterization approaches is therefore essential to establish reliable structure–function relationships and to tailor polysaccharides for food, pharmaceutical, and biomedical applications.

4.3.4 Resistance to Digestion and Fermentation Profile

Microalgal polysaccharides typically resist enzymatic digestion in the human gastrointestinal tract due to their structural complexity. The presence of uronic acids, sulfate groups, and β-glycosidic linkages renders them largely indigestible by salivary, gastric, and pancreatic enzymes, allowing these polymers to reach the colon intact (Lv et al. 2022). Within the anaerobic environment of the colon, resident microbiota selectively ferments these polysaccharides, producing short-chain fatty acids (SCFAs) such as acetate, propionate, and butyrate. This fermentation process lowers the luminal pH and decreases residual carbohydrate levels, reflecting the targeted utilization of these compounds by beneficial gut microorganisms (Han et al. 2020). Recent studies also indicate that variations in the degree of sulfation, branching, and molecular weight influence the extent and rate of fermentation, thereby shaping the spectrum of SCFAs generated. Figure 4.4 illustrates the resistance of microalgal polysaccharides to simulated gastrointestinal digestion and their subsequent fermentation by colonic microbiota, highlighting enhanced SCFA production. Collectively, their resistance to upper gastrointestinal digestion coupled with selective microbial fermentation underscores the prebiotic potential of microalgal polysaccharides and their contribution to host gut health.

4.3.5 Comparison of Microalgal and Conventional Prebiotics

Conventional prebiotics such as inulin, fructooligosaccharides (FOS), and galactooligosaccharides (GOS) are well-established, with robust human clinical evidence demonstrating selective stimulation of *Bifidobacterium* and *Lactobacillus*, enhanced short-chain fatty acid (SCFA) production, improved bowel function, and metabolic benefits including calcium absorption and glycemic control (Hughes et al. 2022; Bouhnik et al. 2007).

In contrast, evidence for microalgal-derived prebiotics is still emerging. Polysaccharides and oligosaccharides from species such as *Chlorella vulgaris*, *Arthrospira (Spirulina) platensis*, and *Scenedesmus obliquus* have demonstrated selective stimulation of *Bifidobacterium* and *Lactobacillus* in vitro, with enhanced acetate and butyrate production (Gouda et al. 2022). *Spirulina* extracts, for example, have been shown to increase beneficial gut taxa and modulate immune responses in both animal and preliminary human studies (de Jesus Raposo et al. 2016). Similarly, exopolysaccharides (EPS) from *Chlorella* and *Nannochloropsis* promote SCFA

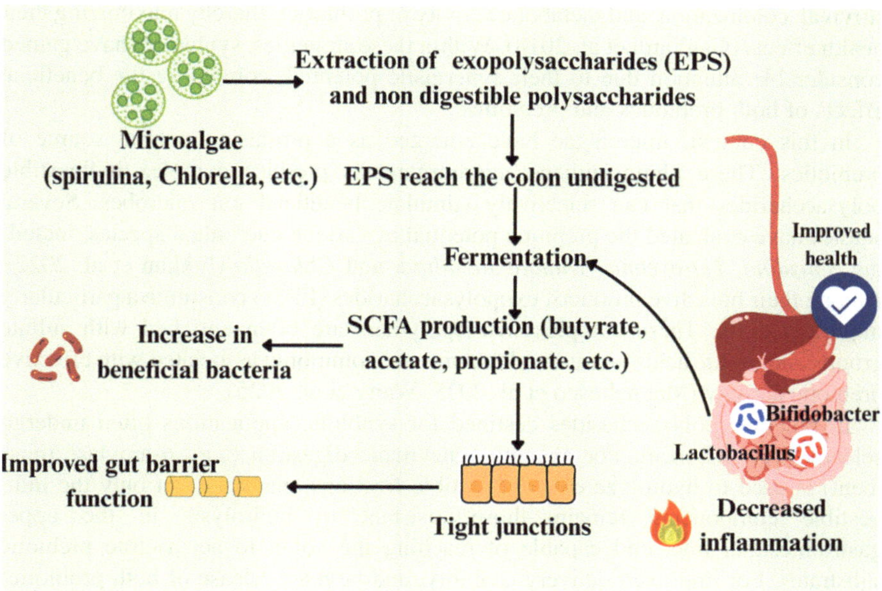

Fig. 4.4 Schematic representation of prebiotic mechanism of microalgae in human gut health, showing resistance to digestion, fermentation by colonic microbiota, SCFA production, modulation of beneficial bacteria, and host health benefits

production and suppress potential pathogens in fecal fermentation assays (Laroche 2022). Compared to conventional prebiotics, microalgal polysaccharides often exhibit slower, sustained fermentability due to structural complexity, which provides prolonged SCFA release (Lv et al. 2022). Their additional antioxidant and immunomodulatory properties also suggest multifunctional health benefits beyond the microbiota. However, clinical validation is scarce, and large, well-controlled human trials are still required. Overall, inulin, FOS, and GOS remain the benchmark, but microalgal polysaccharides, particularly from *Chlorella* and *Spirulina*, show promising potential as next-generation prebiotics.

4.4 Synbiotics: Combining Prebiotics and Probiotics for Gut Health Enhancement

Numerous treatment approaches have been developed to restore and maintain eubiosis given the critical role of gut microbiota in health and disease. Among these, the most widely recognized and commonly employed strategies include the following: (1) probiotics which can suppress pathogenic bacteria and promote the growth of beneficial bacteria; (2) prebiotics, typically fibers or complex carbohydrates that selectively stimulate the activity of health-promoting microbes; and (3) synbiotics, combinations of probiotics and prebiotics that enhance the survival, colonization, and metabolic activity of probiotics, thereby maximizing their health effects (Gagliardi et al. 2018). Within these strategies, synbiotics have gained considerable attention due to their synergistic potential, combining the beneficial effects of both probiotics and prebiotics.

In this context, microalgae have emerged as a promising natural source of prebiotics. These photosynthetic microorganisms produce complex indigestible polysaccharides that can selectively stimulate beneficial gut microbes. Several studies have evaluated the prebiotic potential of various microalgal species, including *Spirulina*, *Tetraselmis*, *Dunaliella salina*, and *Chlorella* (Ilıkkan et al. 2022). Among their bioactive products, exopolysaccharides (EPSs) constitute a particularly important class. These complex heteropolymers are often enriched with sulfate groups and uronic acids—structural features are commonly associated with effective prebiotic activity (Magnabosco et al. 2025; Wang et al. 2025).

In practice, polysaccharides destined for synbiotic applications often undergo selective pre-treatment. For example, enzymatic digestion (e.g., α-amylase treatment) is used to hydrolyze easily digestible fractions, ensuring that only the indigestible components remain—those resistant to hydrolysis in the upper gastrointestinal tract and capable of reaching the colon to act as true prebiotic substrates. For improved delivery, stability, and targeted release of both probiotics and polysaccharides, a number of formulation strategies have been explored, including encapsulation methods (spray-drying, freeze-drying, coacervation, extrusion), polysaccharide-based matrices and films, lipid-based carriers, and nano- and microparticle systems. Although these approaches expand the scope of synbiotic development, they remain supplementary. In most cases, the intrinsic resistance of microalgal polysaccharides to digestion already provides sufficient protection and functionality, allowing them to act effectively without elaborate carriers.

4.5 Evidence of Prebiotic Activity in Microalgae

Evidence from preclinical studies supports the prebiotic activity of microalgal polysaccharides. For example, *Chlorella pyrenoidosa* polysaccharides (CPP) remained intact in gastrointestinal and saliva simulation models but were moderately fermented by human fecal microbiota (Lv et al. 2022). Fermentation led to suppression of *Escherichia schreibella*, *Fusobacterium*, and *Klebsiella*, while enriching *Parabacteroides distasonis*. Compared to traditional prebiotics such as FOS and IMO, CPP promoted higher levels of acetate, propionate, butyrate, and valerate, indicating distinct fermentation dynamics.

Similarly, *C. pyrenoidosa* was tested in a continuous pH-controlled gut model simulating the human colon microbiota from both healthy and coeliac donors. The treatment reduced *Enterobacteriaceae* while enriching *Faecalibacterium*, *Bifidobacterium*, and *Megasphaera* (coeliac donor) or *Prevotella* and *Ruminococcus* (healthy donor). While acetate responses varied by donor, butyrate and propionate levels rose significantly ($p < 0.0001$), suggesting strong implications for inflammation control and mucosal health (van der Linde et al. 2021).

In vivo findings further strengthen this evidence. Two preparations, CPS from *C. pyrenoidosa* and SPS from *Spirulina platensis*, were administered to mice on a high-fat diet (HFD). Both reduced weight gain, glucose intolerance, inflammation, dyslipidemia, and liver fat accumulation, performing comparably to β-glucan as a positive control. Mechanistic studies revealed downregulation of PPARγ and SREBP-1c and increased CPT-1 expression, consistent with inhibited lipogenesis and enhanced lipid breakdown. Moreover, HFD-induced dysbiosis was corrected, with declines in *Actinobacteria* and *Verrucomicrobia* and increases in *Clostridium*, *Bacteroidia*, and *Mollicutes*. SCFA and bile acid metabolism pathways were restored, highlighting gut microbial mediation as a key regulatory mechanism (Guo et al. 2021). Taken together, these findings demonstrate that microalgal polysaccharides—particularly from *Chlorella* and *Spirulina*—exhibit significant prebiotic-like effects, modulating both microbial ecology and host metabolic pathways. This dual action, combined with the option of advanced delivery systems where required, underlines the potential of microalgae as a valuable component in next-generation synbiotic formulations.

4.6 Applications in Food and Nutraceutical Industries

Microalgae are increasingly valued as functional food and nutraceutical ingredients because of their polysaccharides, proteins, essential fatty acids, pigments, and antioxidants. Among these, microalgal polysaccharides—especially exopolysaccharides (EPSs)—stand out for their prebiotic potential and ability to modulate gut microbiota. Food formulations incorporating microalgae demonstrate both nutritional and digestive benefits. *Spirulina* has been added to pasta, biscuits, and health bars, while *Chlorella* has been integrated into dairy and bakery products. For example, *Chlorella* biomass (0.5–2%) in Appenzeller cheese improved lactic acid bacteria viability and sensory attributes (Lousada Falcão et al. 2023). Similarly, microalgae addition enhanced probiotic LAB activity during fermentation, opening avenues for synbiotic dairy innovations (Albuquerque et al. 2024). Bakery applications such as *Chlorella*-enriched breads have shown increased phenolic content and antioxidant activity (Garzon et al. 2021). Beyond solids, microalgae contribute to fortified beverages, including infant formulas and health drinks, enriched with EPA, DHA, and ALA to support cognitive, cardiovascular, and gut functions (Geppert et al. 2005).

4.7 Regulatory Considerations for Microalgae-Based Prebiotics

The translation of microalgae into prebiotic foods is shaped by regulatory frameworks that differ globally. Prebiotics from traditional sources such as inulin, FOS, and GOS are already widely accepted, whereas microalgal polysaccharides often fall under "novel food" categories. In the European Union, EFSA regulates

Table 4.4 Comparative overview of regulatory frameworks relevant to microalgal prebiotics

Regulatory pathway	Jurisdiction	Key requirements	Challenges for microalgal prebiotics
Novel food (EU)	European Union	Safety assessment by EFSA; demonstration of history of use or comprehensive toxicological data; compositional analysis; labeling compliance	Lack of long-term human trials on microalgal polysaccharides; regulatory delays; high cost of dossier preparation
GRAS (generally recognized as safe)	USA (FDA)	Evidence of safety through published literature and/or expert panel review; self-affirmation possible; must be intended for use in food	Limited precedent for microalgal prebiotics compared to algal pigments (e.g., astaxanthin, DHA); need for more toxicological evidence
FOSHU (foods for specified health uses)	Japan	Requires human clinical evidence demonstrating functional benefits (gut health, immunity, etc.); approval by consumer affairs agency	Few clinical trials on microalgal polysaccharides; difficulty in meeting rigorous endpoint requirements; competition with well-established prebiotics (FOS, GOS, inulin)

novel foods under Regulation (EU) 2015/2283, requiring assessment of any microalgae-derived ingredient not consumed prior to 1997 (Ververis et al. 2020). In the United States, the FDA permits marketing of microalgae-based foods under the Generally Recognized as Safe (GRAS) pathway, as with *Spirulina* and *Chlorella*, though prebiotic claims require supporting evidence and may trigger New Dietary Ingredient (NDI) evaluation (Commissioner Office of the FDA 2025). In Japan, microalgae are sometimes approved under the Foods for Specified Health Uses (FOSHU), contingent on consistent quality and robust clinical data (Cherry et al. 2019). To better illustrate these differences, Table 4.4 provides a comparative overview of regulatory frameworks in key jurisdictions. This highlights both common requirements (safety, compositional analysis, and labeling) and unique challenges for microalgal prebiotics, such as proof of clinical efficacy under FOSHU or dossier costs under EU Novel Food rules.

Key safety considerations include potential accumulation of toxins or heavy metals, allergenicity, or generation of unwanted metabolites, especially in novel or genetically modified strains (Pulz and Gross 2004). Labeling claims such as "prebiotic" or "supports gut health" also require validation. A persistent challenge remains the lack of harmonized prebiotic definitions across international markets, despite ISAPP's efforts to standardize criteria (Gibson et al. 2017). In short, while GRAS recognition and FOSHU approval demonstrate regulatory openness, microalgal prebiotics still face hurdles of safety validation, harmonized definitions, and robust clinical substantiation (Nuin Garciarena et al. 2025).

4.8 Future Prospects for Microalgae in Gut Health

Although several microalgal species are already commercialized as dietary supplements, their polysaccharides with potential prebiotic activity remain underexplored. Scaling production is challenged by the trade-off between cost and quality—open ponds are cheaper but vulnerable to contamination, whereas closed photobioreactors ensure purity at higher expense (Sidari and Tofalo 2019). Integrating biorefinery approaches could improve economic feasibility by co-extracting proteins, pigments, and polysaccharides (Patel et al. 2023). The absence of lignin in microalgal cell walls further enhances their appeal for nanocellulose and functional applications.

Technological innovations, including CRISPR-based strain engineering, can boost polysaccharide yield and photosynthetic efficiency, paving the way for precision-designed microalgae with targeted gut health benefits (Ambily et al. 2025). Although in vitro and animal model studies of microalgal polysaccharides demonstrate positive effects on gut microbiota composition, SCFA production, and metabolite profiles, translating these results to humans remains challenging. Key limitations of current human studies include small sample sizes, short durations (often <4 weeks) (Patel et al. 2023; Ambily et al. 2025), lack of mechanistic data, and primary focus on general microbiota shifts rather than clinical endpoints such as inflammation, metabolic health, or symptom relief in disease populations. To establish microalgal PS as clinically relevant prebiotics, future trials should be randomized and controlled and have a sufficient duration (8–12 weeks or more). They should use well-characterized polysaccharides, include disease-relevant populations (e.g., IBS, obesity, IBD), and evaluate multiple endpoints, including microbiome function, metabolite production, host biomarkers, and safety (Gan et al. 2022).

In summary, microalgae offer a sustainable, polysaccharide-rich resource with antioxidant, immunomodulatory, and prebiotic potential. Addressing cultivation, processing, and regulatory challenges through interdisciplinary collaboration will be key to establishing them as mainstream functional ingredients in foods and nutraceuticals.

References

Abdur Razzak S, Bahar K, Islam KMO, Haniffa AK, Faruque MO, Hossain SMZ et al (2024) Microalgae cultivation in photobioreactors: sustainable solutions for a greener future. Green Chem Eng 5(4):418–439

Afzaal M, Saeed F, Shah YA, Hussain M, Rabail R, Socol CT et al (2022) Human gut microbiota in health and disease: unveiling the relationship. Front Microbiol 13:999001

Albuquerque RCV, Silva CEDF, WDS C, Andreola K, da Gama BMV, da Silva AE (2024) Incorporation of cyanobacteria and microalgae in yogurt: formulation challenges and nutritional, rheological, sensory, and functional implications. Appl Microbiol 4(4):1493–1514

Ambily B, Limna Mol VP, Sini H, Nevin KG (2025) CRISPR-based microalgal genome editing and the potential for sustainable aquaculture: a comprehensive review. J Appl Phycol 37(1):265–285

Aminzai MT, Yabalak E, Akay S, Kayan B (2025) Recent developments in subcritical water extraction of industrially important bioactive substances from plants, microorganisms, and organic wastes. Biomass Convers Biorefinery 15(12):17927–17949

Andriamanantoanina H, Rinaudo M (2010) Characterization of the alginates from five madagascan brown algae. Carbohydr Polym 82(3):555–560

Barkia I, Saari N, Manning SR (2019) Microalgae for high-value products towards human health and nutrition. Mar Drugs 17(5):304

Bhardwaj K, Das B, Zuo T, Vrati S, Bhardwaj K, Das B et al (2025) Gut microbiota: role in health and diseases. In: Probiotics, prebiotics, and postbiotics in human health and sustainable food systems. IntechOpen. Available from: https://www.intechopen.com/chapters/1194868

Bouhnik Y, Achour L, Paineau D, Riottot M, Attar A, Bornet F (2007) Four-week short chain fructo-oligosaccharides ingestion leads to increasing fecal bifidobacteria and cholesterol excretion in healthy elderly volunteers. Nutr J 6(1):42

Carvalho AP, Meireles LA, Malcata FX (2006) Microalgal reactors: a review of enclosed system designs and performances. Biotechnol Prog 22(6):1490–1506

Chen Y, Liu X, Wu L, Tong A, Zhao L, Liu B et al (2018) Physicochemical characterization of polysaccharides from *Chlorella pyrenoidosa* and its anti-ageing effects in *Drosophila melanogaster*. Carbohydr Polym 185:120–126

Cherry P, O'Hara C, Magee PJ, McSorley EM, Allsopp PJ (2019) Risks and benefits of consuming edible seaweeds. Nutr Rev 77(5):307–329

Ciric J, Oostland J, de Vries JW, Woortman AJJ, Loos K (2012) Size exclusion chromatography with multi detection in combination with matrix-assisted laser desorption ionization-time-of-flight mass spectrometry as a tool for unraveling the mechanism of the enzymatic polymerization of polysaccharides. Anal Chem 84(23):10463–10470

Commissioner Office of the FDA (2025) U.S. Food and Drug Administration. FDA. Available from: https://www.fda.gov/

Corrêa PS, Morais Júnior WG, Martins AA, Caetano NS, Mata TM (2021) Microalgae biomolecules: extraction, separation and purification methods. Processes 9(1):10

Costa JAV, Freitas BCB, Santos TD, Mitchell BG, Morais MG (2019) Chapter 9: Open pond systems for microalgal culture. In: Pandey A, Chang JS, Soccol CR, Lee DJ, Chisti Y (eds) Biomass, biofuels, biochemicals: biofuels from algae, 2nd edn. Elsevier, pp 199–223. Available from: https://www.sciencedirect.com/science/article/pii/B9780444641922000093

David LA, Maurice CF, Carmody RN, Gootenberg DB, Button JE, Wolfe BE et al (2014) Diet rapidly and reproducibly alters the human gut microbiome. Nature 505(7484):559–563

de Jesus Raposo MF, de Morais RMSC, de Morais AMMB (2013) Health applications of bioactive compounds from marine microalgae. Life Sci 93(15):479–486

De Jesus Raposo MF, De Morais AMMB, De Morais RMSC (2016) Emergent sources of prebiotics: seaweeds and microalgae. Mar Drugs 14(2):27

Dobrinčić A, Balbino S, Zorić Z, Pedisić S, Bursać Kovačević D, Elez Garofulić I et al (2020) Advanced technologies for the extraction of marine brown algal polysaccharides. Mar Drugs 18(3):168

Frick K, Ebbing T, Yeh YC, Schmid-Staiger U, Tovar GEM (2023) Beta-glucan production of *Phaeodactylum tricornutum, Monodopsis subterranea* and *Cylindrotheca fusiformis* during nitrogen depletion. J Appl Phycol 35(6):2607–2618

Gagliardi A, Totino V, Cacciotti F, Iebba V, Neroni B, Bonfiglio G et al (2018) Rebuilding the gut microbiota ecosystem. Int J Environ Res Public Health 15(8):1679

Gan L, Wang J, Guo Y (2022) Polysaccharides influence human health via microbiota-dependent and -independent pathways. Front Nutr 9:1030063. Available from: https://www.frontiersin.org/journals/nutrition/articles/10.3389/fnut.2022.1030063/full

Garzon R, Skendi A, Antonio Lazo-Velez M, Papageorgiou M, Rosell CM (2021) Interaction of dough acidity and microalga level on bread quality and antioxidant properties. Food Chem 344: 128710

Geppert J, Kraft V, Demmelmair H, Koletzko B (2005) Docosahexaenoic acid supplementation in vegetarians effectively increases omega-3 index: a randomized trial. Lipids 40(8):807

Gibson GR, Hutkins R, Sanders ME, Prescott SL, Reimer RA, Salminen SJ et al (2017) Expert consensus document: The International Scientific Association for Probiotics and Prebiotics (ISAPP) consensus statement on the definition and scope of prebiotics. Nat Rev. Gastroenterol Hepatol 14(8):491–502

Gorgônio CMDS, Aranda DAG, Couri S (2013) Morphological and chemical aspects of *Chlorella pyrenoidosa*, *Dunaliella tertiolecta*, *Isochrysis galbana* and *Tetraselmis gracilis* microalgae. Nat Sci 05(07):783–791

Gouda M, Tadda MA, Zhao Y, Farmanullah F, Chu B, Li X et al (2022) Microalgae bioactive carbohydrates as a novel sustainable and eco-friendly source of prebiotics: emerging health functionality and recent technologies for extraction and detection. Front Nutr 9:806692. Available from: https://www.frontiersin.org/journals/nutrition/articles/10.3389/fnut.2022.806692/full

Guo W, Zhu S, Li S, Feng Y, Wu H, Zeng M (2021)Microalgae polysaccharides ameliorates obesity in association with modulation of lipid metabolism and gut microbiota in high-fat-diet fed C57BL/6 mice. Int J Biol Macromol 182:1371–1383

Hallmann A (2015) Algae biotechnology—green cell-factories on the rise. Curr Biotechnol 4(4): 389–415. Available from: https://pub.uni-bielefeld.de/record/2900884

Han R, Pang D, Wen L, You L, Huang R, Kulikouskaya V (2020) In vitro digestibility and prebiotic activities of a sulfated polysaccharide from *Gracilaria lemaneiformis*. J Funct Foods 64:103652

Hughes RL, Alvarado DA, Swanson KS, Holscher HD (2022) The prebiotic potential of inulin-type fructans: a systematic review. Adv Nutr 13(2):492–529

Ilıkkan Ö, Bağdat E, Yalçın D (2022) Evaluation of prebiotic, probiotic, and synbiotic potentials of microalgae. Food Health 8(2):161–171

Kapoore RV, Butler TO, Pandhal J, Vaidyanathan S (2018) Microwave-assisted extraction for microalgae: from biofuels to biorefinery. Biology 7(1):18

Kassim MA, Meng TK, Serri NA, Yusoff SB, Shahrin NAM, Seng KY et al (2020) Sustainable biorefinery concept for industrial bioprocessing. In: Biorefinery production technologies for chemicals and energy. Wiley, pp 15–53. Available from: https://onlinelibrary.wiley.com/doi/abs/10.1002/9781119593065.ch2

Laroche C (2022) Exopolysaccharides from microalgae and cyanobacteria: diversity of strains, production strategies, and applications. Mar Drugs 20(5):336

Lihanová D, Lukáčová A, Beck T, Jedlička A, Vešelényiová D, Krajčovič J et al (2023) Versatile biotechnological applications of *Euglena gracilis*. World J Microbiol Biotechnol 39(5):133

Lousada Falcão R, Pinheiro V, Ribeiro C, Sousa I, Raymundo A, Nunes MC (2023) Nutritional improvement of fresh cheese with microalga *Chlorella vulgaris*: impact on composition, structure and sensory acceptance. Food Technol Biotechnol 61(2):259–270

Lu X (2023) Changes in the structure of polysaccharides under different extraction methods. eFood 4(2):e82

Lv K, Yuan Q, Li H, Li T, Ma H, Gao C et al (2022) *Chlorella pyrenoidosa* polysaccharides as a prebiotic to modulate gut microbiota: physicochemical properties and fermentation characteristics in vitro. Foods 11(5):725

Magnabosco C, Santaniello G, Romano G (2025) Microalgae: a promising source of bioactive polysaccharides for biotechnological applications. Molecules 30(9):2055

Martínez-Ruiz FE, Andrade-Bustamante G, Holguín-Peña RJ, Renganathan P, Gaysina LA, Sukhanova NV et al (2025) Microalgae as functional food ingredients: nutritional benefits, challenges, and regulatory considerations for safe consumption. Biomass 5(2):25

Mishra A, Kavita K, Jha B (2011) Characterization of extracellular polymeric substances produced by micro-algae *Dunaliella salina*. Carbohydr Polym 83(2):852–857

Narang PK, Yashpal M (2024) Microalgae and their diverse applications: a review. J Bioresour 11(2):40–46

Nuin Garciarena I, Ackerl R, García Ruiz E, Glymenaki M, Mendes V, Muñoz-González A et al (2025) The safety assessment of microalgae-derived products as novel foods by the European Food Safety Authority. Future Foods 11:100661
O'Sullivan L, Murphy B, McLoughlin P, Duggan P, Lawlor PG, Hughes H et al (2010) Prebiotics from marine macroalgae for human and animal health applications. Mar Drugs 8(7):2038–2064
Origüela V, Lopez-Zaplana A (2025) Gut microbiota: an immersion in dysbiosis, associated pathologies, and probiotics. Microorganisms 13(5):1084
Palol VV, Waidha K, Moovarkumudalvan B, Valavath Baburajan N, Saravanan SK, Lakshmanan D et al (2024) β-1,3-glucan from *Euglena gracilis*: a promising epidrug targeting epigenetic regulators PRMTs and SIRTs for therapeutic applications in ovarian cancer. J Biomol Struct Dyn 1–16. https://doi.org/10.1080/07391102.2024.2425832
Parakh SK, Tian Z, Wong JZE, Tong YW (2023) From microalgae to bioenergy: recent advances in biochemical conversion processes. Fermentation 9(6):529
Patel AK, Singhania RR, Awasthi MK, Varjani S, Bhatia SK, Tsai ML et al (2021) Emerging prospects of macro- and microalgae as prebiotic. Microb Cell Factories 20(1):112
Patel AK, Vadrale AP, Singhania RR, Michaud P, Pandey A, Chen SJ et al (2023) Algal polysaccharides: current status and future prospects. Phytochem Rev 22(4):1167–1196
Pintea AM, Rugin D (2025) Chapter 14: Antioxidant potential of microalgae. In: Pérez-Gálvez A, Jacob-Lopes E, Queiroz Zepka L, Roca M (eds) Microalgae and one health. Academic Press, pp 197–226. Available from: https://www.sciencedirect.com/science/article/pii/B9780443220807000387
Popper ZA, Michel G, Hervé C, Domozych DS, Willats WGT, Tuohy MG et al (2011) Evolution and diversity of plant cell walls: from algae to flowering plants. Annu Rev Plant Biol 62:567–590
Pulz O, Gross W (2004) Valuable products from biotechnology of microalgae. Appl Microbiol Biotechnol 65(6):635–648
Richmond A, Cheng-Wu Z (2001) Optimization of a flat plate glass reactor for mass production of *Nannochloropsis* sp. outdoors. J Biotechnol 85(3):259–269
Shen Y, Fan N, Ma S, Cheng X, Yang X, Wang G (2025) Gut microbiota dysbiosis: pathogenesis, diseases, prevention, and therapy. MedComm 6(5):e70168
Sidari R, Tofalo R (2019) A comprehensive overview on microalgal-fortified/based food and beverages. Food Rev Int 35(8):778–805
Singh A, Nigam PS, Murphy JD (2011) Mechanism and challenges in commercialisation of algal biofuels. Bioresour Technol 102(1):26–34
Singh SK, Kumar Y, Sasmal S (2022) Enhanced algal biomass production in a novel electromagnetic photobioreactor (E-PBR). Curr Microbiol 79(12):395
Singh SK, Kumar Y, Sasmal S (2023) Perspectives of HMF and LA from microalgal biomass. Algal Res 72:103133
Song H, He M, Gu C, Wei D, Liang Y, Yan J et al (2018) Extraction optimization, purification, antioxidant activity, and preliminary structural characterization of crude polysaccharide from an Arctic *Chlorella* sp. Polymers 10(3):292
Tan JS, Lee SY, Chew KW, Lam MK, Lim JW, Ho SH et al (2020) A review on microalgae cultivation and harvesting, and their biomass extraction processing using ionic liquids. Bioengineered 11(1):116–129
Valdes AM, Walter J, Segal E, Spector TD (2018) Role of the gut microbiota in nutrition and health. BMJ 361:k2179
van der Linde C, Barone M, Turroni S, Brigidi P, Keleszade E, Swann JR et al (2021) An in vitro pilot fermentation study on the impact of *Chlorella pyrenoidosa* on gut microbiome composition and metabolites in healthy and coeliac subjects. Molecules 26(8):2330
Ververis E, Ackerl R, Azzollini D, Colombo PA, de Sesmaisons A, Dumas C et al (2020) Novel foods in the European Union: scientific requirements and challenges of the risk assessment process by the European Food Safety Authority. Food Res Int 137:109515

Wan XZ, Ai C, Chen YH, Gao XX, Zhong RT, Liu B et al (2020) Physicochemical characterization of a polysaccharide from green microalga *Chlorella pyrenoidosa* and its hypolipidemic activity via gut microbiota regulation in rats. J Agric Food Chem 68(5):1186–1197

Wang Y, Li Y, Ma X, Ren H, Fan W, Leng F et al (2018) Extraction, purification, and bioactivities analyses of polysaccharides from *Glycyrrhiza uralensis*. Ind Crops Prod 122:596–608

Wang ZD, Zhang W, Liang TX (2025) Advancements in oral delivery systems for probiotics based on polysaccharides. Polymers 17(2):144

Wassie T, Niu K, Xie C, Wang H, Xin W (2021) Extraction techniques, biological activities and health benefits of marine algae *Enteromorpha prolifera* polysaccharide. Front Nutr 8:747928

Wong JF, Hong HJ, Foo SC, Yap MKK, Tan JW (2022) A review on current and future advancements for commercialized microalgae species. Food Sci Hum Wellness 11(5): 1156–1170

Wu C, Wang X, Wang H, Shen B, He X, Gu W et al (2014) Extraction optimization, isolation, preliminary structural characterization and antioxidant activities of the cell wall polysaccharides in the petioles and pedicels of Chinese herbal medicine Qian (*Euryale ferox* Salisb.). Int J Biol Macromol. 64:458–467

Zhang J, Wen C, Zhang H, Duan Y, Ma H (2020) Recent advances in the extraction of bioactive compounds with subcritical water: a review. Trends Food Sci Technol 95:183–195

The Prebiotic Role of Algal Polysaccharides and Their Impact on Gut Microbiome Composition and Function

5

Saravanan Palanivishwanath, Senthil Nagappan, Komal Tripathi, and Sanjeev Mishra

Abstract

Prebiotics are non-digestible food components that benefit the host by selectively stimulating the growth and activity of beneficial gut microbiota. Algal polysaccharides, such as fucoidan, alginate, agar, carrageenan, and xylan, are complex carbohydrates derived from marine algae and have garnered significant attention for their potential prebiotic effects, as well as other bioactive properties. Generally, these polysaccharides are resistant to human digestive enzymes and thus reach the colon, fermented by specific gut microbiota, promoting the growth of beneficial bacteria such as bifidobacteria and lactobacilli. The fermentation of these non-digested polysaccharides produces short-chain fatty acids (SCFAs) like acetate, propionate, and butyrate, which play pivotal role in maintaining gut homeostasis, enhancing epithelial barrier integrity, modulating immune responses, and reducing inflammation. In addition to promoting beneficial gut bacteria, algal polysaccharides have been found to boost overall microbial diversity, suppress harmful pathogens, and support the body's mucosal immune

S. Palanivishwanath
Biochemical Conversion Division, Sardar Swaran Singh National Institute of Bio-Energy, Kapurthala, India

Department of Chemical Engineering, Indian Institute of Technology Roorkee, Roorkee, Uttarakhand, India

S. Nagappan · S. Mishra (✉)
Biochemical Conversion Division, Sardar Swaran Singh National Institute of Bio-Energy, Kapurthala, India
e-mail: sanjeevmishra@nibe.res.in

K. Tripathi
Department of Chemical Engineering, Indian Institute of Technology Roorkee, Roorkee, Uttarakhand, India

K. Mohanty et al. (eds.), *Advances in Algal Research*,
https://doi.org/10.1007/978-981-95-7234-2_5

defenses. Their natural antioxidant, antimicrobial, and anti-inflammatory properties make them especially promising as ingredients in functional foods aimed at improving gut health and overall well-being. This chapter highlights the role of algal polysaccharides in modulating the gut microbiome and their associated health benefits, particularly in the context of gastrointestinal diseases such as inflammatory bowel disease (IBD), underscoring their promise as functional ingredients for improving gut health and preventing related disorders.

Keywords

Algal polysaccharides · Gut microbiome · Prebiotics · Short-chain fatty acids (SCFAs)

5.1 Introduction

The human body is widely colonized by the microbiome, which is an integral part of normal human function. Among these microbial communities, the human gastrointestinal tract harbors a highly complex ecosystem of microorganisms that play a vital role in maintaining health. These include modulating the immune system, improving nutritional absorption, protecting against pathogens, and supporting the overall immune system. Disruption of the microbial homeostasis affects the normal functions, leading to health complications, including several gastrointestinal diseases, such as irritable bowel syndrome (Li et al. 2020). Mounting evidence suggests that dietary modulation of this gut microbiota is a promising therapeutic strategy for various gastrointestinal tract disorders (Li et al. 2020).

Prebiotics are short-chain carbohydrates and non-digestible ingredients that are resistant to human digestive enzymes and can significantly influence the composition and activity of the gut microbiota (Al-Sheraji et al. 2013). The intake of prebiotics can beneficially influence the composition and activity of the gut microbiota and thereby support overall health. Among the various natural sources, Marine Algal Polysaccharides (MAP) exhibit prebiotic properties. MAP can be fermented in the colon by gut microbiota and positively influence the gut microbiota (probiotics) (Gouda et al. 2022). Generally, these non-digestible carbohydrates are not hydrolyzed in the upper part of the intestine and are transported to the colon, where they will ferment by probiotic/gut microbiota (Gomez-Zavaglia et al. 2019). These photosynthetic organisms reside in the marine environment, including both microalgae (unicellular) and macroalgae. Based on the pigments, these marine algae are divided into three categories they are Chlorophyta (green algae), Ochrophyta (brown algae), and Rhodophyta (Red algae) (Ni et al. 2017). From a nutritional point of view, marine algae contain polysaccharides (15–76%), protein (5–47%), and minerals (7–36%) (Rattigan et al. 2020), along with bioactive components including pycocyanin, allophycocyaninastaxanthin, leutin, zeaxanthin, canthaxanthin, β-carotene, and oleic acid (Ni et al. 2017).

The polysaccharides derived from algae exhibit unique properties and functional versatility that make them valuable. These complex saccharides derived from algae and found applications in the food industry, cosmetics, pharmaceuticals, biomedical, nanotechnology, environmental engineering, and water treatment (Ahmed et al. 2024). Particularly in the food industry, it is used as emulsifying and gelling agents, as flocculants and hydrating agents, emulsifiers, stabilizers, and thickening agents (De Jesus Raposo et al. 2015). The use of marine algal polysaccharides as functional ingredients offers promising avenues for the management of gastrointestinal disorders such as inflammatory bowel diseases (IBD) (Zheng et al. 2025). As such, their incorporation into functional foods underscored their promise as functional ingredients for improving gut health and preventing related disorders (Thiviya et al. 2022; Ricós-Muñoz et al. 2025).

With the above perspective, this chapter discusses the potential of algal polysaccharides as prebiotics and their impact on shaping the gut microbiome composition and functions. It emphasizes the structural diversity of marine algal polysaccharides and their selective fermentation by beneficial microbes, and the resulting health-promoting effects mediated through microbial metabolites.

5.2 Marine Algal Polysaccharides

Marine algal polysaccharides (MAP) are diverse and complex saccharides derived from marine macroalgae (red, brown, and green algae) as well as microalgae (De Jesus Raposo et al. 2015) (Fig. 5.1). They are composed of repeating units of

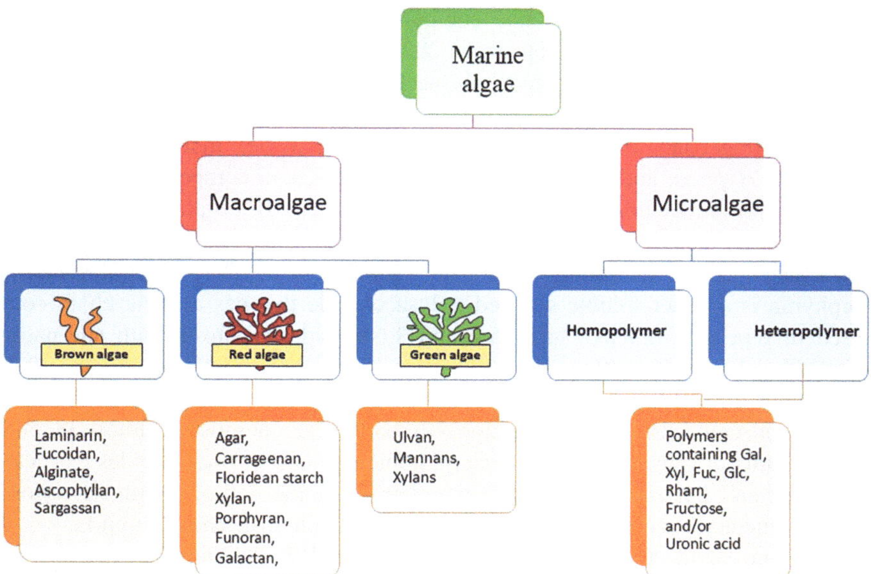

Fig. 5.1 Key polysaccharides found in marine algae

neutral and/or acidic sugars linked via glycosidic bonds, often modified by sulfate, methyl, or pyruvate groups (Ni et al. 2017). A significant portion of marine algal biomass comprises polysaccharides, including mucopolysaccharides, storage polysaccharides, and structural polysaccharides (Usman et al. 2017). Generally, the algal polysaccharide composition, as well as the physical and chemical properties of these polysaccharides, are influenced by several factors, including season, species, age, population density, and geographical location (Lim and Aida 2017).

Brown algae (Phaeophyceae) include 265 genera and 2040 species, of which 95% of the species are found in the marine environment (Mekinić et al. 2019). They are composed of water-soluble polysaccharides, including Alginate, laminarin, and fucoidans, which are widely studied polysaccharides for their health benefits (Sardari and Karlsson 2018; Xiong et al. 2022). Fucoidan is a highly sulfated and branched polymer, found in *Cladsiphon okamuranus, Saccharina japonica, Undaria pinnatifida, and Fucus vesiculosus*. Fucoidan is characterized structurally by the presence of α- fucopyranose backbone with sugar (De Jesus Raposo et al. 2015). Alginate is an acidic linear polysaccharide found in the cell wall and intercellular matrix of *Macrocystis pyrifera, Laminaria digitata, Laminaria japonica, Ascophyllum nodosum*, and *Laminaria hyperborean*. It is a co-polymer composed of 1,4 linked β-D-mannuronic acid (M) and α-L-guluronic acid (G) (Mensah et al. 2023; Lee and Mooney 2011). Laminarin is a storage polysaccharide extracted from the brown seaweeds, including *Laminaria japonica*, *Ecklonia kurome,* or *Eisenia bicyclis*. It is a water-soluble linear chain consisting of β-(1-3) linkages but also containing β-(1-6) linkages (Zhang et al. 2019).

Red algae are the oldest phylogenetic lineage among the marine macrophytes and are an essential source of bioactive compounds. Red algae contain mostly carrageenan and porphyran (Shannon et al. 2021; Pei et al. 2021; Wei et al. 2012). Carrageenan is a linear, sulphated polysaccharide which is made up of galactose units and their derivatives with alternating α-(1-3) and β-(1-4) bonds (Déléris et al. 2016). Carrageenan is commonly found in red algae, including *Chondrus*, *Gigartina*, *Hypnea*, and *Eucheuma*, where it occurs as a component of the outer cell wall and intracellular matrix (Besednova et al. 2020). Porphyran is a primary polysaccharide component found in the cell wall and intercellular space of several *Porphyra* species, including *Porphyra yezoensis* and *Porphyra haitanensis*. Porphyran is a water-soluble sulfated polysaccharide that has a linear chain composed of repeating units of galactose and 3,6-anhydrogalactose, with alternating units of galactose-6-sulfate and 6-O-methyl-galactose (Shannon et al. 2021; Liyanage et al. 2023).

In green seaweeds, Ulvan is a major polysaccharide, comprising almost 38–54% of the thallus dry mass. They are water-soluble sulfated polysaccharides extracted from *Ulva* and *Enteromorpha* sp. They are mainly composed of rhamnose, xylose, and iduronic acid, and they also contain a characteristic disaccharide unit known as albodiuronic acid (→4)-D-glucuronic acid-(1→4)-L-rhamnose-3-sulfate-(1→), which is a common structural feature among Ulvans (Usman et al. 2017; Déléris et al. 2016).

Polysaccharides extracted from microalgae display considerable structural diversity, ranging from species-specific homopolysaccharides to complex heteropolymers. For example, *Gymnodinium impudicum* and *Chlorella vulgaris* produce homopolysaccharides composed exclusively of galactose and glucose, respectively, whereas most other species synthesize heteropolymers of galactose, xylose, and glucose, often containing rhamnose, fucose, fructose, and uronic acids. Linkage analyses reveal distinctive patterns, such as the predominantly 1,3-linked backbone in *Aphanothece halophytica* with terminal glucose and glucuronic acid, alongside branched mannose residues (1,2,4- and 1,3,6-linkages). Unique features include acofriose (3-O-methyl-rhamnose) in *Chlorella*, *Botryococcus braunii*, and calcium-spirulan from *Arthrospira platensis*. Sulphated polysaccharides from *Phaeodactylum tricornutum* are characterized as ramified glucuronomannans with β-(1,3)-linked mannose backbones, decorated with side chains containing triuronic and aldobiuronic acids and β-(1,3)-linked glucans. In *Porphyridium cruentum* and related species (*Porphyridium*, *Rhodella*), distinctive aldobiuronic acids, hetero-oligosaccharides, and repeating disaccharide units further highlight the complexity and functional potential of these biopolymers (De Jesus Raposo et al. 2015).

5.3 Prebiotic Properties of Algal Polysaccharides

As outlined in the introduction, prebiotics are short-chain carbohydrates and non-digestible ingredients that are resistant to human digestive enzymes and can significantly influence the composition and activity of the gut microbiota (Al-Sheraji et al. 2013). The established criteria for considering a food ingredient as a prebiotic include the following: (a) resistance to gastric acidity and enzymatic degradation, and to gastrointestinal absorption; (b) fermentability by microorganisms colonizing the large intestine; and (c) selective stimulation of the growth and/or activity of beneficial bacteria in the gut, which can contribute to health and well-being (Roberfroid 2007).

Marine algal polysaccharides (MAPs) meet these criteria to a considerable extent. Their structural complexity, along with the presence of sulfate groups and uronic acids, imparts acid-binding and cation exchange properties. Furthermore, their high fiber content and properties of fecal bulking also make them a promising prebiotic candidate (De Jesus Raposo et al. 2015). Notably, due to the absence of specific hydrolytic enzymes in the upper gastrointestinal tract, they remain undigested until they reach the colon, where they undergo fermentation by the gut microbiota (Zheng et al. 2020; Li et al. 2017). Detailly, the digestion process of MAPs begins with oral processing, which includes mechanical breakdown and salivary impregnation. Though this process did not digest the MAPs, it affects the behavior. Followed by oral processing, the stomach digestion begins, and the indigested MAPs enter the stomach through the esophagus, where they are mixed with acidic gastric juices and subject to mechanical agitation. But without substantial degradation, the processed

MAPs pass through the stomach and reach the small intestine for the third part of digestion. After entering the small intestine, the MAPs are mixed with sodium bicarbonate, phospholipids, bile salts, and enzymes secreted by the liver, pancreas, and gall bladder. Due to the absence of specific enzymes in the small intestines, MAPs are not digested and reach the colon. These indigested MAPs are subject to fermentation by the gut microbiota found in the large intestine and secrete metabolites like Short Chain Fatty Acids (SCFAs), and will be discussed in detail in the following section (Zheng et al. 2020).

5.4 Gut Microbiome Modulation

The undigested MAPs reach the colon, where they are metabolized by the microorganisms present in this region (Fig. 5.2). The human colon harbors 10 trillion microorganisms that influence host health through various activities, including metabolic transformations, interaction with host cells, and the host immune system in a wide range (Zheng et al. 2020; Li et al. 2016). The dominant bacterial phyla in the healthy human gastrointestinal tract include Bacteroidetes and Firmicutes, while Proteobacteria, Actinobacteria, and Verrucomicrobia are considered subdominant phyla (Li et al. 2016; Madhogaria et al. 2022). Gut microbiota varies both between individuals and within the same individual over time, and its diversity and composition are closely associated with the host's health. The gut microbiome possesses a

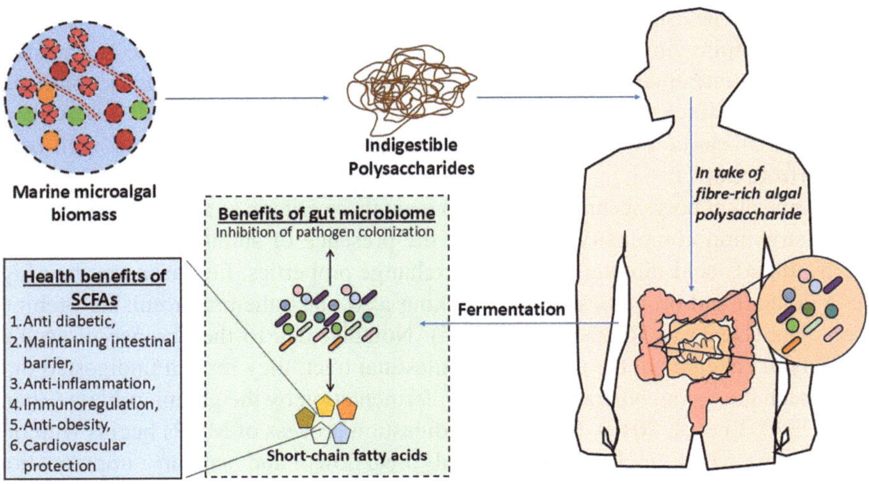

Fig. 5.2 The intake MAPs are undigested till they reach the colon and are fermented inside the colon by the gut microorganisms. These gut microbiome metabolites contribute to the health of humans

diverse array of genes encoding enzymes that enable it to break down complex carbohydrates, including MAPs. These genes are typically clustered in Polysaccharide Utilization Loci (PUL). PUL includes genes encoding for Carbohydrate-Active enzymes (CAZymes) which can lyse the complex carbohydrates into simpler carbohydrate units. By catalytic mechanisms, these CAZymes are classified into three types: Glycoside Hydrolases (GHs), Polysaccharide Lyases (PLs), and Carbohydrate Esterases (CEs) (Madhogaria et al. 2022; Gotteland et al. 2020). GH hydrolyses the glycosidic linkage between carbohydrate units, whereas PL cleaves the acidic sugar unit linkage, and CE breaks the ester linkage. There are notable differences in the CAZymes profiles of various gut microbiota. For instance, *Bacteroides thetaiotaomicron* 269GHs, 17PLs, and 18 CEs, whereas *Campylobacter hominis* contains only 6GHs, 3CEs, and surprisingly lacks PLs. This suggests species-specific abilities for polymer degradation and selective carbohydrate utilization, which in turn supports the enrichment of specific microbial communities within the gut. Such selective pressure enables specific bacteria to thrive based on the available polysaccharide substrates (including MAPs) in the gut, thereby modulating the gut microbiome (Zheng et al. 2020). For example, in vitro fermentation studies on polysaccharides derived from brown seaweed *Ecklonia radiate* increase the total bacterial numbers, especially Bacteroides (Moreira et al. 2023). Similarly, the intake of sulfated polysaccharides derived from *Gracilaria lemaeiformis* alters the microbiome composition, particularly Proteobacteria and Firmicutes (Cantarel et al. 2012). These gut microbiome modulation shows a positive impact on the host's health by producing metabolites like SCFAs and inhibiting pathogen colonization in the colon. Prebiotic-derived SCFAs, including acetate, propionate, and butyrate, not only act as an energy source but also exert properties including anti-inflammatory, immunoregulatory, anti-cancer, anti-obesity, cardiovascular protection, hepatoprotective, and neuroprotective activities (Han et al. 2019).

5.5 Inflammatory Bowel Diseases

Inflammatory bowel diseases (IBD) is a group of idiopathic chronic inflammatory intestinal diseases caused by chronic and excessive inflammation of gastrointestinal tract involving a complex interplay of genetic susceptibility, disturbance of the immune system, damage to the epithelial linings, and/or imbalanced interactions with gut microbes, and environmental variables (alcohols, chemicals, medications, and diets). Crohn's disease (CD) and ulcerative colitis (UC) are the two primary forms of IBD (Besednova et al. 2020).

In the early stages of IBD, the intestinal mucosa often shows redness, swelling, immune cell accumulation, and weakening of the gut's protective layer, which could lead to colorectal cancer and other serious consequences. Additionally, the abnormal immune system is an additional factor in the IBD pathogenesis. Under normal conditions, the gut microbiota helps in maintaining the intestinal barrier and produces SCFAs. But in the abnormal conditions, where the relative abundance of the beneficial microorganisms goes very low and the harmful microorganism goes very high, it eventually leads to the destruction of the tight junction of the intestinal mucosa, causing bacteria and antigens to invade the submucosa, and inducing the inflammation (De Borba Gurpilhares et al. 2018).

In both CD and UC, T cells are activated, but the nature of T cell differentiation and effector function differs. CD is linked to an increase in Th1 and Th17 and their cytokines IL-12, IL-23, IFN-γ, and IL-17, whereas UC is associated with an increase in Th2 with their cytokines IL-12, IL-5, and IL-9 production (Besednova et al. 2020).

In addition to the disturbance of the immune system, genetic susceptibility also plays a vital role in the pathogenesis of IBD. Advanced sequence technologies revealed the role of genes linked to IBD, including NOD2, Atg16L1, and CARD9 (De Borba Gurpilhares et al. 2018). Though several drugs are administered to treat IBD, the adverse side effects show the need for the discovery of novel compounds or a way to treat IBD.

5.6 Algae for Treating Gastrointestinal Disease: Focus on IBD

Marine algae are a promising organism to fulfill the human nutraceutical need and are packed with rich biologically active substances as well. Several algal polysaccharides have anticoagulant, antitumor, anti-inflammatory, antiviral, antihyperlipidemic, and antioxidant properties. Hence, the intake of algal polysaccharides can modulate gut microbiota and limit the occurrence of diseases as well (Qiu et al. 2021). Several studies have explored the potential of algae and their bioactive components for IBD treatment. In a study conducted on Dextron Sodium Sulfate (DSS)-induced ulcerative colitis mouse model, it was shown that the oral administration of a polysaccharide isolated from *Ulva pertusa* significantly alleviated disease symptoms by suppressing the pro-inflammatory cytokines and downregulating MAPK and NF-κB signaling. It also regulates the gut microbiota and improves the tight junction proteins (Occludin, claudin, mucin) (Yuan et al. 2025). Similarly, polysaccharides extracted from red alga Digenea simplex, which show anti-inflammatory and antinociceptive activity, reduce the carrageenan-induced edema in a dose-dependent manner and suppress inflammation triggered by dextran, histamine, serotonin, and bradykinin, and inhibit the transfer of neutrophils to the peritoneal cavity and paws of mice (Son et al. 2024). In another study, Sulphated polysaccharides extracted from the green algae Calulerpa Mexicana have shown antinociceptive and anti-inflammatory activities by reducing paw edema induced by dextran, histamine, and carrageenan, likely through targeting histamine and lowering myeloperoxidase activity in rodent models (Pereira et al. 2014). In a study conducted on a Dextron Sodium Sulfate (DSS)-induced ulcerative colitis mouse model, it was shown that the oral administration of 100–400 kg/day of fucoidan significantly prevents the weight loss, colon shortening, strengthens the epithelial barrier, and reduces the serum inflammatory factors concentrations. Apart from that, it increases the relative abundance of beneficial microorganisms and lowers the relative abundance of unwanted microbiota (Xiong et al. 2022). Similarly, the supplementation of laminarin in DSS-challenged pigs shows good microbiota profiles by reducing Proteobacteria and increasing SCFAs production (Li et al. 2023).

5.7 Applications in Functional Foods

For a long time, marine algae have been used as a food in East Asian countries due to their rich nutritional value and diverse culinary applications. They are packed with a good source of vitamins, minerals, polysaccharides, and protein. Beyond their nutritional content, algal polysaccharides, including alginate, agar, and carrageenan, are used as food additives (thickener and emulsifier) due to their extraordinary rheological properties, such as high viscosity, shear-thinning, viscoelasticity, gelation, and thermal stability (Madhogaria et al. 2022; Charoensiddhi et al. 2017; Rhein-Knudsen and Meyer 2021).

In recent years, the combination of algal-derived prebiotics with probiotics has led to the preparation of algae-based symbiotic formulations. These formulations are helpful to enhance both the survival of beneficial microorganisms and their functional activity. For example, Fucoidan from *Fucus vesiculosus, Macrocystis pyrifera,* and *Undaria pinnatifida has* been used for promoting the growth of probiotic bacteria added to the yogurt. This novel yogurt formulation can be used as an alternative strategy to promote immunomodulatory beneficial effects in allergic patients (Thiviya et al. 2022). Similarly, Synbiotic candy was prepared using *Enteromorpha compressa* as a prebiotic source and *Lactobacillus plantarum* as a probiotic source, and it could be an option to treat colon cancer (Ricós-Muñoz et al. 2025).

The immunomodulatory and anti-inflammatory roles of algal-derived polysaccharides have been extensively demonstrated in various experimental models of gastrointestinal disorders, as summarized in Table 5.1.

5.8 Conclusion

Exploration of prebiotic properties from marine algae is a wonderful idea, as they are the source of several bioactive compounds and complex polysaccharides. Their structural diversity of polysaccharides, resistance to digestion in the upper gastrointestinal tract, and selective utilization by the beneficial microorganisms in the gut contribute to the improved gut microbiome composition and function. Indirectly, the intake of these polysaccharides is improving the SCFAs production that supports the overall intestinal health, modulating immune responses. Most of the prebiotic research is restricted to the macroalgae, so the research community should come forward and explore the untapped potential of the microalgae polysaccharides.

Table 5.1 Animal studies evaluating the effect of the algal polysaccharides on gastrointestinal diseases

S. no	Algal polysaccharide	Algal source	Benefits	Disease	Animal model	Reference
1.	Sulphated polysaccharides	*Caulerpa mexicana*	Restored glutathione levels; reduced oxidative stress; it exhibits gastroprotective activity	Ethanol-induced gastric damage	Mice	Carneiro et al. (2014)
2.	Fucoidan	*Fucus vesiculosus*	Reduced colitis symptoms, lowered colon/spleen weight, restored colon architecture, reduced infiltration of immune cells and edema, and decreased production of ≥15 pro-inflammatory cytokines	DSS-induced colitis	Mice	Carneiro et al. (2018)
3.	Polysaccharide rich extract	*Eucheuma cottonii*	Protected against weight loss, reduced colon weight/length ratio, decreased TNF-α, IL-1β,IL-6, and increased IL-10, reduced colonic damage	DSS-induced colitis	BALB/c mice	Lean et al. (2015)
4.	Sulfated polysaccharide	*Hypnea musciformis*	Reduced the colitis and glutathione, malonyldialdehyde acid, myeloperoxidase, nitrate/nitrite and cytokines	Trinitrobenzene sulfonic acid induced colitis	Rat	Sudirman et al. (2018)
5.	β-glucans	*Laminaria hyperborea, Liminaria digitata*	Reduced expression of Th17 related cytokines, IL23R, and IL-6, suggesting anti-inflammatory effects	IBD	Pig	Brito et al. (2016)
6.	Low molecular weight Ulvan	*Ulva pertusa*	Reduced DSS-induced disease activity index, prevented colon shortening, alleviated colonic tissue damage, decreased oxidative stress and inflammation, improved tight junction protein expression	DSS-induced colitis	B6 mice	Ryan et al. (2012)
7.	Fucoidan	*Ascophyllum nodosum, Laminaria japonica*	Maintaining a balanced composition of microbiota reduce the antigen load and inflammatory response	Intestinal dysbiosis	Mice	Praveen et al. (2020)
8.	Fucoidan	*Undaria pinnatifida*	Alleviates colitis symptoms in fiber-deficient mice, preserves gut barrier integrity, improved the microbiota and metabolites profile, increased weight gain and reduced disease activity index	Colitis	Mice	Shang et al. (2016)

Acknowledgments The authors would like to acknowledge Sardar Swaran Singh National Institute of Bio-Energy for the support and scholarship (SSS-NIBE Bioenergy promotion fellowship). Authors also acknowledge the organizers of 2nd International Symposium on Advances in Algal Research (AAR 2024).

References

Ahmed N, Sheikh MA, Ubaid M, Chauhan P, Kumar K, Choudhary S (2024) Comprehensive exploration of marine algae diversity, bioactive compounds, health benefits, regulatory issues, and food and drug applications. Measurement Food [Internet] 14:100163. Available from: https://doi.org/10.1016/j.meafoo.2024.100163

Al-Sheraji SH, Ismail A, Manap MY, Mustafa S, Yusof RM, Hassan FA (2013) Prebiotics as functional foods: a review. J Funct Foods [Internet] 5(4):1542–1553. Available from: https://doi.org/10.1016/j.jff.2013.08.009

Besednova NN, Zaporozhets TS, Kuznetsova TA, Makarenkova ID, Kryzhanovsky SP, Fedyanina LN et al (2020) Extracts and marine algae polysaccharides in therapy and prevention of inflammatory diseases of the intestine. Mar Drugs [Internet] 18(6):289. Available from: https://doi.org/10.3390/md18060289

Brito TV, Barros FCN, Silva RO, Júnior GJD, Júnior JSC, Franco ÁX et al (2016) Sulfated polysaccharide from the marine algae Hypnea musciformis inhibits TNBS-induced intestinal damage in rats. Carbohydr Polym [Internet] 151:957–964. Available from: https://doi.org/10.1016/j.carbpol.2016.06.047

Cantarel BL, Lombard V, Henrissat B (2012) Complex carbohydrate utilization by the healthy human microbiome. PLoS One [Internet] 7(6):e28742. Available from: https://doi.org/10.1371/journal.pone.0028742

Carneiro JG, Rodrigues JAG, De Sousa Oliveira Vanderlei E, Souza RB, Quinderé ALG, Coura CO et al (2014) Peripheral Antinociception and anti-inflammatory effects of sulphated polysaccharides from the AlgaCaulerpa mexicana. Basic Clin Pharmacol Toxicol [Internet] 115(4):335–342. Available from: https://doi.org/10.1111/bcpt.12234

Carneiro J, Holanda T, Quinderé A, Frota A, Soares V, Sousa R et al (2018) Gastroprotective effects of sulphated polysaccharides from the Alga Caulerpa mexicana reducing ethanol-induced gastric damage. Pharmaceuticals [Internet] 11(1):6. Available from: https://doi.org/10.3390/ph11010006

Charoensiddhi S, Conlon MA, Vuaran MS, Franco CMM, Zhang W (2017) Polysaccharide and phlorotannin-enriched extracts of the brown seaweed Ecklonia radiata influence human gut microbiota and fermentation in vitro. J Appl Phycol [Internet] 29(5):2407–2416. Available from: https://doi.org/10.1007/s10811-017-1146-y

De Borba Gurpilhares D, Cinelli LP, Simas NK, Pessoa A Jr, Sette LD (2018) Marine prebiotics: polysaccharides and oligosaccharides obtained by using microbial enzymes. Food Chem [Internet] 280:175–186. Available from: https://doi.org/10.1016/j.foodchem.2018.12.023

De Jesus Raposo MF, De Morais AMB, De Morais RMSC (2015) Marine polysaccharides from algae with potential biomedical applications. Mar Drugs [Internet] 13(5):2967–3028. Available from: https://doi.org/10.3390/md13052967

Déléris P, Nazih H, Bard J m (2016) Seaweeds in human health. In: Elsevier eBooks [Internet], pp 319–367. Available from: https://doi.org/10.1016/b978-0-12-802772-1.00010-5

Gomez-Zavaglia A, Lage M a P, Jimenez-Lopez C, Mejuto JC, Simal-Gandara J (2019) The potential of seaweeds as a source of functional ingredients of prebiotic and antioxidant value. Antioxidants [Internet]. 8(9):406. Available from: https://doi.org/10.3390/antiox8090406

Gotteland M, Riveros K, Gasaly N, Carcamo C, Magne F, Liabeuf G et al (2020) The pros and cons of using algal polysaccharides as prebiotics. Front Nutr [Internet] 7. Available from: https://doi.org/10.3389/fnut.2020.00163

Gouda M, Tadda MA, Zhao Y, Farmanullah F, Chu B, Li X et al (2022) Microalgae bioactive carbohydrates as a novel sustainable and eco-friendly source of prebiotics: emerging health functionality and recent technologies for extraction and detection. Front Nutr [Internet] 9. Available from: https://doi.org/10.3389/fnut.2022.806692

Han R, Pang D, Wen L, You L, Huang R, Kulikouskaya V (2019) In vitro digestibility and prebiotic activities of a sulfated polysaccharide from Gracilaria Lemaneiformis. J Funct Foods [Internet] 64:103652. Available from: https://doi.org/10.1016/j.jff.2019.103652

Lean QY, Eri RD, Fitton JH, Patel RP, Gueven N (2015) Fucoidan extracts ameliorate acute colitis. PLoS One [Internet] 10(6):e0128453. Available from: https://doi.org/10.1371/journal.pone.0128453

Lee KY, Mooney DJ (2011) Alginate: properties and biomedical applications. Prog Polym Sci [Internet] 37(1):106–126. Available from: https://doi.org/10.1016/j.progpolymsci.2011.06.003

Li M, Li G, Shang Q, Chen X, Liu W, Pi X et al (2016) In vitro fermen tation of alginate and its derivatives by human gut microbiota. Anaerobe [Internet] 39:19–25. Available from: https://doi.org/10.1016/j.anaerobe.2016.02.003

Li M, Shang Q, Li G, Wang X, Yu G (2017) Degradation of marine algae-derived carbohydrates by bacteroidetes isolated from human gut microbiota. Mar Drugs [Internet] 15(4):92. Available from: https://doi.org/10.3390/md15040092

Li Y, Ye H, Wang T, Wang P, Liu R, Li Y et al (2020) Characterization of low molecular weight sulfate ulva polysaccharide and its protective effect against IBD in mice. Mar Drugs [Internet] 18(10):499. Available from: https://doi.org/10.3390/md18100499

Li S, Qian Q, Yang H, Wu Z, Xie Y, Yin Y et al (2023) Fucoidan alleviated dextran sulfate sodium–induced ulcerative colitis with improved intestinal barrier, reshaped gut microbiota composition, and promoted autophagy in male C57BL/6 mice. Nutr Res [Internet] 122:1–18. Available from: https://doi.org/10.1016/j.nutres.2023.11.009

Lim SJ, Aida WMW (2017) Extraction of sulfated polysaccharides (Fucoidan) from brown seaweed. In: Elsevier eBooks [Internet], pp 27–46. Available from: https://doi.org/10.1016/b978-0-12-809816-5.00003-7

Liyanage NM, Nagahawatta DP, Jayawardena TU, Jeon YJ (2023) The role of seaweed polysaccharides in gastrointestinal health: protective effect against inflammatory bowel disease. Life [Internet] 13(4):1026. Available from: https://doi.org/10.3390/life13041026

Madhogaria B, Bhowmik P, Kundu A (2022) Correlation between human gut microbiome and diseases. Infect Med [Internet] 1(3):180–191. Available from: https://doi.org/10.1016/j.imj.2022.08.004

Mekinić IG, Skroza D, Šimat V, Hamed I, Čagalj M, Perković ZP (2019) Phenolic content of brown algae (Pheophyceae) species: extraction, identification, and quantification. Biomolecules [Internet] 9(6):244. Available from: https://doi.org/10.3390/biom9060244

Mensah EO, Kanwugu ON, Panda PK, Adadi P (2023) Marine fucoidans: structural, extraction, biological activities and their applications in the food industry. Food Hydrocoll [Internet] 142:108784. Available from: https://doi.org/10.1016/j.foodhyd.2023.108784

Moreira JB, Santos TD, Cruz CG, Da Silveira JT, De Carvalho LF, De Morais MG et al (2023) Algal polysaccharides-based nanomaterials: general aspects and potential applications in food and biomedical fields. Polysaccharides [Internet] 4(4):371–389. Available from: https://doi.org/10.3390/polysaccharides4040022

Ni J, Wu GD, Albenberg L, Tomov VT (2017) Gut microbiota and IBD: causation or correlation? Nat Rev Gastroenterol Hepatol [Internet] 14(10):573–584. Available from: https://doi.org/10.1038/nrgastro.2017.88

Pei Y, Yang S, Xiao Z, Zhou C, Hong P, Qian ZJ (2021) Structural characterization of sulfated polysaccharide isolated from red algae (Gelidium crinale) and antioxidant and anti-inflammatory effects in macrophage cells. Front Bioeng Biotechnol [Internet] 9. Available from: https://doi.org/10.3389/fbioe.2021.794818

Pereira JG, Mesquita JX, Aragão KS, Franco ÁX, Souza MHLP, Brito TV et al (2014) Polysaccharides isolated from Digenea simplex inhibit inflammatory and nociceptive responses. Carbohydr Polym [Internet] 108:17–25. Available from: https://doi.org/10.1016/j.carbpol.2014.01.105

Praveen MA, Parvathy KRK, Patra S, Khan I, Natarajan P, Balasubramanian P (2020) Cytotoxic and pharmacokinetic studies of Indian seaweed polysaccharides for formulating raindrop synbiotic candy. Int J Biol Macromol [Internet] 154:557–566. Available from: https://doi.org/10.1016/j.ijbiomac.2020.03.086

Qiu Y, Jiang H, Fu L, Ci F, Mao X (2021) Porphyran and oligo-porphyran originating from red algae Porphyra: preparation, biological activities, and potential applications. Food Chem [Internet] 349:129209. Available from: https://doi.org/10.1016/j.foodchem.2021.129209

Rattigan R, O'Doherty JV, Vigors S, Ryan MT, Sebastiano RS, Callanan JJ et al (2020) The effects of the marine-derived polysaccharides laminarin and chitosan on aspects of colonic health in pigs challenged with dextran sodium sulphate. Mar Drugs [Internet] 18(5):262. Available from: https://doi.org/10.3390/md18050262

Rhein-Knudsen N, Meyer AS (2021) Chemistry, gelation, and enzymatic modification of seaweed food hydrocolloids. Trends Food Sci Technol [Internet] 109:608–621. Available from: https://doi.org/10.1016/j.tifs.2021.01.052

Ricós-Muñoz N, Maicas S, Tortajada-Girbés M, Pina-Pérez MC (2025) Novel synbiotic yogurt formulation supplemented with fucoidan from phaeophyceae algae to promote limosilactobacillus reuteri and lacticaseibacillus rhamnosus GG. Foods [Internet] 14(15):2589. Available from: https://doi.org/10.3390/foods14152589

Roberfroid M (2007) Prebiotics: the concept revisited1. J Nutr [Internet] 137(3):830S–837S. Available from: https://doi.org/10.1093/jn/137.3.830s

Ryan MT, O'Shea CJ, Collins CB, O'Doherty JV, Sweeney T (2012) Effects of dietary supplementation with Laminaria hyperborea, Laminaria digitata, and Saccharomyces cerevisiae on the IL-17 pathway in the porcine colon. J Anim Sci [Internet] 90(suppl_4):263–265. Available from: https://doi.org/10.2527/jas.53802

Sardari RRR, Karlsson EN (2018) Marine poly- and oligosaccharides as prebiotics. J Agric Food Chem [Internet] 66(44):11544–11549. Available from: https://doi.org/10.1021/acs.jafc.8b04418

Shang Q, Shan X, Cai C, Hao J, Li G, Yu G (2016) Dietary fucoidan modulates the gut microbiota in mice by increasing the abundance ofLactobacillusandRuminococcaceae. Food Funct [Internet] 7(7):3224–3232. Available from: https://doi.org/10.1039/c6fo00309e

Shannon E, Conlon M, Hayes M (2021) Seaweed components as potential modulators of the gut microbiota. Mar Drugs [Internet] 19(7):358. Available from: https://doi.org/10.3390/md19070358

Son SU, Suh HJ, Shin KS (2024) Characterization of a novel sulfated-rhamnoglucuronan isolated from Korean seaweed Ulva pertusa and its efficacy for treatment of inflammatory bowel disease in mice. Carbohydr Polym [Internet] 342:122373. Available from: https://doi.org/10.1016/j.carbpol.2024.122373

Sudirman S, Hsu YH, He JL, Kong ZL (2018) Dietary polysaccharide-rich extract from Eucheuma cottonii modulates the inflammatory response and suppresses colonic injury on dextran sulfate sodium-induced colitis in mice. PLoS One [Internet] 13(10):e0205252. Available from: https://doi.org/10.1371/journal.pone.0205252

Thiviya P, Gamage A, Liyanapathiranage A, Makehelwala M, Dassanayake RS, Manamperi A et al (2022) Algal polysaccharides: structure, preparation and applications in food packaging. Food Chem [Internet] 405:134903. Available from: https://doi.org/10.1016/j.foodchem.2022.134903

Usman A, Khalid S, Usman A, Hussain Z, Wang Y (2017) Algal polysaccharides, novel application, and outlook. In: Elsevier eBooks [Internet], pp 115–153. Available from: https://doi.org/10.1016/b978-0-12-812360-7.00005-7

Wei N, Quarterman J, Jin YS (2012) Marine macroalgae: an untapped resource for producing fuels and chemicals. Trends Biotechnol [Internet] 31(2):70–77. Available from: https://doi.org/10.1016/j.tibtech.2012.10.009

Xiong RG, Zhou DD, Wu SX, Huang SY, Saimaiti A, Yang ZJ et al (2022) Health benefits and side effects of Short-Chain fatty acids. Foods [Internet] 11(18):2863. Available from: https://doi.org/10.3390/foods11182863

Yuan Z, Zhang Y, Zhang X, Wang L, Ling G, Zhang P (2025) Microalgae and bioactive substances therein: unveiling therapeutic promise against inflammatory bowel disease. J Agric Food Chem [Internet] 73(30):18497–18510. Available from: https://doi.org/10.1021/acs.jafc.5c04341

Zhang C, Show PL, Ho SH (2019) Progress and perspective on algal plastics – a critical review. Bioresour Technol [Internet] 289:121700. Available from: https://doi.org/10.1016/j.biortech.2019.121700

Zheng LX, Chen XQ, Cheong KL (2020) Current trends in marine algae polysaccharides: the digestive tract, microbial catabolism, and prebiotic potential. Int J Biol Macromol [Internet] 151:344–354. Available from: https://doi.org/10.1016/j.ijbiomac.2020.02.168

Zheng W, Tang S, Ren X, Song S, Ai C (2025) Fucoidan alleviated colitis aggravated by fiber deficiency through protecting the gut barrier, suppressing the MAPK/NF-κB pathway, and modulating gut microbiota and metabolites. Front Nutr [Internet] 11. Available from: https://doi.org/10.3389/fnut.2024.1462584

Astaxanthin from *Haematococcus pluvialis*: A Natural Key to Future Therapeutics and Functional Foods

6

N. Vinuba and Jibu Thomas

Abstract

With the rise in global demand for natural, sustainable, and preventive healthcare solutions, the nutraceutical industry is undergoing a rapid transformation. However, there remains a significant gap in consolidated knowledge about the role of microalgae, which are recognized as one of the most diverse and sustainable natural resources and have become crucial in the development of therapeutics and functional foods. Among them, *Haematococcus pluvialis* is recognized as a scalable source of high-value compounds— astaxanthin which has the highest antioxidant property. This chapter provides an in-depth exploration of microalgal astaxanthin with special emphasis on its therapeutic application and incorporation into functional food systems. It discusses the growth phases of *H. pluvialis*, triggers for the astaxanthin biosynthesis pathway, optimization of culture conditions for enhanced biomass and pigment production while highlighting the biochemical composition of microalgal species. The economical demand of microalgal production systems is addressed to highlight commercialization challenges and opportunities. By bridging the gap between algal biotechnology and health sciences, this review underscores the potential of *H. pluvialis*-derived astaxanthin as a key component of future therapeutic and functional food products, supporting a shift toward preventive, natural, and sustainable healthcare solutions.

Keywords

H. pluvialis · Optimization · Astaxanthin · Functional foods

N. Vinuba · J. Thomas (✉)
Division of Biotechnology, Karunya Institute of Technology and Sciences, Coimbatore, Tamil Nadu, India
e-mail: vinuban@karunya.edu.in; jibuthomas@karunya.edu

K. Mohanty et al. (eds.), *Advances in Algal Research*,
https://doi.org/10.1007/978-981-95-7234-2_6

6.1 Introduction

In the current era, the rise in chronic diseases such as diabetes, obesity, and heart disorders has made people more health-conscious. Globally, consumers are actively seeking natural and safe compounds that not only provide nutrition but also help prevent or manage various health issues. This has created a strong demand for sustainable bioactive ingredients that combine therapeutic benefits with nutritional value, such as antioxidants, immune-boosters, and anti-inflammatory agents (Bishop and Zubeck 2012; Eze et al. 2023). Microalgae have long been part of human diets, with historical examples such as the use of *Nostoc* by ancient Chinese populations, *Aphanotheca* in other parts of Asia, *Spirulina* by the Aztecs, and as a staple for the Kanembu people in Africa. Today, only a fraction of algal biodiversity is exploited for human consumption, with *Chlorella vulgaris* and *Spirulina pacifica* being the most prominent species, while others like *Haematococcus* and *Dunaliella* gaining biotechnological relevance (Galasso et al. 2019; Kratzer and Murkovic 2021; Osathanunkul et al. 2025). From a protein nutrition perspective, freshwater algae present a valuable source of amino acids, demonstrating high essential amino acid index (Mišurcová et al. 2014). Based on these properties, screening studies on various freshwater microalgal strains, including *Scenedesmus*, *Chlamydomonas debariana*, and *Chlorella sorokiniana*, have identified promising candidates for simultaneous biomass production and extraction of nutraceutical compounds under optimized cultivation and extraction conditions (Chiellini et al. 2022). The bioactive compounds include primary metabolites such as polysaccharides (phycocolloids) and phycobiliproteins, secondary metabolites like carotenoids and xanthophylls (astaxanthin, fucoxanthin), offering antioxidant, antimicrobial, and anti-inflammatory properties (Garcia-Perez et al. 2023; Guil-Guerrero and Prates 2025). A comparative assessment between *Arthrospira platensis* and *C. vulgaris* have revealed differences in EPA (Eicosapentaenoic Acid) and DHA (Docosahexaenoic Acid), polyphenols, antioxidant capacity, and cytotoxicity, allowing clearer identification of species-specific nutraceutical applications (Matos et al. 2020). Among them, *Haematococcus pluvialis* is well known for its ability to produce pigments under stress conditions. When the vegetative, green, motile cells are exposed to stress, they transform into red, non-motile cysts that accumulate the red pigment astaxanthin. This pigment, renowned for its potent antioxidant activity, is often referred to as the King of Carotenoids (Nair et al. 2023). Astaxanthin is a keto-carotenoid (3,3′-dihydroxy-β, β-carotene-4,4′-dione) distinguished by its long polyene chain with conjugated double bonds and functional hydroxyl (–OH) and keto (C=O) groups at both ends (Davan et al. 2023). It is naturally produced by a range of organisms, including plants, yeast, lichens, and bacteria, but is most abundantly synthesized in microalgae (Cunningham and Gantt 2011; Andrewes and Starr 1976; Yu and Liu 2020; Guerin et al. 2003). Unlike non-polar carotenoids such as β-carotene, the presence of polar groups makes astaxanthin both fat-soluble and relatively hydrophilic, enabling efficient integration into cell membranes and enhancing its antioxidant defence (Santos-Sanchez et al. 2021). Due to these structural and functional attributes, astaxanthin has been widely studied in recent years in both animal and human models, demonstrating neuroprotective, cardioprotective, and antitumour effects, as well as benefits for skin and eye health (Nishida et al. 2023). Its therapeutic potential has been highlighted in the prevention or co-management of conditions such as Alzheimer's disease, Parkinson's disease, cardiovascular disorders, and other oxidative stress-related illnesses (Aneesh et al. 2022). Furthermore, astaxanthin derived from microalgae is considered highly suitable for incorporation into functional foods and therapeutic formulations, as it supports immune defence, tissue repair, and the accumulation of essential nutrients and bioactive compounds (Ambrozova et al. 2014). However, despite these promising properties, significant research gaps remain in optimizing large-scale cultivation of *Haematococcus pluvialis*, improving cost-effective extraction and purification strategies, and conducting extensive clinical trials to validate its efficacy and bioavailability in humans. This review highlights the bioactive potential of *H. pluvialis*, with particular emphasis on astaxanthin. It discusses the upstream and downstream processing strategies for astaxanthin production, its therapeutic applications, and its incorporation into functional foods.

6.2 *Haematococcus pluvialis* as a Primary Source of Astaxanthin

Haematococcus pluvialis is a unicellular, freshwater green microalga from the Chlorophyceae family, known for being the richest natural source of astaxanthin (Johnson and An 1991). It was described by J. Von Flotow (1844) and later explored by T.E. Hazen (1899). *H. pluvialis* is commonly found in freshwater environments such as rain pools, artificial ponds, and birdbaths (Basiony et al. 2022). It thrives under a broad range of environmental conditions and has been isolated from many regions across Europe, North America, and Asia (Boussiba 2000). It has also been found across diverse environmental and climate conditions, such as a freshwater basin in the rock filled with melted snow on Blomstrandhalvøya Island (Norway), dried fountain near Rozhen, Blagoevgrad in Bulgaria (Gacheva et al. 2015), freshwater fishpond in Bihor, Romania (Dragos et al. 2010), rooftop surface of a building of KIOST in Seoul, Korea (Kim et al. 2015), freshwater lakes in Hoa Binh province, Vietnam, Palampur (Ha et al. 2024), Indian state of Himachal Pradesh (Karuppan et al. 2022). During the life cycle of this organism, it undergoes an ultrastructural change. When it is subjected to any stress condition, it changes from the green vegetative stage to the red cyst (aplanospore) stage. This transition is a critical phase in the algae in response to environmental stress, such as high light intensity, nutrient limitation, or salinity and the red cyst is rich in carotenoids, particularly Astaxanthin.

6.2.1 Morphology and Life Cycle of the Organism

The life cycle of *Haematococcus pluvialis* is complex, consisting of multiple distinct morphological stages that are adapted to specific environmental conditions. These stages reflect the algae's ability to adjust rapidly between growth and survival

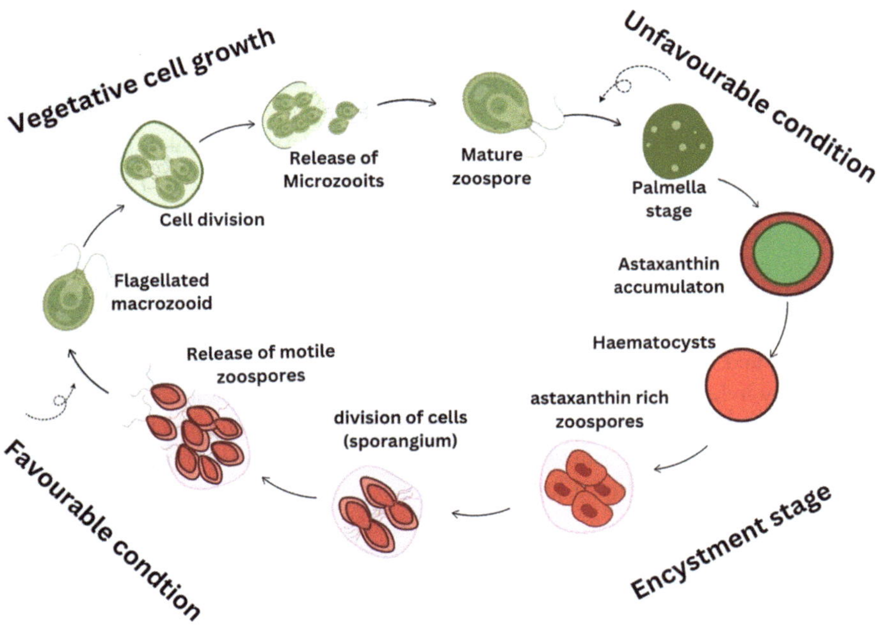

Fig. 6.1 Life cycle of *H. pluvialis*

modes, particularly as it transitions between vegetative (green) and dormant (red) states (Luo et al. 2019). The complete life cycle of *H. pluvialis* is illustrated in Fig. 6.1. It comprises four distinguishable cellular morphologies: macrozooids (zoospores), microzooids, palmella, and hematocysts (aplanospores). Macrozooids (zoospores) are spherical, ellipsoidal, or pear-shaped cells with two flagella of equal length emerging from the anterior end, and a cup-shaped chloroplast with numerous, scattered pyrenoids (Kobayashi et al. 1997). They are between 8 and 20 μm in length and are surrounded by a distinct gelatinous extracellular matrix of variable thickness. Numerous contractile vacuoles are irregularly distributed near the protoplast surface of the cell and are surrounded by a gelatinous extracellular matrix of variable thickness (Shah et al. 2016; Marinho et al. 2021). In contrast, microzooids, which originate through reproductive processes or successive cell divisions, are considerably smaller, generally measuring less than 10 μm (Li et al. 2022a). When these cells are exposed to unfavourable environmental or culture conditions, they gradually lose their flagella and undergo an increase in cell size. This transformation leads to the development of the non-motile palmella stage, a resting vegetative form that plays a crucial role in the survival strategy of *Haematococcus pluvialis*. During the palmella stage, the motile cells lose their flagella and become embedded within a thickened extracellular matrix, which provides enhanced protection against environmental stressors such as nutrient depletion, high light intensity, salinity fluctuations, or desiccation. The palmella stage is characterized by the formation of an amorphous,

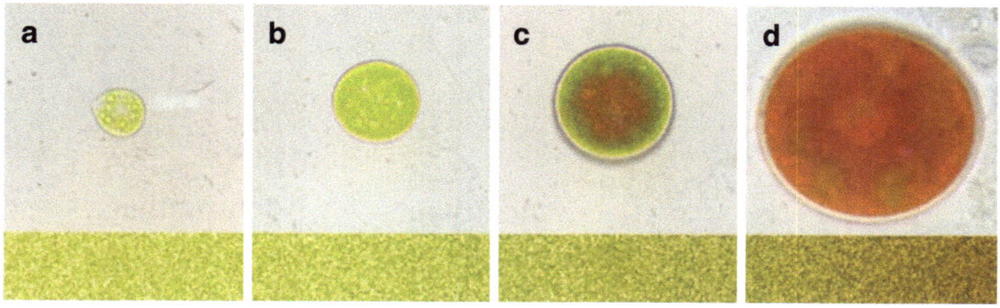

Fig. 6.2 The pigment production process in *Haematococcus pluvialis* progresses through four stages: (**a**) the flagellated zoospore stage, (**b**) the palmella stage, (**c**) the astaxanthin accumulation stage, and (**d**) the haematocyst stage

multilayered structure within the inner regions of the extracellular matrix or the primary cell wall. These structural modifications not only safeguard the cells from physical and chemical stress but also enable them to remain dormant until favourable growth conditions reappear, allowing them to revert to an active, motile state (Hang 2018; Christabel et al. 2024). With continued environmental stress (i.e. nutrient deprivation, high light irradiance, and high salinity), the "palmella" transform into the asexual "aplanospores". At this stage, cells contain two distinct structures: a thick and rigid trilaminar sheath and a secondary cell wall of acetolysis-resistant material. Such cells become resistant to prevailing extreme environmental conditions (Yu et al. 2022; Muniz Souza et al. 2020). Mature aplanospores accumulate large amounts of secondary carotenoids, predominantly astaxanthin, in lipid droplets deposited in the cytoplasm, resulting in a characteristic bright red colour of these cells (Zhang et al. 2017). Figure 6.2 highlights the developmental stages directly involved in pigment production.

6.2.2 Biosynthesis Pathway of Astaxanthin from *H. pluvialis*

Astaxanthin is synthesized via two distinct pathways, such as the mevalonate (MVA) pathway and the 2-C-methyl-D-erythritol 4-phosphate (MEP) pathway, also known as the 1-deoxy-D-xylulose-5-phosphate (DOXP) pathway or non-mevalonate pathway. The MVA pathway is widely reported in higher plants, fungi, and animals (Grünewald et al. 2000). Transcriptomic and genomic evidence indicate that *H. pluvialis* either lacks or has incomplete MVA pathway machinery, with several key enzymes of the classical cytosolic route missing, thereby limiting its contribution to isoprenoid biosynthesis. In contrast, the MEP pathway is fully functional within the chloroplast and provides the precursors necessary for downstream carotenoid biosynthesis, including astaxanthin (Liang et al. 2016). A schematic representation of this pathway is shown in Fig. 6.3.

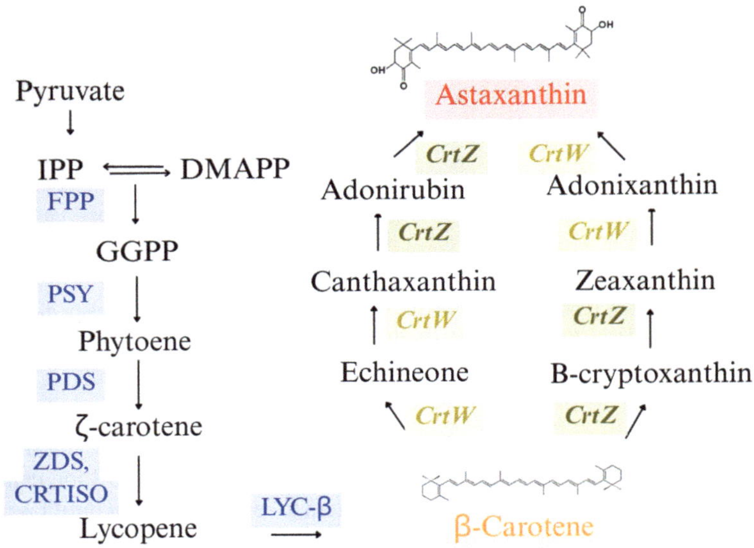

Fig. 6.3 Metabolic pathway of astaxanthin from microalgae

The biosynthesis pathway begins with the condensation of isopentenyl pyrophosphate (IPP) and dimethylallyl pyrophosphate (DMAPP) into geranylgeranyl pyrophosphate (GGPP) by farnesyl pyrophosphate (FPP) and GGPP synthases. GGPP serves as the immediate precursor for carotenoid biosynthesis. The next step is catalysed by phytoene synthase (PSY), which condenses two molecules of GGPP to form phytoene, a colourless carotenoid. Phytoene then undergoes a series of desaturation reactions mediated by phytoene desaturase (PDS) and ζ-carotene desaturase (ZDS), followed by isomerization by carotenoid isomerase (CRTISO), resulting in the formation of lycopene, which serves as a central point in the pathway. Through the activity of lycopene β-cyclase (LCY), lycopene is cyclized to produce β-carotene (Gao et al. 2013; Vidhyavathi et al. 2008; Paniagua-Michel et al. 2012; Göttl et al. 2024).

The final stage of the pathway involves the conversion of β-carotene into astaxanthin via the enzymes β-carotenoid ketolase (CrtW) and β-carotene hydroxylase (CrtZ). These two alternative routes are achieved by the Hydroxylation pathway and Ketolation pathway, wherein the hydroxylation pathway, CrtZ converts β-cryptoxanthin into zeaxanthin, which is then ketolated by CrtW to form adonixanthin, and ultimately astaxanthin. In the Ketolation pathway, CrtW oxidizes β-carotene to canthaxanthin via echinenone as an intermediate, followed by hydroxylation by CrtZ to produce adonirubin, which is finally converted into astaxanthin (Wan et al. 2021; Shah et al. 2016).

6.2.3 Cultivation System in Different Phases

During the biomass accumulation phase of *H. pluvialis*, moderate light intensity (30–80 μmol photons m^{-2} s^{-1}) is maintained to support photosynthesis without inducing stress. Temperature is controlled between 20 and 25 °C to promote optimal metabolic activity, while adequate nutrient availability, including nitrogen (as nitrate or urea), phosphorus, and essential micronutrients are ensured for healthy growth. CO_2 supplementation at 1–2% enhances photosynthetic efficiency and stabilized pH levels in culture medium are also a crucial parameter (Samara et al. 2025). Continuous agitation is maintained to facilitate gas exchange and prevent cell sedimentation. During the growth phase, cells predominantly exist as macrozooid or palmella forms, exhibiting active chloroplasts, high chlorophyll content, and rapid mitotic division. Typically cultivated in either open raceway ponds, which are cost-effective but vulnerable to contamination, or in closed photobioreactors (PBRs), which offer superior environmental control and culture purity (Jannel et al. 2024).

Once optimal biomass is achieved in the cultivation of *H pluvialis*, the culture is subjected to induction of stress conditions that prompt the cells to transition into the red aplanospore stage, thereby initiating the biosynthesis of astaxanthin. These stress conditions include high light intensity (150–400 μmol photons m^{-2} s^{-1} or exposure to natural sunlight), nitrogen and phosphorus starvation, salt stress through the addition of NaCl (0.1–0.2 M), and temperature (25–30 °C). These stress inducers degrade the chloroplasts and stimulate the accumulation of secondary carotenoids like astaxanthin in the cytoplasmic lipid droplets (Zhao et al. 2025). However, this method also presents challenges, including longer cultivation times, higher infrastructure costs, increased risk of microbial contamination during the stress phase, and difficulties in harvesting due to the thick-walled nature of red cysts (Azizi et al. 2024). Recent innovations have addressed these limitations through the use of LED lighting with specific wavelengths (red or blue) to enhance productivity, nutrient pulse strategies to boost pigment accumulation, and immobilized cultivation systems are also employed (Lee et al. 2018).

6.3 Optimization of Cultural Conditions and Downstream Processing

6.3.1 Impact of Stress on Astaxanthin Accumulation

As already mentioned, astaxanthin accumulation is primarily triggered under stress condition that redirect the cellular metabolism towards carotenoid biosynthesis. Studies have mentioned that high light intensity and specific light-quality combinations, such as red and blue light, stimulate photosynthetic electron transport and reactive oxygen species (ROS) formation, which act as a signalling cause for astaxanthin synthesis. Also, salinity and nitrogen limitation are the most effective triggering units, with low salinity promoting antioxidant responses and nitrogen deficiency significantly increasing both pigment content and productivity, especially when combined with high light or fed-batch nutrient strategies (Jin et al. 2025; Dou et al. 2025). Apart from that, comparative metabolomic analyses indicated that osmotic stress induces distinct metabolic markers, such as proline and succinate, and reconfigures several metabolic pathways to achieve higher pigment accumulation than nitrogen stress alone (Ma et al. 2025). In addition, moderate physical stresses like depressurization and electro treatment enhances astaxanthin content while preserving cell viability (Kim et al. 2024). The use of chemical modulators such as dopamine, melatonin, arginine, and phenolic compounds further supports biomass growth, improves photosynthetic efficiency, and directs carbon flux toward carotenoid synthesis, leading to enhanced astaxanthin accumulation even under mild conditions (Sathiyavahisan et al. 2023; Zhao et al. 2025; Acheampong et al. 2024). Table 6.1 presents the various stress conditions induced and the resulting outcomes.

6.3.2 Harvesting and Extraction Methods

Harvesting of *H. pluvialis* cells is a crucial step before the recovery of astaxanthin, as it involves separating the algal biomass from the culture medium. Several methods are employed for this purpose, each with specific advantages and limitations. Centrifugation is the most common and efficient laboratory-scale method, where high centrifugal force sediments the dense algal cells, although it is energy-intensive and costly for large-scale operations (Hagen et al. 2001). Flocculation is another widely used approach in which chemicals (such as alum or ferric chloride), bio-flocculants (like chitosan), or pH adjustment are applied to induce aggregation of algal cells, making them easier to settle and recover (Galán-González et al. 2024; Marinho et al. 2022). However, chemical residues may pose challenges for downstream processing, especially for food and nutraceutical applications. Filtration and membrane separation techniques, such as microfiltration and ultrafiltration, provide continuous harvesting and high recovery efficiency, but they are limited by issues like membrane fouling and high setup costs (Zhao et al. 2022; Gherabli et al. 2023). Overall, centrifugation and flocculation remain the most practical methods, with flocculation being preferred for large-scale cultivation due to its lower energy requirements, while advanced membrane-based systems are being explored for sustainable and scalable harvesting. Following harvesting, the extraction of astaxanthin is accomplished through diverse methodologies, which can be broadly categorized as follows.

6.3.2.1 Mechanical Cell Disruption

Various mechanical cell disruption techniques have been employed to break the tough outer wall. Mechanical methods, such as bead milling, high-pressure homogenization, and French-press treatments, are widely used to break the rigid cyst walls. High-pressure homogenization at 20,000–30,000 psi achieved up to 91% disruption efficiency, with an astaxanthin yield of 15.2 mg/L (Praveenkumar et al. 2020). Similarly, bead milling and ultrasonic-assisted extraction (UAE) have been reported to significantly enhance pigment recovery by creating acoustic cavitation that

Table 6.1 Stress-induced pigment production

Culture condition (green stage)	Induction of stress (red stage)	Outcome	Reference
BG-11 medium;25 °C under continuous 24-h white light (LED) with light intensity of 30 μmol photons m^{-2} s^{-1} for 12 days in fed batch culture	Algae 25 μM dopamine combined with 1 g NaCl under salinity and high light	Biomass of 1.41 g L^{-1}, astaxanthin content to 32.37 mg g^{-1}	Zhao et al. (2025)
C/RM medium, modified BG-11, and OHM media each in 250 ml Erlenmeyer flasks 25 ± 2 °C under 12:12 h light/dark with light intensity 40 μmol photons m^{-2} s^{-1}	Raising bicarbonate con. from 60 to 160 mM for every 2 days	Biomass of 0.511 g L^{-1} after 22 days and astaxanthin content to 48.8 mg g^{-L} achieved at 100 mM bicarbonate	Kobayashi et al. (1997)
NIES-C medium, pH:7.0 ± 0.2, 25.0 ± 1.0 °C; light intensity 200 ± 10 μmol photons m^{-2} s^{-1} for 14 days in 3 L photobioreactor	Electrostimulation was induced at 2, 4, 6, 8, and 10 V for 60 min	Productivity 13.2 mg L^{-1} after 14 d of subsequent photobioreactor cultivation; 2 V for 60 min enhanced astaxanthin content 44.9 mg g^{-L}	Lakshmi Narasimhan et al. (2024)
OHM medium with a pH of 7, 25 ± 1 °C under 12:12 h light/dark with light 50 μmol·s^{-1}·m^{-2} in Erlenmeyer flasks	SiO_2 nanoparticles and silicon salts	SiO_2 nanoparticles stimulated 194 mg $g^{-1}$of astaxanthin	Fazelinejad et al. (2024)
Jaworski's medium (JM), Walne/Conway, Optimal *Haematococcus* Medium (OHM), Blue Green-11, modified Chu-13, and Bold's Basal Medium (BBM) with N and specific organic compounds	Nitrogen starvation	Astaxanthin content, 20.484 mg/L, was obtained in JM without nitrogen	Rayamajhi et al. (2024)
BG-11 medium, pH 7–7.5, 24 °C, 16:8-h and 150 μmol/m^2/s under varying aeration rates (1.0, 1.5, 2.0 L/L/min), light intensities (150–1050 μmol/m^2/s), and CO_2 flow rates (0.6–300 mL/L/min)	Increase light intensity to 150 μmol/m^2/s to 1050 μmol/m^2/s,24:0-h photoperiod, pH 7, and 24 °C	With light 300 μmol/m^2/s & 900 μmol/m^2/s 1.5 L/L/min aeration rate, 225 mL/L/min CO_2 supply highest astaxanthin	Azizi et al. (2024)

(continued)

Table 6.1 (continued)

Culture condition (green stage)	Induction of stress (red stage)	Outcome	Reference
BG11 culture medium,24 ± 1 °C, 16: 8 L/D in white light in erlenmeyer flasks	Monochromic red (peak wavelength 660 nm) and blue light (peak wavelength 455 nm)	Red: blue = 3:4 astaxanthin content were 1.58% and 15.28 g L^{-1}, which were 2.1 and 2.2 times greater than at white light	Jin et al. (2024)
Selenite enrichment (SE) media 25 °C, 12: 12 L/D 100 μmol photons m^{-2} s^{-1}	Light (100, 200, 300, 400, 500 μmol photons m^{-2} s^{-1}); CH_3COONa(0 to 1.0 g L^{-1}); $C_7H_6O_3$ (0 to 12 mg L^{-1}) NaCl (0 to 4.0 g L^{-1}) $FeSO_4$ (0 to 10 mg L^{-1})	1 g L^{-1} CH_3COONa,10 mg L^{-1} $C_7H_6O_3$ con. of 10 mg L^{-1} increased astaxanthin to 6.50 mg L^{-1} at 489 μmol photons m^{-2} s^{-1} based on response surface methodology (RSM)	Wang et al. (2019)
BG11 culture medium, 25 ± 0.5 °C, 18:6 L/D in white light 25 μmol photons m^{-2} s^{-1}, aeration- 0.2 μm filter at a rate of 250 mL min^{-1}	UV mutagenesis, four mutants choose for high pigment production	MT-3-7-2 mutant strain of *H. pluvialis* can serve as a more efficient model for large-scale astaxanthin production	Telli and Ünlü (2023)
BB medium,21 °C, 14.43 ± 0.8 μmol·m^{-2}·s^{-1}, silver nanoparticles (AgNPs) in different concentration (0 mg/L to 8 mg/L)	BB medium without $NaNO_3$, 112.6 ± 3.21 μmol·m^{-2}·s^{-1}, silver nanoparticles (AgNPs) in different concentration (0 mg/L to 8 mg/L)	High con. of AgNPs had a negative impact on productivity in both green and red stage, productivity was reduced by up to 85% and shows changes in cell morphology	Venckus et al. (2023)

destabilizes cell membranes (Pei et al. 2025). These methods are efficient for large-scale processing but may lead to pigment degradation due to excessive shear and heat generation, making them less ideal for preserving astaxanthin stability in sensitive nutraceutical applications.

6.3.2.2 Enzymatic Cell Disruption

Enzymatic digestion represents a milder approach, employing hydrolytic enzymes such as cellulase, lysozyme, and chitosanase to selectively weaken the carbohydrate-rich cell wall (Ye et al. 2020; Machado et al. 2016). This method improves permeability without extensive damage to astaxanthin's chemical structure, thus maintaining antioxidant integrity. However, enzymatic extraction is costly, relatively slow, and less suitable for industrial-scale operations, but it is highly advantageous for pharmaceutical and cosmetic applications where purity and bioactivity are prioritized.

6.3.2.3 Solvent-Based Extraction

Traditional organic solvents such as acetone, ethanol, methanol, hexane, and dimethyl sulfoxide (DMSO) remain among the most widely employed extraction media due to their efficiency in dissolving astaxanthin. Acetone is particularly effective for high recovery, with studies reporting up to 57.42% recovery using microwave-assisted extraction (Dewati et al. 2023). Ethanol, a food-grade solvent, is favoured for functional food and nutraceutical applications due to safety and regulatory acceptance. DMSO, on the other hand, has demonstrated remarkable efficiency (yield of 97.3 mg g^{-1} with 92% pigment retention), though it is limited by toxicity concerns for food and pharma-grade products (Khadilkar et al. 2023). The main drawback of solvent extraction is the potential for residual solvents and pigment degradation. However, the choice of solvent allows for tailoring the extraction to specific downstream applications.

6.3.2.4 Green Solvent and Advanced Extraction Methods

Recent innovations have focused on environmentally friendly and advanced technologies for astaxanthin recovery, moving beyond conventional solvents. Green solvent approaches such as deep eutectic solvents (DES), ionic liquids (ILs), and natural flocculants have shown promising results. A study achieved 99.64% recovery using dipotassium hydrogen phosphate and choline chloride–urea-based DES, offering a non-toxic and biodegradable alternative aligned with sustainable bioprocessing (Nemani et al. 2024). Likewise, ultrathin α-quartz nanoplates combined with ionic liquids significantly improved cell-wall disruption, enhancing extraction efficiency by up to 80% (Lee et al. 2022). In parallel, edible oil-based recovery and salt-assisted emulsion breaking have shown promise in generating astaxanthin-enriched oils suitable for direct food and nutraceutical applications (Myint et al. 2023) . These methods comply with green chemistry principles; however, challenges such as scalability and regulatory approval for food-grade applications remain to be addressed. Complementing these solvent innovations, supercritical fluid extraction (SFE) with CO_2 has gained prominence as a clean, selective, and residue-free technique (Duan et al. 2024). These green and advanced methods represent a shift toward sustainable, high-quality astaxanthin production, bridging environmental responsibility with industrial scalability.

6.3.3 Purification and Quantification

The crude astaxanthin must be purified to separate it from other compounds such as lipids, chlorophylls, and other carotenoids like β-carotene and lutein to ensure the productivity of right quality astaxanthin suitable for industrial and other application. The purification process commonly employs techniques like liquid-liquid extraction, column chromatography, and membrane filtration, based on the desired purity and end use application (Zhang et al. 2023). Conventional purification methods include solvent extraction by DMSO and saponification to release astaxanthin from esterified forms to remove residual solvents, unwanted impurities and recovered 97.3 mg g^{-1}

and 163 mg g^{-1}, respectively in order to yield the free-form compound that offers greater bioavailability (Khadilkar et al. 2023). To enhance recovery rates, optimization techniques are also been integrated into the extraction process. For instance, applied response surface methodology (RSM) using a central composite design to evaluate the combined effects of time, temperature, and acid concentration, achieving an astaxanthin recovery efficiency of up to 97% (Abdel-Hameed et al. 2023). The purity of astaxanthin is quantified by high-performance liquid chromatography (HPLC) which remains as the gold standard for quantification and ester profile analysis, often complemented by FTIR spectroscopic analysis and GC-MS/MS metabolomics to monitor pathway intermediates and stress-induced metabolites (Machado et al. 2016). Advances in stabilization strategies, such as liposome encapsulation, further protect astaxanthin against oxidative degradation and extend its shelf life, enabling its incorporation into functional foods, pharmaceuticals, and therapeutic uses (Luo et al. 2024).

6.4 Application in Therapeutics and Functional Foods

6.4.1 Astaxanthin-Based Therapeutics and Nutraceutical Supplements

Astaxanthin displays exceptional antioxidant and anti-inflammatory properties, primarily due to its conjugated double bonds and polar end groups. It has a unique dual natural property that is both hydrophilic and lipophilic, which enables it to protect both inside and outside the cell and effectively scavenge free radicals and prevent lipid peroxidation, protein oxidation, and DNA damage (Fan et al. 2021). In preclinical models, astaxanthin from microalgae has demonstrated hepatoprotective effects by reducing oxidative stress and liver injury caused by carbon tetrachloride (CCl_4), supporting its use in pharmaceutical and nutraceutical formulations. It also shows gastroprotective activity, preventing ethanol-induced gastric ulcers in mice without signs of acute toxicity up to 500 mg kg^{-1} (Meryem et al. 2022). Beyond gastrointestinal protection, astaxanthin also enhances immune responses, increasing lymphocyte proliferation, natural killer cell activity, and phagocytosis, while showing no genotoxic effects in animal studies. It downregulates key inflammatory pathways such as NF-κB, COX-2, and IL-6, and significantly reduces inflammatory markers like C-reactive protein (CRP) and tumour necrosis factor-alpha (TNF-α), highlighting its potential in managing chronic inflammation (Radzun et al. 2022). Moreover, in terms of cardiovascular protection, astaxanthin has been shown to lower the oxidation of low-density lipoprotein (LDL), enhance high-density lipoprotein (HDL) function, and reduce blood pressure and arterial stiffness. It also improves lipid profiles by increasing HDL and reducing triglyceride levels, all of which contribute to a decreased risk of atherosclerosis and ischemic heart disease (Chou et al. 2020).

As a neuroprotective agent, astaxanthin is capable of crossing the blood-brain barrier, where it inhibits neuroinflammation, prevents the accumulation of β-amyloid plaques associated with Alzheimer's disease, enhances mitochondrial function in neurons, and reduces oxidative stress in models of Parkinson's disease and dementia. Clinical studies further suggest that astaxanthin supplementation may improve cognitive function in elderly individuals experiencing mild memory impairment (Patil et al. 2022). Apart from therapeutic benefits nutritionally, astaxanthin is gaining prominence in both human health and aquaculture. Astaxanthin bioavailability is influenced by the gut microbiota, with studies indicating that probiotic strains such as *Lactobacillus* enhance intestinal uptake by modulating lipid metabolism and facilitating isomer conversion (Li et al. 2022b).

In aquaculture, dietary supplementation with natural astaxanthin improves fish pigmentation, growth performance, and antioxidant defence mechanisms, while simultaneously reducing inflammation and oxidative stress. For example, rainbow trout fed with astaxanthin enriched diets showed enhanced muscle pigmentation, improved serum antioxidant enzyme activity, and favourable changes in lipid composition. Similarly, in zebrafish, dietary astaxanthin alleviated microcystin induced oxidative stress by lowering ROS and lipid peroxidation markers while boosting antioxidant enzymes (Dong et al. 2021). Both the US Food and Drug Administration (FDA) and the European Food Safety Authority (EFSA) have approved natural astaxanthin derived from *Haematococcus pluvialis* as a safe nutraceutical ingredient (Turck et al. 2023). An overview of the production process and its end-use applications is provided in Fig. 6.4.

6.4.2 Astaxanthin as a Functional Food

According to recent studies, consumer interest in microalgae-derived ingredients is increasing, with astaxanthin-rich supplements and fortified foods. Clinical trials have validated the safety and effectiveness of astaxanthin supplementation at daily doses between 4 and 12 mg per day (Chou et al. 2020). In international markets, astaxanthin is commonly incorporated into capsules, beverages, and functional snacks, typically providing around 1–8 mg per serving (Capelli et al. 2019). From a nutritional perspective, astaxanthin plays a key role in modulating metabolic health. Recent findings demonstrate its ability to regulate glucose and lipid homeostasis by improving insulin sensitivity, lowering blood glucose, and reducing triglycerides and LDL (low-density lipoprotein) cholesterol, thereby contributing to its protection against obesity, type 2 diabetes, and cardiovascular diseases (Villaró et al. 2021). These effects are largely attributed to its strong antioxidant capacity, which is reported to be several times higher than that of vitamin C, vitamin E, β-carotene, lutein, and zeaxanthin. Significantly, astaxanthin also prevents lipid peroxidation and protects pancreatic β-cells from oxidative stress, highlighting its role as a dietary intervention for metabolic syndrome (Yamashita 2013). Moreover, studies proposed astaxanthin as a medical food, citing clinical applications in reducing lifestyle-related diseases such as atherosclerosis, hyperlipidemia, and diabetes, while also improving quality of life in patients with chronic conditions. The addition of *H. pluvialis* extract (0.3–0.45 g/kg) to raw ground pork significantly enhanced antioxidant activity, delayed lipid oxidation, and improved colour

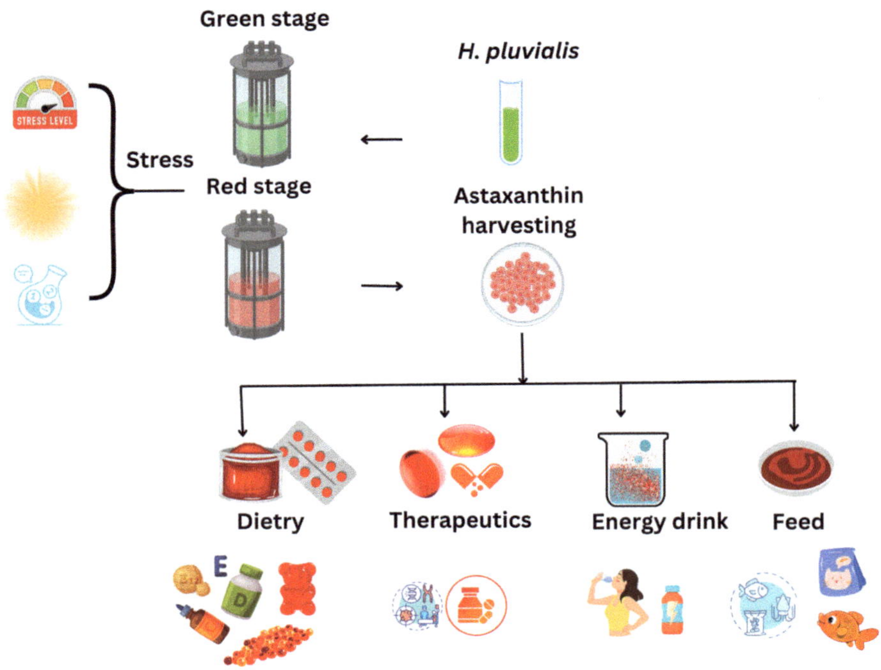

Fig. 6.4 Two-stage cultivation of *Haematococcus pluvialis* enhances astaxanthin accumulation, which is then applied in multiple sectors. It is marketed as capsules, tablets, and gummies as a dietary supplement; used therapeutically for cardiovascular health, diabetes, cancer, and related conditions; added to energy drinks due to its role in reducing muscle fatigue; and incorporated into aquaculture feed to improve fish health and pigmentation

stability, leading to better consumer acceptance after 7 days of refrigerated storage (Espinosa-Ramírez et al. 2023). Research also emphasized that esterified natural astaxanthin from *H. pluvialis* is significantly more effective than synthetic or yeast-derived forms, showing superior results in reducing oxidative stress, improving exercise performance, and enhancing resistance to stress in animal and human studies (Medoro et al. 2024). Based on these studies, astaxanthin can be considered as a promising natural compound for the development of a functional food ingredient since it is sustainable and has high bioactive compounds, ideally positioned to meet the growing demand for functional foods that combine nutrition with disease prevention (Mont'alverne et al. 2010).

6.4.3 Market Trends and Demand

Currently, more than 95% of the market is dominated by synthetic astaxanthin, priced around USD 1000/kg, and is mainly used in aquaculture for pigmentation (Mussagy et al. 2023). However, synthetic astaxanthin, derived from petrochemicals, poses safety and sustainability concerns, and is also not permitted for human consumption in food or supplements in many countries due to its structural difference and weak antioxidant properties. In contrast, natural astaxanthin from *Haematococcus pluvialis* commands a much higher price from USD 2500–7000/kg due to its superior antioxidant activity, health benefits, and consumer preference for natural, safe, and sustainable bioactives (Rahman 2020). Despite contributing less than 5% of the market share, natural astaxanthin demand is rapidly growing due to rising health awareness, expansion of functional foods, and regulatory shifts toward natural ingredients (Panis and Carreon 2016). The global astaxanthin market value in 2022 is USD 864.22 million and it is expected to reach USD 167.5 million to USD 2.13 billion by 2032, with a CAGR of 9.46%, mainly due to the rising popularity in human health products (Mussagy et al. 2023). Global producers such as Cyanotech (USA), Algatech (Israel), Atacama Bio (Chile), and Algalif (Iceland) are leading industries that produce axtaxanthin on large scale, increasingly relying on hybrid systems (photobioreactors + open raceway ponds) and renewable energy sources to reduce costs and improve environmental sustainability. Other than them several countries such as India, China, Japan, and Sweden have established industrial scale production of astaxanthin from microalgae with a capacity of 50–100 tons per year employing different cultivate systems (Shah et al. 2016). Techno-economic studies suggest that hybrid cultivation systems (photobioreactors + open ponds) combined with advanced extraction methods (e.g. supercritical CO_2) can pointedly reduce costs and improve scalability (Villaró et al. 2021). Regions with high solar radiation, like Santa Marta in Colombia, or Mediterranean areas such as Greece, have shown strong economic feasibility (Ambati et al. 2014). Overall, market trends clearly indicate a shift from synthetic to natural astaxanthin, fuelled by consumer demand for clear label products, regulatory restrictions on synthetic antioxidants, and expansion of the functional foods and nutraceuticals sector. This makes *H. pluvialis-derived* astaxanthin as a high valuable and sustainable alternative with strong commercial growth potential.

6.5 Conclusion

Microalgae with its promising bioactive compounds is emerging as a next generation nutraceuticals, associating traditional dietary practices with modern functional foods. Its thriving ability under different environmental conditions followed by the production of primary and secondary metabolites, makes them the most valuable source with health promoting compounds such as polysaccharides, phycobiliproteins, carotenoids, and polyunsaturated fatty acids. Among them, astaxanthin plays a vital role with high antioxidant and anti-inflammation properties that strengthen human defence system, repair mechanism and overall health benefits. At the same time the lower productivity and high production cost are the major drawbacks in the cultivation and extraction process. To overcome these, innovation and improvement in cultural conditions, harvesting, extraction by different methods like solvent free, deep eutectic solvent, and supercritical CO_2 approaches are making production more environmentally friendly and cost-effective. Downstream purification technologies coupled with quantification platforms such as HPLC and LC-MS ensure product safety, stability, and standardization, which are essential for industrial and clinical applications. Therapeutically, astaxanthin has shown remarkable potential as an antioxidant, anti-inflammatory, photoprotective, and hepatoprotective agent, with applications spanning dermatology, gastroprotection, immunomodulation, and radioprotection. Likewise, nutritionally, it enhances gut health, supports systemic antioxidant defence, and has been widely applied in aquaculture to improve pigmentation, growth, and stress tolerance in fish. It is also formulated into functional foods and dietary supplements, and encapsulated delivery systems are steadily bridging the gap between laboratory discovery and consumer health products. With the increasing customer demand for natural astaxanthin, expanding clinical validation of astaxanthin's benefits in metabolic health, cardiovascular protection, skin health, and cognitive function will strengthen its position as a premium nutraceutical, opening new opportunities in personalized nutrition and medical foods. Furthermost with its wide range of biological properties, astaxanthin has a promising opportunity for future applications beyond earth. In the unique conditions of space, astronauts are exposed to elevated oxidative stress, immune suppression, metabolic imbalances, and muscle atrophy due to prolonged exposure to microgravity. The multifunctional benefits of astaxanthin make it an attractive candidate for use as a nutritional supplement in space science, where maintaining physiological balance is critical. Its role in neutralizing reactive oxygen species, supporting glucose metabolism, and preserving muscle health could help mitigate health risks faced by astronauts during long-duration missions. In fact, preliminary research is already exploring the use of natural antioxidants like astaxanthin in space nutrition to ensure sufficient nutrient intake and to protect against the detrimental effects of zero-gravity environments. This opens a novel frontier for astaxanthin as not only a functional food on Earth but also as a nutraceutical solution for human space exploration.

References

Abdel-Hameed M, Sestric R, Hardy B, Levin DB, Sorensen JL (2023) Astaxanthin stability in intracellular and extracellular conditions from the green alga Haematococcus pluvialis. Food Chem Adv 3:100432

Acheampong A, Wang R, Elsherbiny SM, Bondzie-Quaye P, Huang Q (2024) Exogenous arginine promotes the coproduction of biomass and astaxanthin under high-light conditions in Haematococcus pluvialis. Bioresour Technol 393:130001

Ambati R, Phang SM, Ravi S, Aswathanarayana R (2014) Astaxanthin: sources, extraction, stability, biological activities and its commercial applications—a review. Mar Drugs 12(1): 128–152

Ambrozova J, Misurcova L, Vicha R, Machu L, Samek D, Baron M et al (2014) Influence of extractive solvents on lipid and fatty acids content of edible freshwater algal and seaweed products, the green microalga Chlorella kessleri and the cyanobacterium Spirulina platensis. Molecules 19(2):2344–2360

Andrewes AG, Starr MP (1976) (3R,3′R)-astaxanthin from the yeast Phaffia rhodozyma. Phytochemistry 15(6):1009–1011

Aneesh PA, Ajeeshkumar KK, Lekshmi RGK, Anandan R, Ravishankar CN, Mathew S (2022) Bioactivities of astaxanthin from natural sources, augmenting its biomedical potential: a review. Trends Food Sci Technol 125:81–90

Azizi M, Moteshafi H, Hashemi M (2024) Novel paradigm of light and CO2 feeding to improve the co-production of astaxanthin and polyunsaturated fatty acids in Haematococcus pluvialis. J CO_2 Util 84:102855

Basiony M, Ouyang L, Wang D, Yu J, Zhou L, Zhu M et al (2022) Optimization of microbial cell factories for astaxanthin production: biosynthesis and regulations, engineering strategies and fermentation optimization strategies. Synth Syst Biotechnol 7(2):689–704

Bishop WM, Zubeck H (2012) Evaluation of microalgae for use as nutraceuticals and nutritional supplements. J Nutr Food Sci 02(05):1–6

Boussiba S (2000) Carotenogenesis in the green alga *Haematococcus pluvialis* : cellular physiology and stress response. Physiol Plant 108(2):111–117

Capelli B, Talbott S, Ding L (2019) Astaxanthin sources: suitability for human health and nutrition. Funct Food Health Dis 9(6):430–445

Chiellini C, Serra V, Gammuto L, Ciurli A, Longo V, Gabriele M (2022) Evaluation of nutraceutical properties of eleven microalgal strains isolated from different freshwater aquatic environments: perspectives for their application as nutraceuticals. Foods 11(5):654

Chou HY, Ma DL, Leung CH, Chiu CC, Hour TC, Wang HMD (2020) Purified Astaxanthin from *Haematococcus pluvialis* promotes tissue regeneration by reducing oxidative stress and the secretion of collagen *in vitro* and *in vivo*. Oxidative Med Cell Longev 2020(1):1–13

Christabel C, Kim B, Lakshmi Narasimhan A, Sathiyavahisan LP, Ilhamsyah DPA, Kim EJ et al (2024) Enhanced cell growth and Astaxanthin production in Haematococcus lacustris by mechanostimulation of seed cysts. Appl Sci 14(22):10434

Cunningham FX, Gantt E (2011) Elucidation of the pathway to Astaxanthin in the flowers of *Adonis aestivalis*. Plant Cell 23(8):3055–3069

Davan I, Fakurazi S, Alias E, ‘Izzah IN, Hwei NM, Hassan H (2023) Astaxanthin as a potent antioxidant for promoting bone health: an up-to-date review. Antioxidants 12(7):1480

Dewati PR, Rochmadi RA, Budiman A (2023) Mathematical model of astaxanthin purification process using the low-pressure column chromatography method. S Afr J Chem Eng 45:256–268

Dong J, Li C, Dai D, Zhang M, Gao Y, Li X et al (2021) Protective effects of astaxanthin from Haematococcus pluvialis on the survival and oxidative stress of zebrafish embryos induced by microcystin-LR. J Appl Phycol 33(4):2261–2271

Dou Y, Cheng L, Wang Y, Yu G, Zhou W (2025) Comparative metabolomic analysis of Haematococcus pluvialis during hyperaccumulation of astaxanthin under the high salinity and nitrogen deficiency conditions. World J Microbiol Biotechnol 41(2):37

Dragos N, Hegedus A, Druga B, Coman C (2010) Astaxanthin production from a new strain of haematococcus pluvialis grown in batch culture [Internet]. Ann Rom Soc Cell Biol 40. Available from: https://www.researchgate.net/publication/228480563

Duan M, Zhou X, Fang T, Zhou D, Shi Q, Ge F et al (2024) Efficient supercritical carbon dioxide extraction of astaxanthin from Hematococcus pluvialis at high pressure. J Supercrit Fluids 205: 106145

Espinosa-Ramírez J, Mondragón-Portocarrero AC, Rodríguez JA, Lorenzo JM, Santos EM (2023) Algae as a potential source of protein meat alternatives. Front Nutr. Frontiers Media SA 10: 1254300

Eze CN, Onyejiaka CK, Ihim SA, Ayoka TO, Aduba CC, Ndukwe JK et al (2023) Bioactive compounds by microalgae and potentials for the management of some human disease conditions. AIMS Microbiol 9(1):55–74

Fan Q, Chen Z, Wu Y, Zhu J, Yu Z (2021) Study on the enhancement of immune function of Astaxanthin from Haematococcus pluvialis. Foods 10(8):1847

Fazelinejad A, Behbahani M, Harsij Z (2024) Utilization of silicon dioxide nanoparticles and silicon salts to enhance astaxanthin production in Haematococcus Pluvialis. Algal Res 82: 103633

Flotow JV (1844) Beobachtungen über Haematococcus pluvialis. Verhandlungen der Kaiserlichen Leopoldinisch-Carolinischen Deutschen Akademie der Naturforscher 20:413–606

Gacheva G, Draganova PH, Pilarski P. New strain Haematococcus cf. pluvialis Rozhen-12-growth, biochemical characteristics and future perspectives [Internet]. 2015. Available from: http://www.ifrg-bg.com

Galán-González J, Quintero-Zapata I, Elías-Santos M, Galán-Wong LJ, López-Chuken UJ, Guajardo-Barbosa C et al (2024) Harvest by autoflocculation, biomass, and carotenoid production in sequential batch culture of Haematococcus Pluvialis under high ionic strength and macroelements content. J Mar Sci Technol 32(4):6

Galasso C, Gentile A, Orefice I, Ianora A, Bruno A, Noonan DM et al (2019) Microalgal derivatives as potential nutraceutical and food supplements for human health: a focus on cancer prevention and interception. Nutrients 11(6):1226

Gao Z, Meng C, Gao H, Zhang X, Xu D, Su Y et al (2013) Analysis of mRNA expression profiles of carotenogenesis and astaxanthin production of Haematococcus pluvialis under exogenous 2, 4-epibrassinolide (EBR). Biol Res 46:201

Garcia-Perez P, Cassani L, Garcia-Oliveira P, Xiao J, Simal-Gandara J, Prieto MA et al (2023) Algal nutraceuticals: a perspective on metabolic diversity, current food applications, and prospects in the field of metabolomics. Food Chem 409:135295

Gherabli A, Grimi N, Lemaire J, Vorobiev E, Lebovka N (2023) Extraction of valuable biomolecules from the microalga Haematococcus pluvialis assisted by electrotechnologies. Molecules 28(5):2089

Göttl VL, Meyer F, Schmitt I, Persicke M, Peters-Wendisch P, Wendisch VF et al (2024) Enhancing astaxanthin biosynthesis and pathway expansion towards glycosylated C40 carotenoids by Corynebacterium glutamicum. Sci Rep 14(1):8081

Grünewald K, Eckert M, Hirschberg J, Hagen C (2000) Phytoene desaturase is localized exclusively in the chloroplast and up-regulated at the mRNA level during accumulation of secondary carotenoids in *Haematococcus pluvialis* (Volvocales, Chlorophyceae). Plant Physiol 122(4):1261–1268

Guerin M, Huntley ME, Olaizola M (2003) Haematococcus astaxanthin: applications for human health and nutrition. Trends Biotechnol 21(5):210–216

Guil-Guerrero JL, Prates JAM (2025) Microalgae bioactives for functional food innovation and health promotion. Foods 14(12):2122

Ha NC, Tam LT, Hien HTM, Thu NTH, Hong DD, Thom LT (2024) Optimization of culture conditions for high cell productivity and Astaxanthin accumulation in Vietnam's green microalgae Haematococcus pluvialis HB and a neuroprotective activity of its Astaxanthin. Bioengineering 11(12):1176

Hagen C, Grünewald K, Xyländer M, Rothe E (2001) Effect of cultivation parameters on growth and pigment biosynthesis in flagellated cells of Haematococcus pluvialis. J Appl Phycol 13(1):79–87

Hang HY (2018) Maximization of astaxanthin production from green microalga Haematococcus pluvialis using internally-illuminated photobioreactor. Adv Biosci Bioeng 6(2):10

Hazen TE (1899) The life history of Sphaerella lacustris (Haematococcus pluvialis). Mem Torrey Bot Club 6(3):211–246. https://www.jstor.org/stable/43392147

Jannel S, Caro Y, Bermudes M, Petit T (2024) Exploration of the biotechnological potential of two newly isolated Haematococcus strains from Reunion Island for the production of natural Astaxanthin. Foods 13(22):3681

Jin C, Zhou Z, Zhu Y, Liu Q, Zhou X (2024) Photosynthetic activity and astaxanthin production of Haematococcus pluvialis regulated by manipulated light quality. Aquac Int 32(3):3617–3635

Jin C, You J, Zhou Z, Liu Q, Zhou X (2025) Novel insights into saline stress on photosynthetic activity and astaxanthin production of Haematococcus pluvialis. J Oceanol Limnol 43(3):921–938

Johnson EA, An GH (1991) Astaxanthin from microbial sources. Crit Rev Biotechnol 11(4):297–326

Karuppan R, Javee A, Gopidas SK, Pathmanapan A, Kattusamy K, Narayanan V et al (2022) Impact of cultivation parameters on astaxanthin accumulation in the green alga Haematococcus lacustris RRGK isolated from Himachal Pradesh, India. Energy Nexus 6:100083

Khadilkar J, Karande V, Prakash G, Pandit R (2023) Simultaneous extraction and purification of natural astaxanthin from Haematococcus pluvialis using adsorptive chromatography. Bioresour Technol Rep 24:101691

Kim JH, Affan MDA, Jang J, Kang MH, Ko AR, Jeon SM et al (2015) Morphological, molecular, and biochemical characterization of Astaxanthin-producing green microalga Haematococcus sp. KORDI03 (Haematococcaceae, Chlorophyta) isolated from Korea. J Microbiol Biotechnol 25(2):238–246

Kim S, Mahadi R, Narasimhan AL, Christabel C, Yu H, Kim EJ et al (2024) Effect of reduced atmospheric pressures on the morphology and astaxanthin biosynthesis of microalga Haematococcus lacustris. Biotechnol Bioprocess Eng 29(6):1131–1140

Kobayashi M, Kurimura Y, Kakizono T, Nishio N, Tsuji Y (1997) Morphological changes in the life cycle of the green alga Haematococcus pluvialis. J Ferment Bioeng 84(1):94–97

Kratzer R, Murkovic M (2021) Food ingredients and nutraceuticals from microalgae: main product classes and biotechnological production. Foods 10(7):1626

Lakshmi Narasimhan A, Lee N, Kim S, Kim YE, Christabel C, Yu H et al (2024) Enhanced astaxanthin production in Haematococcus lacustris by electrochemical stimulation of cyst germination. Bioresour Technol 411:131301

Lee C, Ahn JW, Kim JB, Kim JY, Choi YE (2018) Comparative transcriptome analysis of Haematococcus pluvialis on astaxanthin biosynthesis in response to irradiation with red or blue LED wavelength. World J Microbiol Biotechnol 34(7):96

Lee N, Narasimhan AL, Moon G, Kim YE, Park M, Kim B et al (2022) Room-temperature cell disruption and Astaxanthin recovery from Haematococcus lacustris cysts using ultrathin α-quartz nanoplates and ionic liquids. Appl Sci 12(4):2210

Li Q, You J, Qiao T, Zhong DB, Yu X (2022a) Sodium chloride stimulates the biomass and astaxanthin production by Haematococcus pluvialis via a two-stage cultivation strategy. Bioresour Technol 344:126214

Li C, Gao Y, Huan Y, Ren P, Zhi J, Wu A et al (2022b) Colon and gut microbiota greatly affect the absorption and utilization of astaxanthin derived from Haematococcus pluvialis. Food Res Int 156:111324

Liang C, Zhang W, Zhang X, Fan X, Xu D, Ye N et al (2016) Isolation and expression analyses of methyl-d-erythritol 4-phosphate (MEP) pathway genes from Haematococcus pluvialis. J Appl Phycol 28(1):209–218

Luo Q, Bian C, Tao M, Huang Y, Zheng Y, Lv Y et al (2019) Genome and transcriptome sequencing of the Astaxanthin-producing green microalga, *Haematococcus pluvialis*. Genome Biol Evol 11(1):166–173

Luo F, Wang S, Zhang X, Liu Z, Zhu R, Xue W (2024) Extraction of Astaxanthin from Haematococcus pluvialis and preparation of Astaxanthin liposomes. Molecules 29(14):3320

Ma R, Tao X, Wang B, Ho SH, Chen J, Xie Y (2025) Hyper-production of astaxanthin from Haematococcus pluvialis by a highly efficient nitrogen feeding strategy accompanied with high light induction. Algal Res 85:103865

Machado FRS, Trevisol TC, Boschetto DL, Burkert JFM, Ferreira SRS, Oliveira JV et al (2016) Technological process for cell disruption, extraction and encapsulation of astaxanthin from Haematococcus pluvialis. J Biotechnol 218:108–114

Marinho YF, Malafaia CB, de Araújo KS, da Silva TD, dos Santos APF, de Moraes LB et al (2021) Evaluation of the influence of different culture media on growth, life cycle, biochemical composition, and astaxanthin production in Haematococcus pluvialis. Aquac Int 29(2):757–778

Marinho YF, Oliveira CYB, Malafaia CB, Cahú TB, Oliveira APS, Napoleão TH et al (2022) A circular approach for the efficient recovery of astaxanthin from Haematococcus pluvialis biomass harvested by flocculation and water reusability. Sci Total Environ 841:156795

Matos J, Cardoso CL, Falé P, Afonso CM, Bandarra NM (2020) Investigation of nutraceutical potential of the microalgae *Chlorella vulgaris* and *Arthrospira platensis*. Int J Food Sci Technol 55(1):303–312

Medoro A, Intrieri M, Passarella D, Willcox DC, Davinelli S, Scapagnini G (2024) Astaxanthin as a metabolic regulator of glucose and lipid homeostasis. J Funct Foods. Elsevier Ltd 112:105937

Meryem S, Sarah B, Nabil B, Meriem M, Ali R (2022) Gastric ulcer prevention, harmlessness and antioxidant activity of astaxanthin extracted from a new Algerian strain of Haematococcus pluvialis. Asian J Agric Biol. (Online First)

Mišurcová L, Buňka F, Vávra Ambrožová J, Machů L, Samek D, Kráčmar S (2014) Amino acid composition of algal products and its contribution to RDI. Food Chem 151:120–125

Mont'alverne L, Seabra J, Fátima L, Pedrosa C (2010) Astaxanthin: structural and functional aspects 1 Astaxantina: aspectos estruturais e funcionais. Rev Nutr 23:1041

Muniz Souza RL, Ramalho Viana WK, Martins Moreira MS, Soares Filho AA, Lisboa V, Bezerra Vidigal RCDA et al (2020) Technological innovations in the cultivation of Haematococcus pluvialis microalgae: review. Sist Gest 15(3):223–234

Mussagy CU, Kot A, Dufossé L, Gonçalves CNDP, Pereira JFB, Santos-Ebinuma VC et al (2023) Microbial astaxanthin: from bioprocessing to the market recognition. Appl Microbiol Biotechnol. Springer Science and Business Media Deutschland GmbH 107:4199–4215

Myint AA, Wulandari S, Choi J, Sim SJ, Kim J (2023) High-yield dimethyl ether-based recovery of astaxanthin and fatty acids directly from wet Haematococcus pluvialis. Sep Purif Technol 320: 124226

Nair A, Ahirwar A, Singh S, Lodhi R, Lodhi A, Rai A et al (2023) Astaxanthin as a king of ketocarotenoids: structure, synthesis, accumulation, bioavailability and antioxidant properties. Mar Drugs 21(3):176

Nemani N, Dehnavi SM, Pazuki G (2024) Extraction and separation of astaxanthin with the help of pre-treatment of Haematococcus pluvialis microalgae biomass using aqueous two-phase systems based on deep eutectic solvents. Sci Rep 14(1):5420

Nishida Y, Berg P, Shakersain B, Hecht K, Takikawa A, Tao R et al (2023) Astaxanthin: past, present, and future. Mar Drugs 21(10):514

Osathanunkul M, Thanaporn S, Karapetsi L, Nteve GM, Pratsinakis E, Stefanidou E et al (2025) Diversity of bioactive compounds in microalgae: key classes and functional applications. Mar Drugs 23(6):222

Paniagua-Michel J, Olmos-Soto J, Ruiz MA (2012) Pathways of carotenoid biosynthesis in bacteria and microalgae. Methods Mol Biol 892:1–12

Panis G, Carreon JR (2016) Commercial astaxanthin production derived by green alga Haematococcus pluvialis: a microalgae process model and a techno-economic assessment all through production line. Algal Res 18:175–190

Patil AD, Kasabe PJ, Dandge PB (2022) Pharmaceutical and nutraceutical potential of natural bioactive pigment: astaxanthin. Nat Prod Bioprospect 12(1):25

Pei H, Bao K, Ye Y, Song Y, Liu P, Han Y et al (2025) Development of a ready-to-use astaxanthin esters extract from Haematococcus pluvialis using hydrophobic NADES: optimization, mechanistic insights, and stability evaluation. Microchem J 209:112794

Praveenkumar R, Lee J, Vijayan D, Lee SY, Lee K, Sim SJ et al (2020) Morphological change and cell disruption of Haematococcus pluvialis cyst during high-pressure homogenization for Astaxanthin recovery. Appl Sci 10(2):513

Radzun K, Rusmidi M, Norisam I, Iran N, Pardi F, Ismail A et al (2022) Anti-inflammatory effects of Astaxanthin extracted from microalgae Hematococcus pluvialis and combinations with palm Tocotrienol rich-fraction in RAW 264.7 macrophages. Pharm J 14(1):205–215

Rahman KM (2020) Food and high value products from microalgae: market opportunities and challenges. In: Microalgae biotechnology for food, health and high value products. Springer Singapore, Singapore, pp 3–27

Rayamajhi V, An Y, Byeon H, Lee J, Kim T, Choi A et al (2024) A study on the effect of various media and the supplementation of organic compounds on the enhanced production of Astaxanthin from Haematococcus lacustris (Girod—Chantrans) Rostafinski (Chlorophyta). Microorganisms 12(6):1040

Samara C, Biziouras A, Papapanagiotou G, Chatzidoukas C (2025 Aug) Extending the green stage of Haematococcus pluvialis as a crucial precursor for efficient astaxanthin production: optimization via Taguchi design. Biochem Eng J 220:109763

Santos-Sanchez NF, Hernández-Carlos B, Torres-Arino A, Salas-Coronado R (2021) Astaxanthin and its formulations as potent oxidative stress inhibitors. Pharmacogn Rev 14(27):8–15

Sathiyavahisan LP, Lakshmi Narasimhan A, Mahadi R, Kim S, Christabel C, Yu H et al (2023) Rapid induction of Astaxanthin in Haematococcus lacustris by mild electric stimulation. Appl Sci 13(23):12959

Shah MDMR, Liang Y, Cheng JJ, Daroch M (2016) Astaxanthin-producing green microalga Haematococcus pluvialis: from single cell to high value commercial products. Front Plant Sci 7:531

Telli M, Ünlü ES (2023) Comparative de novo transcriptome analysis and random UV mutagenesis: application in high biomass and astaxanthin production enhancement for Haematococcus pluvialis. Mol Biol Rep 50(10):8133–8143

Turck D, Bohn T, Castenmiller J, De Henauw S, Hirsch-Ernst KI, Maciuk A, et al (2023) Safety of a change in specifications of the novel food oleoresin from Haematococcus pluvialis containing astaxanthin pursuant to regulation (EU) 2015/2283. EFSA J 21(11):e08338

Venckus P, Endriukaitytė I, Čekuolytė K, Gudiukaitė R, Pakalniškis A, Lastauskienė E (2023) Effect of biosynthesized silver nanoparticles on the growth of the green microalga Haematococcus pluvialis and Astaxanthin synthesis. Nanomaterials 13(10):1618

Vidhyavathi R, Venkatachalam L, Sarada R, Ravishankar GA (2008) Regulation of carotenoid biosynthetic genes expression and carotenoid accumulation in the green alga Haematococcus pluvialis under nutrient stress conditions. J Exp Bot 59(6):1409–1418

Villaró S, Ciardi M, Morillas-españa A, Sánchez-zurano A, Acién-fernández G, Lafarga T (2021) Microalgae derived astaxanthin: research and consumer trends and industrial use as food. Foods. MDPI 10:2303

Wan X, Zhou XR, Moncalian G, Su L, Chen WC, Zhu HZ et al (2021) Reprogramming microorganisms for the biosynthesis of astaxanthin via metabolic engineering. Prog Lipid Res 81:101083

Wang Q, Quan S, Xiao H (2019) Towards efficient terpenoid biosynthesis: manipulating IPP and DMAPP supply. Bioresour Bioprocess 6(1):6

Yamashita E (2013) Astaxanthin as a medical food. Funct Foods Health Dis 3:254

Ye Z, Tan X, Liu Z, Aadil RM, Tan Y, Inam-ur-Raheem M (2020) Mechanisms of breakdown of *Haematococcus pluvialis* cell wall by ionic liquids, hydrochloric acid and multi-enzyme treatment. Int J Food Sci Technol 55(9):3182–3189

Yu W, Liu J (2020) Astaxanthin isomers: selective distribution and isomerization in aquatic animals. Aquaculture 520:734915

Yu C, Wang HP, Yu X (2022) The associative induction of succinic acid and hydrogen sulfide for high-producing biomass, astaxanthin and lipids in Haematococcus pluvialis. Bioresour Technol 358:127397

Zhang C, Liu J, Zhang L (2017) Cell cycles and proliferation patterns in Haematococcus pluvialis. Chin J Oceanol Limnol 35(5):1205–1211

Zhang L, Hu T, Yao S, Hu C, Xing H, Liu K et al (2023) Enhancement of astaxanthin production, recovery, and bio-accessibility in Haematococcus pluvialis through taurine-mediated inhibition of secondary cell wall formation under high light conditions. Bioresour Technol 389:129802

Zhao Z, Blockx J, Muylaert K, Thielemans W, Szymczyk A, Vankelecom IFJ (2022) Exploiting flocculation and membrane filtration synergies for highly energy-efficient, high-yield microalgae harvesting. Sep Purif Technol 296:121386

Zhao Y, Li Q, Chen D, Yang M, Huang F, Liu J et al (2025) Exploiting synergy of dopamine and stressful conditions in enhancing Haematococcus lacustris biomass and astaxanthin yield. Bioresour Technol 417:131848

Part II

Algal Biotechnology, Bioengineering and Sustainable Bioproducts

7 Beyond Carbon Capture: Trehalose Transforms Algae into Biofactories

Makaranga Abdalah, Bharmjeet Singh, Asha Arumugam Nesamma, and Pannaga Pavan Jutur

Abstract

Algae-based lutein is sustainable and another eco-friendly alternative to plant and synthetic sources. Lutein is a carotenoid pigment and naturally occurring antioxidant commonly used as a supplement for ocular health. In this study, the co-culture of microalgae and bacteria, *Chlorella saccharophila* and *Exiguobacterium* sp., was investigated for enhanced lutein production. The co-culture, particularly under nitrate-depleted conditions with 3% CO_2 supplementation, exhibited the highest lutein productivity (238.31 $\mu g.L^{-1}$ d^{-1}), a 1.33-fold and 3.54-fold increase compared to the co-culture nitrate-depleted and mono-culture, respectively. Additionally, the co-culture showed an elevated chlorophyll a level, indicating a positive influence of both bacteria and carbon dioxide (CO_2) on algal photosynthesis. Metabolomics analysis revealed a significant decrease in bacterial trehalose and an increase in algal sucrose, suggesting beneficial metabolic interactions. These findings highlight the potential of co-culture strategies for inducing new metabolite pathways and increasing the yield of valuable compounds, contributing to a circular bioeconomy through CO_2 valorization.

M. Abdalah
Omics of Algae Group, Industrial Biotechnology, International Centre for Genetic Engineering and Biotechnology (ICGEB), New Delhi, India

Department of Agriculture, Earth and Environmental Sciences (DAES), Faculty of Science, Mwenge Catholic University (MWECAU), Moshi, Kilimanjaro, Tanzania

B. Singh · A. A. Nesamma · P. P. Jutur (✉)
Omics of Algae Group, Industrial Biotechnology, International Centre for Genetic Engineering and Biotechnology (ICGEB), New Delhi, India
e-mail: Pavan.Jutur@icgeb.org

K. Mohanty et al. (eds.), *Advances in Algal Research*,
https://doi.org/10.1007/978-981-95-7234-2_7

Keywords

Algae · Bacteria · Metabolomics · Sucrose · Trehalose

7.1 Introduction

Photosynthetic eukaryotic microalgae and prokaryotic bacteria offer immense potential in the circular bioeconomy due to their metabolic versatility and high yields of biomass, biofuels, and biorenewables (Jutur 2025; Makaranga and Jutur 2025). Among microalgae, *Chlorella* species are particularly valuable for their ability to produce high-value carotenoids, including lutein, without competing with food crops or harming the environment (Kareya et al. 2022; Fu et al. 2023). The continuous search for sustainable industrial feedstocks has positioned microalgae as key players in valorization, transforming industrial flue gases directly into valuable chemicals and biomass (Kareya et al. 2022). Compared to mitigation via geologic sequestration, this biotechnological approach offers both an environmental solution and an economic incentive, directly linking carbon capture to the production of high-value compounds (Sørensen et al. 2022). Lutein, a yellow dihydroxy xanthophyll, is essential for human health, particularly in preventing age-related macular degeneration (Eun et al. 2025; Dang Mai et al. 2025). The market, valued at approximately $150 million in 2025, is projected to exhibit a Compound Annual Growth Rate (CAGR) of 7% from 2025 to 2033, driven by increasing demand in the pharmaceutical, nutraceutical, and food industries (Makaranga and Jutur 2025; Fariz-Salinas et al. 2024). Certain microalgae exhibit lutein levels (up to 1.7%) surpassing those of common dietary sources like kale, spinach, and egg yolk (approximately 0.7–0.9%). Predominantly associated with photosystem II light-harvesting complex, lutein is crucial in photosynthesis and cellular defense mechanisms (Fu et al. 2023; Iwamoto et al. 2024). However, current production methods, which are largely dependent on marigold flower extraction, are limited by seasonal variability, high costs, and land requirements (Fu et al. 2023; Eun et al. 2025; Lin et al. 2015). Microalgae present a superior alternative due to their rapid growth and higher lutein yields, yet challenges in productivity and scalability persist (Makaranga and Jutur 2025; Fu et al. 2023; Dang Mai et al. 2025; Fariz-Salinas et al. 2024).

To overcome these limitations, microbial co-culture strategies are emerging as a promising solution. Algae-bacteria interactions can enhance metabolite production through mutualistic exchanges of nutrients, vitamins, and growth-promoting factors (Jutur 2025; Makaranga and Jutur 2025; Carrasco Flores et al. 2024; Kim et al. 2025; Henshaw et al. 2024). For instance, bacteria such as *Stenotrophomonas maltophilia* and algae *Monoraphidium* sp. have been shown to improve microalgal lutein synthesis (Manoel et al. 2025). *Chlorella saccharophila* UTEX247, a freshwater microalga, exhibits high lutein accumulation potential, especially when co-cultured with symbiotic bacteria like *Exiguobacterium* sp. (Makaranga and Jutur 2023, 2025), known for its plant-growth-promoting properties. The organism maintains a substantial investment in other photoprotective pigments in addition to increasing lutein synthesis. While many co-culture strategies focus on generic nutrient recycling (e.g., exchange) or the action of general growth-promoting factors, our work is distinct in identifying the specific cross-kingdom exchange of the disaccharides trehalose and sucrose as a key metabolic signal and carbon flux regulator for lutein synthesis.

Recent studies suggest that microbial co-cultures activate silent metabolic pathways, increasing the yield of high-value-added products (Guo et al. 2024). In previous work, *Exiguobacterium* sp. AMK1 found to stimulate *C. saccharophila* growth via trehalose and sucrose metabolism, inducing novel metabolites linked to carotenoid biosynthesis (Makaranga and Jutur 2025). However, the precise mechanisms, such as nutrient exchange, carbon flux, and stress-induced lutein production, remain unclear. Key questions include the following: (i) How do sucrose and trehalose dynamics differ in mono- vs. co-culture? (ii) How does nitrate depletion or CO_2 supplementation influence symbiosis? (iii) What metabolic biomarkers correlate with enhanced lutein yield? Building upon the established roles of trehalose as a bacterial stress-protectant and sucrose as an algal photosynthate, we hypothesize a novel metabolic exchange: The algal partner consumes trehalose, stimulating bacterial nutrient services, while the bacteria promote algal sucrose exudation, enhancing carbon flow into high-value carotenoid pathways under nitrogen-limited stress. This mutualistic exchange, tracked by key sugar dynamics, offers a scalable strategy for enhancing bioproducts. This study investigates the co-culture of *C. saccharophila* UTEX247 and *Exiguobacterium* sp. under autotrophic conditions to decipher their metabolic crosstalk and optimize lutein production. By leveraging their symbiotic relationship, we aim to develop a sustainable, high-yield bioprocess that aligns with circular economy principles, offering a viable alternative to conventional lutein production methods.

7.2 Materials and Methods

7.2.1 Microalgae and Symbiotic Bacteria Strain(s)

The C. saccharophila UTEX247 and its associated bacteria, *Exiguobacterium* sp. was chosen as study organisms in this investigation. The algal species *C. saccharophila* UTEX247 was obtained from the University of Texas at Austin (UTEX) culture collection of algae and cultured using BG11 agar plates. The strain *Exiguobacterium* sp. was isolated based on a previous study screening algal growth-promoting bacteria from an in-house strain library at the Omics of Algae group, Industrial Biotechnology, International Centre for Genetic Engineering and Biotechnology (ICGEB), India, and cryopreserved in 10% sterile DMSO and stored at −80 °C (Makaranga and Jutur 2023, 2025). This specific pairing is considered a synthetic co-culture in the sense that the two organisms were individually isolated, characterized, and then intentionally combined and optimized in a controlled system to maximize a specific output (lutein), distinguishing it from a naturally occurring, unoptimized environmental consortium. The absence of bacteria in axenic algal cultures was monitored periodically using PCR amplification with selected universal

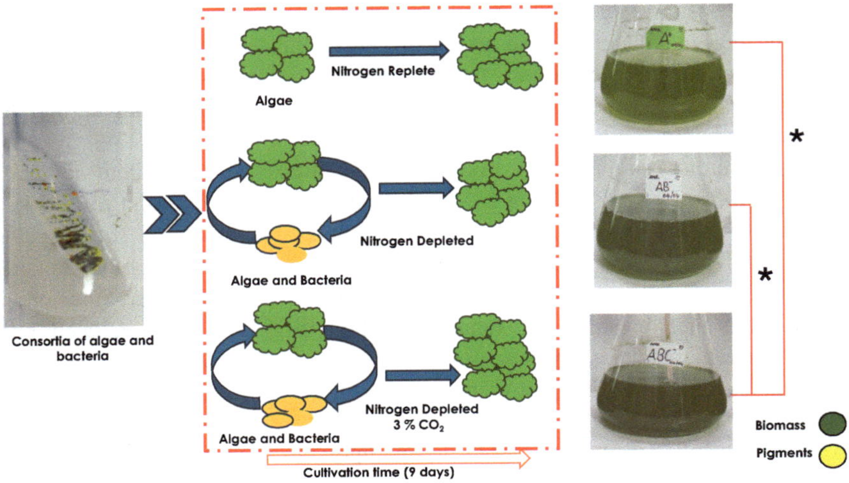

Fig. 7.1 Schematic representation of experimental design: From isolation of algae and bacteria to mono-culture and co-culture systems

primers for 16S rRNA gene amplification in the genomic DNA (Makaranga and Jutur 2023, 2025). Previously investigated interactions between *C. saccharophila* UTEX247 and *Exiguobacterium* sp. established a valuable foundation for our comparative analysis (Makaranga and Jutur 2023, 2025). The experiment design consisted of Mono-culture nitrogen-replete (M_{ANR}), Co-culture nitrate-depleted (C_{ND}), and Co-culture nitrate-depleted with 3% CO_2 supplementation (C_{NDC}), and each was conducted in 1.0 L Erlenmeyer flasks (Fig. 7.1).

7.2.2 Microalgae Growth Profiling in Mono- and Co-Cultures

The cell number of microalgae was counted under a light microscope using a hemocytometer (Makaranga and Jutur 2023, 2025). The growth of microalgae with the addition of bacteria was compared to the growth of axenic microalgae without the addition of bacteria, using a cell count of algae. Cell numbers were monitored over time to determine growth rates. Growth rates (μ) were obtained using Eq. (7.1):

$$\mu = \ln\ (N_1/N_0)/(t_1 - t_0) \tag{7.1}$$

N_1 represents the number of cells at a given sampling time, while N_0 represents the initial number of cells at the beginning of the experiment (t_0). The specific growth rate (μ) was used to calculate the doubling time using the following Eq. (7.2):

$$\text{Doubling time} = \ln\ (2)/\mu \tag{7.2}$$

Samples were collected at 0, 3, 6, and 9 days for all experiments. These four time points were specifically chosen to capture key physiological phases: lag (Day 0), exponential growth (Day 3), the transition into the stationary/stress phase (Day 6), and the final peak accumulation (Day 9). These samples were analyzed to determine pigment content and generate metabolic profiles for both mono-cultures and co-cultures (Makaranga and Jutur 2025).

7.2.3 Pigment Extraction, Identification, and Quantification by HPLC-UV

Pigment quantification in *C. saccharophila* UTEX247 co-cultures was performed with slight modifications to the established protocol by (Makaranga and Jutur 2023, 2025). Briefly, 1×10^7 cells were resuspended in 1 mL of absolute methanol. The cells were then lysed by vortexing with glass beads for 30 min and sonicated in a sonication bath at 4 °C for 15 min. The mixture was centrifuged, and the supernatant was collected for pigment analysis. Agilent Infinity series 1260 HPLC system (Agilent Technologies, Santa Clara, California (CA), USA) coupled with UV detector using C-30 Acclaim column (4.6 × 250 mm, 5 μm) was used for pigment estimation where the separation was carried out using a binary solvent system with methanol as the primary solvent (A) and methyl tert-butyl ether (MTBE) as the secondary solvent (B) at 35 °C. The gradient conditions listed below were used to isolate the six carotenoids: 2–20% B for the first 10 min, then 20–20% B (10–12 min), 20–80% B (12–30 min), 80–2% B (32–35 min). All reagents used in pigment extraction and chromatography analysis were HPLC grade. By contrasting the retention times of the standards, acquired from DHI (Hørsholm, Denmark), pigments were identified and quantified at the UV detector at 450 nm by comparing their retention times and absorption spectra with those of the samples. All analyses were performed in biological triplicates.

7.2.4 Metabolomic Analysis

Untargeted metabolomics was performed using gas chromatography-tandem mass spectrometry with a triple quadrupole (QQQ) MS/MS system (GC-MS/MS) (Agilent Technologies, Santa Clara, California, USA) (Fig. 7.2). The metabolome detection was conducted through three distinct treatments: M_{ANR}, C_{ND}, and C_{NDC}. Metabolite extraction of the mono-cultures and co-cultures was performed (Makaranga and Jutur 2023, 2025). Approximately 1×10^8 cells were centrifuged at 10,000× *g* for 10 min at 4 °C and immediately quenched in liquid nitrogen to obtain steady-state metabolite concentrations. For resuspension, 1 mL of ice-cold methanol/ethanol/chloroform (in a 1:3:1 volumetric ratio) was added to the cells, followed by 15 min of sonication in an ice bath. These samples were then centrifuged at 10,000× *g* for 15 min at 4 °C, and the supernatant was filtered using a 0.2 μm filter. Five microliters of ribitol (adonitol, 5 mg/mL) was added as an internal standard (IS) in the sample.

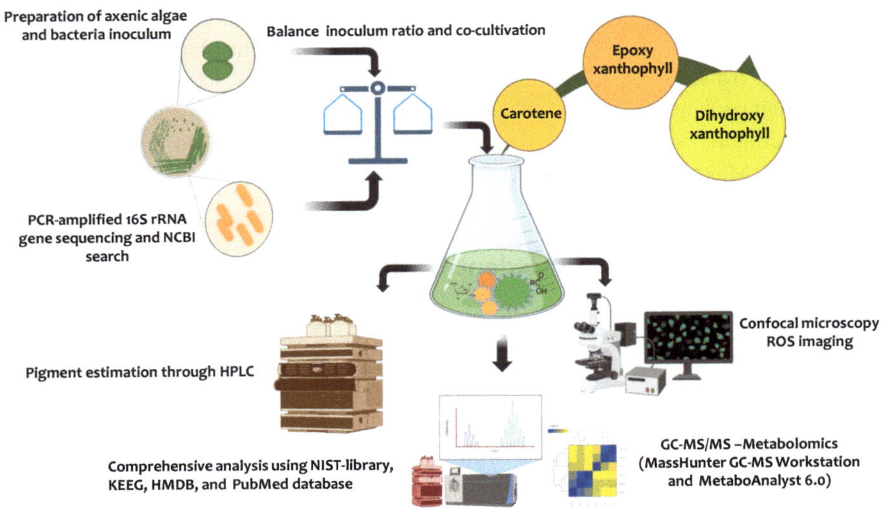

Fig 7.2 A comprehensive flowchart of experimental methodology

The supernatant was collected and dried under a stream of nitrogen gas. The samples were derivatized to be prepared for GC-MS/MS analysis by following steps: The dried residue was dissolved in 10 μL of methoxyamine hydrochloride solution (4% w/v in pyridine), then incubated at 30 °C for 90 min. For derivatization, N-methyl-N-(trimethylsilyl) trifluoroacetamide (90 μL) was added to the abovementioned solution and incubated at 37 °C for 30 min. After thorough vortexing, the samples were centrifuged for 3 min at 14,000× *g* and transferred to 2 mL amber micro-vials with 250 μL glass inserts (Agilent Technologies Pvt. Ltd., USA). Every sample was analyzed in triplicate to ensure accuracy. The detailed GC/MS parameters setup for metabolite detection was reported in our previous studies (Makaranga and Jutur 2025).

The targeted trehalose and sucrose standard was detected as described previously (Makaranga and Jutur 2025; Mayack et al. 2020). Sucrose and trehalose, ranging from 0.5 to 3.0 mg. mL^{-1} for sugar, were prepared according to authentic standards (Sigma-Aldrich Pvt. Ltd., Cat. 190675-25G). These solutions were diluted and derivatized as described above for the mono-cultures (M_{ANR}) mono-culture bacteria under nitrate-depleted in BG11 medium (M_{BND}) and co-culture (C_{ND} and C_{NDC}) samples. The typical retention times and prominent ion numbers of the targeted sugar and internal standard were used to identify each peak. Sample peaks were identified and aligned using the NIST 14 Mass Spectral Library collection of the U.S. National Institute of Standards and Technology (NIST). Trehalose and sucrose peaks were manually integrated and normalized using the internal standard peak area. Standard curves for each run were generated from the peak areas of the two standards with known sugar concentrations (Sigma Aldrich, St. Louis, MO). The standard was used for each mono- and co-culture sample to convert peak areas into mg. mL^{-1} concentrations.

7.3 Results and Discussion

7.3.1 Growth Performance

The growth of microalgae estimation 0, 3, 6, and 9 days was significantly higher ($p < 0.05$) under nitrate-depleted conditions supplemented with 3% CO_2, particularly on days 6 and 9, compared to the co-culture nitrate-depleted condition, and followed by the mono-culture nitrogen-replete condition (Fig. 7.3). Biomass yield under nitrate-depleted conditions with 3% CO_2 supplementation reached 912.17 ± 29.20, which was significantly greater ($p < 0.05$) than that of the co-culture nitrate-depleted (546.00 ± 21.20), and next to the mono-culture nitrogen-replete (392.17 ± 3.06). Furthermore, in co-culture nitrate-depleted conditions with 3% CO_2 supplementation, biomass increase corresponded with higher specific growth rates and reduced doubling time when compared to the co-culture nitrate-depleted, followed by the mono-culture algae under nitrogen-replete conditions (Table 7.1). This enhanced productivity is likely linked to adjustments in the photosynthetic apparatus, particularly the accumulation of lutein in the light-harvesting complexes, which improves photoprotection and photosystem stability under stress. Lutein accumulation is a key adaptive response to combined nutrient

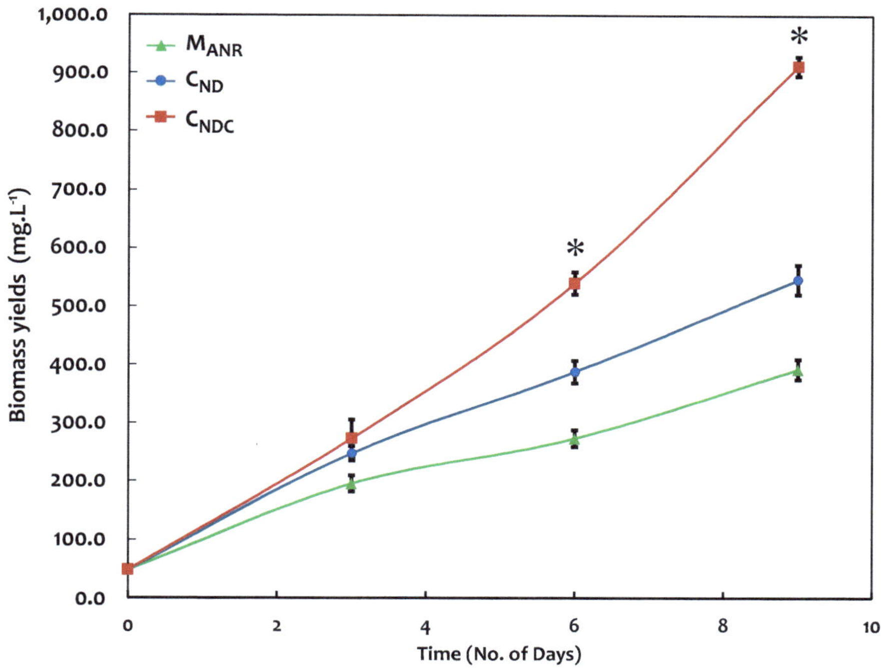

Fig. 7.3 Line graph depicting the biomass yield (mg L^{-1}) of *C. saccharophila* UTEX247 in mono-culture and co-culture under 150 μE light intensity for 9 days. Bars indicate mean values ± standard error ($n = 3$). One-way ANOVA ($p < 0.05$) determined statistical significance

Table 7.1 Growth parameters, biomass yields, and lutein productivity of *C. saccharophila* UTEX247 cultivated in mono-culture and co-cultures

Treatments	Specific growth rate (day^{-1})	Doubling time (h)	Biomass yield (mg. L^{-1})	Lutein productivity (μg.L^{-1} day^{-1})
M_{ANR}	(0.35 ± 0.01)	(47.57 ± 0.43)	(392.17 ± 3.06)	(33.65 ± 5.40)
C_{ND}	(0.40 ± 0.00)	(41.08 ± 1.08)	(546.00 ± 21.20)	(89.81 ± 12.61)
C_{NDC}	(0.49 ± 0.03)*	(33.92 ± 0.78)*	(912.17 ± 29.20)*	(238.31 ± 30.04)*

*Stars indicate mean values ± standard error ($n = 3$). One-way ANOVA ($p < 0.05$) determined statistical significance

and photo-oxidative stress, acting to quench reactive oxygen species (ROS) and facilitate non-photochemical quenching (NPQ) to safely dissipate excess light energy, a process that is maximized under a CO_2-enriched, nitrate-depleted environment.

7.4 Photosynthetic Activity Assessment

Using HPLC, chlorophyll a and b were quantified over a 9-day co-cultivation period (Fig. 7.3). Photosynthetic activity showed a direct proportional relationship with lutein productivity from the alpha-carotene branch (Figs. 7.2 and 7.3). Chlorophyll a demonstrated substantially higher productivity (1763.72 μg·L^{-1}·d^{-1}) compared to chlorophyll b (314.11 μg·L^{-1}·d^{-1}). This pigment productivity pattern, with chlorophyll a being highest, followed by chlorophyll b, and then total carotenoids, remained consistent across all tested conditions: co-culture nitrate-depleted with 3% CO_2 supplementation, co-culture nitrate-depleted, and mono-culture algae under nitrogen-replete environments.

7.5 Lutein Production Enhancement

Under co-culture nitrate-depleted conditions with 3% CO_2 supplementation, HPLC analysis revealed a significant reallocation of carbon flux into the α-carotene branch of the carotenoid pathway, resulting in lutein the primary downstream product being the most abundant carotenoid at 238.31 μg.$L^{-1}d^{-1}$, markedly exceeding levels in both co-culture nitrate-depleted and mono-culture algae under nitrogen-replete conditions; substantial production was also observed in the β-carotene branch, with zeaxanthin and violaxanthin accumulating at 220.72 μg.$L^{-1}d^{-1}$ and 185.42 μg.$L^{-1}d^{-1}$, respectively, while antheraxanthin 84.07 μg.$L^{-1}d^{-1}$ canthaxanthin, and β-carotene itself were produced in comparatively low amounts across all conditions (Fig. 7.4).

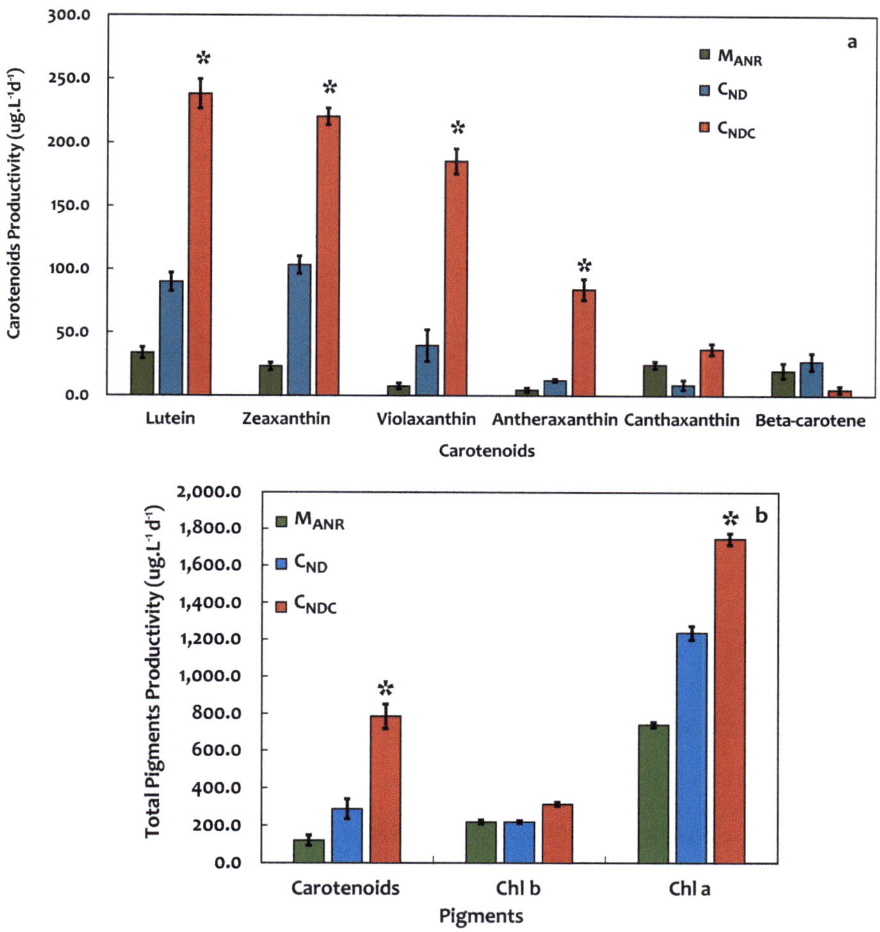

Fig. 7.4 Comparative pigment productivity across experimental conditions. Mono-culture nitrogen-replete (M_{ANR}), co-culture nitrate-depleted (C_{ND}), co-culture nitrate-depleted with 3% CO_2 supplementation (C_{NDC}). Data show the productivity ($\mu g \cdot L^{-1} \cdot d^{-1}$) of major carotenoids and chlorophylls quantified via HPLC analysis over a 9-day cultivation period

7.6 Metabolomic Profiling

7.6.1 Untargeted GC-MS/MS Metabolomics

Using untargeted GC-MS/MS metabolomics, we developed a comprehensive profile of metabolite classes, including amino acids, fatty acids, sugars and alcohols, and organic acids. The expression of individual metabolites within each class is visualized in the heatmap (Fig. 7.5), where color indicates both the class and the

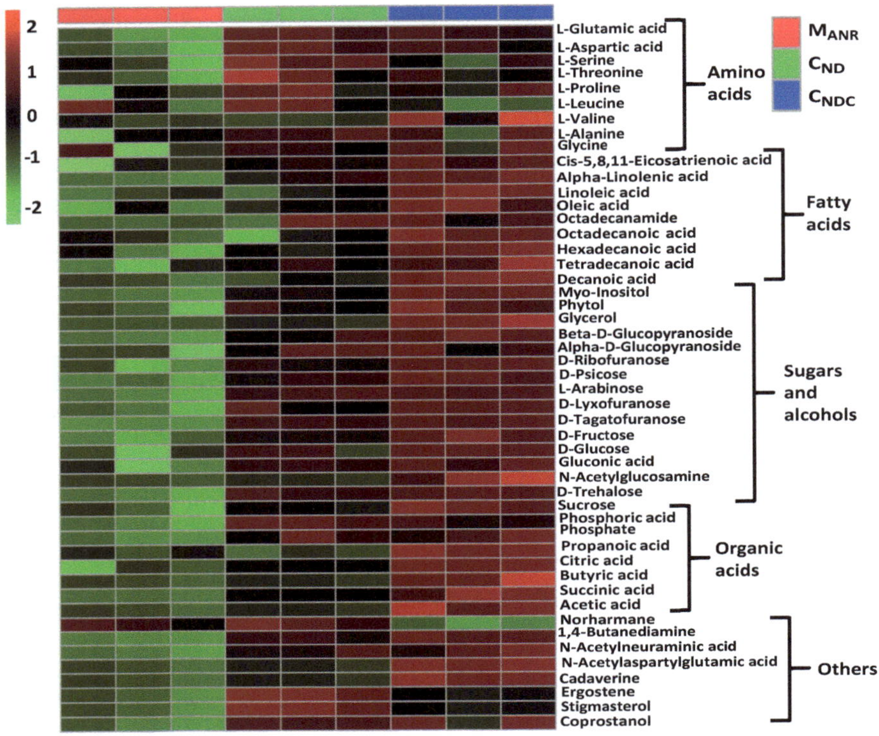

Fig. 7.5 The GC-MS/MS heatmap of untargeted metabolites class expression: comparison of mono-culture algae under nitrogen-replete (M_{ANR}), co-culture nitrate-depleted (C_{ND}), and co-culture nitrate-depleted with 3% CO_2 supplementation (C_{NDC}). Statistical analysis of color intensity: expression on a $\pm$ 2.0 scale and significance determined by one-way ANOVA ($p < 0.05$)

relative quantity of expression on a normalized intensity scale from −2 to +2. The analysis revealed a clear hierarchy of metabolic activity across the experimental conditions. While advanced pathway enrichment and flux analysis using tools could provide deeper system-level insights, the scope of the current study focused on identifying and quantifying the most highly regulated metabolites in central carbon metabolism. The expression of metabolites across all major classes was significantly higher in the co-culture nitrate-depleted supplemented with 3% CO_2, followed by the co-culture nitrate-depleted, with the mono-culture algae under nitrogen-replete conditions showing the lowest overall expression (Fig. 7.5). Among the most prominently expressed classes were sugars and alcohols. This class was dominated by monosaccharides, such as glucose and fructose and their conjugates, which showed statistically validated upregulation in the C_{NDC} condition (Fig. 7.5). Furthermore, we identified and quantified disaccharide sugars, including sucrose and trehalose, which were also significantly elevated. Due to their potential critical role in microbial interactions, these specific disaccharides were selected for subsequent targeted metabolomics analysis to precisely quantify their levels and further elucidate their impact on the symbiotic relationship between the algae and bacteria.

7.6.2 Targeted GC-MS/MS Metabolomics

Using targeted metabolomics, we quantified the disaccharides trehalose and sucrose at three time points (days 3, 6, and 9) across the different culture conditions (Fig. 7.6a). Chromatographic analysis revealed a distinct presence of these sugars depending on the culture: Sucrose (eluting at 40.4-40.5 min) was undetectable in bacteria-alone cultures, while trehalose (eluting at 41.3-41.5 min) was only a minor peak in algae-alone cultures but constituted a major peak in bacteria-alone cultures. Furthermore, each culture condition exhibited a unique phenotypic appearance, ranging from deep green (mono-culture algae, nitrogen-replete) and deep cream (mono-culture bacteria, nitrogen-replete) to green orange (co-culture, nitrate-depleted) and deep green, orange (co-culture nitrate-depleted with 3% CO_2 supplementation) (Fig. 7.6).

Quantification of these disaccharides revealed clear temporal and condition-dependent trends. Trehalose, primarily produced by the bacteria, peaked in bacteria-alone cultures on day 6 (360.90 $\mu g{\cdot}mL^{-1}$) and day 9 (297.76 $\mu g{\cdot}mL^{-1}$), but its concentration was significantly reduced in both co-culture conditions (Fig. 7.6b). Conversely, sucrose accumulation was highest in the co-culture with nitrate depletion and CO_2 supplementation, reaching a maximum of 799.48 $\mu g{\cdot}mL^{-1}$ (Fig. 7.6c). Overall, the productivity of sucrose in co-cultures was substantially higher over the time course compared to the algae-alone mono-culture, which only reached 487.92 $\mu g{\cdot}mL^{-1}$, demonstrating a clear synergistic effect of the co-cultivation strategy.

7.7 Discussion

The metabolomic data, particularly the trehalose-sucrose exchange, provide the mechanistic validation for the enhanced pigment productivity observed under the C_{NDC} condition.

7.8 Mechanisms of Enhanced Lutein Production

High-value tetraterpenoid carotenoid lutein is utilized extensively in medications, food, feed, and cosmetics (Makaranga and Jutur 2025; Dang Mai et al. 2025; Eun et al. 2025). In the current study, the observed reallocation of carbon flux into the α-carotene branch under high-CO_2, nitrate-depleted conditions point to a targeted metabolic response to environmental stress. While direct validation using gene

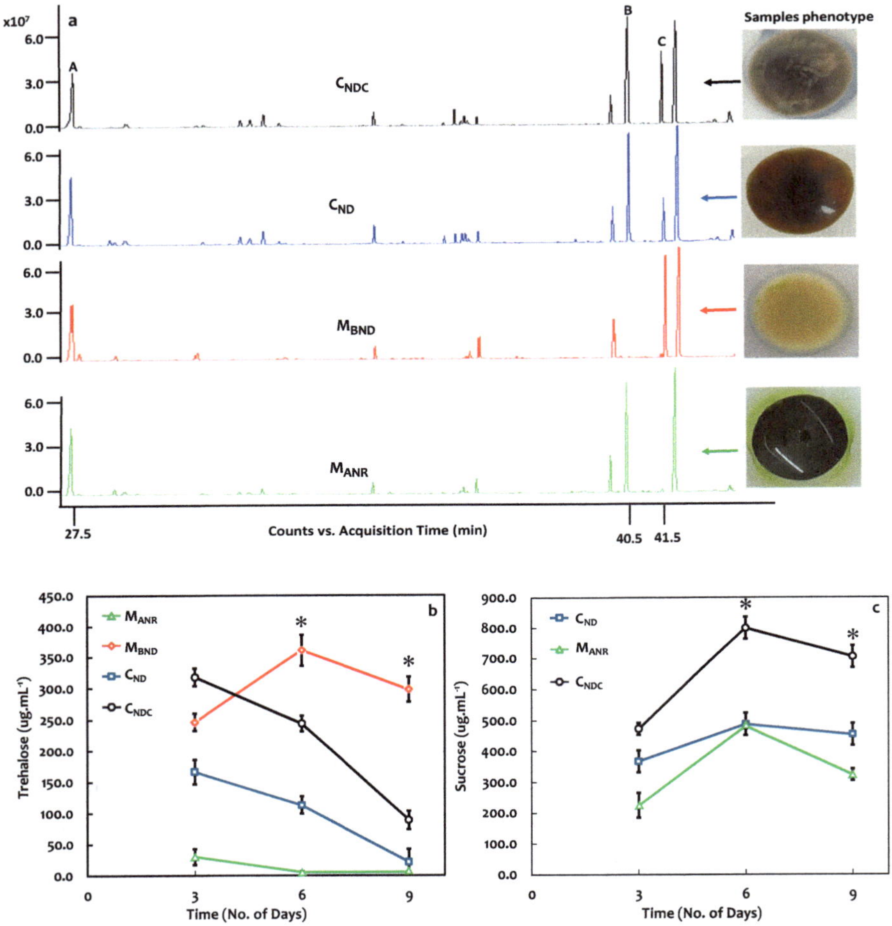

Fig. 7.6 (**a**) GC-MS/MS targeted metabolite chromatograms of extracts from mono-cultures and co-cultures. Abbreviations: "A" represents the internal standard ribitol retention time: 27.3–27.5 min, target ion: 206), "B" denotes sucrose (retention time: 40.4–40.5 min, target ion: 361), and "C" represents trehalose (retention time: 41.3–41.5 min, target ion: 361). Chromatograms indicate that trehalose is a minor disaccharide, while sucrose is the major disaccharide in *C. saccharophila* mono-culture under nitrogen replete, 150 μmol photons m^{-2} s^{-1} light intensity and 3% CO_2 supplementation. (**b**) Trehalose in algal and bacterial, and (**c**) sucrose in algal (statistical significance by one-way ANOVA, $p < 0.05$)

expression analysis (e.g., LCYB/LCYE ratio) is necessary for full biological confirmation, the robust phenotypic accumulation of lutein (terminal α-carotene product) and low levels of β-carotene (competing terminal product) strongly suggest LCYE/LCYB flux control is exerted at the enzymatic level under the co-culture nitrate-depleted with 3% CO_2 supplementation. This particular carotenoid may be essential for photoprotection and cellular stability when nitrogen is scarce and carbon is plentiful, as evidenced by the notable accumulation of lutein (238.3 $\mu g.L^{-1}d^{-1}$), which is far higher than levels under other circumstances. This is probably because lutein plays a crucial role in the xanthophyll cycle, which helps to dissipate excess light energy and prevent photo-oxidative damage (Jahns and Holzwarth 2012). This process is especially important when nutrient stress impairs the body's ability to grow and repair itself.

Additionally, the competitive β-carotene branch's significant production of zeaxanthin and violaxanthin demonstrates a coordinated, albeit divided, metabolic strategy. The organism maintains a substantial investment in other photoprotective pigments in addition to increasing lutein synthesis. Carbon is preferentially directed toward the pathways' terminal products (lutein, zeaxanthin, and violaxanthin) rather than their precursors, as evidenced by the consistently low yield of β-carotene, antheraxanthin, and canthaxanthin under all conditions. This effective channelling suggests a complex regulatory system that maximizes the synthesis of the most functionally relevant photoprotective compounds at the critical times, optimizing the photosynthetic apparatus for survival under stress (Jahns and Holzwarth 2012).

7.9 Bacterial-Algal Metabolic Interactions

The targeted metabolomics data show a complex metabolic conversation between the bacterial and algal partners, marked by resource optimization and cross-feeding that is heightened in stressful environments (Fig. 7.7). A basic metabolic interdependence is highlighted by the different sugar profiles, where trehalose is a major bacterial product and sucrose is undetectable in bacterial mono-cultures. Despite being produced in large quantities in bacterial mono-cultures, trehalose was significantly reduced in co-cultures. This suggests that the microalgae either consume this molecule as a carbon source or that its production is downregulated when microalgal metabolites are present. On the other hand, the production of sucrose by microalgae, which is significantly increased in co-culture under nitrate depletion and high CO_2, suggests that the bacteria may promote microalgal carbon fixation and exudation.

This interaction appears to be a synergistic adaptation to nutrient stress (Fig. 7.7). The microalgae appear to adjust their growth strategy to produce carbon-rich compounds, such as sucrose, when nitrate levels are low. The bright orange pigmentation in co-cultures, a sign of increased carotenoid production, suggests that the heterotrophic bacteria may offer vital services like nutrient remineralization or stress protection in exchange for this organic carbon. When the algal partner is given plenty of carbon to compensate for nitrogen limitation, resource exchange is maximized, as evidenced by the highest sucrose yield in the CO_2-supplemented, nitrate-depleted co-culture. The potential of synthetic co-cultures for sustainable bioproduction is demonstrated by this mutualistic interaction, which not only enhances survival under stress but also significantly increases overall productivity.

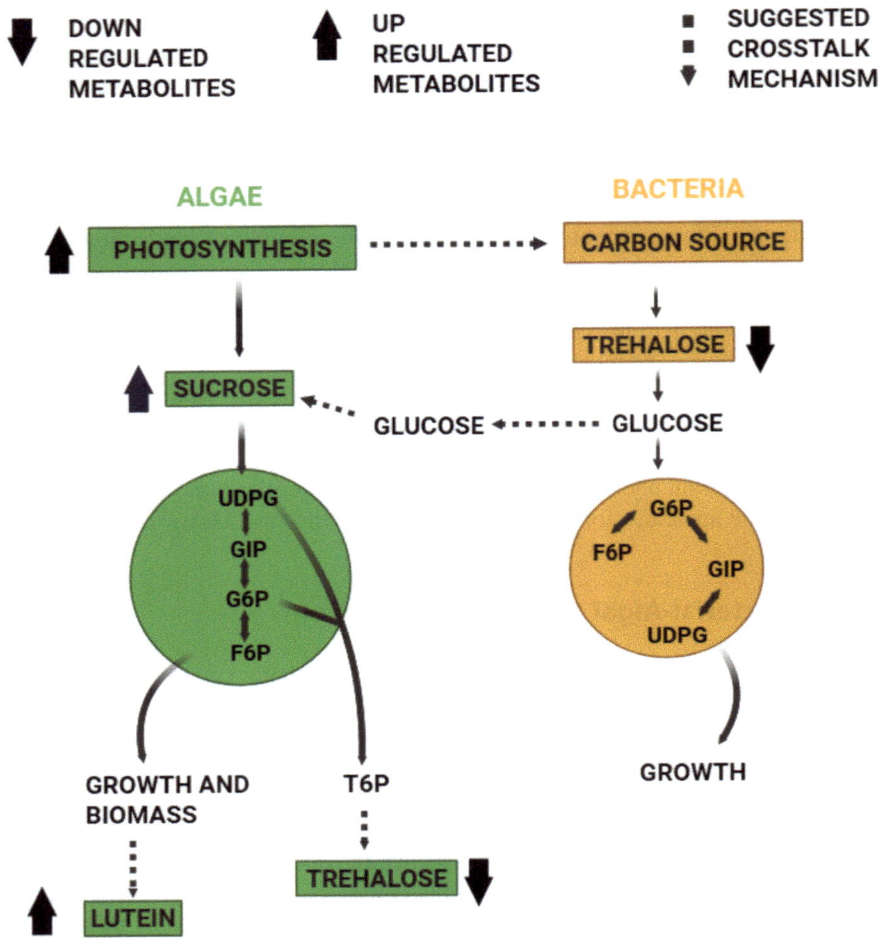

Fig. 7.7 A simplified schematic diagram illustrating key metabolic pathways in the interaction between algae and bacteria. Highlighting the interconnected biosynthesis of sucrose, trehalose, and the pigment lutein from central carbon intermediates in algae

7.10 Implications for Bioprocess Design

Crucial metabolic biomarkers for maximizing the performance of algae-bacteria co-cultures are trehalose and sucrose. The different patterns of sucrose as an algal product and trehalose as a bacterial product suggest a cross-kingdom exchange, in which bacteria probably eat algal sucrose and algae may use bacterial trehalose to boost mutual growth under stress. Trehalose levels in co-cultures are significantly lower than those in bacterial mono-cultures, suggesting active consumption or downregulation and dynamic resource allocation. Stress conditions can be designed to optimize carbon exchange, as evidenced by the synergistic increase in sucrose productivity under nitrate depletion with CO_2 supplementation. This realization supports the use of gas composition (e.g., CO_2 enrichment) and nutrient stress (e.g., nitrogen depletion) to improve the production of specific compounds. Real-time process control, made possible by monitoring these biomarkers, enables the implementation of precision fermentation techniques that take advantage of mutualistic interactions to produce carotenoids, sugars, and other valuable metabolites in a sustainable manner. This study validates the principle that the sucrose/trehalose ratio acts as a robust ex vivo metabolic indicator of the co-culture's symbiotic state. Moving forward, integrating these biomarkers into a real-time process control system for automated nutrient and CO_2 delivery represents a crucial next step for industrial implementation.

7.11 Applications and Future Perspectives

7.11.1 Industrial Implementation Strategies

This research provides a transformative framework for industrial biotechnology by using stress-induced algae-bacteria co-cultures. Immediate applications include multi-product biorefineries for the simultaneous production of carotenoids and biofuel, carbon capture from industrial emissions, and wastewater treatment with value-added metabolite generation. These approaches enhance economic viability while promoting circular carbon and nutrient economies. Future implementation strategies focus on real-time process control using sucrose/trehalose biomarkers as performance indicators. Advancements will require engineered strains for enhanced productivity, novel photobioreactor designs for scale-up, and market-driven selection of algal-bacterial pairs for specific high-value products, moving from laboratory validation to sustainable industrial processes.

7.11.2 Circular Bioeconomy Integration

This algae-bacteria co-culture system exemplifies circular bioeconomy principles by integrating waste streams into value-added production. Using nitrate-depleted wastewater and industrial CO_2 emissions as inputs, the process simultaneously treats waste and generates high-value outputs (lutein, sucrose), eliminating pollution while creating economic value. A full Life Cycle Assessment (LCA) would likely show significantly reduced environmental footprints in water, carbon, and nutrient cycles compared to traditional agriculture or fermentation. Future integration requires scaling these synergies into industrial ecosystems. While a preliminary LCA and pilot-scale outlook are beyond the scope of this fundamental study, the high volumetric productivity (238.3 $\mu g.L^{-1}$ Day^{-1}) and dual-product output (lutein and sucrose) provide a strong techno-economic justification for future pilot trials and integrated biorefinery design. Factories could become closed-loop systems where their CO_2 and nutrient-rich wastewater feed adjacent photobioreactors, producing raw materials for nutraceuticals, bioplastics, and biofuels on-site. This model transforms linear waste disposal into circular productivity, demanding new LCA metrics that value ecosystem services and multi-output efficiency alongside traditional productivity measures.

7.11.3 Expansion to Other Valuable Compounds

The demonstrated success in enhancing lutein and sucrose production through algae-bacteria synergy under stress conditions provides a robust platform for expanding this strategy to other high-value compounds. The same principles of nutrient limitation and CO_2 supplementation can be applied to stimulate production of omega-3 fatty acids (EPA/DHA), astaxanthin, or phycobiliproteins by pairing specialized algal strains with complementary bacteria. This approach could unlock new pathways for compounds typically produced in low yields, transforming their economic feasibility. Future applications should explore consortia designed for specific compound classes, such as pairing cyanobacteria with heterotrophic bacteria to enhance hydrogen production or using stress-induced lipid synthesis for biodiesel precursors. The metabolic crosstalk evidenced by sucrose/trehalose exchange suggests that similar interactions could be engineered to overproduce virtually any algal metabolite. This paradigm shift from axenic to synergistic cultivation could revolutionize industrial biotechnology, enabling sustainable production of diverse bioactive compounds within integrated biorefineries.

7.12 Future Directions

Future research must prioritize translating these lab-scale synergies into industrial reality. Key directions include engineering algal strains to suppress competitive carbon sinks, such as starch, and enhancing sucrose excretion for bacterial partners. Concurrently, bacteria should be engineered to overproduce and export stress-protectants, such as trehalose, in response to algal signals. Crucially, to mechanistically validate the proposed sugar exchange, future studies should employ stable isotope tracing (e.g., using ^{13}C-labeled CO_2 or sugars) to confirm the carbon flow from bacterial trehalose to algal biomass, and integrate enzymatic gene expression analysis (e.g., qPCR) to determine whether trehalose activity in the algae, or trehalose synthase activity in the bacteria, is the primary driver of the observed sugar dynamics. This requires the development of robust genetic tools for non-model industrial strains. Research must also focus on dynamic process control, using real-time metabolomic sensors to monitor sucrose/trehalose ratios as biomarkers for automated nutrient and CO_2 delivery, optimizing the symbiosis continuously. Scaling these interactions demands novel photobioreactor designs that facilitate intimate microbial partnership while ensuring light penetration and gas exchange.

Furthermore, LCA studies are crucial for quantifying the environmental and economic benefits of using waste CO_2 and wastewater in comparison to the energy costs of reactor operation. Ultimately, research should expand this principle to diverse algal-bacterial pairs for producing various high-value compounds, creating a versatile platform for the circular bioeconomy.

7.13 Conclusions

This study conclusively demonstrates that co-culturing *C. saccharophila* with bacteria *Exiguobacterium* sp. under nitrate-depleted, CO_2-supplemented stress conditions are a metabolically responsive strategy for enhancing lutein production. The significant 3.54-fold increase in productivity underscores the profound synergistic effect of this algae-bacteria partnership. The metabolomic shifts, notably the reduction in bacterial trehalose and increase in algal sucrose, reveal a beneficial exchange that redirects carbon flux toward valuable metabolites. These findings validate the potential of engineered synthetic co-cultures as a process-efficient, sustainable platform for the circular bioeconomy, enabling high-value compound production through CO_2 valorization and stress-induced biosynthesis.

Acknowledgments We gratefully acknowledge the support for this research provided by the Department of Biotechnology (DBT), India. AM acknowledges the ICGEB-Arturo Falaschi fellowship, and BS and AAN acknowledge the fellowships received from the DBT.

References

Carrasco Flores D, Hotter V, Vuong T, Hou Y, Bando Y, Scherlach K, Burgunter-Delamare B, Hermenau R, Komor AJ, Aiyar P et al (2024) A mutualistic bacterium rescues a green alga from an antagonist. Proc Natl Acad Sci U S A 121(15):e2401632121

Dang Mai T, Bui Thu Uyen T, Ba Duong T, Nguyen Ngoan T, Nguyen Thu Phuong T, Do Cam Van T, Luu Ngoc S, Tran Dang T (2025) Sequential optimization of biomass and lutein production by Dictyosphaerium sp. HT3. Asia-Pac J Chem Eng:e70058. https://doi.org/10.1002/apj.70058

Eun H, Prabowo CPS, Lee SY (2025) Gram-per-litre-scale production of lutein by engineered Corynebacterium. Nat Synth 4:1200–1211

Fariz-Salinas EA, Limón-Rodríguez B, Beltrán-Rocha JC, Guajardo-Barbosa C, Cantú-Cárdenas ME, Martínez-Ávila GCG, Castillo-Zacarías C, López-Chuken UJ (2024) Effect of light stress on lutein production with associated phosphorus removal from a secondary effluent by the autoflocculating microalgae consortium BR-UANL-01. 3 Biotech 14(1):23

Fu Y, Wang Y, Yi L, Liu J, Yang S, Liu B, Chen F, Sun H (2023) Lutein production from microalgae: a review. Bioresour Technol 376:128875

Guo L, Xi B, Lu L (2024) Strategies to enhance production of metabolites in microbial co-culture systems. Bioresour Technol 406:131049

Henshaw RJ, Moon J, Stehnach MR, Bowen BP, Kosina SM, Northen TR, Guasto JS, Floge SA (2024) Metabolites from intact phage-infected Synechococcus chemotactically attract heterotrophic marine bacteria. Nat Microbiol 9(12):3184–3195

Iwamoto H, Soccol CR, Molina-Aulestia DT, Cardoso J, de Melo Pereira GV, de Souza Vandenberghe LP, Manzoki MC, Ambati RR, Ravishankar GA, de Carvalho JC (2024) Lutein from microalgae: an industrial perspective of its production, downstream processing, and market. Fermentation 10(2):106

Jahns P, Holzwarth AR (2012) The role of the xanthophyll cycle and of lutein in photoprotection of photosystem II. Biochim Biophys Acta Bioenerg 1817(1):182–193

Jutur PP (2025) The phycosphere's promise: algae and bacteria rewrite biotechnology's rules. J Phycol 61(3):440–442

Kareya MS, Mariam I, Rajacharya GH, Nesamma AA, Jutur PP (2022) Valorization of carbon dioxide (CO2) to enhance production of biomass, biofuels, and biorenewables (B3) in Chlorella saccharophila UTEX247: a circular bioeconomy perspective. Biofuels Bioprod Biorefin 16(3): 682–697

Kim H, Brisson VL, Casey JR, Swink C, Rolison KA, McCall N, Golini AN, Northen TR, Veličković D, Weber PK et al (2025) Spatially structured bacterial interactions alter algal carbon flow to bacteria. ISME J 19(1):wraf096

Lin J-H, Lee D-J, Chang J-S (2015) Lutein production from biomass: marigold flowers versus microalgae. Bioresour Technol 184:421–428

Makaranga A, Jutur PP (2023) Dynamic metabolomic crosstalk between Chlorella saccharophila and its new symbiotic bacteria enhances lutein production in microalga without compromising its biomass. Enzym Microb Technol 170:110291

Makaranga A, Jutur PP (2025) Nutrient stress triggers sugar-mediated carotenoid production in algal-bacterial interactions. World J Microbiol Biotechnol 41(3):93

Manoel JAC, Šterbová K, Saini MK, Bárcenas-Pérez D, Cheel J, Grivalský T, Lakatos GE, Lukeš M, Urajová P, Ferreira A et al (2025) Microalgae-associated Stenotrophomonas maltophilia enhances lutein production and biostimulant activity in Monoraphidium sp. Algal Res 91:104250

Mayack C, Carmichael K, Phalen N, Khan Z, Hirche F, Stangl GI, White HK (2020) Gas chromatography – mass spectrometry as a preferred method for quantification of insect hemolymph sugars. J Insect Physiol 127:104115

Sørensen M, Andersen-Ranberg J, Hankamer B, Møller BL (2022) Circular biomanufacturing through harvesting solar energy and CO(2). Trends Plant Sci 27(7):655–673

Strategies for Enhanced Terpenoid Production in Microalgae for Therapeutic Applications

8

Sana Parveen, Simran Maratha, Ruchika Agarwal, Aryan Singh, Kumkum Sharma, and Shailesh Kumar Patidar

Abstract

Terpenoids are the largest group of natural compounds. Many classes of terpenes are known to have high antioxidant, anticancer, hypoglycemic, and antibacterial activity. Microalgae are also considered as a good source of terpenoid in addition to various other valuable compounds of applications. Microalgae possess biosynthesis pathways for both eukaryotic (Mevalonic acid pathway) and prokaryotic (methyl-D-erythritol 4-phosphate pathway) organisms, enabling them to produce wide range of terpenes. Various genetic engineering technologies for improved terpenoid production have been developed till date such as overexpression of rate limiting enzymes and inhibition of competing pathways, but these technologies are not economically feasible when operated alone. Therefore, there is a need to understand the effect of chemical and biological additives, which could enhance the accumulation of terpenoids in microalgae in a more ecological friendly and efficient manner. Modulating culture condition by providing high light, optimize carbon dioxide, appropriate nutrient limitation and optimize reactive oxygen species generation could ameliorate the terpenoid biosynthesis. Further, exogenous addition of chemical regulators in microalgae such as trimethylamine and terbinafine may enhance lycopene and squalene, respectively. Besides understanding only, the up-stream process of terpenoids accumulation, downstream processes are also essential for effective extraction of the desired compounds for industrial scale production. Recent downstream technologies used different solvent-based extraction, supercritical fluid extraction, and solvent free physical; extraction or combining all of these methodologies for effective industrial scale

S. Parveen · S. Maratha · R. Agarwal · A. Singh · K. Sharma · S. K. Patidar (✉)
Department of Environmental Science, School of Earth Sciences, Central University of Rajasthan, Kishangarh, Ajmer, India
e-mail: shailesh.patidar@curaj.ac.in

K. Mohanty et al. (eds.), *Advances in Algal Research*,
https://doi.org/10.1007/978-981-95-7234-2_8

production. This chapter will explore the biosynthetic pathways of terpenoids production, genetic engineering strategies for improved terpenoids productivity and limitations. In addition, this chapter will cover chemical and biological regulators for upstream production and downstream process for efficient terpenoid extraction/production at an industrial scale for therapeutic applications.

Keywords

Microalgae · Terpenoids · Therapeutic agent · Chemical regulators · Biological additives · Nutrient starvation

8.1 Introduction

Terpenoids, also referred to as isoprenoids, represent one of the most diverse and functionally important classes of natural products, synthesized through the condensation and modification of isoprene units via the mevalonate (MVA) and methylerythritol phosphate (MEP) pathways (Zhang and Lu 2017; Böttger et al. 2018). They are ubiquitous across living organisms, where they play fundamental roles in growth regulation, photosynthesis, stress response, and defence. Their chemical diversity arises from complex enzymatic transformations of isopentenyl pyrophosphate (IPP) and dimethylallyl pyrophosphate (DMAPP), giving rise to monoterpenes, sesquiterpenes, diterpenes, triterpenes, and other subclasses with wide-ranging biological functions (Jiang et al. 2016). From an evolutionary perspective, terpenoid biosynthesis predates the divergence of Bacteria and Archaea, reflecting their ancient and indispensable role in life processes (Hoshino and Villanueva 2023). Beyond their intrinsic biological roles, terpenoids are indispensable in modern industry, where their structural diversity and bioactivity underpin applications in pharmaceuticals, nutraceuticals, cosmetics, food, and renewable energy. In medicine, iconic examples include the anticancer agent *Taxol® (paclitaxel)* and the antimalarial compound *Artemisinin*, which exemplify the immense therapeutic relevance of plant-derived terpenoids. These compounds, along with numerous others, display anticancer, antimicrobial, anti-inflammatory, and neuroprotective activities that continue to drive drug discovery and development (Wang et al. 2005; Powder-George 2024). Their aromatic and volatile properties have secured their extensive use in cosmetics and perfumery, where they contribute fragrance, flavor, and pigmentation (Lowe et al. 2024). While in the food industry, terpenoids serve as natural additives, preservatives, and flavoring agents, enhancing both taste and safety of consumable products (Lowe et al. 2024). More recently, their potential as biofuels has attracted attention due to their high energy density, renewable nature, and compatibility with existing fuel infrastructures. Therefore, advances in synthetic biology enabling microbial and algal production systems that offer sustainable alternatives to fossil fuels (Jahangeer et al. 2021; Wang et al. 2018). Besides their uses in industries, terpenoids are highly beneficial with therapeutic potentials as they have anti-cancer effects (able to limit cancerous growth spread), blood supply, stimulate apoptosis, and interfere with oncogenesis pathways (Tripathi et al. 2023; Kumar 2024). Their addition to antimicrobial and antiviral effect also widens their applicability as therapeutic agents at time when resistance to antibacterial drugs is on the rise (Lowe et al. 2024; Wang et al. 2005). Nevertheless, terpenoids have been in use since time immemorial and in very large amounts, however their high prices as compared to other drugs at large-scale production is still a big challenge. Conventional methods of extracting plants are limited by poor yield, tedious growth periods, and environmental destruction, whereas chemical synthesis is uneconomical owing to structural complexity of the terpenoids in most cases. This has been necessitating the move towards microbial and algal production platforms although they also have their techno-economic hurdles. In eukaryotes like *Saccharomyces cerevisiae* and *Escherichia coli*, there are specific challenges to the balance of precursor availability, enzyme activity and cofactor supply, along with expression and stability of cytochrome P450 enzymes and compartmentalization of the intermediates in the terpenoid pathway (Wang et al. 2025; Zhang et al. 2017; Xia et al. n.d.; Mani et al. 2021). Simultaneously, industrial scale-up of such processes is also costly because downstream processing technologies also must be eco-friendly and efficient enough to match the conventional technologies (Zhang et al. 2017; Wang et al. 2024; Chen et al. 2021). The complexity of the structure of terpenoids also complicates the solution, and the bioavailability of the latter is especially difficult regarding pharmaceutical applications (He 2025). The host selection remains one of the current issues as traditional microbial chassis are not natural terpenoid producers and require extensive pathway reconstruction (Lin and Pakrasi 2019; Moser and Pichler 2019). Against this backdrop, microalgae have emerged as highly promising platforms for sustainable terpenoid production, offering advantages such as rapid growth, photosynthetic carbon fixation, and the ability to thrive on non-arable land with minimal freshwater input. Unlike higher plants, microalgae can be cultivated under controlled conditions, allowing for more predictable yields and lower environmental impact, while their possession of both MVA and MEP pathways provides multiple entry points for genetic engineering (Huang et al. 2021). For example, targeted overexpression of key enzymes such as 1-deoxy-D-xylulose-5-phosphate synthase (DXS) in species like *Chlamydomonas reinhardtii* and *Nannochloropsis gaditana* has been shown to significantly enhance terpenoid yields (Hoqani 2017), while the introduction of the isopentenol utilization pathway (IUP) in *C. reinhardtii* increased limonene production 23-fold compared to strains expressing only limonene synthase (Zhao et al. 2022) . Cyanobacteria have also been engineered to produce a wide range of terpenoids by integrating plant-specific terpene synthases into their native MEP pathway (Rautela and Kumar 2022) demonstrating the potential of photosynthetic microbial systems. Besides genetic methods, cultivation approaches are also important: Dense cultivation of *Synechocystis* sp. PCC 6803 has been used to increase terpenoid productivity (Dienst et al. 2019). Other physico-chemical parameters such as light intensity, nutrient supply, and the addition of specific chemical inducers have all been found to positively affect yields (Herold and Bryan 2020; Huang et al. 2021). The economics of downstream recovery of industrial terpenoids is critically important.

Advanced extraction strategies, such as supercritical fluid extraction (SFE), enable more efficient recovery of carotenoids compared to conventional solvent-based methods (Huang et al. 2021), In addition, the application of in situ extractants—such as dodecane overlays during high-density cultivation—circumvents feedback inhibition and significantly enhances terpenoid production (Dienst et al. 2019). Besides genetic and process optimization, one strategy that is very essential to scalable microalgal production is strain selection/improvement, where mutageneses as well as adaptive laboratory evolution are used to identify robust strains that can thrive under industrial conditions and high-throughput screening is used to identify the desired production strain. The fact that the artificial intelligence is being integrated into the process of cultivation optimization and strain selection is an encouraging venture to negate the existing drawbacks of microalgal biotechnology (Alzahmi et al. 2024). Even after such developments, scaling microalgal terpenoid yields to an industrial level and their cost-effectiveness as well as other issues relating to environmental impact and long-term stability of genetically engineered traits still remain a challenge (Balasubramaniam et al. 2021; Simeis et al. 2021). However, the emergence of genetic engineering, metabolic engineering, cultivation engineering, and eco-friendly processing can significantly change microalgae into effective bio-factories of terpenoids. Due to the worldwide increased demand in these compounds in the areas of pharmaceuticals, nutraceuticals, cosmetics, and renewable energy, the research behind microalgal production of terpenoids is of great science and industrial interest. A combination of synthetic biology with microalgal production and harvesting technologies can not only yield increased production, but also yield more sustainable and cost-effective production systems, placing microalgae as the backbone of sustaining future markets with terpenoids in medicinal and industrial usage.

8.2 Classification of Terpenoids and Their Biosynthetic Pathways

Terpenoids are a diverse group of bioactive natural compounds derived from various sources, including plants, animals, and microorganisms. Their wide structural diversity, indicated by different carbon skeletons and functional groups, supports their extensive range of biological activities. The classification of terpenoids depends on the number of carbon atoms in the isoprene units, which generally have the formula (C_5H_8) n.

8.2.1 Classification of Terpenoids

Based on the number of isoprene units present, it can be classified as monoterpenoids, sesquiterpenes, diterpenes, triterpenes, tetraterpenes, and hyper-polyterpenes. They are discussed in detail as follows:

(a) *Hemiterpenoids*: Isoprene is classified as a hemiterpenoid because it contains only one isoprene unit. However, some oxygen-containing isoprene derivatives, such as isovaleric acid and prenol, are also considered hemiterpenoids (Chanda and Ramachandra 2019).
(b) *Monoterpenoids*: Monoterpenoids consist of two isoprene units, resulting in a molecular formula of $C_{10}H_{16}$. Based on their structure, they can be categorized as having no ring, monocyclic (single ring), or bicyclic (two rings). These compounds have extensive market value across different industrial sectors, like medicinal, agricultural, and food, due to their versatility in biological and chemical activities (Gao and Xiao 2025). Previous studies show that monoterpenoids have broad bioactivities, including antitumor, pathogen-resistant, germicidal, and hypersensitivity-reducing effects (Vyhlídalová et al. 2025). Various examples of monoterpenoids include eucalyptol, citral, limonene, and camphor.
(c) *Sesquiterpenoids*: These terpenoids consist of three isoprene units, having 15 carbon atoms in their structure. A carbon atom can be arranged in a linear form or a cyclic form. These compounds are extensively distributed across several plant orders, such as Magnoliales, Rutinales, Cornales, and Chrysanthemum. They are typically present as volatile constituents, including alcohols, ketones, and lactones (Shataer et al. 2025). Guaianolide, eudesmanolides, and germacranolides are the three main varieties of sesquiterpene lactones. The three main varieties of sesquiterpene containing a lactone ring are guaianolide, eudesmanolides, and germacranolides. Different examples of sesquiterpenes are β-farnesene, α-humulene, zingiberene, and caryophyllene (Das et al. 2024).
(d) *Diterpenoids*: These are non-volatile compounds containing 20 carbon atoms with structural variability, derived from the substrate geranylgeranyl diphosphate (GGPP) with the help of the enzyme diterpene synthase. Their molecular formula is $(C_{10}H_{16})_2$, and they exhibit various structural forms, such as linear and cyclic compounds. Their structures often contain keto and hydroxyl groups attached to the terminal positions. They also serve as precursors for many biologically important compounds, such as phytol, phytoretinal, and retinol (Lanzotti 2013).
(e) *Sesterterpenoids*: Sesterterpenoids are a relatively small group of natural products. Around 1000 natural sesterterpenoids have been reported. The basic carbon skeletons of many sesterterpenoids are composed of 25 carbon atoms. However, it should also be noted that the basic carbon skeletons of some sesterterpenoids are composed of fewer than 25 carbons, due to a degradation reaction during their biosynthesis. For example, even though ircinin-3 (22) from the sponge Ircinia oros possesses only 21 carbon atoms, 22 is a sesterterpenoid (Kinghorn et al. 2020).
(f) *Triterpenoids*: Triterpenoids constitute a diverse class of naturally occurring phytochemicals, typically comprising approximately 30 carbon atoms arranged as a sequence of 6 isoprene monomers. These compounds are widely distributed among various plant species. They may be present in their native, unaltered free forms or may form conjugates with sugar groups or other molecular entities,

thereby giving rise to specialized metabolites with enhanced biological activities. The biosynthesis of triterpenes and sterols predominantly occurs through the cyclization of 2,3-oxidosqualene, a straightforward and universally utilized linear isoprenoid substrate, a process catalyzed by 2,3-oxidosqualene cyclases (OSCs) (Zeng et al. 2025).

(g) *Tetraterpenoid*: Tetraterpenoids, commonly called carotenoids, are naturally occurring pigments composed of 40 carbon atoms, formed through the head-to-head condensation of 2 geranylgeranyl diphosphate (GGPP) molecules. These lipid-soluble molecules consist of eight isoprene units (C40H64) and are known for their bright colors—ranging from yellow to dark red—in many fruits, vegetables, and animals that obtain them through their diet. Among these, β-carotene is the most prevalent and is an essential precursor to vitamin A. Found in colorful produce like spinach, peppers, kale, and apricots, carotenoids are vital for plant functions and human health. Their striking colors result from a long polyene chain of conjugated double bonds, called the chromophore, which absorbs light effectively. In addition to β-carotene and lycopene, the group includes oxygen-containing derivatives called xanthophylls, such as lutein, zeaxanthin, and astaxanthin. These compounds are especially common in higher plants and fungi, contributing not only to pigmentation but also to antioxidant activity and other biological roles (Ludwiczuk et al. 2017).

(h) *Polyterpenoids*: Polyterpenes are polymeric compounds that are formed by more than eight isoprene units and include natural rubbers, which contain isoprene units in the *cis*-configuration (latex elastomer), and gutta-percha and balata, which contain isoprene units with trans double bonds (inelastic) (Mosquera et al. 2021).

8.2.2 Biosynthetic Pathways of Terpenoid Production

Terpenoids are an enormous family of organics that are classified on the basis of structures of isoprene (C_5H_8), biosynthesized across a diverse taxonomic group, viz. plants (as a secondary metabolite), animal, and micro-organisms (Hillier and Lathe 2019). Microalgae holds two important biosynthetic pathways for production of terpenoids, that is, Mevalonic Acid Pathway (MVA) for eukaryotic and methyl-D-esrythritol-4-phosphate pathway (MEP) for prokaryotic origin (Huang et al. 2021). The pathways are shown in Fig. 8.1:

The MVA pathway is the sole terpenoid pathway that exclusively works in the cytosol and is present in various organisms, whereas autotrophs employ the MEP pathway in the plastid or chloroplast for the production of terpenoids (Lange et al. 2000). It has also been seen that in various unicellular algae, viz. *Chlamydomonas* and *Nannochloropsis* sp. deprecated the MVA pathway after adopting the MEP pathway (Hale et al. 2012).

The raw material is acetyl-CoA for MVA pathway and its two molecules trigger the MVA pathway which results in the formation of one acetoacetyl-CoA molecule,

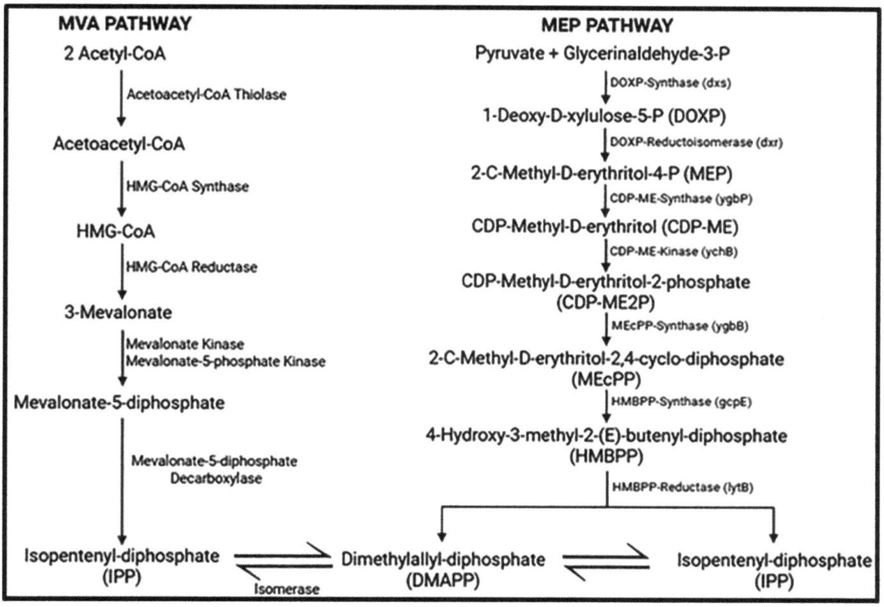

Fig. 8.1 MVA and MEP pathways of terpenoid production in microalgae

and in turn, one MVA molecule is generated when 3-hydroxy-3-methyl-glutaryl-coenzyme A (HMG-CoA) is reduced by the enzyme HMG-CoA reductase (HMGR). Therefore, HGMR expression can be modulated at successive levels of gene expression (Burg and Espenshade 2011). Ultimately, a bottleneck step in which isopentenyl pyrophosphate (IPP) is interconverted to dimethylallyl pyrophosphate (DMAPP), that is catalyzed by isopentenyl diphosphate isomerase (IDI), which also equilibrates the IPP and DMAPP ratio (Berthelot et al. 2012; Withers et al. 2007).

In the MEP pathway, glyceraldehyde 3-phosphate (GAP) and pyruvate serve as raw materials, and their one molecule is combined to produce 1-deoxy-D-xylulose 5-phosphate (DXP) which is catalyzed by DXP synthase, serving as an essential step in the MEP pathway (DeColli et al. 2018). Hence, in some literature, it is documented that embryophytes contain numerous copies of DXP synthase that are segmented into three groups of isoforms used in terpenoid synthesis (Cordoba et al. 2011). At the final stage, IPP or DMAPP was formed by the reduction of 4-hydroxy 3-methyl-butenyl 1-phosphate (HMBPP), alike the MVA pathway. In this case as well, the IDI serves as a key function to equilibrate the IPP and DMAPP (Huang et al. 2021).

After the yield of IPP and DMAPP (i.e., building blocks) from both pathways, the downstream biosynthesis of terpenoids is initiated with a set of enzymes known as Prenyltransferases, responsible for catalyzing the successive addition of IPP units to DMAPP substrates by stepwise condensation reaction which leads to formation of linear 5-carbon Prenyl pyrophosphate ($C_5H_{10}O_7P_2$) molecules (i.e., building blocks for various larger molecules like terpenes and steroids) (Davies et al. 2015). In turn, both IPP and DMAPP units are aggregated to form geranyl pyrophosphate (GPP), a 10 C molecule and also act as a biosynthetic precursor of monoterpenes. Therefore, sequential addition of IPP units forming farnesyl pyrophosphate (FPP) precursors of sesquiterpenes and geranylgeranyl pyrophosphate (GGPP) the precursors of diterpenes (Vavitsas et al. 2018).

8.2.3 Terpenoid Extraction Technologies

There are various extraction technologies for efficient and enhanced terpenoid yield which are given as follows:

(a) *Green extraction technology*: Various techniques have been employed to extract terpenoids. Traditional methods like solvent extraction, hot extraction, and percolation are used to obtain bioactive compounds from plants, microbes, and other organisms. However, these methods are limited by long extraction times, low efficiency, and thermal degradation, whereas modern eco-friendly techniques improve yield and shorten extraction. Industries, including cosmetics, are increasingly adopting green solvents to minimize reliance on hazardous chemicals like chloroform and phenol. Green extraction, as defined by Chemat et al. (2012), involves developing extraction processes that minimize energy consumption, utilize alternative and renewable natural products, and guarantee a safe, high-quality extract. The six principles emphasize renewable plant resources, green solvents, energy efficiency, co-product generation, simplified yet safe processes, and biodegradable, contaminant-free extracts. Innovative methods include steam distillation, solvent extraction, and supercritical fluid extraction.
(b) *Steam distillation*: Steam distillation techniques are widely employed to extract volatile compounds, relying on Dalton's Law of Partial Pressures and the immiscibility of these compounds with water. Currently, four variants are recognized: hydro distillation, direct steam stripping, indirect steam stripping, and water-bubble distillation. A key benefit of these methods is the use of water, which significantly minimizes environmentally harmful waste. Additionally, their design enables easy scalability without requiring complex technologies.
(c) *Supercritical fluid (SCF) extraction technology*: SCF extraction shows potential for improving molecule recovery with promising industrial uses. It provides greater selectivity and quicker processing times compared to conventional methods. In SCF – CO_2 extraction, the solvent exists as a gas at atmospheric conditions, enabling significant CO_2 removal from the extracts and resulting in a solvent-free product. If implemented industrially, CO_2 could also be reused and recycled. Different factors affect the composition of the extract as well as the recovery efficiency, for example, heat level, applied force, auxiliary agents, solvent velocity, and processing time, all of which must be fine-tuned to achieve optimal recovery of bioactive compounds (Molino et al. 2020).

(d) *Ultrasound-assisted extraction (UAE)*: Ultrasound-assisted extraction has become a promising green technology because of its rapid process, high efficiency, and low solvent requirement. It works by using high-frequency ultrasonic waves to efficiently extract bioactive compounds. These waves create pressure changes that produce bubbles, which form and burst repeatedly. The energy released from these bubble disruptions breaks open cells, facilitating compound extraction (Siddiqui et al. 2025). Previously, Ferreira et al. (2023) investigated the performance of UAE for the extraction of bioactive compounds in *Dunaliella salina* and found that UAE outperformed Soxhlet and maceration methods, yielding up to 41.4% extract, compared to 10.4% and 3.1%, respectively. A 1:10 solid-to-liquid ratio and 30 min continuous sonication gave the highest yield. Ultrasound-assisted extraction (UAE) proved to be a highly effective and eco-friendly method for extracting valuable compounds from *Spirulina platensis*. Under optimal conditions (25 kHz, 30 minutes, 1:20 ratio), UAE boosted the extraction of proteins (up to 57%), carbohydrates (41%), and chlorophyll (4.15 mg/g). It also enhanced antioxidant activity, with DPPH inhibition reaching 70%. Additionally, the UAE reduced particle size, improving the bioavailability of the extracts. These results highlight UAE as a promising technique for obtaining functional ingredients from microalgae in a faster and more sustainable way (Tavakoli et al. 2021).

(e) *Microwave-assisted extraction (MAE)*: Microwave-assisted extraction (MAE) is an alternative technique used to isolate valuable compounds from microalgae biomass. Unlike traditional extraction methods, MAE employs microwave energy that is swiftly absorbed by the biomass and converted into heat. This rapid heating occurs via mechanisms like dipole rotation and ionic conduction. The extent of microwave energy absorption—and thus heat production—is evaluated using the dissipation factor (tan δ), which is directly linked to the dielectric loss (ε'') and inversely related to the dielectric constant (ε') of the solvent. Because of this relationship, polar solvents are essential for effective extraction. Compared to conventional solid-liquid extraction, MAE significantly minimizes thermal gradients and enables instant, uniform heating. This leads to quicker and more efficient extraction processes with reduced solvent use and improved yields (Rehman et al. 2020). Microwave-assisted extraction (MAE) significantly enhanced the extraction of bioactive compounds from *Chlorella vulgaris*. Under optimal conditions (0.5 g biomass, 120 °C, 5 min, 1:16 solid–liquid ratio), the extract showed high protein content (50.2%), carbohydrates (28.1%), and chlorophyll (2.6 mg/g). Antioxidant activity was also notably high, with 81% DPPH scavenging. MAE drastically reduced extraction time and improved yield and bioactivity compared to conventional methods, confirming its efficiency as a rapid and sustainable extraction technique for microalgal applications (Georgiopoulou et al. 2023a, b). Georgiopoulou et al. (2023a, b) highlighted that MAE turns out to be a promising, sustainable way for the extraction of bioactive compounds from algal biomass, as high amounts of chlorophyll and carotenoids were obtained at the end of extraction.

(f) *Solvent-based extraction*: Bio-based solvents derived from renewable sources like crops and plant waste are gaining traction as greener alternatives to traditional petrochemical solvents. They're safer, eco-friendly and chemically compatible but often costly due to expensive biomass processing. In the context of extracting carotenoids, bio-based solvents such as ethyl lactate, 2-MeTHF, natural deep eutectic solvents (NaDESs), deep eutectic solvents (DESs), and ionic liquids (ILs) have shown strong environmental advantages over petroleum-based options (Kopp and Lauritano 2025).

8.3 Effect of Culture Conditions of Microalgae for Enhanced Terpenoid Production

The development in molecular biology, biotechnology, and genetic engineering are the best strategy for enhancing the production of terpenoids in microalgae. The controlled environmental and nutritional conditions of microalgae are strongly associated to photosynthetic activity, carbon flux, and biosynthesis pathways of terpenoids (MVA and MEP) (Arendt et al. 2016). The production of terpenoids are influenced by both physical and biological culture conditions. Physical conditions include such as light intensity and CO_2 concentration, nutrient availability (nitrogen and phosphorus), temperature, pH, and salinity (Shukla and Phulara 2021). However, biological conditions such as co-culturing, genetic engineering, rate-limiting enzymes, and media optimization play a promising role in enhancing yield of terpenoids (Huang et al. 2021).

8.3.1 Various Physical Parameters of Culture Conditions to Enhance Terpenoid Productivity

(a) *Light intensity and CO_2 concentration*: Microalgae are photoautotrophic organisms that can absorb energy from solar radiation and perform photosynthesis to fix inorganic carbon (CO2) as carbon sources (Grechanik et al. 2020). Therefore, Light intensity and CO_2 concentration are two most important factors that influence the photosynthesis efficiency and stimulate the microalgal cell growth and carotenoid production (Huang et al. 2021). The lighting conditions such as light intensity, light duration, and photoperiod regime are essential for the growth, functional state, and reproduction of algae. The lower intensity of light can have limiting effect while too higher intensity can have an inhibitory effect on the growth of microalgae (Maltsev et al. 2021). During high intensity of light until the saturation point, the photosynthesis rate of microalgae also increases. Therefore, more pigments such as carotenoids, astaxanthin, and zeaxanthin associated with photosynthesis increase photosystems as well. However, a further increase in light intensity leads to photooxidative damage to microalgae and therefore, does not increase photosynthetic efficiency and rate. The wavelength of light also affects the growth of microalgae. Generally, microalgae consist high amount of pigment that absorbed blue and red light comparing to other colors (Mulders et al. 2014). The biosynthesis of terpenoid is coupled with the increasing photosynthetic rate and dry cell weight. For example, in *Chlamydomonas* sp. under high lighting lutein productivity increased from 1.61 to 5.08 mg/L/day and increased in cell biomass from 536.26 to 1821.51 mg/L/day (Ma et al. 2019). High CO_2 concentration in between 5 and 20% is beneficial for the photosynthetic activity and also promotes carboxylase activity which helps to increase biomass (Su 2021). It was also reported that increased chlorophyll content in *Dunaliella salina* was observed under periodical aeration of CO_2, which is more than 26 mg/L (Ying et al. 2015). Another study found that when *Haematococcus pluvialis* was exposed to 15% CO_2, the concentration of astaxanthin increased to 36.23 mg / L, which was two to three times higher than the control. In addition, the biomass of *Haematococcus pluvialis* increased from 200 mg/L to 878 mg/L. It is believed that high CO_2 concentrations increased the C/N ratio, which caused nutritional deficiencies and increased the biomass and astaxanthin levels in *Haematococcus pluvialis* (Zhang et al. 2018).

(b) *Temperature*: Besides the light intensity and nutrient availability, temperature is also an important parameter that influenced the growth efficiency of microalgae. Below the optimal condition between 20 and 25 °C, increase in temperature positively increased photosynthesis, growth, and cell division (Ras et al. 2013). In some microalgae, such as *Murillopsis* sp. and *Scenedesmus almeriensis*, lutein accumulates under high temperature conditions. In contrast, during the night, slightly lower temperature is effective for cell growth and astaxanthin accumulation in the microalgae *Haematococcus pluvialis* (Shi et al. 2020). It has been also observed that increased temperature results in increased photooxidative stress, which is associated with increased carotenoid synthesis in microalgae (Tripathi et al. 2002). Lower temperatures reduce the nutrient and slow down the accumulation of lutein. In *Trentepohlia* sp., accumulation of carotenoids progressively increased by decreasing the temperature. The optimum production of carotenoids was observed at 5 °C (Kharkongor and Ramanujam 2015).

(c) *pH*: Algal biomass production efficiency depends on the pH level, which affects photosynthesis activity, carbon fixation, and the distribution of carbon in different molecules, as well as algae growth and cellular metabolism. By controlling the distribution of carbonate species in the medium, culture pH affects the solubility and availability of carbon. The absorption of inorganic carbon influenced to enhance pH of the medium (Filali et al. 2021). The optimal range of pH is between 7 and 8 for most of the microalgae. It was observed that maximum carotenoid production occurred at pH 6, and stress conditions in the culture medium also had an effect on the increase in carotenoid production (Kumar 2018). The higher temperature with intermediate pH favors carotenoid accumulation and biomass productivity (Assunção et al. 2023). According to Khalil et al. (2010), *Dunaliella bardawil* and *Chlorella ellipsoidea* are microalgae that grow in the pH range of 4–9 and 4–10, respectively.

D. bardawil could accumulate β-carotene, vitamin E, and vitamin C maximum at pH of 9. In contrast to the control, β-carotene content in *C. ellipsoidea* increased at pH 6 and pH 10, but it was significantly higher at pH 7.

(d) *Salinity*: High salt concentrations lead to osmotic and oxidative stress. Reactive oxygen species (ROS) induced due to NaCl stress generate oxidative stress and synthesize antioxidant enzymes, terpenoid pigments such as carotenoids. At a high salt concentration of 30%, the accumulation of carotenoid and β-carotene showed in *D. salina* and *Dunaliella bardawil* (Elloumi et al. 2020). The microalgae *D. salina* was grown under the high salt stress that contributed in increased lipid content (up to 70%) (Takagi and Yoshida 2006). The growth rate and salinity tolerance of microalgae is dependent on the species and its habitat. Kumar (2018) has observed that *Nannochloropsis salina* cultured in a wide range of salinity between 0 ppt and 50 ppt and found maximal growth at 40 ppt and no growth at 0 ppt and 10 ppt. Salinity has a positive effect on the production of astaxanthin and β-carotene, but not on the biomass growth. In another example, *Haematococcus pluvialis* treated with different concentrations of NaCl, the maximum astaxanthin content occurred at 0.2 g/L NaCl treatment, which was 5.01 mg/L, significantly 42.74% higher than the control (Ding et al. 2019).

8.3.2 Effect of Chemical Modulators for Improved Terpenoid Productivity

There are various methods available for the production and accumulation of bioactive compounds. Certain chemical modulators have been proposed that directly modulate cellular metabolism and are applied to various microalgae at a commercial level without delving into molecular biology. In addition, the researchers took a further step to verify these chemicals in large-scale cultures and concluded that propyl gallate and butylated hydroxyanisole could be used in large-scale applications, considering the low cost of the chemicals and enhanced lipid content (Yu et al. 2015).

Alterations in nitrogen and phosphorus in the growth media can also play a major role in terpenoid biosynthesis. Microalgae can fix inorganic carbon by using sunlight, instead they also require other macro elements for their growth and production of terpenoids. Nitrogen and phosphorus are both elements that help to influence biosynthesis of terpenoids (Huang et al. 2021). Nitrogen limitation might promote the production of carotenoids, lutein, astaxanthin, and β-carotene in different species. After nitrogen starvation in *D. salina,* overproduction of total lipids and beta-carotene was observed. Under nitrogen limitation level, growth diminishes and carbon and energy from photosynthesis are stored in β-carotene and triglycerides (Bonnefond et al. 2017). In *H. pluvialis,* nitrogen starvation inhibits the growth. Thus, sufficient nutrients are usually required for higher algal growth. Although in H. pluvialis, under starvation of nitrogen at high light conditions of 200 μmol/m^2.s and 400 μmol/m^2.s, the biomass yield reached up to 2.01 ± 0.06 and 2.19 ± 0.08 g/

L, respectively (Zhang et al. 2018). Phosphorus is also an important component of nucleic acids, proteins, and several metabolites. It is an essential constituent of cellular membranes which provide structure and protection to cell, and are indispensable for the fixation of carbon and cellular metabolism during photosynthesis and respiration (Bossa et al. 2024). Phosphorus supplementation with nitrogen starvation contributes 10–100% enhancement of biomass production and also increased 15–50% of lipid accumulation. In *C. regularis*, the growth rate essentially showed that the effect of phosphorus depended on its concentration ($P < 0.05$). The excessive increment of its supply ($P \geq 150$ mg/L) could result into the growth inhibition and reduced lipid productivity (Fu et al. 2020). Phosphorus deficiency also has a positive correlation with the synthesis of carotenoids. Under nutrient, limited condition is expected to increase accumulation of carotenoid content (Dahmen-Ben Moussa et al. 2017).

8.3.3 Impact of Biological Modifications for Improved Terpenoid Yield

8.3.3.1 Influence of Co-culturing Techniques on Terpenoid Biosynthesis

Co-culturing is a setup of cell cultivation of two or more different microorganisms such as bacteria, microalgae, and fungi, when grown together synergistically to improve the survival, effective increment in biomass, and bioactive metabolites along with lipid production (Das et al. 2022). Co-culturing of green microalgae *H. pluvialis* with antibiotic-synthesizing fungus *Simplicillium lanosoniveum* DT06 under non-aseptic and aseptic pure culture enhanced the biochemical metabolites and growth of microalgal cells. Non-aseptic culture condition led to increase in biomass production, lipid productivity and astaxanthin as comparative to aseptic culture by 119%, 279%, and 175% respectively (Yang et al. 2017).

8.3.3.2 Optimization of Growth Media

The growth of microalgae and lipid is significantly dependent on the composition of the media. According to the previous study, researchers optimized the culture media to enhance β-carotene, biomass growth, and lipid production in *D. salina* by using the Box Behnken design model. Result showed that the experimental optimized media was efficient to increase biomass production, β-carotene, and lipid productivity than their original production by 2.17 folds, 1.45 folds, and 1.56 folds, respectively (Ahuja et al. 2020). In another study, *Chlorella vulgar is* cultivated with or without the replacement of culture media under the effect of lighting and salt (NaCl and $MgCl_2$) stress conditions. The result showed that the supplement of 10 g/L NaCl with a media replacement under nitrogen limitation and high light intensity helps to enhance total carotenoids (4.37 ± 0.33 μg/mL). The growth of microalgae and lipid is significantly dependent on the composition of the media (Ali et al. 2021).

8.3.3.3 Genetic Engineering Approaches for Enhanced Biosynthesis of Terpenoid Production

Terpenoids are highly diverse natural products with broad industrial applications, but low natural yields and unsustainable extraction have driven the use of genetic engineering to optimize pathways and enable sustainable, commercially viable production.

(a) *Metabolic pathway engineering*: One of the most effective approaches to boosting terpenoid biosynthesis is metabolic pathway engineering. By reprogramming carbon flux, genetic engineers can enhance precursor availability and maximize terpenoid accumulation.
(b) *Gene overexpression* is a direct strategy whereby rate-limiting enzymes in the MEP and MVA pathways are upregulated. Overexpression of genes including DXS and HMGR has been shown to substantially expand precursor pools, thereby leading to increase in the terpenoid yields (Lu et al. 2016).
(c) *Pathway optimization* enhances even more on the biosynthesis by diverting the metabolic pathways towards terpenoid biosynthesis. Recombinant DNA technologies have been used to design and construct more competitive pathways in microbial system as well as plants (Qingping 2006). In algae, optimization often involves carbon channelling towards terpenoid biosynthesis and away from storage lipid.
(d) *Silencing competing pathways* is an additional route of modulate terpenoid biosynthesis. Suppressing branches such as sterol or carotenoid synthesis ensures that available metabolic resources are preferentially allocated to terpenoid production. Post-translational gene silencing, antisense RNA, and CRISPR-Cas technologies have been extensively used to perform this (Lu et al. 2016).
(e) *Genetic transformation and tissue culture*: Transformation techniques remain central to genetic engineering for terpenoid enhancement. Among them agrobacterium-mediated transformation is widely used in plant systems to introduce novel genes or modify existing ones, leading to improved terpenoid yields, especially for pharmaceutical applications (Lu et al. 2016). Another important technique generated by *Agrobacterium rhizogenes* is Hairy Root Cultures, which have emerged as efficient biotechnological platforms for secondary metabolite production. These genetically stable cultures are capable of sustained terpenoid biosynthesis and are already being considered for industrial applications. Other methods such as electroporation, particle bombardment, and CRISPR-based genome editing have been successfully applied to various microalgae to enhance terpenoid biosynthesis pathways. These strategies leverage the photosynthetic nature of algae, coupling CO_2 fixation with terpenoid biosynthesis to create a sustainable production model.
(f) *Regulation and omics-driven approaches*: Besides structural genes, several levels of regulation have been described to operate on terpenoid biosynthesis. Transcriptional and Post-Transcriptional regulation can be taken as advantage to increase production. Terpenoid biosynthesis can be upregulated by

manipulation of transcription factors, promoters, and phytohormone signaling pathways (Abbas et al. 2017; Lu et al. 2016). Other omics-based approaches, including genomics, transcriptomics, proteomics, and metabolomics, have furnished significant information on terpenoid metabolism. Integrated omics (iOMICS) analyses such as genome wide construction are being used for the development of metabolic network that have been used since last two decades. Integration of omics data and synthetic biology can provide the ability to rationally design new biosynthetic pathways, determine bottlenecks and create better production in host cells (Kumari et al. 2024). These strategies, when combined with computational modelling and systems biology, allow the development of predictive frameworks for pathway optimization and large-scale implementation. Although genetic engineering has provided promising potent avenues in the improvement of terpenoids, some issues still exist. Plant systems have complicated regulation as well as poor productivity which makes mass production challenging. Microbial hosts are very tractable, but they are not natural terpenoid producers and thus require a lot of engineering to get these complex pathways working in these hosts (Shukla et al. 2023). With their potential, algal systems are limited by the transformation efficiency and stable gene expression. However, innovation in CRISPR-Cas editing, synthetic pathway, and systemic biology integration will be critical to overcoming these barriers. A combination of factors of algal biotechnology, metabolic engineering, and omics (high-throughput measurements) will be beneficial to the realization of high yield of terpenoids using algal biotechnology pursuing sustainability. Ultimately, genetic engineering is poised to transform terpenoid biosynthesis from a constrained natural process into a scalable, industrially viable biomanufacturing system. Various strategies for the terpenoid production in different microalgae are given in Table 8.1:

8.4 Various Therapeutic Applications of Terpenoids

There has been a growing interest in terpenoids due to their diverse and distinct biological features, which enable a wide range of applications across various industries. The growing interest in terpenoids is due to their versatility in chemical forms which is attributed to their wide range of biological activities, such as antioxidant, antimicrobial, and anti-inflammatory effects, making them valuable in therapeutical applications. These compounds, are integral to numerous sectors, including pharmaceuticals, food, cosmetics, and agriculture. Several microalgae are rich source of terpenoids, for example, *Dunaliella salina* for β-Carotene, *Chlorella. Sorokiniana* and *Chromochloris zoofingiensis* for lutein, *Nannochloropsis oculate* and *Dunaliella* sp. for zeaxanthin, *Haematococcus lacustris* for astaxanthin, and *Isochrysis* sp. for fucoxanthin (Ávila-Román et al. 2021). Monoterpenes like prefuroplocamioid, cyclohexane, pirene furoplocamioid C and lactone derived from algae like *Plocamium* sp. and *Sargassum* sp. have shown selective cytotoxicity against cancer cells as well as HeLa and Vero cells (Ferdous et al. 2021). The

Table 8.1 List of metabolic engineering strategies, their mode of action, type of terpenoid produce, and the product

Strategy	Mode of action	Strain	Type of terpenoid	Product	Result	Reference
Metabolomic engineering	Episomal expression of CrGES was driven either by the phosphate-responsive inducible promoter AP1p or the constitutive endogenous promoter Phatr3_J492020p	*P. tricornutum*	Monoterpene	Geraniol	Production of geraniol increases in 3 days up to 48.6μgmL^{-1} by the application of the inducible promoter AP1p	Fabris et al. (2020)
Metabolomic engineering	Upregulation of endogenic gene DXS or *PSY*	*P. tricornutum*	Tetraterpenoid	Fucoxanthin	The gene transformant increases 2 1nd 1.8-fold increase in fucoxanthin	Eilers et al. (2016)
Genetic engineering	Overexpression of the *OR* gene in the presence of the promoter CaMV35S	*Chlamydomonas reinhardtii*	Tetraterpenoid	α and βcarotene, lutein, and violaxanthin	4,3.1, 3.2, and 3.6 times increased	Yazdani et al. (2021)
Metabolomic engineering	Upregulation of the gene *PSY* in transformants by an inducible promoter RBCS2 and HSP70A	*Chlamydomonas reinhardtii*	Tetraterpenoid	Violaxanthin and lutein	2.0- and 2.2-multifold increased	Velmurugan and Kodiveri Muthukaliannan (2022)
Genetic engineering	Upregulated expression of *HpDGAT1*	*Haematococcus pluvialis*	Carotenoid	Multifold increase in esterified astaxanthin (EAST)	–	Cui et al. (2021)
Random mutagenesis	Chemical mutation through ethyl methyl sulfonate	*Cyanobacterium Arthrospira*	Monoterpenoid	Limonene	1.5-fold times multiplied	Guan et al. (2017)
Genetic engineering	Enhanced expression and antisense-based downregulation	*Dunaliella tertiolecta*	Carotenoid	Zeaxanthin	Multifold increase by 2.22-fold	Zhang et al. (2015)

green microalga *Chlorococcum humicola* demonstrates anticancer and chemopreventive properties, attributed to its rich content of bioactive compounds including terpenoids (Matin et al. 2024). Diterpenoids such as eleganolone and eleganonal, have shown to exert neuroprotective effect on neuroblastoma due to their standard antioxidant properties (Silva et al. 2019). Oxidative stress is a key factor to carcinogenesis, antioxidant properties of terpenoids thereby helps enhancing the efficacy of cancer treatments. Diterpenes downregulates AK2/STAT3, PI3K/Akt, and NF-κB regulatory pathways which lead to apoptosis in cancer cells. Furthermore, diterpenoids can also inhibit metastasis and angiogenesis by hindering PI3K/Akt/mTOR and VEGFR-2 signaling pathways (Chen et al. 2024; Ferdous et al. 2021). Fucoxanthin, abundantly produced by species such as *Phaeodactylum tricornutum*, *C. calcitrans*, and others, demonstrates strong antioxidative and anticancer activities (Neumann et al. 2019). Other tetraterpenoids such as zeaxanthin, violaxanthin, neoxanthin, astaxanthin, and β-carotene are some common algal terpenoids that possess impressive antioxidant properties (Ferdous et al. 2021). Triterpenoid derivatives like betulinic acid and ursolic acid have demonstrated neuroprotective effects by mitigating oxidative stress and downregulating pro-apoptotic proteins and inflammatory cytokinin. These multifaceted pharmacological properties make terpenoids promising candidates for therapeutic applications in conditions such as Alzheimer's disease, Parkinson's disease, and other neurodegenerative disorders (Bankar et al. 2024; Farzan et al. 2024). Tetraprenyltoluquinol (meroterpenoid), lutein, benzene dicarboxylic acid, diisooctyl ester (triterpenoids) also displayed excellent radical scavenging and reducing activity (Antony and Chakraborty 2020a, b; Ferdous et al. 2021; Zbakh et al. 2020). These properties are also beneficial in cosmetic applications, where terpenoids are used for their antiaging effects. Major classes of mono-, di-, and tri-terpenes specifically dominate as the potentiators of antimicrobial and antiseptic activities. Terpenoids exhibit potential antimicrobial activity by inhibiting two crucial processes of microbial survival: oxygen uptake and oxidative phosphorylation. Carbonylation of terpenoids is believed to enhance bacteriostatic activity (Mahizan et al. 2019). When used in combination, terpenoids exhibit enhanced antimicrobial activity compared to their individual effects. In one study, nootkatone was effective against Gram-positive bacteria, while eugenol inhibited *Corynebacterium diphtheriae* and *Acinetobacter baumannii*. However, when combined with other terpenoids such as carvacrol and thymol, the mixture demonstrated potent inhibitory activity against *C. diphtheriae, A. baumannii*, *Listeria monocytogenes*, *Escherichia coli*, and *Salmonella enteritidis* as well *Listeria monocytogenes* (Yamaguchi 2022). Found in various microalgal species, these compounds act through multiple biological pathways to mitigate inflammation. Their anti-inflammatory action is primarily linked to the inhibition of key pro-inflammatory enzymes like cyclooxygenase-2 (COX-2) and 5-lipoxygenase (5-LOX), which are involved in the biosynthesis of inflammatory mediators such as prostaglandins and leukotrienes (Antony and Chakraborty 2020a, b). Algal extracts and diverse algae-derived polysaccharides, phytosterols, and terpenoids have been revealed to induce Nrf2 activation, leading to an increase in the expression of the HO-1 protein in macrophages. This, in part, contributes to the observed anti-inflammatory and cytoprotective effects (Matin et al. 2024). Lutein alleviates inflammation in osteoarthritis, a common bone disease related to aging. It also exhibits anti-aging effects in mesenchymal stem cells and reduces senescence by downregulating inflammatory processes in vitro and in vivo (Zhao et al. 2023). Cartenoids have versatile roles in combating inflammation. β-Carotene reduces vascular inflammation and oxidative stress in diabetic and healthy HUVECs by downregulating adhesion molecules and NF-κB, suggesting its role in preventing diabetes-related cardiovascular issues (Ucci et al. 2019). Besides these, a number of in vivo and in vitro models have evidenced the anti-inflammatory and anti-cancer effects of carotenoids which are reviewed by (Ávila-Román et al. 2021). Diterpenes such as pachydictyol A, isopachydictyol A, and dichotomanol, have shown potent anticoagulant and antiplatelet effects by inhibiting thrombin activity and platelet aggregation—key processes involved in thrombus formation associated with cardiovascular diseases (De Andrade Moura et al. 2014). Terpenoids have been identified as potential agents for the prophylaxis and treatment of atherosclerosis and metabolic syndrome. These compounds can lower total lipids, cholesterol, and triglycerides, while also normalizing blood pressure and glucose levels, thereby reducing the risk of CVDs. Sesquiterpenoids, diterpenoids, triterpenoids, and tetraterpenoids show potential as agents across a broad dosage range (Liu et al. 2019). A list of microalgal based terpenoids and their therapeutics applications are described in Table 8.2.

Microalgal terpenoids offer distinct advantages over terrestrial plant-derived terpenoids, including faster growth rates, ability to be cultured under controlled conditions, and higher metabolic plasticity enabling diverse compound synthesis. Microalgae also provide more sustainable and scalable production platforms with lower land and water requirements, reducing environmental impact (Matin et al. 2024). Despite therapeutic promise, technical and regulatory challenges hinder the large-scale commercialization of microalgal terpenoid-based drugs. More research in this area is required to improve the drug production and enhance therapeutic applications. The overall mechanism of applied stresses for improved terpenoid biosynthesis and their therapeutic applications is given in Fig. 8.2.

8.5 Conclusion

Terpenoids are among the most useful and varied classes of natural products with a range of applications in pharmaceutical, nutraceutical and cosmetic industries, foods, and renewable energy. Their pharmacological importance, indicated using such drugs as paclitaxel and artemisinin, allows them to be used in the treatment of cancer, inflammations, metabolic diseases, and microbial diseases. The main concern in their commercial production is their restricted natural occurrence, and more complicated production process. The recent developments in metabolic engineering and synthetic biology have offered alternative ways of overcoming such challenges. The microalgae are an emerging platform as these systems can grow rapidly, have photosynthetic capability to fix CO_2, and harbor both the MVA and MEP pathways

Table 8.2 List of microalgal based terpenoids and their therapeutic applications

Microalgal species	Major terpenoid (s) produced	Type of terpenoid (class)	Therapeutic application(s)	Reference(s)
Chlamydomonas reinhardtii	Limonene, patchoulol, other monoterpenes	Monoterpenes, sesquiterpenes	Anticancer, antimicrobial, antioxidant; potential use in metabolic and inflammatory disorders	Zhao et al.(2022); Rautela and Kumar (2022)
Nannochloropsis gaditana	Carotenoids (β-carotene, astaxanthin), diterpenes	Carotenoids, diterpenes	Antioxidant, anti-inflammatory, cardioprotective; prevention of oxidative stress–related diseases	Hoqani (2017); Huang et al. (2021)
Dunaliella salina	β-carotene (high concentration)	Carotenoid (tetraterpene)	Antioxidant, anticancer, immune modulation, skin protection (UV defense, anti-aging in cosmetics)	Rajesh et al. (2017); Simeis et al. (2021)
Haematococcus pluvialis	Astaxanthin	Carotenoid (xanthophyll)	Strong antioxidant, neuroprotective, cardioprotective, anti-inflammatory; potential in cancer prevention	Balasubramaniam et al. (2021); Lowe et al. (2024)
Synechocystis sp. *PCC 6803*	Sesquiterpenes (e.g., bisabolene, farnesene)	Sesquiterpenes	Antimicrobial, anticancer, anti-inflammatory properties	Dienst et al. (2019); Herold and Bryan (2020)
Chlorella vulgaris	Lutein, carotenoids	Carotenoids	Antioxidant, anti-inflammatory; benefits in eye health and prevention of age-related macular degeneration (AMD)	Câmara et al. (2024); Huang et al. (2021)
Spirulina platensis (*Arthrospira*)	β-carotene, tocopherols (isoprenoid derivatives)	Carotenoid, terpenoid derivative	Immunomodulatory, anticancer, antioxidant; metabolic regulation	Yao and Liu (2022); Kumar (2024)

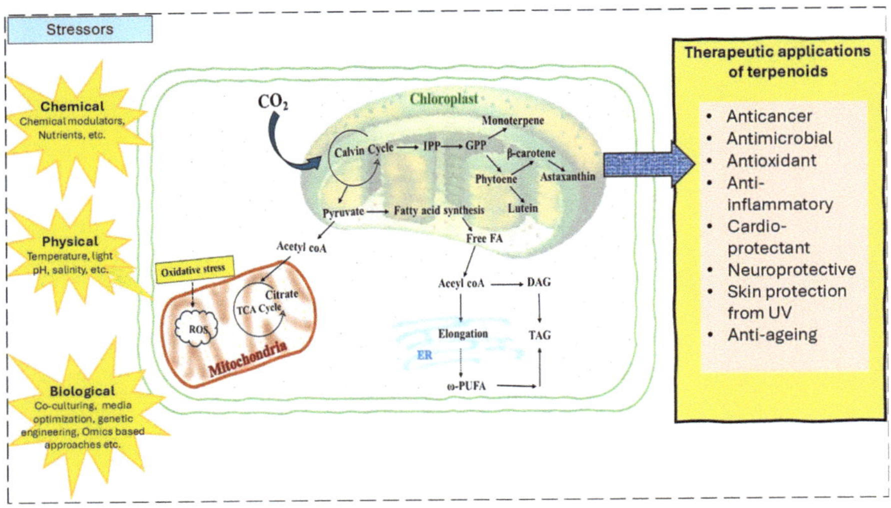

Fig. 8.2 The overall mechanism of applied stresses for improved terpenoid biosynthesis and their therapeutic applications

to produce terpenoids. Genetic tools including overexpression of enzymes, rate-limiting steps, and heterologous pathway introduction, as well as the incorporation of plant-specific terpene synthase have already increased terpenoid yields. In parallel, cultivation improvements, high-density fermentation, and extraction techniques including supercritical fluid extraction have also developed microalgal production systems. Many other metabolic bottlenecks involve enzyme-stability, strain robustness, and the cultivation and downstream processing that are also too costly. Such barriers will need to be overcome by systems biology, state of the art tech applications, coupled with environmentally benign processing technologies. In summary, increased terpenoid in microalgae will integrate sustainable answers to handle increased world demand. Microalgal platforms will achieve this and further innovation will go a long way to fill the space between therapeutic potential and commercial feasibility, becoming a definitive asset of the bioeconomy.

Acknowledgments The author(s) extend special thanks to the DBT for the grant BT/PR54521/BSA/33/308/2024.

References

Abbas F, Ke Y, Yu R, Yue Y, Amanullah S, Jahangir MM, Fan Y (2017) Volatile terpenoids: multiple functions, biosynthesis, modulation and manipulation by genetic engineering. Planta 246(5):803–816

Ahuja S, Roy A, Kumar L, Bharadvaja N (2020) Media optimization using box Behnken design for enhanced production of biomass, beta-carotene and lipid from Dunaliella salina. Vegetos 33(1): 31–39

Ali HEA, El-Fayoumy EA, Rasmy WE, Soliman RM, Abdullah MA (2021) Two-stage cultivation of Chlorella vulgaris using light and salt stress conditions for simultaneous production of lipid, carotenoids, and antioxidants. J Appl Phycol 33(1):227–239

Alzahmi AS, Daakour S, Nelson D, Al-Khairy D, Twizere JC, Salehi-Ashtiani K (2024) Enhancing algal production strategies: strain selection, AI-informed cultivation, and mutagenesis. Front Sustain Food Syst 8:1331251

Antony T, Chakraborty K (2020a) Anti-inflammatory polyether triterpenoids from the marine macroalga Gracilaria salicornia: newly described natural leads attenuate pro-inflammatory 5-lipoxygenase and cyclooxygenase-2. Algal Res 47:101791. https://doi.org/10.1016/j.algal.2020.101791

Antony T, Chakraborty K (2020b) First report of antioxidant abeo-labdane type diterpenoid from intertidal red seaweed Gracilaria salicornia with 5-lipoxygenase inhibitory potential. Nat Prod Res 34(10):1409–1416. https://doi.org/10.1080/14786419.2018.1508150

Arendt P, Pollier J, Callewaert N, Goossens A (2016) Synthetic biology for production of natural and new-to-nature terpenoids in photosynthetic organisms. Plant J 87(1):16–37. https://doi.org/10.1111/tpj.13138

Assunção J, Amaro HM, Tavares T, Malcata FX, Guedes AC (2023) Effects of temperature, pH, and NaCl concentration on biomass and bioactive compound production by Synechocystis salina. Life 13(1):187. https://doi.org/10.3390/life13010187

Ávila-Román J, García-Gil S, Rodríguez-Luna A, Motilva V, Talero E (2021) Anti-inflammatory and anticancer effects of microalgal carotenoids. Mar Drugs 19(10):10. https://doi.org/10.3390/md19100531

Balasubramaniam V, Gunasegavan RDN, Mustar S, Lee JC, Noh MFM (2021) Isolation of industrial important bioactive compounds from microalgae. Molecules. https://doi.org/10.3390/MOLECULES26040943

Bankar AA, Nagulwar VP, Kotagale NR, Inamdar NN (2024) Neuroprotective prospectives of triterpenoids. Explorat Neurosci 3(3):Article 3. https://doi.org/10.37349/en.2024.00047

Berthelot K, Estevez Y, Deffieux A, Peruch F (2012) Isopentenyl diphosphate isomerase: a checkpoint to isoprenoid biosynthesis. Biochimie 94(8):1621–1634. https://doi.org/10.1016/j.biochi.2012.03.021

Bonnefond H, Moelants N, Talec A, Mayzaud P, Bernard O, Sciandra A (2017) Coupling and uncoupling of triglyceride and beta-carotene production by Dunaliella salina under nitrogen limitation and starvation. Biotechnol Biofuels 10(1). https://doi.org/10.1186/s13068-017-0713-4

Bossa R, Di Colandrea M, Salbitani G, Carfagna S (2024) Phosphorous utilization in microalgae: physiological aspects and applied implications. Plants 13(15):2127. https://doi.org/10.3390/plants13152127

Böttger A, Vothknecht U, Bolle C, Wolf A (2018) Terpenes and terpenoids. In: Lessons on caffeine, Cannabis & co: plant-derived drugs and their interaction with human receptors. Springer International Publishing, Cham, pp 153–170

Burg JS, Espenshade PJ (2011) Regulation of HMG-CoA reductase in mammals and yeast. Prog Lipid Res 50(4):403–410. https://doi.org/10.1016/j.plipres.2011.07.002

Câmara RB et al (2024) Antioxidant and anti-inflammatory effects of terpenoids: current evidence and future perspectives. Front Pharmacol 15:116723

Chanda S, Ramachandra TV (2019) A review on some therapeutic aspects of phytochemicals present in medicinal plants. Int J Pharm Life Sci 10(1)

Chemat F, Vian MA, Cravotto G (2012) Green extraction of natural products: concept and principles. Int J Mol Sci 13(7):8615–8627. https://doi.org/10.3390/ijms13078615

Chen L, Pang Y, Luo Y, Cheng X, Lv B, Li C, Li C (2021) Separation and purification of plant terpenoids from biotransformation. Eng Life Sci. https://doi.org/10.1002/ELSC.202100014

Chen B, Li Y, Li W, Ye S, Zhu L, Ding Y (2024) Antitumor activity and mechanism of Terpenoids in seaweeds based on literature review and network pharmacology. Adv Biol 8(3):2300541. https://doi.org/10.1002/adbi.202300541

Cordoba E, Porta H, Arroyo A, San Román C, Medina L, Rodríguez-Concepción M, León P (2011) Functional characterization of the three genes encoding 1-deoxy-D-xylulose 5-phosphate synthase in maize. J Exp Bot 62(6):2023–2038. https://doi.org/10.1093/jxb/erq393
Cui H, Xu W, Zhu X, Zhao C, Cui Y, Ji C, Zhang C, Xue J, Qin S, Jia X, Li R (2021) Characterization of a Haematococcus pluvialis Diacylglycerol acyltransferase 1 and its potential in unsaturated fatty acid-rich triacylglycerol production. Front Plant Sci 12:771300. https://doi.org/10.3389/fpls.2021.771300
Dahmen-Ben Moussa I, Chtourou H, Karray F, Sayadi S, Dhouib A (2017) Nitrogen or phosphorus repletion strategies for enhancing lipid or carotenoid production from Tetraselmis marina. Bioresour Technol 238:325–332. https://doi.org/10.1016/j.biortech.2017.04.008
Das PK, Rani J, Rawat S, Kumar S (2022) Microalgal co-cultivation for biofuel production and bioremediation: current status and benefits. Bioenergy Res 15(1):1–26. https://doi.org/10.1007/s12155-021-10254-8
Das DD, Sharma N, Chawla V, Chawla PA (2024) Current trends of analytical techniques for bioactive Terpenoids: A review. Crit Rev Anal Chem 54(8):2984–3000. https://doi.org/10.1080/10408347.2023.2219757
Davies FK, Jinkerson RE, Posewitz MC (2015) Toward a photosynthetic microbial platform for terpenoid engineering. Photosynth Res 123(3):265–284. https://doi.org/10.1007/s11120-014-9979-6. Kluwer Academic Publishers
De Andrade Moura L, Marqui de Almeida AC, Domingos TFS, Ortiz-Ramirez F, Cavalcanti DN, Teixeira VL, Fuly AL (2014) Antiplatelet and anticoagulant effects of diterpenes isolated from the Marine Alga, Dictyota menstrualis. Marine Drugs 12(5):Article 5. https://doi.org/10.3390/md12052471
DeColli AA, Nemeria NS, Majumdar A, Gerfen GJ, Jordan F, Freel Meyers CL (2018) Oxidative decarboxylation of pyruvate by 1-deoxy-D-xyulose 5-phosphate synthase, a central metabolic enzyme in bacteria. J Biol Chem 293(28):10857–10869. https://doi.org/10.1074/jbc.RA118.001980
Dienst D, Wichmann J, Mantovani O, Rodrigues JS, Lindberg P (2019) High density cultivation for efficient sesquiterpenoid biosynthesis in Synechocystis sp. PCC 6803. bioRxiv. https://doi.org/10.1101/834499
Ding W, Cui J, Zhao Y, Han B, Li T, Zhao P, Xu J-W, Yu X (2019) Enhancing Haematococcus pluvialis biomass and γ-aminobutyric acid accumulation by two-step cultivation and salt supplementation. Bioresour Technol 285:121334. https://doi.org/10.1016/j.biortech.2019.121334
Eilers U, Bikoulis A, Breitenbach J, Büchel C, Sandmann G (2016) Limitations in the biosynthesis of fucoxanthin as targets for genetic engineering in Phaeodactylum tricornutum. J Appl Phycol 28(1):123–129. https://doi.org/10.1007/s10811-015-0583-8
Elloumi W, Jebali A, Maalej A, Chamkha M, Sayadi S (2020) Effect of mild salinity stress on the growth, fatty acid and carotenoid compositions, and biological activities of the thermal freshwater microalgae Scenedesmus sp. Biomolecules 10(11):1515. https://doi.org/10.3390/biom10111515
Fabris M, George J, Kuzhiumparambil U, Lawson CA, Jaramillo-Madrid AC, Abbriano RM, Vickers CE, Ralph P (2020) Extrachromosomal genetic engineering of the marine diatom Phaeodactylum tricornutum enables the heterologous production of Monoterpenoids. ACS Synth Biol 9(3):598–612. https://doi.org/10.1021/acssynbio.9b00455
Farzan M, Farzan M, Shahrani M, Navabi SP, Vardanjani HR, Amini-Khoei H, Shabani S (2024) Neuroprotective properties of Betulin, Betulinic acid, and Ursolic acid as triterpenoids derivatives: a comprehensive review of mechanistic studies. Nutr Neurosci 27(3):223–240. https://doi.org/10.1080/1028415X.2023.2180865
Ferdous UT, Yusof ZNB, Ferdous UT, Yusof ZNB (2021) Algal terpenoids: a potential source of antioxidants for cancer therapy. In: Terpenes and terpenoids—recent advances. IntechOpen. https://doi.org/10.5772/intechopen.94122

Ferreira JPA, Grácio M, Sousa I, Pagarete A, Nunes MC, Raymundo A (2023) Tuning the bioactive properties of Dunaliella salina water extracts by ultrasound-assisted extraction. Mar Drugs 21(9):472. https://doi.org/10.3390/md21090472

Filali R, Tian H, Micheils E, Taidi B (2021) Evaluation of the growth performance of microalgae based on fine pH changes. Austin J Biotechnol Bioeng 8(1)

Fu L, Yan G, Li Y, Li Q, Zhou D (2020) Phosphorus supply via a fed-batch strategy improves lipid heterotrophic production of Chlorella regularis. Environ Sci Pollut Res 27:31677–31685. https://doi.org/10.1007/s11356-020-09495-5

Gao Q, Xiao W (2025) Advances in the biosynthesis of monoterpenes by yeast. Synth Biol J 6(2): 357–372

Georgiopoulou I, Louli V, Magoulas K (2023a) Comparative study of conventional, microwave-assisted and supercritical fluid extraction of bioactive compounds from microalgae: the case of Scenedesmus obliquus. Separations 10(5):290. https://doi.org/10.3390/separations10050290

Georgiopoulou I, Tzima S, Louli V, Magoulas K (2023b) Process optimization of microwave-assisted extraction of chlorophyll, carotenoid and phenolic compounds from Chlorella vulgaris and comparison with conventional and supercritical fluid extraction. Appl Sci 13(4):2740. https://doi.org/10.3390/app13042740

Grechanik V, Romanova A, Naydov I, Tsygankov A (2020) Photoautotrophic cultures of Chlamydomonas reinhardtii: sulfur deficiency, anoxia, and hydrogen production. Photosynth Res 143(3):275–286. https://doi.org/10.1007/s11120-019-00701-1

Guan J, Shen S, Wu H, Liu X, Shen W, He Y, Duan R (2017) Biomass and terpenoids produced by mutant strains of Arthrospira under low temperature and light conditions. World J Microbiol Biotechnol 33(2):33. https://doi.org/10.1007/s11274-016-2199-9

Hale I, O'Neill PM, Berry NG, Odom A, Sharma R (2012) The MEP pathway and the development of inhibitors as potential anti-infective agents. MedChemComm 3(4):418–433. https://doi.org/10.1039/c2md00298a

He H (2025) Potential and challenges of Terpenoids in cancer therapy: mechanistic review and future perspectives. Theoret Natural Sci. https://doi.org/10.54254/2753-8818/2024.la19342

Herold RA, Bryan SJ (2020) Engineered Terpenoid production in Synechocystis sp. PCC 6803 under different growth conditions. bioRxiv. https://doi.org/10.1101/2020.08.19.256149

Hillier SG, Lathe R (2019) Terpenes, hormones and life: isoprene rule revisited. J Endocrinol 242(2):R9–R22. https://doi.org/10.1530/JOE-19-0084. BioScientifica Ltd

Hoqani UA (2017) Metabolic engineering of the algal chloroplast for terpenoid production

Hoshino Y, Villanueva L (2023) Four billion years of microbial terpenome evolution. FEMS Microbiol Rev 47(2):fuad008

Huang PW, Wang LR, Geng SS, Ye C, Sun XM, Huang H (2021) Strategies for enhancing terpenoids accumulation in microalgae. Appl Microbiol Biotechnol 105(12):4919–4930. https://doi.org/10.1007/s00253-021-11368-x. Springer Science and Business Media Deutschland GmbH

Jahangeer M, Fatima R, Ashiq M, Basharat A, Qamar SA, Bilal M, Iqbal HMN (2021) Therapeutic and biomedical potentialities of Terpenoids—a review. J Pure Appl Microbiol. https://doi.org/10.22207/JPAM.15.2.04

Jiang Z, Kempinski C, Chappell J (2016) Extraction and analysis of terpenes/terpenoids. Curr Protoc Plant Biol 1(2):345–358

Khalil ZI, Asker MMS, El-Sayed S, Kobbia IA (2010) Effect of pH on growth and biochemical responses of Dunaliella bardawil and Chlorella ellipsoidea. World J Microbiol Biotechnol 26(7): 1225–1231. https://doi.org/10.1007/s11274-009-0292-z

Kharkongor D, Ramanujam P (2015) Spatial and temporal variation of carotenoids in four species ofTrentepohlia (Trentepohliales, Chlorophyta). J Bot 2015:1–8. https://doi.org/10.1155/2015/201641

Kinghorn, A. D., Falk, H., Gibbons, S., Kobayashi, J., Asakawa, Y., & Liu, J.-K. (Eds.). (2020). Progress in the chemistry of organic natural products 111. Springer International Publishing https://doi.org/10.1007/978-3-030-37865-3

Kopp G, Lauritano C (2025) Greener extraction solutions for microalgal compounds. Mar Drugs 23(7):269. https://doi.org/10.3390/md23070269
Kumar S (2018) Effect of salinity and pH ranges on the growth and biochemical composition of marine microalga- nannochloropsis Salina. Int J Agric Environ Biotechnol 11(4). https://doi.org/10.30954/0974-1712.08.2018.6
Kumar A (2024) Terpenoids as anticancer agents: mechanisms and clinical perspectives. Cancer Lett 567:112–126
Kumari M, Checker VG, Kathpalia R, Srivastava V, Singh IK, Singh A (2024) Metabolic engineering for enhanced terpenoid production: leveraging new horizons with an old technique. Plant Physiol Biochem. https://doi.org/10.1016/j.plaphy.2024.108511
Lange BM, Rujan T, Martin W, Croteau R (2000) Isoprenoid biosynthesis: the evolution of two ancient and distinct pathways across genomes. Proc Natl Acad Sci USA 97(24):13172–13177. https://doi.org/10.1073/pnas.240454797
Lanzotti V (2013) Diterpenes for therapeutic use. In: Ramawat KG, Mérillon J-M (eds) Natural products. Springer Berlin Heidelberg, pp 3173–3191. https://doi.org/10.1007/978-3-642-22144-6_192
Lin P-C, Pakrasi HB (2019) Engineering cyanobacteria for production of terpenoids. Planta. https://doi.org/10.1007/S00425-018-3047-Y
Liu H, Zhang Y, Sun S, Wang S (2019) Efficacy of Terpenoid in attenuating aortic atherosclerosis in Apolipoprotein-E deficient mice: a meta-analysis of animal studies. Biomed Res Int 2019: 2931831. https://doi.org/10.1155/2019/2931831
Lowe H, Ali A, Steele B, Gordon L, Grant J (2024) The potential therapeutic value of terpenes. INNOSC Theranost Pharmacol Sci. https://doi.org/10.36922/itps.0332
Lu X, Tang K, Li P (2016) Plant metabolic engineering strategies for the production of pharmaceutical Terpenoids. Front Plant Sci. https://doi.org/10.3389/FPLS.2016.01647
Ludwiczuk A, Skalicka-Woźniak K, Georgiev MI (2017) Terpenoids. In: Pharmacognosy. Elsevier, pp 233–266. https://doi.org/10.1016/B978-0-12-802104-0.00011-1
Ma R, Zhao X, Xie Y, Ho S-H, Chen J (2019) Enhancing lutein productivity of Chlamydomonas sp. via high-intensity light exposure with corresponding carotenogenic genes expression profiles. Bioresour Technol 275:416–420. https://doi.org/10.1016/j.biortech.2018.12.109
Mahizan NA, Yang S-K, Moo C-L, Song AA-L, Chong C-M, Chong C-W, Abushelaibi A, Lim S-HE, Lai K-S (2019) Terpene derivatives as a potential agent against antimicrobial resistance (AMR) pathogens. Molecules 24(14):2631. https://doi.org/10.3390/molecules24142631
Maltsev Y, Maltseva K, Kulikovskiy M, Maltseva S (2021) Influence of light conditions on microalgae growth and content of lipids, carotenoids, and fatty acid composition. Biology 10(10):1060. https://doi.org/10.3390/biology10101060
Mani V, Park S, Kim JA, Lee SI, Lee K (2021) Metabolic perturbation and synthetic biology strategies for plant Terpenoid production—an updated overview. Plants. https://doi.org/10.3390/PLANTS10102179
Matin M, Koszarska M, Atanasov AG, Król-Szmajda K, Jóźwik A, Stelmasiak A, Hejna M (2024) Bioactive potential of algae and algae-derived compounds: focus on anti-inflammatory, antimicrobial, and antioxidant effects. Molecules 29(19):Article 19. https://doi.org/10.3390/molecules29194695
Molino A, Mehariya S, Di Sanzo G, Larocca V, Martino M, Leone GP, Marino T, Chianese S, Balducchi R, Musmarra D (2020) Recent developments in supercritical fluid extraction of bioactive compounds from microalgae: role of key parameters, technological achievements and challenges. J CO2 Util 36:196–209. https://doi.org/10.1016/j.jcou.2019.11.014
Moser S, Pichler H (2019) Identifying and engineering the ideal microbial terpenoid production host. Appl Microbiol Biotechnol. https://doi.org/10.1007/S00253-019-09892-Y
Mosquera MEG, Jiménez G, Tabernero V, Vinueza-Vaca J, García-Estrada C, Kosalková K, Sola-Landa A, Monje B, Acosta C, Alonso R, Valera MÁ (2021) Terpenes and Terpenoids: building blocks to produce biopolymers. Sustain Chem 2(3):467–492. https://doi.org/10.3390/suschem2030026

Mulders KJM, Lamers PP, Martens DE, Wijffels RH (2014) Phototrophic pigment production with microalgae: biological constraints and opportunities. J Phycol 50(2):229–242. https://doi.org/10.1111/jpy.12173

Neumann U, Derwenskus F, Flaiz Flister V, Schmid-Staiger U, Hirth T, Bischoff SC (2019) Fucoxanthin, A carotenoid derived from Phaeodactylum tricornutum exerts Antiproliferative and antioxidant activities in vitro. Antioxidants 8(6):Article 6. https://doi.org/10.3390/antiox8060183

Powder-George YL (2024) Terpenoids. https://doi.org/10.1016/b978-0-443-18657-8.00006-2

Qingping Z (2006) Genetic manipulation on biosynthesis of Terpenoids. China Biotechnol

Rajesh T et al (2017) Carotenoid production by Dunaliella salina: industrial and therapeutic significance. J Appl Phycol 29(1):1–12

Ras M, Steyer J-P, Bernard O (2013) Temperature effect on microalgae: a crucial factor for outdoor production. Rev Environ Sci Biotechnol 12(2):153–164. https://doi.org/10.1007/s11157-013-9310-6

Rautela A, Kumar S (2022) Engineering plant family TPS into cyanobacterial host for terpenoids production. Plant Cell Rep. https://doi.org/10.1007/s00299-022-02892-9

Rehman MU, Abdullah, Khan F, Niaz K (2020) Introduction to natural products analysis. In: Recent advances in natural products analysis. Elsevier, pp 3–15. https://doi.org/10.1016/B978-0-12-816455-6.00001-9

Shataer D, Chang Y, Obul M, Aierken K, Liu H (2025) An up-to-date review on the classification, pharmacology, and production of terpenes and Terpenoids. Curr Org Chem 29(19):1508–1522. https://doi.org/10.2174/0113852728373242250319054222

Shi T-Q, Wang L-R, Zhang Z-X, Sun X-M, Huang H (2020) Stresses as first-line tools for enhancing lipid and carotenoid production in microalgae. Front Bioeng Biotechnol 8. https://doi.org/10.3389/fbioe.2020.00610

Shukla V, Phulara SC (2021) Impact of culture condition modulation on the high-yield, high-specificity, and cost-effective production of Terpenoids from microbial sources: a review. Appl Environ Microbiol 87(4):e02369–e02320. https://doi.org/10.1128/AEM.02369-20

Shukla V, Gupta P, Phulara SC (2023) Metabolic engineering for the biosynthesis of terpenoids from microbial cell factories. https://doi.org/10.1007/978-3-031-40082-7_4

Siddiqui SA, Ali Redha A, Salauddin M, Harahap IA, Rupasinghe HPV (2025) Factors affecting the extraction of (poly)phenols from natural resources using deep eutectic solvents combined with ultrasound-assisted extraction. Crit Rev Anal Chem 55(1):139–160. https://doi.org/10.1080/10408347.2023.2266846

Silva J, Alves C, Freitas R, Martins A, Pinteus S, Ribeiro J, Gaspar H, Alfonso A, Pedrosa R (2019) Antioxidant and neuroprotective potential of the Brown seaweed Bifurcaria bifurcata in an in vitro Parkinson's disease model. Mar Drugs 17(2):Article 2. https://doi.org/10.3390/md17020085

Simeis DD, Serra S, Fonzo AD, Secundo F (2021) Oxidation of Terpenoids to achieve high-value flavor and fragrances—questioning microalgae oxidative capabilities in the biotransformation of the Sesquiterpene Valencene and of selected natural Apocarotenoids. Chemistry. https://doi.org/10.3390/CHEMISTRY3030059

Su Y (2021) Revisiting carbon, nitrogen, and phosphorus metabolisms in microalgae for wastewater treatment. Sci Total Environ 762:144590. https://doi.org/10.1016/j.scitotenv.2020.144590

Takagi M, Karseno, Yoshida T (2006) Effect of salt concentration on intracellular accumulation of lipids and triacylglyceride in marine microalgae Dunaliella cells. J Biosci Bioeng 101(3):223–226. https://doi.org/10.1263/jbb.101.223

Tavakoli S, Hong H, Wang K, Yang Q, Gahruie HH, Zhuang S, Li Y, Liang Y, Tan Y, Luo Y (2021) Ultrasonic-assisted food-grade solvent extraction of high-value added compounds from microalgae Spirulina platensis and evaluation of their antioxidant and antibacterial properties. Algal Res 60:102493. https://doi.org/10.1016/j.algal.2021.102493

Tripathi U, Sarada R, Ravishankar GA (2002) Effect of culture conditions on growth of green alga—Haematococcus pluvialis and astaxanthin production. Acta Physiol Plant 24(3):323–329. https://doi.org/10.1007/s11738-002-0058-9

Tripathi SK, Biswal S, Panda M, Biswal BK (2023) Terpenoids A potential scaffold for cancer therapy: a mechanistic approach. https://doi.org/10.2174/9789815124699123010014

Ucci M, Di Tomo P, Tritschler F, Cordone VGP, Lanuti P, Bologna G, Di Silvestre S, Di Pietro N, Pipino C, Mandatori D, Formoso G, Pandolfi A (2019) Anti-inflammatory role of carotenoids in endothelial cells derived from umbilical cord of women affected by gestational diabetes mellitus. Oxidative Med Cell Longev 2019(1):8184656. https://doi.org/10.1155/2019/8184656

Vavitsas K, Fabris M, Vickers CE (2018) Terpenoid metabolic engineering in photosynthetic microorganisms. Genes 9(11):520. https://doi.org/10.3390/genes9110520. Multidisciplinary Digital Publishing Institute

Velmurugan A, Kodiveri Muthukaliannan G (2022) Genetic manipulation for carotenoid production in microalgae an overview. Curr Res Biotechnol 4:221–228. https://doi.org/10.1016/j.crbiot.2022.03.005

Vyhlídalová B, Ondrová K, Zůvalová I (2025) Dietary monoterpenoids and human health: unlocking the potential for therapeutic use. Biochimie 228:89–100. https://doi.org/10.1016/j.biochi.2024.09.002

Wang G, Tang W, Bidigare RR (2005) Terpenoids as therapeutic drugs and pharmaceutical agents. https://doi.org/10.1007/978-1-59259-976-9_9

Wang C, Liwei M, Park J-B, Jeong S-H, Wei G, Wang Y, Kim S-W (2018) Microbial platform for Terpenoid production: Escherichia coli and yeast. Front Microbiol. https://doi.org/10.3389/FMICB.2018.02460

Wang S, Meng D, Feng M, Li C, Wang Y (2024) Efficient plant triterpenoids synthesis in Saccharomyces cerevisiae: from mechanisms to engineering strategies. ACS Synth Biol. https://doi.org/10.1021/acssynbio.4c00061

Wang J, Ji X, Rong Y, Li D, Shen X, Liu Z, Xia Y, Shi S (2025) Heterologous biosynthesis of Terpenoids in Saccharomyces cerevisiae. Biotechnol J. https://doi.org/10.1002/biot.202400712

Withers ST, Gottlieb SS, Lieu B, Newman JD, Keasling JD (2007) Identification of isopentenol biosynthetic genes from Bacillus subtilis by a screening method based on isoprenoid precursor toxicity. Appl Environ Microbiol 73(19):6277–6283. https://doi.org/10.1128/AEM.00861-07

Xia Y, Li C, Cao R, Qin L, Shi S (n.d.) Development and perspective of production of terpenoids in yeast. Synthetic Biol Eng. https://doi.org/10.35534/sbe.2024.10003

Yamaguchi T (2022) Antibacterial effect of the combination of terpenoids. Arch Microbiol 204(8):520. https://doi.org/10.1007/s00203-022-03142-y

Yang Y, Sun X, Zhao Y (2017) Effects of Lugol's iodine solution and formalin on cell volume of three bloom-forming dinoflagellates. Chin J Oceanol Limnol 35(4):858–866. https://doi.org/10.1007/s00343-017-5378-0

Yao X, Liu X (2022) Cardiometabolic health benefits of terpenoids. Front Nutr 9:942560

Yazdani M, Croen MG, Fish TL, Thannhauser TW, Ahner BA (2021) Overexpression of native ORANGE (OR) and OR mutant protein in Chlamydomonas reinhardtii enhances carotenoid and ABA accumulation and increases resistance to abiotic stress. Metab Eng 68:94–105. https://doi.org/10.1016/j.ymben.2021.09.006

Ying K, Gilmour D, Zimmerman W (2015) Periodic CO2 dosing strategy for Dunaliella salina batch culture. Int J Mol Sci 16(5):11509–11521. https://doi.org/10.3390/ijms160511509

Yu X, Chen L, Zhang W (2015) Chemicals to enhance microalgal growth and accumulation of high-value bioproducts. Front Microbiol 6. https://doi.org/10.3389/fmicb.2015.00056

Zbakh H, Zubía E, De Los Reyes C, Calderón-Montaño JM, Motilva V (2020) Anticancer activities of Meroterpenoids isolated from the Brown alga Cystoseira usneoides against the human colon cancer cells HT-29. Foods 9(3):Article 3. https://doi.org/10.3390/foods9030300

Zeng W, Li H, Liu S, Luo Z, Chen J, Zhou J (2025) Biosynthesis and bioactivities of triterpenoids from Centella asiatica: challenges and opportunities. Biotechnol Adv 80:108541. https://doi.org/10.1016/j.biotechadv.2025.108541

Zhang L, Lu S (2017) Overview of medicinally important diterpenoids derived from plastids. Mini Rev Med Chem 17(12):988–1001

Zhang JY, Jeon H, Sim SJ, Lee Y, Jin E (2015) Homologous sense and antisense expression of a gene in Dunaliella tertiolecta. Planta 242(4):1051–1058. https://doi.org/10.1007/s00425-015-2369-2

Zhang Y, Nielsen J, Nielsen J, Nielsen J, Liu Z (2017) Engineering yeast metabolism for production of terpenoids for use as perfume ingredients, pharmaceuticals and biofuels. FEMS Yeast Res. https://doi.org/10.1093/FEMSYR/FOX080

Zhang W, Zhou X, Zhang Y, Cheng P, Ma R, Cheng W, Chu H (2018) Enhancing astaxanthin accumulation in Haematococcus pluvialis by coupled light intensity and nitrogen starvation in column photobioreactors. J Microbiol Biotechnol 28(12):2019–2028. https://doi.org/10.4014/jmb.1807.07008

Zhao M, Cai W, Zheng S-Q, Zhao J, Zhang J, Huang Y, Jia B (2022) Metabolic engineering of the Isopentenol utilization pathway enhanced the production of Terpenoids in Chlamydomonas reinhardtii. Mar Drugs. https://doi.org/10.3390/md20090577

Zhao K, Zhou T, Yang J, Li Y, Qin J, Wang S, Li D, Chen J, Zheng WV (2023) Lutein shows a protective effect against the aging of mesenchymal stem cells by downregulating inflammation. Int Immunopharmacol 116:109749. https://doi.org/10.1016/j.intimp.2023.109749

Diatoms as Hosts and Ghosts for the Engineering of Value-Added Products: Scope, Challenges, and Advances

9

Vaishnavi Newaskar and Gayatri Dave

Abstract

Microalgae have emerged as promising hosts for product development due to their ease of cultivation and minimal nutritional requirements. Among these, diatoms are unicellular algae belonging to the class Bacillariophyceae. They stand out due to their unique nanoporous silica cell walls (frustules), ecological versatility, and capacity for high-value compound synthesis. Diatoms contribute significantly to global oxygen production and serve as ecological indicators, supporting environmental monitoring, paleolimnology, and forensic science research. Due to their inherent synthesis of polyunsaturated fatty acids, carotenoids, sterols, and terpenoids, as well as their capacity to produce bio-silica nanostructures, these organisms are considered highly suitable for applications in drug delivery, biosensing, biofuel development, and recombinant protein production.

This chapter presents an overview of recent advances in using diatoms as both living (hosts) and nonliving (ghosts) systems. It highlights the availability of genomic, transcriptomic, and proteomic resources for model species such as *Phaeodactylum tricornutum* and *Thalassiosira pseudonana,* and gaps in omics data for many other diatom taxa. It examines challenges in genome editing tools like CRISPR-Cas9 and TALENs, gene delivery methods, and the transformation vectors and promoter systems. This chapter further explores applications of genetically engineered frustules in biomedical scaffolding, environmental monitoring, biofuel production, nanotechnological applications, targeted drug delivery, cancer therapy, biodegradation, and biosensing. Also, it explains about the

V. Newaskar · G. Dave (✉)
Department of Biological Sciences, P D Patel Institute of Applied Sciences, CHARUSAT University, Changa, Gujarat, India
e-mail: gayatridave.bt@charusat.ac.in

K. Mohanty et al. (eds.), *Advances in Algal Research*,
https://doi.org/10.1007/978-981-95-7234-2_9

challenges and future perspectives of gene editing, and materials engineering places diatoms at the forefront of algal biotechnology. Their dual role as both living factories and structurally unique organisms opens transformative avenues for sustainable innovation across health, energy, and environmental sectors.

Keywords

Microalgae-based bioproduction · Diatom frustules · CRISPR-Cas9 · Nanobiotechnology · Recombinant protein expression · Diatom engineering

9.1 Introduction

Genetic manipulation of microbial cells to produce valuable products and engineering bioinspired nanomaterials have become essential frontiers in biotechnology. These techniques enable synthesizing smart proteins, bioactive compounds, secondary metabolites, and enzymes. Traditionally, bacteria and fungi have been the primary organisms employed for these purposes. Recently, microalgae have emerged as promising hosts for product development due to their ease of cultivation and minimal nutritional requirements. Diatoms are unicellular microalgae belonging to the class Bacillariophyceae and play a significant role in global ecosystems, contributing approximately 20–25% of the Earth's oxygen supply (Hopes et al. 2016). Notably, diatoms have garnered special attention among all microalgae because of their intricate, nanoporous, silica-based cell walls. They are remarkably adaptable, thriving in diverse environments, including marine, freshwater, brackish areas, soil, and atmospheric habitats (Romero-Guzmán et al. 2021). Their ecological significance extends to their function as sensitive indicators of environmental perturbation, making them invaluable in studies of ecosystem health. Their diversity, sensitivity, and abundance allow for their use in limnological surveys, biomonitoring, and paleolimnological research, providing insights into acidification, eutrophication, and climate change (Anderson 2000). Additionally, they are utilized in forensic investigations, particularly in drowning cases, through the diatom test, which detects their silicified frustules in organs or clothing (Horton 2007).

Diatoms have garnered significant interest in biotechnology due to their potential as hosts for producing polyunsaturated fatty acids and carotenoids. They are capable of producing a diverse set of bioactive compounds, including sterols, hydrocarbons, tocopherols, terpenes, quinines, chlorophylls, and xanthophylls (secondary metabolites) in much shorter durations and with less space than plants. Diatom ghost cells, known as frustules of diatoms, are organic content-free, empty, porous silica shells that are used for drug delivery and other biomedical applications. It can be functionalized chemically or genetically for various applications. The diatom biosilica is considered an eco-friendly alternative for targeted drug delivery of cancer therapeutics. Their diverse biological activities make them valuable for pharmaceutical applications, nutraceuticals, and industrial biotechnology (Marella et al. 2021). Advances in genetic modification techniques have further enhanced their utility, allowing for the exploration of their genomic features and metabolic pathways.

The first section of this chapter presents a brief account of the available omics resources for diatom research. The second segment encapsulates the genetic modifications in diatoms and the scope of frustule functionalization and doping for various applications, along with the previous research efforts, ongoing progress, and the challenges faced in scaling up diatom-based technologies, offering a comprehensive outlook on their future role in biotechnology.

9.2 Availability of "Omics" Resources for Diatoms

The diatom is an eukaryote, a member of the Heterokontophyta, also known as heterokont algae. There are approximately 20,000—two million species of diatoms. More than 27,000 species have been identified, and scientists continue to discover new species yearly. While species richness and widespread presence are compared, the available information is limited.

9.2.1 Genomic Resources

The diversity of diatom species offers specialized models for different research areas, such as *Thalassiosira pseudonana,* which is widely used to study silica biomineralization, *Phaeodactylum tricornutum* for carbohydrate and lipid metabolism, *Cylindrotheca fusiformis* for silicification and ecophysiology, *Pseudonitzschia* and *Chaetoceros* for harmful algal blooms, and *Fragilariopsis cylindrus* for polar marine adaptation (Huang and Daboussi 2017). Table 9.1 summarizes the details on the availability of genomic resources for diatoms.

Phaeodactylum tricornutum and *Thalassiosira pseudonana* have fully sequenced genome and have gained momentum as hosts for protein production. *Paralia guyana*, has the largest known diatom genome to date. It is worth noting that the limited resolution of common molecular markers, such as 18S rDNA, 28S rDNA, ITS, rbcl, and cox1, has hindered a thorough assessment of diatom biodiversity. Organelle genomes, including mitochondrial and chloroplast genomes, are proposed as "super barcodes" for differentiating diatom species due to their extensive genomic content and rapid DNA sequencing technologies (Wang et al. 2022). The NCBI Organelle Database contains organelle genomes for about 151 distinct diatom taxonomic groups. The NCBI genome database and PhycoCosm provide whole-genome sequence data for only 20 diatom taxa.

9.2.2 Transcriptome Data Availability for Diatoms

Diatoms are the most prevalent group in oceanic and freshwater ecosystems, and their functional roles, illuminated through transcriptomics, are essential for global carbon cycling models, climate change studies, and marine food web research. The first global transcriptomic profiling using microarrays was published in 2008,

Table 9.1 Genomic resources on diatoms

Species	Morphology	Size of genome	No of genes	Genes annotated	References
P. tricornutum	Pennate	24.5 Mb	10,402	Yes	Bowler et al. (2008)
T. pseudonana	Centric	34 Mb	11,242	Yes	Armbrust et al. (2004)
C. cryptica	Centric	171 Mb	21,250 (predicted)	No	Roberts et al. (2020)
T. weissflogii	Centric	51 Mb	23,362	Yes	Cheng et al. (2021)
C. gracilis	Centric	35.4 Mb	15,484 (predicted)	No	Kumazawa et al. (2022)
C. fusiformis	Pennate	48.2 Mb	NA	No	Genome assembly ASM1969352v1
S. tropicum	Centric	78.78 Mb	17,757	Yes (86.14%)	Liu and Chen (2024)
E. pelagica	Pennate	60.3 Mb	130.75 Kb chloroplast 40.08 Kb mitochondria	NA	Schvarcz et al. (2024)
F. cylindrus	Pennate	61.1 Mb	NA	NA	Mock et al. (2017)
P. guyana	Centric	558.85 Mb	22,904 (predicted)	Yes (84.45%)	Jian et al. (2024)
F. crotonensis	Pennate	NA	26,015 (predicted)	Yes (81%)	Zepernick et al. (2022)
C. weissflogii	Centric	231 Mb	NA	NA	Li et al. (2021)
P. inscriptura	Pennate	NA	38,013 bp mitochondria	NA	Jeong and Lee (2021)

detailing responses of *P. tricornutum* to nitrate starvation and recovery (Maheswari et al. 2010). The Marine Microbial Eukaryote Transcriptome Sequencing Project (MMETSP) generated RNA-Seq datasets for numerous diatom species, expanding taxonomic coverage. DiatOmicBase (Villar et al. 2025) is a web tool that provides gene annotations, transcriptomic profiles, and a genome browser for *P. tricornutum* and *T. pseudonana*.

The transcriptomic and proteomic analyses of various diatom species have illuminated key processes; for instance, Phosphorus (P) limitation triggers diverse diatom responses, with an adaptation in nutrient acquisition and metabolic reprogramming. *Skeletonema costatum* upregulates genes related to P utilization, nucleotide metabolism, glycolysis, and photosynthesis under P deficiency. *Thalassiosira pseudonana* responds with increased polyphosphate production, enhanced expression of P transporters, metalloenzyme activity, and membrane alterations through sulfolipid synthesis. *Phaeodactylum tricornutum* upregulates alkaline phosphatase and proteins related to organic P scavenging, protein degradation, lipid accumulation, and photorespiration, while downregulating energy metabolism, photosynthesis, and nucleic acid synthesis. *P. tricornutum* primarily focuses on metabolic downscaling and activating selective pathways. All three species enhance organic P utilization and modify central metabolism, reflecting their ecological and evolutionary differences.

9.2.3 Protein 3D Structures from Diatoms

Proteomics can be used in marine applications to investigate cellular expression and adaptation in organisms grown under different nutrient conditions, investigate marine dissolved and particulate organic matter pools for source information and insights into long-term carbon preservation, and validate genomic findings by determining which genes are translated and the level of protein expression through proteomic analyses.

Diatom proteins frequently possess repetitive and charged sequences (such as silaffins), complicating the crystallization process. Many of these proteins are intrinsically disordered (IDPS), particularly those related to silica formation, and struggle to form stable crystals. So far, the Protein Data Bank (PDB) contains approximately 54 protein entries from six to seven diatom species. The structural biology of diatom proteins is an emerging field that holds enormous potential. Utilizing the special biomineralization and stress response properties of diatoms in fundamental science and applied nanotechnology requires structural research. Because of technical difficulties, experimental crystal structures are still uncommon, although developments in cryo-EM and computer modeling are quickly bridging the gap.

9.3 Obtaining Monoclonal Host Cells: Opportunities and Hurdles

In nature, diatoms are closely associated with bacteria, which can interfere with molecular and genetic studies by contributing their genomes and influencing diatom physiology. To avoid this, axenic diatom cultures are obtained through filtration, antibiotic treatment, and monoclonal culture techniques (Han et al. 2016). Commonly used antibiotics include ampicillin, streptomycin, amphotericin, ciprofloxacin, neomycin, kanamycin, and penicillin (Shishlyannikov et al. 2011). Genetic manipulation of diatoms requires monoclonal diatom strains; however, not all species can thrive on agar plates (Kimura and Tomaru 2013). Therefore, a direct antibiotic-based recombinant selection method has limited applications. Cultivating a pure culture using a low agar concentration (1%) effectively promotes diatom growth on solid media (Kimura and Tomaru 2013). Monoclonal diatom strains aid in integrative taxonomic studies, but identifying diatoms based on frustule morphological features is challenging due to species variability, prolonged cultivation, and varying cell division rates, raising concerns about genetic marker stability. This morphological variability raises concerns regarding the stability of the genetic markers examined throughout cultivation (Mohamad et al. 2022).

9.4 Diatom Engineering

Over the past 3 decades, significant advancements have been made in the genetic manipulation of diatoms. Figure 9.1 shows timeline for diatom engineering over past 3 decades. The first attempt was carried out in 1995 using the biolistic transformation method in *Cyclotella cryptica* (Dunahay et al. 1995). Despite progress in developing transformation vectors for diatom genetic manipulation in subsequent years, these methods initially failed to achieve efficient transformation. One of the major challenges was the silicified cell wall of diatoms, which acted as a barrier to transformation. To overcome this, researchers employed the electroporation technique in *Thalassiosira pseudonana* (Niu et al. 2012). However, targeted genetic modifications in diatoms remained problematic. In 2014, researchers achieved a breakthrough by demonstrating stable and targeted genetic modifications in the diatom *Phaeodactylum tricornutum* using TALENs (transcription activator-like effector nucleases) and meganucleases (Daboussi et al. 2014). In 2015, the bacterial conjugation technique was successfully employed for gene insertion in both *P. tricornutum* and *T. pseudonana* (Karas et al. 2015). The advent of CRISPR-Cas9 has provided a highly efficient and precise tool for genome editing. Researchers first attempted genetic manipulation using CRISPR-Cas9 in *T. pseudonana* (Hopes et al. 2016). This technique enabled knock-in and knock-

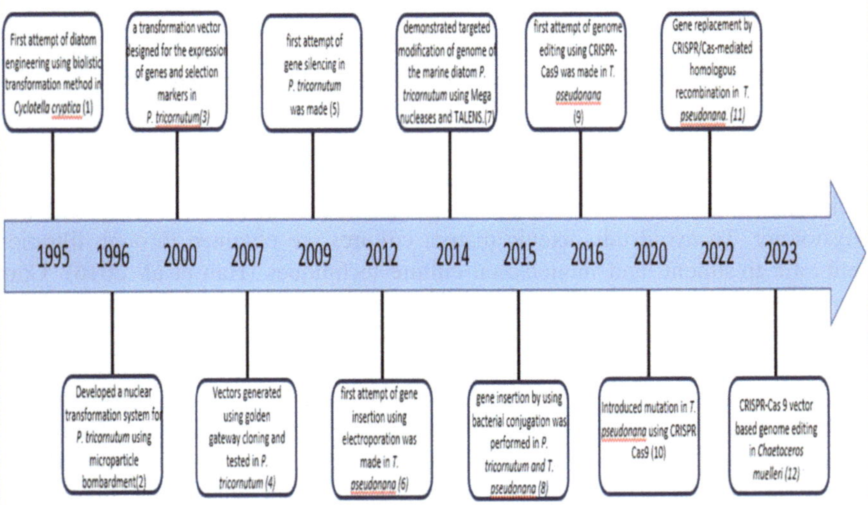

Timeline indicating the development of diatom engineering over past 3 decades: 1. Dunahay et al. 1995; 2. Apt et al. 1996; 3. ; 4. Siaut et al. 2007; 5. De Riso et al 2009; 6. Niu et al. 2012; 7. Daboussi et al. 2014; 8. Karas et al. 2015; 9. Hopes et al. 2016; 10. Nawaly et al 2020; 11. Belshaw et al 2022; 12. Yin et al 2023

Fig. 9.1 Representing the development in diatom gene modification techniques

down experiments in diatoms through RNP-based (ribonucleoprotein) and episome-based approaches (Tanwar et al. 2018). RNP-based method involves delivery of pre-assembled gRNA and Cas9 complex, which provides targeted genome editing. RNPs degrade a few hours after introduction into the cell, reducing cytotoxicity and off-target effects (Velmurugan and Deka 2018). However, their editing efficiency is relatively lower than the episome-based genome editing, as it uses a plasmid carrying Cas9 and sgRNA that replicates independently within the cell without integrating into the genome. Recent research has successfully implemented genetic manipulation in diatoms using CRISPR-Cas9 technology, enabling precise gene editing. One of the key mechanisms facilitating this process is homologous recombination, which allows for targeted gene replacement and ensures accurate modifications within the diatom genome (Yin and Hu 2023; Belshaw et al. 2023).

9.5 Vectors, Promoters, and Selection Markers for Gene Modifications in Diatoms

Diatom-specific vectors are relatively limited in number, which presents challenges for genetic engineering in these organisms. Among the available tools, the ptPUC3 vector, as utilized in *Phaeodactylum tricornutum* by (Karas et al. 2015), has enabled advancements in diatom molecular biology by providing a reliable platform for gene expression. Another significant advancement is the use of DiaCas9 vectors (Nymark et al. 2016), specifically designed to express Cas9 proteins in diatoms, which enables precise genome editing through CRISPR-Cas9 technology (Table 9.2).

Another widely adopted approach involves using Golden Gate cloning for plasmid preparation, allowing efficient assembly of multiple genetic elements in a single reaction. This technique is particularly valuable when incorporating essential features, such as selection markers and promoters tailored for diatom systems (Hopes et al. 2016). Promoters in diatom genetic engineering generally fall into two categories: inducible and constitutive. Inducible promoters offer temporal control over gene expression in response to specific stimuli, providing flexibility for experiments that require precise regulation. In contrast, constitutive promoters drive continuous expression of the target gene, ensuring stable and robust expression across different conditions. The LHCF promoter is widely used to express transgenes in diatoms, and it is derived from the fucoxanthin chlorophyll a/c binding protein gene (Hopes et al. 2016; Nymark et al. 2016; Nam et al. 2022). The strategic selection and combination of these components are critical for optimizing diatom genetic tools, expanding their potential as model organisms and bioengineering platforms.

9.6 Genetic Engineering of Diatoms as a Host for Production of Value-Added Products

Researchers have extensively explored diatom engineering for diverse applications such as green fuel production, i.e., the biofuel production, for nanoporous frustule modification, targeted drug delivery, recombinant protein production and

Table 9.2 The commonly used promoters and selection markers in diatom engineering

Species	Method	Vector	Promoter	Terminator	Selection marker	Efficiency	References
Phaeodactylum tricornutum	Microparticle bombardment	–	FCP	FCP	sh ble	24%	Weyman et al. (2015)
35*Thalassiosira pseudonana*	Microparticle bombardment	–	U6, FCP	FCP	NAT	<61.5%	Hopes et al. (2016)
Phaeodactylum tricornutum	Microparticle bombardment	pKS, diaCas9_sgRNA	LHCF2, U6	LHCF1, U6	Zeocin	31%	Nymark et al. (2016)
Thalassiosira pseudonana	Microparticle bombardment	pMHL_71	SIT	SIT	NAT	NA	Shrestha and Hildebrand (2017)
Cyclotella cryptica	Microparticle bombardment	pMHL_72	SIT	SIT	NAT	NA	Shrestha and Hildebrand (2017)
Phaeodactylum tricornutum	Biolistic particle delivery	ptCC9	FCP, U6, NR	FCP, U6, NR	sh ble (zeocin)	NA	Stukenberg et al. (2018)
Phaeodactylum tricornutum	Bacterial conjugation	pPtPuc3	FCP	FCP	sh ble, NAT	NA	Slattery et al. (2018)
Thalassiosira pseudonana	Microparticle bombardment	pSin1KO.	U6	U6	sh ble	NA	Görlich et al. (2019)
Thalassiosira pseudonana	Biolistic particle delivery	pKSdiaCas9_sgRNA	U6	U6	NAT	39%	Nawaly et al. (2020)
Thalassiosira pseudonana	Microparticle bombardment	pTpNR,pMD19	NA	NA	NAT	NA	Shimakawa et al. (2022)
Thalassiosira pseudonana	Bacterial conjugation	–	FCP	FCP	NAT	58%	Nam et al. (2022)
Thalassiosira pseudonana	Biolistic particle delivery	–	U6, FCP	FCP	NAT	85%	Belshaw et al. (2023)
Phaeodactylum tricornutum	Bacterial conjugation	pPtGE30	40SRPS8	FCP	sh ble	NA	Awwad et al. (2023)

biodegradation, leveraging their unique biological and structural features. In *Phaeodactylum tricornutum*, the ptTES1 gene, which encodes an acyl-CoA thioesterase enzyme involved in lipid breakdown, was knocked out using TALEN technology. This modification resulted in reduced lipid catabolism and significantly higher accumulation of triacylglycerols (TAGs), especially under nitrogen-deprived conditions, thus maximizing the diatom's biofuel potential (Weyman et al. 2015). Further enhancing lipid production, overexpression of the DGAT2 gene in *P. tricornutum*, achieved through the addition of regulatory sequences, resulted in up to a 35% increase in lipid content, underscoring the potential of targeted genetic interventions in optimizing biofuel feedstocks (Gunarathne and Gautam 2019). Naser et al. (2023), developed transformant clones for overexpressing the endogenous fat storage-inducing transmembrane protein for enhanced biofuel production (Table 9.3).

Diatoms have also shown immense promise in targeted drug delivery systems. In *T. pseudonana*, genetic modifications involved incorporating GFP and IgG-binding GB1 proteins into the diatom frustule using plasmid technology introduced via microparticle bombardment. These antibody-functionalized biosilica frustules demonstrated a remarkable ability to selectively target and destroy cancer cells, paving the way for novel, biocompatible drug delivery platforms (Delalat et al. 2015).

Advancements in recombinant protein production have further demonstrated the utility of diatoms as bioengineered tools. By introducing plasmids with silicon starvation-inducible promoters, *T. pseudonana* and *C. cryptica* were engineered to robustly express enhanced green fluorescent protein (eGFP). Under silica starvation conditions, protein expression increased dramatically—up to 500-fold in *T. pseudonana* and even greater in *C. cryptica*—highlighting the efficiency of leveraging diatom-specific metabolic pathways for high-yield protein production (Shrestha and Hildebrand 2017). Awwad et al. (2023) integrated the Cannabis gene, including tetraketide synthase (TKS), and olivetolic acid cyclase (OAC) for the production of olivetolic acid as a precursor for cannabinoid production as a novel biomolecule with potential pharmaceutical application. The first genetically engineered system for ovothiol biosynthesis was developed in *P. tricornutum* by overexpression of the 5-histidyl cysteine sulfoxide synthase OvaA gene, which is involved in ovothiol biosynthesis, which has unique antioxidant and anti-inflammatory properties.

The nanoporous architecture of diatom frustules has been explored for hemostatic applications. Researchers have specifically engineered *Cyclotella* frustules to enhance their hemostatic properties by knocking down the silica-spanning protein (SAP2). This genetic modification resulted in significant changes in frustule morphology, including size, pore dimensions, surface area, and pore volume. Notably, these structural modifications led to an overall improvement in hemostatic efficiency, highlighting the potential of genetically engineered diatom frustules in biomedical applications (Wang et al. 2022).

Table 9.3 Diatom engineering for process and product development

Diatom species	Target	Modification tool	Modification	Result	Reference
Biofuel production					
Phaeodactylum tricornutum	ptTES1	TALEN	Knock out the ptTES1 gene, which encodes an acyl-Coa thioesterase enzyme involved in breaking down lipids.	The diatoms reduced lipid catabolism, resulting in significantly higher TAG accumulation, particularly under nitrogen-deprived conditions.	Weyman et al. (2015)
Phaeodactylum tricornutum	DGAT2	Transformation	Overexpression of the gene by adding regulatory sequences	Up to 35% increased lipid content	Gunarathne and Gautam (2019)
Fistulifera solaris	Endogenous fat storage inducing transmembrane protein (FIT)	Transformation	Introduced a transformant clone	Overexpression of FIT	Naser et al. (2023)
Nanotechnological purposes					
T. pseudonana	Silicic acid transporter genes	Gateway technology	Created SIT genes fused with green fluorescent protein (GFP)	Enabled visualization of SITs within cells	Shrestha and Hildebrand (2017)
T. pseudonana	Silicalemma Associated Proteins (SAPs)	Transformation	Knockdown of TpSAP1 and TpSAP3 genes	Reduced expression of these genes altered the frustule patterning, with mutant cells showing approximately twice the volume of wild-type cells and a mislocated pattern center	Tesson et al. (2017)
T. pseudonana	Silicin-1	CRISPR-Cas9	Sin1 knockout mutants	Knockout mutants exhibited defects in their silica structures, confirming Sin1's role in the silica formation process	Görlich et al. (2019)

Targeted drug delivery					
T. pseudonana	Frustule modification	Transformation	GFP and IgG-binding GB1 were developed using plasmid pTpNR/T8–egfp–pgb1 introduced via microparticle bombardment.	Antibody-functionalized biosilica targeted and killed cancer cells, selectively	Delalat et al. (2015)
Recombinant protein production					
Phaeodactylum tricornutum	Cannabis gene(tetraketide synthase (TKS), and olivetolic acid cyclase (OAC)	Transformation	Integrated the Cannabis gene into the diatom	Production of Olivetolic acid	Awwad et al. (2023)
Phaeodactylum tricornutum	5-histidyl cysteine sulfoxide synthase *ovoA*	Transformation	Introduced overexpressing clones	Overexpression of 5-histidylcysteine sulfoxide synthase	Russo et al. (2023)
Biodegradation					
Phaeodactylum tricornutum	PETase-gfp plasmid	Transformation	Engineered plasmid introduced into the diatom cell	Successful secretion of PETase-gfp in the media and biodegradation of PET	Moog et al. (2019)
Biomedical applications					
Cyclotella cryptica	Silicalemma-spanning protein (SAP2)	Transformation	Knockdown as well as overexpression of SAP2	Change in size, specific pore size, surface area and pore volume on knockdown strains and increased hemostatic property	Wang et al. (2022)

9.7 Ghost Cells

Diatom frustules, commonly referred to as "diatom ghosts," are silica-based structures characterized by nanoscale pores. These unique morphological and chemical features render them highly suitable for diverse applications in nanotechnology, optoelectronics, biomedicine, and solar energy harvesting (Rogato and De Tommasi 2020) (Fig. 9.2).

Such applications are facilitated through approaches including surface immobilization, frustule doping, and genetic engineering. Rogato and Tommasi, (Rogato and De Tommasi 2020) provided a comprehensive review of these developments up to 2020.

Subsequent studies have expanded on this foundation. Researchers investigated the immobilization of industrially significant polypeptides on the frustules of *Thalassiosira pseudonana.* Six polypeptides—protamine, metallothionein, phosphotriesterase, choline oxidase, laccase, and polyamine synthase—were successfully immobilized as GFP-tagged fusion proteins (Shimakawa et al. 2022). Researchers (Dalgic et al. 2023) developed a composite scaffold integrating diatom frustules with poly(hydroxybutyrate-co-hydroxyvalerate) (PHBV) and poly(-ε-caprolactone) (PCL) to enable sustained release of the hydrophilic molecule melatonin, thereby enhancing osteogenic potential. Scientists (Lin et al. 2023) engineered a floatable, micron-scale enzyme platform based on diatom frustules, demonstrating improved enzymatic activity in comparison to conventional silica-based supports. The hemostatic performance of *Cyclotella cryptica* frustules, showing that genetic modification of *Silicalemma Associated Protein 2* enhanced hemophilicity relative to wild-type strains (Wang et al. 2022).

More recently, it is reported that frustules of *Melosira nummuloides* exhibit superior superhydrophilic and hemophilic properties compared to synthetic silica (Lee et al. 2020). Furthermore, CRISPR-Cas9-mediated knockout of *TpSil1*, *TpSil2*, and *TpSil3* genes in *T. pseudonana* resulted in reduced silicification, alongside enhanced cell growth and photosynthetic activity. Conversely, overexpression of silaffin genes led to increased silica deposition, indicating distinct functional roles of these genes in determining frustule morphology (Qiao et al. 2024).

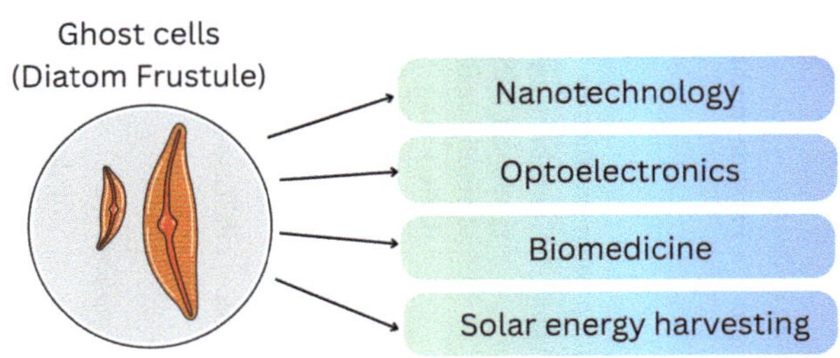

Fig. 9.2 Diatom ghost cell applications

9.8 Challenges and Future Perspective

Diatom biotechnology offers a sustainable and novel way for addressing global challenges in energy, healthcare, and environmental conservation; however, several factors limit its widespread application and industrial scalability. One major challenge is establishing axenic (bacteria-free) cultures, which require careful optimization since commonly used antibiotics may adversely affect diatom growth alongside bacterial contaminants. Additionally, genetic engineering tools such as CRISPR/Cas9 and TALENs exhibit variable efficiency across diatom species and often cause off-target effects, reducing the precision and reproducibility of genetic modifications. The lack of extensive genomic, proteomic, and metabolic data for many diatom species further restricts the development of tailored genetic tools and impedes metabolic pathway analysis. The silicified cell wall, or frustule, of diatoms poses another barrier by making gene delivery through techniques like electroporation and biolistic technically challenging and prone to damaging cells. Despite these constraints, the future of diatom biotechnology is promising. Advances in next-generation sequencing and synthetic biology are expected to drive the development of species-specific CRISPR systems and novel gene delivery methods compatible with the unique structure of diatom cells. By using genetic engineering techniques, the lipid production could be enhanced, which will lead to a sustainable resource for biofuel production. Comprehensive genomic mapping of diverse species will enable rational metabolic engineering for enhanced bio silica-based devices like solar cells, and bioactive compound synthesis. Genetic manipulation and post-synthetic chemical modification of diatom frustules could lead to breakthroughs in drug delivery, biosensing, and catalytic applications. Functionalized biosilica could be used as targeted therapeutic carriers or environmental diagnostic tools. Furthermore, engineered diatoms capable of surviving extreme or polluted conditions could serve in bioremediation efforts. Collaborative research, supported by updated methodologies, ethical oversight, and robust infrastructure, will be key to realizing the full potential of diatoms as versatile agents of sustainable biotechnological innovation.

References

Anderson NJ (2000) Diatoms, temperature and climatic change. Eur J Phycol 35(4):307–314

Armbrust EV, Berges JA, Bowler C, Green BR, Martinez D, Putnam NH et al (2004) The genome of the diatom *Thalassiosira pseudonana*: ecology, evolution, and metabolism. Science 306(5693):79–86

Awwad F, Fantino EI, Héneault M, Diaz-Garza AM, Merindol N, Custeau A, Gélinas SE, Meddeb-Mouelhi F, Li J, Lemay JF, Karas BJ (2023) Bioengineering of the marine diatom *Phaeodactylum tricornutum* with cannabis genes enables the production of the cannabinoid precursor, olivetolic acid. Int J Mol Sci 24(23):16624

Belshaw N, Grouneva I, Aram L, Gal A, Hopes A, Mock T (2023) Efficient gene replacement by CRISPR/Cas-mediated homologous recombination in the model diatom *Thalassiosira pseudonana*. New Phytol 238(1):438–452

Bowler C, Allen AE, Badger JH, Grimwood J, Jabbari K, Kuo A et al (2008) The *Phaeodactylum* genome reveals the evolutionary history of diatom genomes. Nature 456(7219):239–244

Cheng H, Bowler C, Xing X, Bulone V, Shao Z, Duan D (2021) Full-length transcriptome of *Thalassiosira weissflogii* as a reference resource and mining of chitin-related genes. Mar Drugs 19(7):392

Daboussi F, Leduc S, Maréchal A et al (2014) Genome engineering empowers the diatom *Phaeodactylum tricornutum* for biotechnology. Nat Commun 5:3831

Dalgic AD, Atila D, Tezcaner A, Gürses S, Keskin D (2023) Diatom silica frustules-doped fibers for controlled release of melatonin for bone regeneration. Eur Polym J 186:111858

Delalat B, Sheppard VC, Rasi Ghaemi S, Rao S, Prestidge CA, McPhee G et al (2015) Targeted drug delivery using genetically engineered diatom biosilica. Nat Commun 6:8791

Dunahay TG, Jarvis EE, Roessler PG (1995) Genetic transformation of the diatoms *Cyclotella cryptica* and *Navicula saprophila*. J Phycol 31:1004–1012

Görlich S, Pawolski D, Zlotnikov I, Kröger N (2019) Control of biosilica morphology and mechanical performance by the conserved diatom gene Silicanin-1. Commun Biol 2:1–8

Gunarathne SS, Gautam S (2019) Genetically engineered diatoms for the enhanced production of green fuels. Think India J 22(16):4402–4407

Han J, Wang S, Zhang L, Yang G, Zhao L, Pan K (2016) A method of batch-purifying microalgae with multiple antibiotics at extremely high concentrations. Chin J Oceanol Limnol 34(1):79–85

Hopes A, Nekrasov V, Kamoun S, Mock T (2016) Editing of the urease gene by CRISPR-Cas in the diatom *Thalassiosira pseudonana*. Plant Methods 12:49

Horton BP (2007) Diatoms and forensic science. Paleontol Soc Pap 13:181–190

Huang W, Daboussi F (2017) Genetic and metabolic engineering in diatoms. Philos Trans R Soc Lond Ser B Biol Sci 372(1728):20160411

Jeong Y, Lee J (2021) The complete mitochondrial genome of the benthic diatom *Pleurosigma inscriptura*. Mitochondrial DNA Part B 6(9):2584–2586

Jian J, Du F, Wang B, Fang X, Larsen TO, Li Y et al (2024) A high-quality genome of the early diverging tychoplanktonic diatom *Paralia Guyana*. Sci Data 11(1):1–7

Karas BJ, Diner RE, Lefebvre SC et al (2015) Designer diatom episomes delivered by bacterial conjugation. Nat Commun 6:6925

Kimura K, Tomaru Y (2013) A unique method for culturing diatoms on agar plates. Plankton Benthos Res 8(1):46–48

Kumazawa M, Nishide H, Nagao R, Inoue-Kashino N, Shen JR, Nakano T et al (2022) Molecular phylogeny of fucoxanthin-chlorophyll a/c proteins from *Chaetoceros gracilis* and Lhcq/Lhcf diversity. Physiol Plant 174(1):e13598

Lee J, Lee HA, Shin M, Juang LJ, Kastrup CJ, Go GM et al (2020) Diatom frustule silica exhibits superhydrophilicity and superhemophilicity. ACS Nano 14(4):4755–4766

Li L, Wang H, Wang S, Xu Y, Liang H, Liu H, Sonnenschein EC (2021) The draft genome of the centric diatom *Conticribra weissflogii* (Coscinodiscophyceae, Ochrophyta). Protist 172(5–6): 125845

Lin NS, Hirayama K, Kitamura M, Koide S, Kitajima H, Harada T et al (2023) Fabrication of a floatable micron-sized enzyme device using diatom frustules. ACS Omega 8(23):21145–21153

Liu S, Chen N (2024) Chromosome-level genome assembly of marine diatom *Skeletonema tropicum*. Sci Data 11(1):403

Maheswari U, Jabbari K, Petit JL, Porcel BM, Allen AE, Cadoret JP et al (2010) Digital expression profiling of novel diatom transcripts provides insight into their biological functions. Genome Biol 11:1–19

Marella TK, Bansal H, Bhattacharjya R, Parmar N, Chaurasia A, Watanabe MM et al (2021) Deciphering functional biomolecule potential of marine diatoms through complex network approach. Bioresour Technol 342:125927

Mock T, Otillar RP, Strauss J, McMullan M, Paajanen P, Schmutz J, Salamov A, Sanges R, Toseland A, Ward BJ, Allen AE (2017) Evolutionary genomics of the cold-adapted diatom *Fragilariopsis cylindrus*. Nature 541(7638):536–540

Mohamad H, Mora D, Skibbe O, Abarca N, Deutschmeyer V, Enke N et al (2022) Morphological variability and genetic marker stability of 16 monoclonal pennate diatom strains under medium-term culture. Diatom Res 37(4):307–328

Moog D, Schmitt J, Senger J, Zarzycki J, Rexer KH, Linne U et al (2019) Using a marine microalga as a chassis for polyethylene terephthalate (PET) degradation. Microb Cell Factories 18:1–15

Nam O, Grouneva I, Mackinder LC (2022) Endogenous GFP tagging in the diatom *Thalassiosira pseudonana*. bioRxiv. 2022–09

Naser I, Yabu Y, Maeda Y, Tanaka T (2023) Highly efficient genetic transformation methods for the marine oleaginous diatom *Fistulifera solaris*. Mar Biotechnol 25(5):657–665

Nawaly H, Tsuji Y, Matsuda Y (2020) Rapid and precise genome editing in a marine diatom, *Thalassiosira pseudonana* by Cas9 nickase (D10A). Algal Res 47:101855

Niu YF, Yang ZK, Zhang MH et al (2012) Transformation of diatom *Phaeodactylum tricornutum* by electroporation and establishment of inducible selection marker. BioTechniques 52:1–3

Nymark M, Sharma AK, Sparstad T, Bones AM, Winge P (2016) A CRISPR/Cas9 system adapted for gene editing in marine algae. Sci Rep 6(1):24951

Qiao T, Wang L, Zhao Y, Li Y, Yang G, Zhu B, Pan K (2024) Silaffins-driven genetic engineering of diatom cell walls: insight into biosilica morphology and nanomaterial design. bioRxiv:2024–2012

Roberts WR, Downey KM, Ruck EC, Traller JC, Alverson AJ (2020) Improved reference genome for *Cyclotella cryptica* CCMP332, a model for cell wall morphogenesis, salinity adaptation, and lipid production in diatoms (Bacillariophyta). G3 (Bethesda) 10(9):2965–2974

Rogato A, De Tommasi E (2020) Physical, chemical, and genetic techniques for diatom frustule modification: applications in nanotechnology. Appl Sci 10(23):8738

Romero-Guzmán ET, Reyes-Gutiérrez LR, Romero-Guzmán L, Hernández-Mendoza H, Uría-Gómez LC, Gutiérrez-Reyes J (2021) An overview of bioaerosols suspended in the atmosphere of Metropolitan Zone of Toluca Valley. J Mex Chem Soc 65(2):214–224

Russo MT, Rogato A, Jaubert M, Karas BJ, Falciatore A (2023) *Phaeodactylum tricornutum*: an established model species for diatom molecular research and an emerging chassis for algal synthetic biology. J Phycol 59(6):1114–1122

Schvarcz CR, Stancheva R, Turk-Kubo KA, Wilson ST, Zehr JP, Edwards KF, Steward GF, Archibald JM, Oatley G, Sinclair E, Santos C (2024) The genome sequences of the marine diatom *Epithemia pelagica* strain UHM3201 (Schvarcz, Stancheva & Steward, 2022) and its nitrogen-fixing, endosymbiotic cyanobacterium. Wellcome Open Res 9:232

Shimakawa G, Katayama S, Tsuji Y, Yoneda K, Fukuda W, Fujiwara S, Matsuda Y (2022) Immobilization of a broad range of polypeptides on the frustule of the diatom *Thalassiosira pseudonana*. Appl Environ Microbiol 88(21):e01153–e01122

Shishlyannikov SM, Zakharova YR, Volokitina NA, Mikhailov IS, Petrova DP, Likhoshway YV (2011) A procedure for establishing an axenic culture of the diatom Synedra acus subsp. radians (Kütz.) Skabibitsch from Lake Baikal. Limnol Oceanogr Methods 9(10):478–484

Shrestha RP, Hildebrand M (2017) Development of a silicon limitation inducible expression system for recombinant protein production in the centric diatoms *Thalassiosira pseudonana* and *Cyclotella cryptica*. Microb Cell Factories 16:1–14

Slattery SS, Diamond A, Wang H, Therrien JA, Lant JT, Jazey T, Lee K, Klassen Z, Desgagné-Penix I, Karas BJ, Edgell DR (2018) An expanded plasmid-based genetic toolbox enables Cas9 genome editing and stable maintenance of synthetic pathways in *Phaeodactylum tricornutum*. ACS Synth Biol 7(2):328–338

Stukenberg D, Zauner S, Dell'Aquila G, Maier UG (2018) Optimizing CRISPR/Cas9 for the diatom *Phaeodactylum tricornutum*. Front Plant Sci 9:740

Tanwar A, Sharma S, Kumar S (2018) Targeted genome editing in algae using CRISPR/Cas9. Indian J Plant Physiol 23:653–669

Tesson B, Lerch SJ, Hildebrand M (2017) Characterization of a new protein family associated with the silica deposition vesicle membrane enables genetic manipulation of diatom silica. Sci Rep 7: 1–13

Velmurugan N, Deka D (2018) Transformation techniques for metabolic engineering of diatoms and haptophytes: current state and prospects. Appl Microbiol Biotechnol 102:4255–4267

Villar E, Zweig N, Vincens P, Cruz de Carvalho H, Duchene C, Liu S et al (2025) DiatOmicBase: a versatile gene-centered platform for mining functional omics data in diatom research. Plant J 121(6):e70061

Wang Y, Liu S, Wang J, Yao Y, Chen Y, Xu Q et al (2022) Diatom biodiversity and speciation revealed by comparative analysis of mitochondrial genomes. Front Plant Sci 13:749982

Weyman PD, Beeri K, Lefebvre SC, Rivera J, McCarthy JK, Heuberger AL et al (2015) Inactivation of *Phaeodactylum tricornutum* urease gene using transcription activator-like effector nuclease-based targeted mutagenesis. Plant Biotechnol J 13:460–470

Yin W, Hu H (2023) CRISPR/Cas9-mediated genome editing via homologous recombination in a centric diatom *Chaetoceros muelleri*. ACS Synth Biol 12(4):1287–1296

Zepernick BN, Truchon AR, Gann ER, Wilhelm SW (2022) Draft genome sequence of the freshwater diatom *Fragilaria crotonensis* SAG 28.96. Microbiol Resour Announc 11(9): e00289–e00222

Microalgae-Derived Pigments: Sustainable Solution and Their Multifaceted Applications

10

Athulya Babu, Amrutha V. Sunil, Shirin Fathima M, Nikitha Ramalal, and Praveena P

Abstract

Pigments are exceptionally colored substances that are used to color other materials in a variety of industries. Naturally synthesized pigments are in greater demand, and circular bioeconomy strategies can include their manufacturing. Microalgae, macroalgae, cyanobacteria, plants, and animals produce natural pigments. Microalgae, photosynthetic microorganisms, are becoming a key sustainable source of natural pigments, providing an eco-friendly alternative to synthetic dyes. These pigments, including carotenoids, phycobiliproteins, and chlorophylls, are highly valued across industries such as food, cosmetics, pharmaceuticals, and dietary supplements due to their bioactive properties, including antioxidant, anti-inflammatory, antidiabetic, and neuroprotective effects. Unlike synthetic pigments, which have raised safety concerns and are banned in some countries due to their potential risks to human health and food safety, microalgal pigments offer a nontoxic and noncarcinogenic solution. The cultivation of microalgae offers significant advantages, as it is not constrained by seasonal, climatic, or environmental factors and does not require arable land. These microorganisms grow rapidly, possess high pigment concentrations, and can thrive under stress, making them a highly efficient and sustainable resource.

A. Babu · A. V. Sunil · Shirin Fathima M · N. Ramalal
Department of Biotechnology, Sahrdaya College of Engineering and Technology, Kodakara, Thrissur, Kerala, India

Praveena P (✉)
Department of Biotechnology, Sahrdaya College of Engineering and Technology, Kodakara, Thrissur, Kerala, India

Associate Professor, Department of Biotechnology, Sahrdaya College of Engineering and Technology, Thrissur, Kerala, India
e-mail: ppraveena@sahrdaya.ac.in

K. Mohanty et al. (eds.), *Advances in Algal Research*,
https://doi.org/10.1007/978-981-95-7234-2_10

Pigments like phycocyanin, β-carotene, and astaxanthin are particularly gaining popularity for their natural, safe properties. As consumers increasingly demand natural, health-conscious products, microalgal pigments are becoming an attractive alternative, providing an environmentally responsible solution that aligns with global trends toward sustainability. Their wide-ranging applications and eco-friendly attributes position microalgae as a competitive and promising source of natural pigments in the global market.

Keywords

Antidiabetic · Anti-inflammatory · Antioxidant · Eco-friendly · Microalgae · Pigments · Sustainable

Abbreviations

AI	Artificial Intelligence
Chl	Chlorophyll
DSP	Downstream Processes
FDA	U. S. Food and Drug Administration
HPH	High Pressure Homogenization
ML	Machine Learning
OPEX	Operating Expenditure
ORP	Open Raceway Pond
PBP's	Phycobiliproteins
PBR	Photobioreactor
PC	Phycocyanin
PE	Phycoerythrin
PLE	Pressurized Liquid Extraction
ROS	Reactive Oxygen Species
SCF	Subcritical Fluid Extraction
USP	Upstream Processes
UVB	Ultraviolet B Radiation

10.1 Introduction

Global population has increased demand for natural resources, energy, food, and housing. One of the most underutilized resources, microalgae has recently gained attention for its potential to provide a variety of renewable natural metabolites, such as protein, lipids, fuel, biomass, and biochemically active substances like vitamins and pigments. Pigments are intensely colored substances that are used to color other materials in a variety of industries. Usually referred to as "dyes" in the food and textile sectors and "pigments" in the pharmaceutical, ink and cosmetic industries, they are widely utilized to improve the aesthetic appeal of industrial products. Microalgae, with their diverse range of pigments and bioactive compounds, are becoming recognized as nature's undiscovered treasure in the field of bioproducts. The pigments include chlorophylls, carotenoids, and phycobiliproteins. These pigments have unique properties that allow them to be used in a variety of food, cosmetic, and medicinal applications, in addition to playing important roles in the photosynthesis of microalgae. Microalgae have a wide range of pigments, including newly identified polyphenols, which increases their potential as natural colorants and useful constituents in a variety of sectors (Saini et al. 2020; Begum et al. 2016).

Any organism with chlorophyll a and a thallus that has not differentiated into roots, stems, and leaves is considered an alga by phycologists. This definition includes cyanobacteria even though they are prokaryotic species. Thus, in applied phycology, "microalgae" refers to both microscopic algae sensu stricto and oxygenic photosynthetic bacteria, or cyanobacteria, formerly known as Cyanophyceae. These two groups of phototrophic organisms are interesting because, like heterotrophic bacteria, they can use solar energy to produce biomass for fine chemicals, food, and feed. They can be found on the surface of any type of soil, despite the fact that they are mostly located in water (Tomaselli 2004).

They provide one of the most exciting prospects for the provision of next-generation food, feeds, and biofuels, and they serve as a great source of novel bio products with potential applications in numerous biotechnological areas. Simultaneously, it can be combined with environmental management procedures like CO_2 sequestration and wastewater nutrient recycling. Because microalgal biomass contains lipids, pigments, carbohydrates, minerals, enzymes, and vitamins, a wide range of products can be developed from it (Barros et al. 2019).

It has been established that the base of the food chain in aquatic environments is made up of microalgae. Furthermore, they are a vast and diverse group of primitive photosynthetic microorganisms, either eukaryotic or prokaryotic. Because the pigments in algae influence their color, one of their most distinctive features, microalgal biomass is garnering interest from all around the world. These pigments are brightly colored chemical compounds that belong to three different classes within the photosynthetic system of microalgae: phycobiliproteins, carotenoids, and chlorophylls. Microalgal cells can be cultured to accumulate a wide variety of useful pigments throughout the course of their life cycle under various light and nutrition regimes. The establishment of a microalgal bioindustry that is both sustainable and commercially viable depends on the efficient recovery of these pigments (Braga et al. 2020).

Green photosynthetic pigments called chlorophylls are present in every species of cyanobacteria and microalgae. Carotenoids are a class of auxiliary pigments with an absorption range of 400–550 nm that are derived from linear hydrocarbon polyenes, some of which have oxygenated derivatives. On the outside of the thylakoid membranes, colored antennae-protein pigments called phycobiliproteins are attached to supramolecular complexes called phycobilisomes. The pigments serve as accessory pigments that improve cellular light collection for photosynthesis by absorbing energy in the visible light range (450–650 nm) (Halim 2020).

The increasing scrutiny of artificial colorants and health concerns is driving demand for natural pigments. Microalgae, with higher growth rates and photosynthetic efficiencies, offer advantages over traditional methods like fruits and vegetables. The European Food Safety Authority and the FDA have approved microalgae pigments and advised their replacement with synthetic colorants in food, cosmetic, and nutritional markets. They are frequently used as food supplements and coloring agents. They also have anti-inflammatory, antiaging, antioxidant, and anticancer qualities, which make them a potent component of important medications (Rahman et al. 2017).

The majority of synthetic pigments are carbon-based compounds that are chemically produced from acids, petrochemical products, and other substances. Synthetic pigments may not function the same way even when they are exact replicas of natural materials. This is due to the fact that natural items are frequently chiral, but synthetic compounds could be racemic. The need for natural colors has increased due to reports of some synthetic pigments (such as citrus red II, metanil yellow, and rhodamine B) having a variety of toxicological effects, such as cancer, estrogenic activity, and neurotoxicity. Although synthetic pigments are cheap and generally stable, the market has driven the growth of organically produced pigments as part of the growing circular bioeconomy, as health and environmental consciousness have grown (Capelli et al. 2013).

In addition to color, pigments have several possible industrial uses and health-promoting qualities. Dietary consumption of microalgae has been demonstrated to have the capacity to protect organisms against a variety of chronic disorders, including cancer, diabetes, atherosclerosis, coronary disease, ischemic brain development, metabolic syndromes, gastrointestinal and liver diseases, as well as neurodegenerative diseases like Alzheimer's and Parkinson's disease. Consumers are becoming more aware of the relationship between diet, health, and disease prevention (Braga et al. 2020).

Seasonality, climate, and environmental constraints do not limit the industrial production of microalgae as a sustainable natural source of pigments. They can also be grown ecologically and sustainably. They also stand out for having a high pigment content, growing quickly, being able to thrive in unfavorable conditions, and not requiring fertile soils, which makes them one of the most promising and competitive sources of natural pigments (Sun et al. 2023).

10.2 Types of Microalgal Pigments and Their Sources

Photosynthetic organisms, such as cyanobacteria and microalgae, produce different types of pigments to harvest light. These pigments are often classified into three primary classes: carotenoids, chlorophylls, and phycobiliproteins. The kind of pigments that cyanobacteria and microalgae have depends on their phylum. All of these species include chlorophyll, their principal pigment, which is essential for absorbing light and giving them their green color. Astaxanthin, lutein, and β-carotene are a few prominent examples of carotenoids, which can be primary or secondary pigments with varying profiles depending on the species and growing conditions. Moreover, water-soluble pigments known as phycobilins, of which phycocyanin and phycoerythrin are the most well known. When combined, these three pigment classes improve visible-spectrum light harvesting while accommodating different species and growth environments (Pagels et al. 2021).

10.2.1 Chlorophylls

The most common natural pigment that gives microalgae their green hue is chlorophyll, which is also necessary for the photosynthetic process and gives microalgae their energy for metabolism and reproduction. All life on Earth receives its energy from the photosynthetic process, which is supported by chlorophyll. For this reason, chlorophyll is not only the most prevalent pigment but also the most significant one. Chlorophyll (Chl) participates in the processes of energy conversion and light harvesting. The major pigment for photosynthetic processes is chlorophyll *a*, and the concentration of this pigment in microalgae is influenced by exposure to light and the nutrition of minerals in cultural conditions. Its content has been observed to make up roughly 5% of the dry biomass weight in *Chlorella* sp. cells in healthy cells. Chlorophylls are located in intracellular organelles called chloroplasts, where they are attached to proteins forming the chloroplast lamellae. These membranes surround thylakoids, which are compressed "sacs" where light is absorbed during photosynthetic activity. The Chl *a*, *b*, *c*, *d*, and *f* were dispersed among the different groups of microalgae by the evolutionary process, and their presence or absence is significant for taxonomy. Chl *a* is found in Prochlorophyta, Chlorophyta, and other organisms. Green algae, or Chlorophyta, and Euglenophyta contain chlorophyll *b*; Bacillariophyta, Phaeophyta, Pyrrophyta, and Cryptophyta contain chlorophyll *c*; and Rhodophyta is the sole group in which chlorophyll *d* is found. Chl *f* is abundant in Cyanobacteria (da Silva and Lombardi 2020). The distribution of chlorophyll pigments across major algal and cyanobacterial groups is summarized in Table 10.1.

10.2.2 Carotenoids

Carotenoids are a class of auxiliary pigments with an absorption range of 400–550 nm that are derived from linear hydrocarbon polyenes, some of which have oxygenated derivatives. Carotenoids are categorized into distinct classes

Table 10.1 Distribution of chlorophylls in different algal groups (da Silva and Lombardi 2020)

Chl *a*	Cyanobacteria, Prochlorophyta, Glaucophyta, Rhodophyta, and Xanthophyceae
Chl *b*	Euglenophyta, Chlorophyta, and Charophyta
Chl c_1	Chrysophyceae, Raphidophyceae, Phaeophyceae, Bacillarophyta, and Haptophyta
Chl c_2	Chrysophyceae, Raphidophyceae, Phaeophyceae, Bacillarophyta, and Cryptophyta
Chl c_3	Chrysophyceae, Bacillarophyta, and Haptophyta
Chl *d*	Rodophyta
Chl *f*	Cyanobacteria

Table 10.2 Distribution of carotenoids in different algal groups (Duan et al. 2024; Hassaan et al. 2021; Gille et al. 2016; Vadrale et al. 2023; Li et al. 2022; Del Campo et al. 2004; Pelah et al. 2004; Shen et al. 2011; Kim et al. 2017; Granado-Lorencio et al. 2009; Chen et al. 2022; Mohamadnia et al. 2020; Xia et al. 2018)

β-carotene	*Dunaliella salina, Arthrospira platensis,* and *Chlamydomonas reinhardtii*
Lutein	*Chlorella sorokiniana, Scenedesmus* sp., and *Chlorella zofingiensis*
Astaxanthin	*Haematococcus pluvialis* and *Chlorella zofingiensisis*
Zeaxanthin	*Nannochloropsis oculata, Dunaliella tertiolecta,* and *Chromochloris zofingiensis*
Fucoxanthin	*Tisochrysis lutea* and *Odontella aurita*

according to their molecular makeup and functionality. When it comes to their molecular makeup, carotenoids with pure hydrocarbon chains, like β-carotene, are categorized as carotenes, whereas those with molecules that contain oxygenated derivatives, such -OH and -CO replacements, are categorized as xanthophylls, like astaxanthin. Carotenoids are classified as primary or secondary carotenoids based on function. Primary carotenoids, including lutein, violaxanthin, and β-carotene, function as energy harvesters by transferring absorbed light energy to chlorophylls, thereby broadening the cell's range of light absorption capabilities. However, when exposed to intense light, secondary carotenoids (such as astaxanthin and canthaxanthin) create a protective coating over the microalgal cells, shielding the chlorophylls from photodamage. Stressful environments typically cause cells to produce more secondary carotenoids. Strong antioxidant qualities are exhibited by all carotenoids, and they can scavenge reactive oxygen species (ROS) to shield cells from oxidation and assaults by free radicals in stressful environmental conditions (Halim 2020). Major carotenoids and their corresponding microalgal sources are summarized in Table 10.2.

10.2.3 Phycobiliproteins

The water-soluble, brightly colored, auto fluorescent, and auxiliary light—harvesting pigments known as phycobiliproteins (PBPs) are essential to photosynthesis because they carry solar energy to the reaction center in cyanobacteria, red algae, glaucophytes, and cryptomonads. PBPs make up roughly 20–30% of the cell's dry weight, yet in low light conditions, this percentage can increase to 40–50% (Kannaujiya et al. 2020).

They are divided into four primary classes according to their molecular makeup: allophycocyanin (bluish-green in color), phycocyanin (blue), phycoerythrin (red), and phycoerythrocyanin (orange). The color of allophycocyanin is bluish-green. The blue phycobiliprotein known as phycocyanin (PC) has an absorption peak at 610–620 nm and is nearly exclusively present in cyanobacteria, including *Spirulina platensis*. Red phycobiliproteins called phycoerythrin (PE) have an absorption peak at 565 nm and are mainly present in the chloroplasts of red algae (*Porphyridium cruentum*) and cyanobacteria (*Synechococcus* sp., *Leptolyngbya* sp.). Phycobiliproteins, carotenoids, and fucoxanthin are photosynthetic pigments derived from microalgae and cyanobacteria with numerous biotechnological

Table 10.3 Distribution of phycobiliproteins in microalage (Prabakaran et al. 2020; Shanmugam et al. 2017; Patel et al. 2005; Román et al. 2002; Tan et al. 2022; Liu et al. 2012; Liang et al. 2018)

PC	*Spirulina platensis, Oscillatoria* sp., *Phormidium*, and *Lyngbya* spp.
PE	*Porphyridium cruentum, Tolypothrix* sp., and *Pseudanabaena* sp.
APC	*Spirulina platensis* and *Thermosynechococcus elongatus*

applications, including healthy food coloring and cosmetics. Unlike chlorophylls and carotenoids, phycobiliproteins can be extracted from microalgal biomass without the use of organic solvents because they are soluble in water and aqueous buffers (like phosphate buffer) (Pagels et al. 2021; Chini Zittelli et al. 2023). The distribution of major phycobiliproteins and their cyanobacterial and algal sources is summarized in Table 10.3.

10.3 Pigment Production in Microalgae

Upstream (USP) and downstream (DSP) processing procedures are required in the synthesis of carotenoid and phycobiliproteins from microalgae. The USP consists of the variety of microalgae species, the availability of nutrients, the carbon dioxide supply, and the source of light. There have been reports of many methods for the production system, including two-stage, open, closed, heterotrophic, mixotrophic, and photoautotrophic systems. Raceway ponds, as well as flat and tubular type photobioreactors, are them. The two-stage systems technique is the most widely used and has significantly increased in the manufacturing of pigments in industry. After the USP is finished, the DSP starts. Harvesting, extraction, and purification are unit processes that make up the microalgae biomass DSP (Maroneze et al. 2020). The overall downstream processing workflow for microalgal product recovery, from cultivation to purification, is illustrated in Fig. 10.1.

10.3.1 Commercial Microalgae Culture System for Pigment Production

Microalgae-based products require large-scale culture in open systems like ponds or closed systems like photobioreactors. Open systems are widely used for pigment production, but have limitations due to limited control and contamination. Closed systems have advantages but have high costs and scaling difficulties (Hassaan et al. 2021).

10.3.2 Open Systems

Raceway ponds, also known as Oswald lagoons, are a commercial technique for producing microalgae biomass (Hassaan et al. 2021). They are recirculated around a

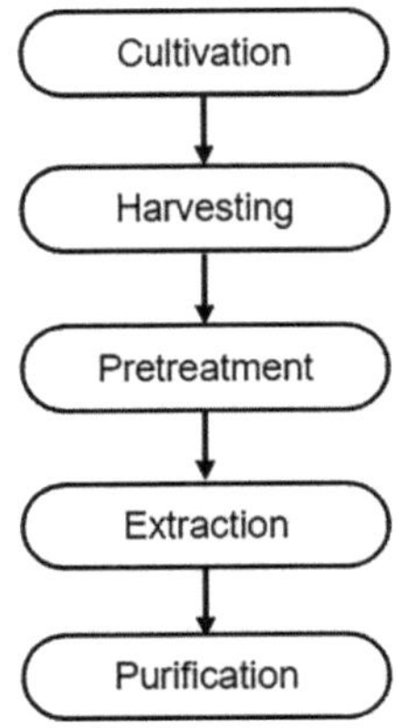

Fig. 10.1 Overview of pigment production in microalgae

racetrack loop, with shallow channels for light (Posten and Walter 2012; Acién et al. 2017). The process involves paddlewheels for energy, nutrient mixing, and cell growth (Kumar et al. 2015). Temperature is a key factor determining biomass productivity (Brusca et al. 2017). Unmixed ponds, with shallow depths and limited mixing mechanisms, are the least expensive and least technical commercial growing techniques. They require microalgae that can withstand harsh conditions or outgrow contaminants (Kumar et al. 2015; Borowitzka 1990).

10.3.3 Closed Systems

Investments in photobioreactors have increased due to high productivity and operational control. Tubular photobioreactors are popular for large-scale commercial applications, but scale-up challenges and high costs remain challenges (Hassaan et al. 2021).

Tubular photobioreactors offer higher yields due to solar light availability, reduced contamination, and better control over growing conditions. They can culture sensible strains like *H. pluvialis*, with tube diameters ranging from 0.03 to 0.12 m. Tube diameter affects solar radiation absorption, temperature, and oxygen concentration (Acién et al. 2017; Torzillo and Chini 2015).

10.4 Downstream Processing

Pigment extraction from microalgae involves dewatering, harvesting cells, biomass pretreatment, or cell rupture, followed by extraction and purification of pigments (Deepika et al. 2022).

Table 10.4 Low energy and cost-effective microalgae harvesting techniques (Zhu et al. 2024)

Emerging technique	Industrial scale advantages
Bioflocculation	High, selective recovery is achieved by this process, which uses living microorganisms or algal biopolymers, without the need for chemicals, thereby offering enhanced sustainability and biocompatibility.
Electroflocculation	Coagulants are electrolytically generated by it, allowing rapid and controlled destabilization and aggregation. Precision dosing is employed in the technique to enhance control over microalgae recovery.
Magnetic separation	Excellent, scalable separation is facilitated by this method using magnetic nanoparticles under low magnetic field gradients, providing an efficient approach for dewatering.
Hybrid systems	Multiple mechanisms (e.g., electrolytic coagulation combined with flotation) are integrated in these systems to achieve improved efficiency and economics through synergies, while minimizing overall processing steps.

10.4.1 Biomass Harvesting

Dewatering is the initial stage of biomass harvesting, separating mixed particles from the liquid phase (Deepika et al. 2022). Techniques like centrifugation, filtration, flocculation, sedimentation, and flotation are used to recover microalgae biomass. Centrifugation is the most effective method, with yields of up to 98%. Filtration is efficient and cost-effective, but requires regular replacement of filtration membranes. Flocculation aggregation can enhance efficiency, while sedimentation separates materials based on density or particle size. Drying, a crucial process for food supplements and lipid extraction, requires high energy consumption and can lead to economic bottlenecks. Spray drying is the primary technique for many microalgal taxa (Junior et al. 2020).

A critical economic blockage that limits large-scale commercial production, particularly for food-grade applications, is through harvesting and dewatering of microalgae biomass from dilute cultures (Zhu et al. 2024). Over 20–30% of total algal biomass yield expenses are often documented for these highly cost-intensive processes (Barros et al. 2014). Intensive separation attempts are called for by the observation that cells are kept dispersed due to the microalgae's small size (typically 2–20 μm) and low concentration (Uduman et al. 2010). Significant hurdles that limit industrial scalability and sustainability are exhibited by conventional techniques broadly adopted for microalgae harvesting, namely centrifugation, filtration, and flocculation. The conventional techniques face barriers in efficiency, cost, scalability, and compatibility with food processing, thereby demanding the development of affordable and effective dewatering methods. By reducing chemical use and energy input, the recent research is focusing on emerging milder technologies that are suitable for food-grade production. Table 10.4 represents the emerging microalgal harvesting techniques and their industrial-scale advantages.

10.4.2 Biomass Pretreatment

Cell disruption or biomass pretreatment is a process that breaks down cell walls to release intracellular components and increase solvent extraction accessibility. This process improves mass transfers in the next extraction step, increasing extraction yields and efficiencies. Three types of cell disruption are enzymatic, chemical, and mechanical.

Mechanical methods include high-pressure homogenization (HPH), bead milling, ultrasonication, pulsed electric field therapy, microwaving, and osmotic shock treatment (Halim 2020). HPH is more effective than ultrasonication and bead milling, but its efficiency varies across species (Dong et al. 2016). Bead milling is used for lipid and pigment extraction, but can lead to temperature denaturation and impair solvent recovery (Doucha and Lívanský 2008). Ultrasonication increases water-insoluble pigment production but has a slower disruption rate and higher energy consumption (Halim 2020). Pulsed electric field treatment increases lutein extraction yields but can lead to emulsion formation (Luengo et al. 2015). Enzymatic treatments hydrolyze algal cell walls for various applications, but they are expensive and require recovery and reuse for economic scalability (Halim 2020; Kobayashi et al. 1997). Chemical hydrolysis targets specific wall linkages and dissolves the entire wall or wall layer, but has slower kinetics (Halim 2020).

10.4.3 Pigment Extraction

The extraction of intracellular pigments from debris following cell rupture can be achieved using various methods, including supercritical carbon dioxide extraction, organic solvent extraction, pressurized solvent extraction, and ionic liquid extraction (Halim 2020; Gogate and Joshi 2020). Organic solvent extraction is a popular technique for water-insoluble pigments like carotenoids and chlorophylls. However, it has drawbacks such as high solvent consumption, selectivity, and extraction efficiency (Halim 2020). Supercritical extraction, which uses a low-polarity solvent like supercritical CO_2, is the most widely used method for extracting carotenoids and chlorophylls from microalgae and cyanobacteria. It is more environmentally friendly and takes less time to extract the material (Pagels et al. 2021). Subcritical fluid extraction (SCF) requires lower pressure and temperature settings and uses liquid fuels like dimethyl ether, CO_2, and 1, 1, 2-tetraflu oroethane (Gogate and Joshi 2020). Pressurized liquid extraction (PLE) uses liquid solvents at greater pressures and temperatures, resulting in higher solubility and mass transfer rates. Ionic liquids, salts of loosely bound anions and cations, maintain their liquid state throughout a broad temperature range and can be regulated through ion exchange or combination. They are an effective extracting solvent and a desirable substitute for organic solvents in various extraction processes, including the extraction of microalgal pigments. However, ionic liquids are costly and have low biodegradability, and their high temperature requirement can potentially harm the integrity of targeted pigments and other important components in the microalgal biomass (Halim 2020).

10.4.4 Pigment Purification

Cyanobacterial or microalgal biomass contains pigments that contain contaminants, which can be removed using various chemical and chromatographic procedures. These include liquid chromatography, spectrophotometric analysis, and thin-layer chromatography for chlorophyll and carotenoids. Chlorophylls in astaxanthin and β-carotene extracts are converted into water-soluble salts that precipitate when reacting with calcium hydroxide or acid. These salts can be removed using solid–liquid separation techniques (Halim 2020; Fujii 2012). The first step in purifying phycobiliproteins is initial lysis, followed by aqueous two-phase purification, gel permeation chromatography, anionic chromatography, and ultrafiltration (Vadrale et al. 2023).

10.5 Applications

Microalgal pigments possess antioxidant, anti-inflammatory, neuroprotective, and hepatoprotective effects, making them ideal for industrial applications. They are already widely utilized in a number of sectors, including cosmetics, food, medicines, nutraceuticals, and aquaculture. Some of the industries are already making use of β-carotene from *Dunaliella,* astaxanthin from *Haematococcus*, and phycocyanin and other phycobiliproteins from *A. platensis*. Because microalgal pigments work well as labels for antibodies and receptors, they have also been used in clinical and research settings. Furthermore, there is growing interest in the use of microalgal pigments in aquaculture for feed composition and supplementation because of the environmentally beneficial aspects of microalgae production.

10.5.1 Nutraceuticals and Pharmaceuticals

The class of microalgal pigments that has been investigated the most in terms of bioactivities are called carotenoids, which include xanthophyll and carotenes. Lycopene has also been shown in experimental and epidemiological studies to function as a chemopreventive agent against breast, stomach, and lung cancer. Numerous research studies have indicated that β-carotene can lower the risk of certain malignancies, including skin and stomach cancer (Honda 2020). Dietary β-carotene may protect against age-related macular degeneration. β-Carotene exhibits a number of health benefits, particularly significant provitamin A activity; however, the right amount must be taken into consideration. Through several molecules and pathways, astaxanthin demonstrates anti-apoptotic, antiproliferative, and anti-invasion properties. Consuming astaxanthin on a daily basis may lessen eye strain, such as that experienced by visual display terminal employees, and it may also enhance the blood flow to the retinal capillaries of the eye. Furthermore, a number of research investigations have documented the accumulation of dietary astaxanthin in the brains of animals, suggesting that the carotenoid may have potential use in the

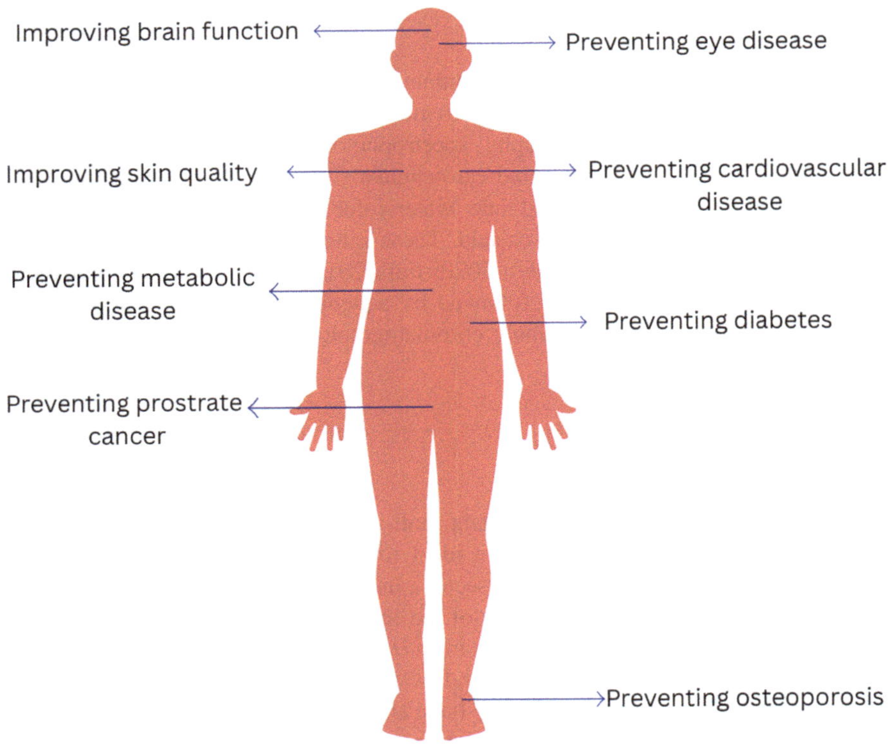

Fig. 10.2 Typical health effects of carotenoids

prevention and/or treatment of Alzheimer's disease (Honda 2020). Figure 10.2 illustrates the potential health benefits of the intervention across multiple organ systems, including brain, skin, eye, cardiovascular, metabolic, endocrine, skeletal, and prostate health.

PBPs are utilized as potent nutraceutical and pharmaceutical ingredients to support the body's natural healing process and to develop innovative pharmaceutical medications that may both prevent and treat recently acquired illnesses. Numerous studies have revealed a variety of biological and pharmacological characteristics of PBPs, such as antioxidant, anti-inflammatory, anticancer, hepatoprotective, and free-radical scavenger activities, that both persistently promote health and impeccably induce it. PBPs have a strong ability to scavenge harmful free radicals, such as peroxynitrile, due to their interaction with apoprotein residues like tryptophan and tyrosine. This ability can lead to the promotion of body health and life maintenance, including the prevention of diseases like cancer that are caused by oxidative stress. The notable antioxidant effect of PBPs, particularly phycocyanin, is associated with their scavenging ability. PBPs have two primary uses in the pharmaceutical and nutraceutical industries: as a natural, nontoxic colorant and, secondly, as a therapeutic agent (Manirafasha et al. 2020; Tounsi et al. 2023).

10.5.2 Food Industry

Chlorophylls, like the majority of natural pigments, are used in food additives for a variety of purposes, with colorants being the most widely used. Natural green colorants are increasingly used by the food industry to impart, intensify, or standardize food color. This allows food to offer the ideal color intensity and hue to meet consumer expectations, requirements, and wants, as opposed to synthetic pigments that may have adverse effects (Sun et al. 2023; Fernandes et al. 2020).

With their exceptional color characteristics and bioactivity, natural carotenoids are the best option available to food producers for colorants, particularly in baked goods, soft drinks, and grilled sausages. Furthermore, because of its exceptional antioxidative and preservative properties, food retains its original flavor and color while FAs and other components are prevented from oxidizing and degrading. Consequently, they have a noticeable impact and increase the desire for foods such as cakes, butter, margarine, milk products, and soft drinks that contain more FAs. To give them a distinctive and striking color, lutein is now widely added to flavorings, tobacco, confections, pastries, baby formula, and a variety of feed items (Sun et al. 2023; Farkas et al. 2020).

PBPs are widely used in the commercial sector as natural colorants. Patents have been awarded for the red or pinkish-red PE food colorant, which is used in confections, syrups, dairy products, baked goods, gelatin desserts, dried goods, fermented milk products, ice cream, and milkshakes (Dufossé et al. 2005; Mishra et al. 2008). By adding its yellow fluorescence properties, PE can also be employed to improve the fluorescence and consumer appeal of alcoholic beverages, soft drinks, and cake decorations. Because cyanobacteria PC naturally has a blue hue, it is used in Japan to color yogurt, ice cream, and soft drinks. It can be found in chewing gum, candies, popsicles, jellies, dairy goods, and soft drinks (Sun et al. 2023; Galetović and Dufossé 2020).

10.5.3 Cosmetics and Cosmeceuticals

Microalgae-derived pigments have been used to create a variety of skin care products, including antiaging, sunscreen, emollient, and anti-irritant creams. The major pigment of microalgae, known as carotenoid pigments (astaxanthin, lutein, zeaxanthin, and canthaxanthin), is found in large quantities in *Haematococcus* and *Dunaliella*. These pigments have antioxidant and UV protection qualities. It has been established that the orange pigment violaxanthin, which is derived from microalgae, provides photoprotective benefits for human skin fibroblasts exposed to UVB radiation. It has been discovered that microalgae pigment has strong antiaging properties for the skin. One of the pigments found in microalgae, β-carotene, has been reported to have great potential for antiaging. Antioxidants that are particularly potent include astaxanthin. Compared to α-tocopherol, astaxanthin has greater promise for use in antiaging therapy due to its exceptional antioxidant properties. Products of microalgae metabolism, such as carotenoids, are frequently utilized in healthcare applications to prevent oxidative stress because of their well-known high antioxidant qualities. Numerous cosmetic products have been developed as a result of research on the cosmeceutical qualities of microalgae pigments. One of such items is a pill form for tanning. Canthaxanthin is the most often found ingredient in tanning tablets. Colorants for cosmetics can be derived from microalgal pigments (Pangestuti et al. 2020).

10.5.4 Aquaculture Feed

In salmon farming, fish color is very important. Usually, the main source of carotenoids is synthetic astaxanthin. As an alternative, salmonid fish, trout, and poultry have been fed canthaxanthin, astaxanthin, and lutein from *Chlorella* sp. on a regular basis to enhance their reddish color or the yellowish hue of their egg yolks. Furthermore, astaxanthin can improve the immunity of shrimp and fish, facilitating effective growth and survival. It should be mentioned that the United States Food and Drug Administration has approved astaxanthin as a color enhancer for fish feed. Additionally, feed contains β-carotene as a colorant to enhance the appearance of fish and shellfish (Yusoff et al. 2020; Fayaazuddin et al. 2023).

10.6 Cost Analysis and Market Value

For a technology to be considered truly sustainable, economic feasibility must coincide with its environmental benefits. A major obstacle for commercialization of microalgal pigment production is high-cost production, especially during energy intensive downstream processing, even though it is a promising green technology. An in-depth financial analysis is essential to demonstrate the long-term capacity of this technology. Table 10.5 compares the cost structure of open raceway ponds (ORPs) and photobioreactors (PBRs), showing that PBRs require substantially higher capital and operating expenditures, with energy as the dominant OPEX component and ORPs yielding a lower biomass production cost per kilogram.

Table 10.5 Comparative financial analysis of microalgae biomass production from open raceway ponds (ORP) and tubular photobioreactors (PBR) (Davis et al. 2011)

Cost type	Open raceway pond (ORP)	Photobioreactor (PBR)	Key takeaway
Capital expenditure	$2,423,000 USD	$12,940,000 USD	PBRs have a significantly higher initial investment.
Total annual operating expenditure (OPEX)	$238,000 USD	$2,257,000 USD	PBRs have significantly higher annual operating costs.
Highest OPEX component	Energy (66%)	Energy (88%)	Primary cost driver for both systems.
Production cost (per kg of biomass)	$0.73/kg	$2.71/kg	ORPs are more cost-effective.

10.7 Industrialization and Future Research Directions

In order to predict the commercial viability of microalgae bioprocess, modern computing methods are vital. The key strategy to optimize the cultivation parameters, yield prediction and to increase the overall efficiency of bioprocess includes artificial intelligence (AI) and machine learning (ML), which helps to overcome persistent challenges in the industry (Yusof et al. 2024).

The rapid growth, high pigment content, and the cultivation method that does not require land make microalgae key producers of natural pigments such as chlorophylls, carotenoids, and phycobiliproteins for the food, nutraceutical, and pharmaceutical industry. The next wave of research is surely about scaling up sustainably and economically. Development of hybrid system which compose of closed photobioreactor and open ponds is included in next-gen cultivation, which balances high productivity and low cost. Integration of wastewater treatment and the microalgae cultivation enables the recycling of nutrients as well as reducing environmental impacts (Benedetti et al. 2018). Desired pigment production can also be obtained by redirecting the metabolic fluxes toward pigment pathways to increase photosynthetic ability with the use of metabolic and genome engineering techniques like CRISPR/Cas9 and chloroplast/nuclear genome editing (Tran et al. 2025).

10.8 Conclusion

Microalgae-derived pigments have shown great promise as sustainable substitutes for synthetic colorants in a range of industries, such as food, pharmaceuticals, and cosmetics. They differ from traditional synthetic alternatives due to their nutritional and bioactive qualities.

Microalgae are known to be abundant suppliers of carotenoids, phycobiliproteins, and chlorophylls. These pigments have several health advantages, such as antioxidant effects, in addition to producing vivid hues. Microalgae pigments are in higher demand in the food industry as colorants and nutraceuticals due to the growing customer desire for natural ingredients. They minimize the impact on the environment because they are a more environmentally friendly substitute for conventional synthetic colors.

In order to improve pigment stability and bioavailability, there are still a number of obstacles to overcome in the process of optimizing extraction and production processes. Scaling up production and increasing economic viability require ongoing research and development activities aimed at strengthening cultivation methods and biorefinery procedures.

Microalgae-derived pigments have a bright prospect ahead of them, with expected improvements in market value and a wide range of uses. This trajectory is consistent with the growing demand for environmentally friendly and health-conscious products, as well as a shift toward more sustainable operations. A more inventive and sustainable industrial environment will surely result from the ongoing research and development of microalgal pigments.

References

Acién FG, Molina E, Reis A, Torzillo G, Zittelli GC, Sepúlveda C, Masojídek J (2017) Photobioreactors for the production of microalgae. In: Microalgae-based biofuels and bioproducts. Woodhead Publishing, United Kingdom, pp 1–44

Barros AI, Gonçalves AL, Simões M, Pires JCM (2014) Harvesting techniques applied to microalgae: a review. Renew Sustain Energy Rev [Internet] 41:1489–1500

Barros A, Pereira H, Campos J, Marques A, Varela J, Silva J (2019) Heterotrophy as a tool to overcome the long and costly autotrophic scale-up process for large scale production of microalgae. Sci Rep 9(1):13935

Begum H, Yusoff FM, Banerjee S, Khatoon H, Shariff M (2016) Availability and utilization of pigments from microalgae. Crit Rev Food Sci Nutr 56(13):2209–2222

Benedetti M, Vecchi V, Barera S, Dall'Osto L. (2018) Biomass from microalgae: the potential of domestication towards sustainable biofactories. Microbial Cell Factor [Internet] 17(1):173

Borowitzka MA (1990) The mass culture of *Dunaliella salina*. In: Technical resource papers regional workshop on the culture and utilization of Seaweeds, vol 2

Braga AR, de Rosso VV (2020) Analytical protocols in the measurement of pigments' bioavailability. In: Pigments from microalgae handbook. Springer International Publishing, Cham, pp 229–240

Brusca S, Famoso F, Lanzafame R, Messina M, Wilson JS (2017) A site selection model to identify optimal locations for microalgae biofuel production facilities in sicily (Italy). Int J Appl Eng Res 12:16058

Capelli B, Bagchi D, Cysewski GR (2013) Synthetic astaxanthin is significantly inferior to algal-based astaxanthin as an antioxidant and may not be suitable as a human nutraceutical supplement. Nutrafoods 12(4):145–152

Chen Q, Chen Y, Xiao L, Li Y, Zou S, Han D (2022) Co-production of lutein, zeaxanthin, and β-carotene by utilization of a mutant of the green alga *Chromochloris zofingiensis*. Algal Res 68: 102882

Chini Zittelli G, Lauceri R, Faraloni C, Silva Benavides AM, Torzillo G (2023) Valuable pigments from microalgae: phycobiliproteins, primary carotenoids, and fucoxanthin. Photochem Photobiol Sci 22(8):1733–1789

da Silva JC, Lombardi AT (2020) Chlorophylls in microalgae: occurrence, distribution, and biosynthesis. In: Pigments from microalgae handbook. Springer International Publishing, Cham, pp 1–18

Davis R, Aden A, Pienkos PT (2011) Techno-economic analysis of autotrophic microalgae for fuel production. Appl Energy [Internet] 88(10):3524–3531

Deepika C, Wolf J, Roles J, Ross I, Hankamer B (2022) Sustainable production of pigments from Cyanobacteria. In: Cyanobacteria in biotechnology: applications and quantitative perspectives. Springer International Publishing, Cham, pp 171–251

Del Campo JA, Rodriguez H, Moreno J, Vargas MA, Rivas J, Guerrero MG (2004) Accumulation of astaxanthin and lutein in *Chlorella zofingiensis* (Chlorophyta). Appl Microbiol Biotechnol 64(6):848–854

Dong T, Knoshaug EP, Pienkos PT, Laurens LM (2016) Lipid recovery from wet oleaginous microbial biomass for biofuel production: a critical review. Appl Energy 177:879–895

Doucha J, Lívanský K (2008) Influence of processing parameters on disintegration of *Chlorella* cells in various types of homogenizers. Appl Microbiol Biotechnol 81(3):431–440

Duan X, Xie C, Hill DR, Barrow CJ, Dunshea FR, Martin GJ, Suleria HA (2024) Bioaccessibility, bioavailability and bioactivities of carotenoids in microalgae: a review. Food Rev Intl 40(1): 230–259

Dufossé L, Galaup P, Yaron A, Arad SM, Blanc P, Murthy KNC et al (2005) Microorganisms and microalgae as sources of pigments for food use: a scientific oddity or an industrial reality? Trends Food Sci Technol [Internet] 16(9):389–406

Farkas Á, Bencsik T, Deli J (2020) Carotenoids as food additives. In: Pigments from microalgae handbook. Springer International Publishing, Cham, pp 421–447

Fayaazuddin T, Prakash P, Shakena Fathima T, Dhanasekaran D (2023) Commercial astaxanthin production from green alga *Haematococcus pluvialis*. In: Food microbiology based entrepreneurship: making money from microbes. Springer Nature Singapore, Singapore, pp 279–304

Fernandes AS, Nass PP, Oliveira Á, Zepka LQ (2020) Chlorophylls as food additives. In: Pigments from microalgae handbook. Springer International Publishing, Cham, pp 391–420

Fujii K (2012) Process integration of supercritical carbon dioxide extraction and acid treatment for astaxanthin extraction from a vegetative microalga. Food Bioprod Process 90(4):762–766

Galetović A, Dufossé L (2020) Phycobiliproteins as food additives. In: Pigments from microalgae handbook. Springer International Publishing, Cham, pp 559–573

Gille A, Trautmann A, Posten C, Briviba K (2016) Bioaccessibility of carotenoids from *Chlorella vulgaris* and *Chlamydomonas reinhardtii*. Int J Food Sci Nutr 67(5):507–513

Gogate PR, Joshi SM (2020) Process intensification aspects of extraction of pigments from microalgae. In: Pigments from microalgae handbook. Springer International Publishing, Cham, pp 309–324

Granado-Lorencio F, Herrero-Barbudo C, Acién-Fernández G, Molina-Grima E, Fernández-Sevilla JM, Pérez-Sacristán B, Blanco-Navarro I (2009) In vitro bioaccesibility of lutein and zeaxanthin from the microalgae *Scenedesmus almeriensis*. Food Chem 114(2):747–752

Halim R (2020) Industrial extraction of microalgal pigments. In: Pigments from microalgae handbook. Springer International Publishing, Cham, pp 265–308

Hassaan MS, Mohammady EY, Soaudy MR, Sabae SA, Mahmoud AM, El-Haroun ER (2021) Comparative study on the effect of dietary β-carotene and phycocyanin extracted from *Spirulina platensis* on immune-oxidative stress biomarkers, genes expression and intestinal enzymes, serum biochemical in Nile tilapia, Oreochromis niloticus. Fish Shellfish Immunol 108:63–72

Honda M (2020) Nutraceutical and pharmaceutical applications of carotenoids. In: Pigments from microalgae handbook, Switzerland, pp 449–469

Junior WG, Gorgich M, Corrêa PS, Martins AA, Mata TM, Caetano NS (2020) Microalgae for biotechnological applications: cultivation, harvesting and biomass processing. Aquaculture 528: 735562

Kannaujiya VK, Singh PR, Kumar D, Sinha RP (2020) Phycobiliproteins in microalgae: occurrence, distribution, and biosynthesis. In: Pigments from microalgae handbook. Springer International Publishing, Cham, pp 43–68

Kim M, Ahn J, Jeon H, Jin E (2017) Development of a *Dunaliella tertiolecta* strain with increased zeaxanthin content using random mutagenesis. Mar Drugs 15(6):189

Kobayashi M, Kurimura Y, Sakamoto Y, Tsuji Y (1997) Selective extraction of astaxanthin and chlorophyll from the green alga *Haematococcus pluvialis*. Biotechnol Tech 11(9):657–660

Kumar K, Mishra SK, Shrivastav A, Park MS, Yang JW (2015) Recent trends in the mass cultivation of algae in raceway ponds. Renew Sust Energ Rev 51:875–885

Li J, Zhao X, Chang JS, Miao X (2022) A two-stage culture strategy for *Scenedesmus* sp. FSP3 for CO2 fixation and the simultaneous production of lutein under light and salt stress. Molecules 27(21):7497

Liang Y, Kaczmarek MB, Kasprzak AK, Tang J, Shah MM, Jin P, Klepacz-Smółka A, Cheng JJ, Ledakowicz S, Daroch M (2018) Thermosynechococcaceae as a source of thermostable C-phycocyanins: properties and molecular insights. Algal Res 35:223–235

Liu Y, Feng Y, Lun J (2012) Aqueous two-phase countercurrent distribution for the separation of c-phycocyanin and allophycocyanin from *Spirulina platensis*. Food Bioprod Process 90(2): 111–117

Luengo E, Martínez JM, Bordetas A, Álvarez I, Raso J (2015) Influence of the treatment medium temperature on lutein extraction assisted by pulsed electric fields from *Chlorella vulgaris*. Innovative Food Sci Emerg Technol 29:15–22

Manirafasha E, Guo L, Jing K (2020) Nutraceutical and pharmaceutical applications of phycobiliproteins. In: Pigments from microalgae handbook. Springer International Publishing, Cham, pp 575–584

Maroneze MM, Dias RR, Severo IA, Queiroz MI (2020) Microalgae-based processes for pigments production. In: Pigments from microalgae handbook. Springer International Publishing, Cham, pp 241–264

Mishra SK, Shrivastav A, Mishra S (2008) Effect of preservatives for food grade C-PC from Spirulina platensis. Proc Biochem [Internet] 43(4):339–345

Mohamadnia S, Tavakoli O, Faramarzi MA, Shamsollahi Z (2020) Production of fucoxanthin by the microalga *Tisochrysis lutea*: a review of recent developments. Aquaculture 516:734637

Pagels F, Pereira RN, Vicente AA, Guedes AC (2021) Extraction of pigments from microalgae and cyanobacteria—a review on current methodologies. Appl Sci 11(11):5187

Pangestuti R, Suryaningtyas IT, Siahaan EA, Kim SK (2020) Cosmetics and cosmeceutical applications of microalgae pigments. In: Pigments from microalgae handbook. Springer International Publishing, Cham, pp 611–633

Patel A, Mishra S, Pawar R, Ghosh PK (2005) Purification and characterization of C-Phycocyanin from cyanobacterial species of marine and freshwater habitat. Protein Expr Purif 40(2):248–255

Pelah D, Sintov A, Cohen E (2004) The effect of salt stress on the production of canthaxanthin and astaxanthin by *Chlorella zofingiensis* grown under limited light intensity. World J Microbiol Biotechnol 20(5):483–486

Posten C, Walter C (eds) (2012) Microalgal biotechnology: potential and production. Walter de Gruyter

Prabakaran G, Sampathkumar P, Kavisri M, Moovendhan M (2020) Extraction and characterization of phycocyanin from *Spirulina platensis* and evaluation of its anticancer, antidiabetic and antiinflammatory effect. Int J Biol Macromol 153:256–263

Rahman DY, Sarian FD, Van Wijk A, Martinez-Garcia M, Van der Maarel MJ (2017) Thermostable phycocyanin from the red microalga *Cyanidioschyzon merolae*, a new natural blue food colorant. J Appl Phycol 29(3):1233–1239

Román RB, Alvárez-Pez JM, Fernández FA, Grima EM (2002) Recovery of pure B-phycoerythrin from the microalga *Porphyridium cruentum*. J Biotechnol 93(1):73–85

Saini DK, Chakdar H, Pabbi S, Shukla P (2020) Enhancing production of microalgal biopigments through metabolic and genetic engineering. Crit Rev Food Sci Nutr 60(3):391–405

Shanmugam A, Sigamani S, Venkatachalam H, Jayaraman JD, Ramamurthy D (2017) Antibacterial activity of extracted phycocyanin from *Oscillatoria* sp. J Appl Pharm Sci 7(3):062–067

Shen CT, Chen PY, Wu JJ, Lee TM, Hsu SL, Chang CM, Young CC, Shieh CJ (2011) Purification of algal anti-tyrosinase zeaxanthin from *Nannochloropsis oculata* using supercritical anti-solvent precipitation. J Supercrit Fluids 55(3):955–962

Sun H, Wang Y, He Y, Liu B, Mou H, Chen F, Yang S (2023) Microalgae-derived pigments for the food industry. Mar Drugs 21(2):82

Tan HT, Yusoff FM, Khaw YS, Noor Mazli NA, Nazarudin MF, Shaharuddin NA, Katayama T, Ahmad SA (2022) A review on a hidden gem: Phycoerythrin from blue-green algae. Mar Drugs 21(1):28

Tomaselli L (2004) The microalgal cell. In: Handbook of microalgal culture: biotechnology and applied phycology, UK, vol 1, pp 3–19

Torzillo G, Chini ZG (2015) Tubular photobioreactors. In: Algal biorefineries: volume 2: products and refinery design. Springer International Publishing, Cham, pp 187–212

Tounsi L, Ben Hlima H, Hentati F, Hentati O, Derbel H, Michaud P, Abdelkafi S (2023) Microalgae: a promising source of bioactive phycobiliproteins. Mar Drugs 21(8):440

Tran QG, Le TT, Choi DY, Cho DH, Yun JH, Choi HI et al (2025) Progress and challenges in CRISPR/Cas applications in microalgae. J Microbiol [Internet] 63(3):e2501028

Uduman N, Qi Y, Danquah MK, Forde GM, Hoadley A (2010) Dewatering of microalgal cultures: a major bottleneck to algae-based fuels. J Renew Sustain Energy [Internet] 2(1):012701

Vadrale AP, Dong CD, Haldar D, Wu CH, Chen CW, Singhania RR, Patel AK (2023) Bioprocess development to enhance biomass and lutein production from *Chlorella sorokiniana* Kh12. Bioresour Technol 370:128583

Xia S, Gao B, Fu J, Xiong J, Zhang C (2018) Production of fucoxanthin, chrysolaminarin, and eicosapentaenoic acid by *Odontella aurita* under different nitrogen supply regimes. J Biosci Bioeng 126(6):723–729

Yusof Z, Tong YW, Selvarajoo K, Parakh SK, Foo SC (2024) Overcoming challenges in microalgal bioprocessing through data-driven and computational approaches. Curr Opin Food Sci [Internet] 61:101253

Yusoff FM, Banerjee S, Nagao N, Imaizumi Y, Shariff M, Toda T (2020) Use of microalgae pigments in aquaculture. In: Pigments from microalgae handbook. Springer International Publishing, Cham, pp 471–513

Zhu J, Wakisaka M, Omura T, Yang Z, Yin Y, Fang W (2024) Advances in industrial harvesting techniques for edible microalgae: recent insights into sustainable, efficient methods and future directions. J Clean Prod [Internet] 436:140626

11 Microalgae-Based Biofuel Generation, Value-Added Production and Bioremediation: An Implication for Sustainability and Circular Economy

Sudha Sahay, Brigita Jain, Devanshi Trivedi, Antony Suresh, and Shailesh Dave

Abstract

Microalgae have several advantages over traditional oil-producing crops, such as a high rate of photosynthesis leading to faster cell growth and development, and high potential to produce biofuels and high-value by-products. Microalgae have robust environmental rigidity, no competition with food or agricultural land, rapid sequestration of environmental carbon, cultivation on wastewater, and time-round cultivation and harvesting. There are different groups of microalgae, colonized in different geographical areas, that retain different characteristics. Selection of an implicit microalgae species to produce its bulk biomass and fulfil

S. Sahay (✉)
Loyola Centre for Research and Development, Xavier Research Foundation, Navrangpura, Ahmedabad, Gujarat, India

Department of Microbiology and Biotechnology, University School of Science, Gujarat University, Ahmedabad, Gujarat, India
e-mail: Sudha.sahay@xrf.res.in

B. Jain
Loyola Centre for Research and Development, Xavier Research Foundation, Navrangpura, Ahmedabad, Gujarat, India

Department of Life Science, University School of Science, Gujarat University, Ahmedabad, Gujarat, India

D. Trivedi · S. Dave
Loyola Centre for Research and Development, Xavier Research Foundation, Navrangpura, Ahmedabad, Gujarat, India

A. Suresh
Loyola Centre for Research and Development, Xavier Research Foundation, Navrangpura, Ahmedabad, Gujarat, India

St. Xavier's College, Navrangpura, Ahmedabad, Gujarat, India

K. Mohanty et al. (eds.), *Advances in Algal Research*,
https://doi.org/10.1007/978-981-95-7234-2_11

the product target of a specific by-product is a challenge. To establish a microalgae production technology, there are several obstacles across the upstream and downstream processes. Researchers are searching for results to overcome the challenges in microalgae technology, especially for the biofuel product. In this endeavor, we have cultivated and characterized several microalgae species isolated from the Sabarmati River water, Ahmedabad, Gujarat, India. Among all, one of the microalgae shown to produce loftiest biomass and crude oil yield, ~2 g/L and ~1.3 g/L, respectively, in a modified artificial saline medium (ASM). The microalgae are identified as *Micractinium reisseri* after the molecular characterization. The analysis based on the FTIR, TLC, GC, and ELISA techniques demonstrated the importance of *M. reisseri* not only as an implicit feedstock for biofuels production but also for the generation of value-added by-products which have several advantages in areas like medicinals, neutraceuticals, bioplastics, bioethanol, and husbandry. The by-products generated from these processes can increase the eventuality for indirect frugality with zero destruction. The integration of *M. reisseri* cultivation in primarily treated municipal wastewater with anaerobic digestion, composting, and ethanol production processes provides a new model for climate change mitigation of biogenic CO_2 and indirect frugality.

Keywords

Microalgae · Biofuel · Photobioreactor · Raceway pond · Lipid accumulation

11.1 Introduction

Microalgae are the phytoplanktons that include both prokaryotic and eukaryotic organisms. The prokaryotes are mainly blue-green algae, also known as cyanobacteria, which belong to the Kingdom Monera and the Division Cyanophyta. The eukaryotic microalgae include diatoms and green microalgae, also known as chlorophyta, which belong to the Kingdom Protista. These organisms are oxygen-producing, photosynthetic, and non-cohesive in nature (Heimann and Huerlimann 2015; Kumar et al. 2021; Rahman 2020). There are about 72,500 microalgal species that have been identified, encompassing both cyanobacteria and eukaryotic microalgae like diatoms and green microalgae (Kang et al. 2023). According to Norton et al. (1996), there may be around 200,000 or more microalgal species; however, only 70,000 to 80,000 have been identified globally. Of these, only about fifteen strains are currently cultivated on an industrial scale. There are several difficulties reported while cultivating the microalgae at a commercial scale (Kumar et al. 2024). Three cultivation methods widely employed at the research and commercial levels are closed, open, and hybrid systems (Jeevanandam et al. 2020). Each method has its own pros and cons. For the adoption of a successful microalgae cultivation technique at a pilot scale, there are several factors one has to consider in terms of economic feasibility. The important environmental factor influencing microalgal growth and development are the light, temperature, CO_2, salinity, pH, and available nutrients such as nitrogen, potassium, phosphorus, and other elements (Venckus et al. 2021).

Microalgae are known for their distinct chemical makeup, which includes a wide array of extracellular and intracellular compounds, such as lipids, proteins, carbohydrates, vitamins, minerals, and other organic substances (Sun et al. 2022; Chen et al. 2020). These bioactive factors have attracted worldwide attention due to their implicit health benefits, including anti-antipathetic (Fleurence and Levine 2018), anti-inflammatory (Prabakaran et al. 2019), antibacterial (Jena 2019), anti-tumor (Zhang et al. 2019), antiviral (Reynolds et al. 2021), neuroprotective (Gallego et al. 2022), nutraceuticals, medicinals, and cardiovascular protective (Kumar et al. 2022). Cyanobacteria, diatoms, and green microalgae are increasingly viewed as promising organisms for the sustainable production of biofuels as well as CO_2 sequestration and bioremediation (Borowitzka 2013; Mishra et al. 2008). Despite the vast diversity of microalgae in nature, only a few strains are employed commercially. Among these, *Arthrospira platensis* (spirulina) and *Chlorella vulgaris* dominate the global demand, particularly in health and nutrition sectors (Koyande et al. 2019). Several strains fail to achieve marketable-scale cultivation due to factors analogous as low biomass productivity, slow growth, lack of rigidity, or less value-added products, all of which contribute to high production costs (Borowitzka 2013; Sahay and Braganza 2016). Another limiting factor is the continued use of outdated technologies in large-scale microalgae processing. Therefore, advancements in sustainable technologies for both upstream (mass cultivation) and downstream (harvesting, extraction, and fractionation) processes are essential for exploring better operations of microalgae. Selection of a potential microalgae species to fulfil the demand and supply is equally important. As mentioned earlier, several strains failed to achieve marketable-scale production due to factors associated with in-built character or externally controlled. Finding a microalgae species with a potential growth rate and oil yield is the demand of today. Many researchers are in a process of isolating new microalgae species or trying to improve the potential of available microalgae by using classical or molecular approaches. Sahay and Braganza (2022) have reported *Micractinium reisseri* for the first time in India, isolated in the laboratory from Sabarmati river water, Ahmedabad, Gujarat. In the modified artificial saline medium, the *M. reisseri* is able to produce ~2 g/L dry biomass and ~1.3 g/L crude oil after 24 days of incubation. Further, multiple efforts have been made to optimize the parameters in the primarily treated municipal wastewater for the successful cultivation of *M. reisseri.* Recently, the scientists have succeeded in growing *M. reisseri* in wastewater and helped to remove various pollutants to improve the quality of water (unpublished data). Besides, it also contributes to produce value-added products. There were about eight pigments, including neoxanthin, carotenoids, and chlorophylls detected in *M. reisseri* biomass (Fig. 11.1). Thin-layer chromatography of the crude solvent extracts revealed the presence of 3–4 pigment fractions extracted from the green *M. reisseri* strains. Chlorophyll a and β-carotene are the dominant pigments present in the bench-scale microalgae culture extract, whereas pheophytin was clearly observed in the large-

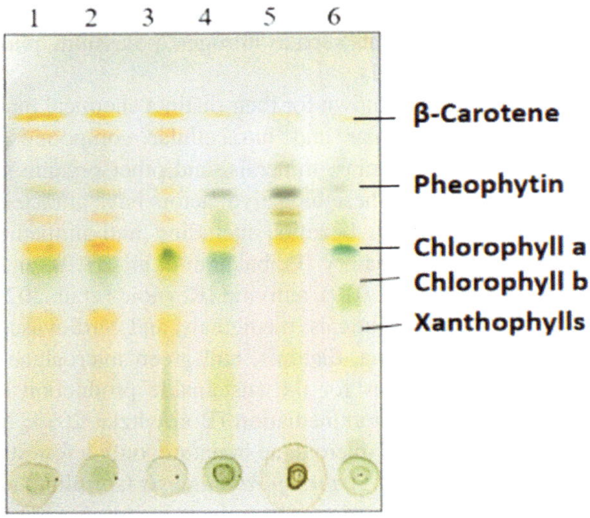

Fig. 11.1 Pigment assay of *M. reisseri* solvent extract by using thin layer chromatography (TLC) technique. Lanes 1, 2, and 3: Dry biomass of bench-scale culture (lab conditions) and Lane 4, 5, and 6: Dry biomass of raceway pond culture (outdoor conditions)

scale raceway pond culture extract. The identified pigments were xanthophylls, chlorophyll b, chlorophyll a, and β-carotene, and their Rf values were 0.45, 0.52, 0.61, and 0.94, respectively. Rf values were compared with the literature data. The study also included the crude oil production bioenergy conversion followed by an essential analysis in the de-oiled biomass for implicit use as a soil nutrient enhancer.

11.2 Microalgae Cultivation and Harvesting

Microalgae are generally cultivated in either open raceway ponds or closed photobioreactors (Nayak et al. 2016). The microalgae needed water, carbon dioxide, and a nutrient-rich culture medium containing essential elements for their growth and development (Abou-Shanab et al. 2014; Baicha et al. 2016). The substantial volume of water needed frequently becomes a major contributor to the overall cost and energy consumption of the cultivation process. Other functional factors that significantly contribute to the cost and energy consumption for biomass yield are design and fabrication of the photobioreactor, artificial light source, aeration, temperature control, agitation, and pH (Cheng et al. 2020; Gutiérrez-Casiano et al. 2022; Javed et al. 2022). The response of photosynthetic organisms like microalgae increases with an increase in light intensity. A group of researchers showed a significant increase in proteins, pigments, and lipids in *Chlorella vulgaris* when the light intensity was increased (Gutiérrez-Casiano et al. 2022). The researchers explained that the increase in color conflation is due to the increase in vacuity of carbon sources. There are four modes of cultivation of microalgae—autotrophic, heterotrophic, photoautotrophic, and purely heterotrophic. A mixotrophic approach is applied where one has to combine the autotrophic and heterotrophic modes of development while using wastewater for the microalgae cultivation. This can enhance the overall microalgae biomass yield, increased fatty acid methyl ester (FAME) content, and reduce chemical oxygen demand (COD) (Javed et al. 2022).

The microalgae cultivation is followed by dewatering the biomass to use it for biofuels and value-added products. There are several challenges in the upstream and downstream processes of microalgae technology. A significant loss of biomass occurs while harvesting the microalgae, leading to hampered algal technology. A microalgal dewatering process that is affordable and effective is pivotal. The self-flocculation method has been applied to determine the fruitful harvesting system that gives a high probability of implicit de-watering and less loss of biomass. Also, if the flocculation is chemical or biological agent-based, then it is less poisonous and also ecofriendly (Zhang et al. 2020). Presently, the entrepreneurs are floundering not only with the query of the algal business for production, but also with high prospects for businesses to be run in a sustainable way (Chen et al. 2016a). The large-sized microalgal cells are occasionally self-flocculated and settled at the bottom of the vessel. Thus, easy to filter and separate. The major difficulty arises with the small to tiny microalgal cells that show a homogenous pattern of growth, as we have found in the case of *M. reisseri*. The harvesting of *M. reisseri* biomass is a major challenge. Because of its small cell size makes it pivotal in the downstream processing of biomass recovery. Thus, we have applied several flocculation methods on *M. reisseri* culture and found a good success in the auto-flocculation method to recover the maximum biomass and oil yield (Sahay et al. 2024). Figure 11.2 shows the auto-flocculation system by aeration applied to *M. reisseri* culture to separate the biomass from the liquid.

Fig. 11.2 Flocculation of *M. resserie* biomass by aeration: (**a**) fresh culture before flocculation, (**b**) culture after 48 h of aeration and auto-flocculation (Sahay et al. 2024)

There is no extra addition of chemical or biological flocculants required in the auto-flocculation method. This approach is not only simple and efficient for industrial-scale applications, but also sustainable and economical, as it eliminates the need for any pre-treatment of the biomass before utilization.

11.3 Microalgae Cultivation Using Wastewater

It is well reported that a large volume of water resources is lost due to the growing population and fast economic development (Fito and Van-Hulle 2021). As a result, the need for effective treatment of wastewater has become an urgent global issue. Wastewater is generated from various sources, including households, municipalities, and industries (such as dye, textile, breweries, tanneries, and dairy industries), as well as from agriculture and aquaculture operations (Natasha et al. 2023; Jiao et al. 2021). The wastewater contains high pollutants like nitrogen, phosphorus (both organic and inorganic forms), heavy metals, and substances contributing to high chemical oxygen demand (COD). These have prompted significant research into effective removal strategies (Chen et al. 2020; Cheng et al. 2020). Traditional treatment methods have helped, but they come with drawbacks such as greenhouse gas emissions and high energy and nutrient demands (Chai et al. 2021).

To promote sustainability and align with circular economy principles, waste should be transformed into valuable resources rather than discarded. Microalgae offer a promising solution in this context—they can remove a broad spectrum of pollutants from wastewater (Altowayti et al. 2022; Dolganyuk et al. 2020), while also capturing carbon dioxide, thus reducing net emissions. Cultivating microalgae in wastewater not only cleans the water but also results in the production of biofuels, valuable biomaterials, and co-products (Goh et al. 2022). However, there is a need for a comprehensive review of current strategies used in microalgal cultivation and the generation of value-added products, to support ongoing research and development across academia and industry. Extensive research is underway to develop greener and more efficient methods for producing microalgal biomass, biofuels, and high-value products using wastewater. For instance, the *Chlorella* species are used to treat municipal wastewater at various stages, such as before and after the primary settling and post-activated sludge treatment. Many studies on *Chlorella, Scanedesmus*, and other microalgae species show that microalgae can help remove heavy metals, carbon, nitrogen, phosphorus, and COD (Annapoorna et al. 2022) from wastewater. Interestingly, bacteria co-cultured with the microalgae enhanced the breakdown of glucose, nitrogen, phosphorus, and COD, although they also reduced the microalgal lipid content (Matter et al. 2019). Some microalgae also form consortia with aerobic or anaerobic microbes, aiding in nutrient recovery (Lisa et al. 2022). Microalgae cultivated in wastewater produce large amounts of hydrocarbons, which can be converted into biofuels such as biogas, biodiesel, biohydrogen, bioethanol, and biobutanol using both thermochemical and biological methods (Lisa et al. 2022). Utilizing wastewater for cultivating microalgae presents an innovative and eco-friendly approach that supports the circular economy and contributes to long-term sustainability. Instead of discarding the wastewater, the wise idea is to clean it and reuse for valuable purposes. Growing microalgae in such nutrient-rich effluents not only facilitates effective nutrient recovery but also results in the simultaneous production of biofuels, bioproducts, and valuable co-products (Cheirsilp et al. 2022; Chen et al. 2016b). Although microalgae can grow in the 100% volume of wastewater, the preliminary optimization of physico-chemical parameters of wastewater is required to make it more compatible with microalgae cultivation (Chen et al. 2016b). Many researchers have recommended pretreating or diluting wastewater prior to microalgal inoculation to improve microalgal growth. Cheng et al. (2020) demonstrated that pre-treatment of swine wastewater using titanium dioxide and intense pulsed light (TiO_2 + IPL) improved the growth of *Tribonema* sp. and *Synechocystis* sp. Similarly, Javed et al. (2022) showed that diluting textile wastewater by 50% significantly enhanced *C. vulgaris* biomass production, which is beneficial for biofuel applications. Pancha et al. (2019) have done an experiment with acid pretreatment of potato juice-based wastewater. The treated wastewater was used to culture *Haematococcus pluvialis* and found to achieve higher yields compared to growth in standard nutrient medium. The *Arthrospira platensis* cultivated in outdoor raceway ponds achieved up to 96% removal of ammonical nitrogen (NH_4^+-N), 87% phosphorus, 82.19% COD, and 60.92% Cr(VI) metal from wastewater (Chew et al. 2017). Markou et al. (2012) reported 73.18% COD removal and 91.19% carbohydrate reduction, along with complete removal of phenols, phosphorus, and nitrates from olive mill wastewater. For economic prospects, the studies suggest that 25% wastewater concentration favors protein accumulation, while 50% concentrations are ideal for lipid and carbohydrate production (Chini et al. 2023). Additionally, decolorizing colored wastewater has been shown to nearly double the growth rate and increase biomass yield of *A. platensis* by 50%. A group of researchers (Markou et al. 2012) used carbon-rich wastewater from sago starch processing, which has been reported to boost microalgal biomass as well as crude protein, lipid, and carbohydrate content, along with reductions of 98% in COD, 99.9% in ammoniacal nitrogen, and 99.4% in phosphate levels in the open raceway pond set up.

Microalgae often coexist with aerobic and anaerobic microbes like bacteria and fungi, which assist in nutrient recycling. Maintaining axenic (pure) microalgal cultures in open systems is challenging, as surrounding environmental factors and water quality significantly influence biomass composition and attract other microbes too. Well-studied genera such as *Chlorella, Scenedesmus, Chlamydomonas*, and natural microalgal consortia are widely recognized for their potential in wastewater treatment and biorefinery applications due to their resilience and valuable by-products (Table 11.1) (Chinnasamy et al. 2010a). In poultry wastewater, *C. vulgaris* effectively removed nitrogen and organic content, while the native bacteria broke down waste to release CO_2, which was then used by the microalgae for photosynthesis, simultaneously producing dissolved oxygen (Gutiérrez-Casiano et al. 2022). All these examples highlight the multiple applications of microalgae, which contribute to making microalgal technology more financially viable. Although capital and operational expenses (CAPEX and OPEX) can be significant, coupling biomass production with wastewater treatment increases economic feasibility.

Table 11.1 Recent updates on microalgae used for the treatment of different wastewater

S.No.	Microalgae sp. used for bioremediation	Source of wastewater
1.	*Chlorella* sp. L38, L166	Organic (soybean)
2.	*Arthrospira* sp.	Dairy, aquaculture, and tannery
3.	*Chlorella variabilis* and *Scenedesmus obliquus*	Dairy
4.	*Tribonema* sp. and *Synechocystis* sp.	Swine
5.	*Coelastrella* sp.	Swine
6.	*Parachlorella kessleri* and *Chlorella vulgaris*	Brewery
7.	*Haematococcus pluvialis*	Organic (Potato)
8.	*Chlorella sorokiniana*	Brewery
9.	*Chlorella vulgaris*	Brewery
10.	*Chlorella vulgaris*	Textile
11.	*Arthrospira platensis*	Dyeing
12.	*Chlorella vulgaris*	Poultry
13.	*Chlorella* sp.	Brewery
14.	*Chlorella* sp. and *Neochloris* sp.	Poultry
15.	*Nanochloropsis*	Dairy
16.	*Tetradesmus obliquus* TS03	Tannery
17.	*Scenedesmus* sp. ASK22	Dairy
18.	*Arthrospira platensis* and *Dunaliella salina*	Saline aquaculture
19.	*Chlorella vulgaris*	Aquaculture
20.	*Microcystis aeruginosa* and *Chroococcus minutus,*	Aquaculture
21.	*Chaetoceros gracilis*, and *Thalassiosira weissflogii*	Dairy

11.4 Extraction of High-Value Compounds

Microalgae have been used in wastewater treatment for decades. The US Department of Energy's Aquatic Species Program emphasized that integrating microalgae-based wastewater treatment with the production of biofuels and coproducts offer a cost-effective approach to generating biodiesel and pharmaceuticals. The increasing relevance of microalgal biotechnology and circular bioeconomy has driven growing interest in utilizing wastewater-cultivated microalgae to produce a broad array of value-added products and by-products (Chinnasamy et al. 2010b). Microalgae are widely utilized as both food and feed in aquaculture. Exploration of microalgae for the biorefinery concept—modeled after traditional petroleum refineries—has emerged as an efficient framework for converting microalgal biomass into a range of high-value end products. Microalgae can generate numerous bioactive compounds suitable for use in food, pharmaceuticals, nutraceuticals, and chemical industries (Table 11.2).

Notably, by-products from microalgae cultivation are rich in nutraceutical components such as polyunsaturated fatty acids (PUFAs), carotenoids, phytosterols, vitamins, and polyphenols—key ingredients in health-promoting products (Su 2021). To enhance the production of such value-added compounds while

Table 11.2 Microalgae-based value products and their application

Microalgae	Product	Application area
Arthrospira platensis	Phycocyanin and biomass	Health food and cosmetics
Chlorella vulgaris	Biomass	Health food supplement
Dunaliella salina	Carotenoides and β-carotene	Health food supplement
Haematococcus pluvialis	Carotenoides and Astaxanthin	Health food and pharmaceuticals
Isochrysis galbana	Fatty acids	Animal nutrition

contributing to a sustainable circular economy, wastewater-based microalgal cultivation has become an increasingly adopted strategy. Moreover, to maximize productivity with zero discharge, the advancements in biotechnology techniques hold great promise. Techniques in metabolic engineering and synthetic biology—including the development of gene promoters, vectors, selectable markers, and gene-editing tools—have enabled precise modification of biosynthetic pathways in microalgae (Otondo et al. 2018). The microalgal-derived compounds offer superior bioactivity and broader industrial applications. Their primary and secondary metabolites serve as robust feedstocks for various bio-products, for example, proteins can be converted into bioplastics (Zhang et al. 2012), carbohydrates are potential sources for bio-hydrogen production, and lipids are precursors for PUFAs like omega-3 fatty acids (Qv et al. 2023). Many industries are now investing in microalgal production platforms to harvest valuable compounds from cultures grown in wastewater. This innovative approach not only mitigates environmental pollution but also transforms waste into valuable bio-resources, aligning with the "waste-to-wealth" model (Premaratne et al. 2022).

11.5 Lipid Accumulation in Microalgae Cultivated in Various Wastewater Types

Lipids are one of the key biochemical components synthesized by microalgae. Several studies have indicated that under certain stress conditions, microalgae can accumulate triacylglycerol (TAG) levels as high as 60%, making them a promising feedstock for biodiesel production (Magro et al. 2012). Stress factors such as nutrient limitation, increased salinity, temperature fluctuations, and variations in light intensity have all been shown to promote lipid accumulation in microalgal cells (Jiang et al. 2015). Using different forms of wastewater as nutrient sources—rich in both organic and inorganic matter—is considered a sustainable and efficient strategy for cultivating microalgae at scale. Research by Chinnasamy et al. (2010b) estimates that biodiesel yields from wastewater-grown microalgae can range from 0.40 to 0.78 tonnes per hectare per year. However, lower lipid concentrations in certain growth conditions remain a limiting factor for optimal biodiesel output. This challenge can potentially be overcome by cultivating microalgae in nutrient-rich wastewater, which encourages higher lipid synthesis. Table 11.3 (adapted from Pancha et al. (2014)) highlights lipid production by different microalgal strains grown in various wastewater types.

Table 11.3 Lipid content of various microalgae grown in various wastewaters (Pancha et al. 2014)

Microalgae	Type of wastewater	Lipid content
Selenastrum minutum	Mix wastewater	37%
Consortia of microalgae	Municipal wastewater	28.50%
Scenedesmus sp.	Domestic wastewater	23.10%
Mix microalgae	Domestic wastewater	26.20%
Coelastrella sp.	Swine wastewater	24.80%
Consortia of microalgae	Dairy farm wastewaters	16.89%
Scenedesmus bijuga	Food-based effluent	35.06%
Chlorella sp.	Piggery wastewater	29.30%
Chlorella vulgaris	Domestic wastewater	32.70%
Chlorella vulgaris	Brewery wastewater	18%
Chlorella zofingiensis	Piggery wastewater	45.81%
Chlorella pyrenoidosa	Municipal wastewater	63.45%
Micractinium reisseri	Municipal wastewater	40%
Chlamydomonas mexicana	Piggery wastewater	33%
Chlamydomonas polypyrenoideum	Dairy wastewaters	42%
Ourococcus multisporus	Municipal wastewater	31%
Euglena sp.	Municipal wastewater	25.60%
Monoraphidium sp.	Municipal wastewater	26%
Scenedesmus sp.	Molasses wastewater	28.90%

Ji et al. (2015) demonstrated that *Scenedesmus obliquus* cultivated in 1% food-processing wastewater removed 38.9 mg/L of total nitrogen (TN) and 12.1 mg/L of total phosphorus (TP), while also storing significant amounts of lipids and carbohydrates—suitable for biodiesel and bioethanol production. These results suggest that the specific microalgal strain and its origin play a crucial role in both wastewater remediation efficiency and the generation of value-added products. Cyanobacteria are another group of microalgae known for producing lipids with important bioactivities. In particular, their lipid profiles are rich in essential fatty acids such as gamma-linolenic acid (GLA), which has been associated with antioxidant and antimicrobial properties. Although borage oil is traditionally the primary source of GLA, cyanobacterial species like *Arthrospira platensis* have also shown high GLA content, accounting for approximately 18–21% of total fatty acids (Phang et al. 2000; Mühling et al. 2005). Strains of *Arthrospira* are especially valued for their unique GLA profiles, which vary among different genera and offer significant biotechnological potential (Phang et al. 2000).

11.6 Economic and Environmental Sustainability

Microalgae are poised to become an advanced bioremediation strategy, effectively treating pollutants from diverse sources such as domestic sewage and industrial effluents. In addition to producing biomass, these organisms are highly efficient in removing contaminants, supporting cleaner ecosystems, and enabling sustainable wastewater treatment solutions. The contribution of microalgae to biorefinery systems is expected to grow significantly. As a renewable feedstock, microalgae support the production of a wide spectrum of high-value compounds—ranging from biofuels to pharmaceuticals and cosmetics. Their utility extends beyond lipids and proteins, fostering innovation in bio-based product development and broadening the application of biorefinery technologies. Microalgal systems are increasingly being integrated into circular bioeconomy frameworks, which emphasize the efficient recycling and reuse of resources within closed-loop systems (Gupta et al. 2024; Hossain and Mahlia 2019). This transition not only reduces environmental burden and waste output but also reinforces sustainable, low-impact industrial practices. These synergistic partnerships enhance nutrient cycling and resilience, enabling more robust and efficient production platforms for bio-based outputs.

11.7 Challenges and Future Perspectives

Future innovations will focus on establishing co-culture systems that combine microalgae with bacteria or fungi to improve nutrient assimilation, biomass productivity, and system stability. The genetic tools—particularly CRISPR and other gene-editing technologies—will facilitate the creation of next-generation microalgal strains with optimized traits for targeted applications (Grama et al. 2022). These customized strains will support enhanced pigment synthesis, pollutant degradation, and biofuel generation, redefining the role of microalgae in sustainable biotechnology. The potential of microalgae largely hinges on progress in metabolic and genetic engineering. Currently, most high-value compounds are derived from a limited number of microalgae genera. However, with the integration of advanced OMICS technologies—such as genomics, proteomics, and metabolomics—there is substantial opportunity to boost the synthesis of specialized metabolites (Chong et al. 2022). Synthetic biology approaches, including pathway engineering and precision gene editing, can significantly improve yield and product specificity. Despite these advancements, slow adoption rates persist, mainly due to limited public awareness and underestimation of the commercial and environmental benefits offered by microalgal biotechnology. To support industrialization, future research should address key challenges in process efficiency and cost reduction.

11.8 Conclusion

Addressing global water scarcity has become an urgent priority, demanding innovative and sustainable wastewater treatment solutions. Microalgae-based technologies offer a promising dual-purpose approach—simultaneously treating wastewater and generating biomass for a range of high-value products within an integrated biorefinery framework. This approach represents a "win–win" scenario, effectively combining environmental management with the production of renewable resources.

The findings of the present study have presented the potential to produce biodiesel from the microalgae *M. reisseri*—a model for neutral lipid production. These findings emphasize the profitable eventuality of *M. reisseri* in producing biofuels and high-value products while promoting waste reduction and resource effectiveness. It is important to remember the major obstacles in microalgae growing methods and their constraints, with a focus on the financial aspect. To address these issues, attention should be directed toward developing a novel and affordable biorefinery-based technology. It is clear that significant progress has been made in the field of microalgal biotechnology for the environmentally friendly production of green chemicals and third-generation biofuels. The goal of economically viable lipid synthesis could be achieved via metabolic engineering genetically accessible microalgal strains or by bioprospecting more suitable host species. Future studies must also concentrate on microalgae that show promise for high-value lipids, as they are most likely to achieve economic viability quickly. Additionally, it aids in CO_2 mitigation, wastewater treatment at zero cost, and the production of desirable value-added byproducts. Adopting different cost-cutting measures, like using industrial flue gas and wastewater as carbon and nutrient sources, respectively, could enhance these economic processes. The path ahead involves refining cultivation strategies and advancing biorefinery techniques to boost overall efficiency and reduce production costs. To realize the full potential of microalgae in large-scale wastewater remediation, coordinated efforts among researchers, industry stakeholders, and policy-makers are essential. Lastly, in order to compete with fossil fuels and create a sustainable and renewable bioenergy supply, the microalgae-based biorefinery method appears to be the most practical strategy in the years to come. Strengthening investment in research and technological development will be key to unlocking the capabilities of microalgal systems—ultimately contributing to a sustainable future for water security, environmental protection, public health, and clean energy.

References

Abou-Shanab RA, El-Dalatony MM, EL-Sheekh MM, Ji MK, Salama ES, Kabra AN, Jeon BH (2014) Cultivation of a new microalga, *Micractinium reisseri*, in municipal wastewater for nutrient removal, biomass, lipid, and fatty acid production. Biotechnol Bioproc E 19(3): 510–518. https://doi.org/10.1007/s12257-013-0485-z

Altowayti WAH, Shahir S, Othman N, Eisa TAE, Yafooz WM, Al-Dhaqm A, Soon CY, Yahya IB, Che Rahim NANB, Abaker M, Ali A (2022) The role of conventional methods and artificial intelligence in the wastewater treatment: a comprehensive review. PRO 10(9):1832. https://doi.org/10.3390/pr10091832

Annapoorna BR, Vasudevan S, Sindhu K, Vani V, Nivya V, Venkateish VP, Kumar PM (2022) Hepatoprotective marine phytochemicals: recent perspectives. In: Kim S-K (ed) Marine biochemistry: isolations and techniques, 1st edn. CRC Press, pp 207–230. https://doi.org/10.1201/9781003303909

Baicha Z, Salar-García M, Ortiz-Martínez V, Hernández-Fernández F, de los Ríos A, Labjar N, Lotfi E, Elmahi M (2016) A critical review on microalgae as an alternative source for bioenergy production: a promising low cost substrate for microbial fuel cells. Fuel Process Technol 154: 104–116. https://doi.org/10.1016/j.fuproc.2016.08.017

Borowitzka MA (2013) High-value products from microalgae—their development and commercialisation. J Appl Phycol 25(3):743–756. https://doi.org/10.1007/s10811-013-9983-9

Chai WS, Tan WG, Munawaroh HSH, Gupta VK, Ho SH, Show PL (2021) Multifaceted roles of microalgae in the application of wastewater bio-treatment: a review. Environ Pollut 269: 116236. https://doi.org/10.1016/j.envpol.2020.116236

Cheirsilp B, Wantip K, Chai-issarapap N, Maneechote W, Pekkoh J, Duangjan K, Ruangrit K, Pumas C, Pathom-aree W, Srinuanpan S (2022) Enhanced production of astaxanthin and co-bioproducts from microalga *Haematococcus* sp. integrated with valorization of industrial wastewater under two-stage LED light illumination strategy. Environ Technol Innov 28: 102620. https://doi.org/10.1016/j.eti.2022.102620

Chen J, Wang Y, Benemann JR, Zhang X, Hu H, Qin S (2016a) Microalgal industry in China: challenges and prospects. J Appl Phycol 28:715–725. https://doi.org/10.1007/s10811-015-0720-4

Chen YH, Chang GK, Kuo SM, Huang SY, Hu I, Lo YL, Shih SR (2016b) Well-tolerated *Spirulina* extract inhibits influenza virus replication and reduces virus-induced mortality. Sci Rep 6(1): 24253. https://doi.org/10.1038/srep24253

Chen YD, Liu F, Ren NQ, Ho SH (2020) Revolutions in algal biochar for different applications: state-of-the-art techniques and future scenarios. Chin Chem Lett 31(10):2591–2602. https://doi.org/10.1016/j.cclet.2020.08.019

Cheng P, Chen D, Liu W, Cobb K, Zhou N, Liu Y, Liu H, Wang Q, Chen P, Zhou C, Ruan R (2020) Auto-flocculation microalgae species *Tribonema* sp. and *Synechocystis* sp. with T-IPL pre-treatment to improve swine wastewater nutrient removal. Sci Total Environ 725:138263. https://doi.org/10.1016/j.scitotenv.2020.138263

Chew KW, Yap JY, Show PL, Suan NH, Juan JC, Ling TC, Lee DJ, Chang JS (2017) Microalgae biorefinery: high value products perspectives. Bioresour Technol 229:53–62. https://doi.org/10.1016/j.biortech.2017.01.006

Chini ZG, Lauceri R, Faraloni C, Silva Benavides AM, Torzillo G (2023) Valuable pigments from microalgae: phycobiliproteins, primary carotenoids, and fucoxanthin. Photochem Photobiol Sci 22(8):1733–1789. https://doi.org/10.1007/s43630-023-00407-3

Chinnasamy S, Bhatnagar A, Hunt RW, Das KC (2010a) Microalgae cultivation in a wastewater dominated by carpet mill effluents for biofuel applications. Bioresour Technol 101(9): 3097–3105. https://doi.org/10.1016/j.biortech.2009.12.026

Chinnasamy S, Bhatnagar A, Claxton R, Das KC (2010b) Biomass and bioenergy production potential of microalgae consortium in open and closed bioreactors using untreated carpet industry effluent as growth medium. Bioresour Technol 101(17):6751–6760. https://doi.org/10.1016/j.biortech.2010.03.094

Chong JWR, Tan X, Khoo KS, Ng HS, Jonglertjunya W, Yew GY, Show PL (2022) Microalgae-based bioplastics: future solution towards mitigation of plastic wastes. Environ Res 206:112620. https://doi.org/10.1016/j.envres.2021.112620

Dolganyuk V, Belova D, Babich O, Prosekov A, Ivanova S, Katserov D, Patyukov N, Sukhikh S (2020) Microalgae: a promising source of valuable bioproducts. Biomolecules 10(8):1153. https://doi.org/10.3390/biom10081153

Fito J, Van-Hulle SWH (2021) Wastewater reclamation and reuse potentials in agriculture: towards environmental sustainability. Environ Dev Sustain 23(3):2949–2972. https://doi.org/10.1007/s10668-020-00732-y

Fleurence J, Levine IA (2018) Anti-allergic and allergic properties. In: Levine IA, Fleurence J (eds) Microalgae in health and disease prevention. Academic Press, pp 307–315. https://doi.org/10.1016/B978-0-12-811405-6.00014-1

Gallego R, Valdés A, Suárez-Montenegro ZJ, Sánchez-Martínez JD, Cifuentes A, Ibáñez E, Herrero M (2022) Anti-inflammatory and neuroprotective evaluation of diverse microalgae extracts enriched in carotenoids. Algal Res 67:102830. https://doi.org/10.1016/j.algal.2022.102830

Goh PS, Lau WJ, Ismail AF, Samawati Z, Liang YY, Kanakaraju D (2022) Microalgae-enabled wastewater treatment: a sustainable strategy for bioremediation of pesticides. Water 15(1):70. https://doi.org/10.3390/w15010070

Grama SB, Liu Z, Li J (2022) Emerging trends in genetic engineering of microalgae for commercial applications. Mar Drugs 20(5):285. https://doi.org/10.3390/md20050285

Gupta R, Mishra N, Singh G (2024) Microalgae cultivation and value-based products from wastewater: insights and applications. Blue Biotechnol 1:20. https://doi.org/10.1186/s44315-024-00019-1

Gutiérrez-Casiano N, Hernández-Aguilar E, Alvarado-Lassman A, Méndez-Contreras JM (2022) Removal of carbon and nitrogen in wastewater from a poultry processing plant in a photobioreactor cultivated with the microalga *Chlorella vulgaris*. J Environ Sci Health, Part A 57(7):620–633. https://doi.org/10.1080/10934529.2022.2096986

Heimann K, Huerlimann R (2015) Microalgal classification: major classes and genera of commercial microalgal species. In: Kim S-K (ed) Handbook of marine microalgae. Academic Press, pp 25–41. https://doi.org/10.1016/B978-0-12-800776-1.00003-0

Hossain N, Mahlia TMI (2019) Progress in physicochemical parameters of microalgae cultivation for biofuel production. Crit Rev Biotechnol 39:835e859. https://doi.org/10.1080/07388551.2019.1624945

Javed F, Rehman F, Khan AU, Fazal T, Hafeez A, Rashid N (2022) Real textile industrial wastewater treatment and biodiesel production using microalgae. Biomass Bioenerg J 165: 106559. https://doi.org/10.1016/j.biombioe.2022.106559

Jeevanandam J, Harun MR, Lau SY, Sewu DD, Danquah MK (2020) Microalgal biomass generation via electroflotation: a cost-effective Dewatering technology. Appl Sci 10(24):9053. https://doi.org/10.3390/app10249053

Jena J, Subudhi E (2019) Microalgae: an untapped resource for natural antimicrobials. In: Sukla L, Subudhi E, Pradhan D (eds) The role of microalgae in wastewater treatment. Springer, Singapore, pp 99–114. https://doi.org/10.1007/978-981-13-1586-2_8

Ji MK, Yun HS, Park S, Lee H, Park YT, Bae S, Ham J, Choi J (2015) Effect of food wastewater on biomass production by a green microalga *Scenedesmus obliquus* for bioenergy generation. Bioresour Technol 179:624–628. https://doi.org/10.1016/j.biortech.2014.12.053

Jiang L, Pei H, Hu W, Ji Y, Han L, Ma G (2015) The feasibility of using complex wastewater from a monosodium glutamate factory to cultivate *Spirulina subsalsa* and accumulate biochemical composition. Bioresour Technol 180:304–310. https://doi.org/10.1016/j.biortech.2015.01.019

Jiao GJ, Ma J, Li Y, Jin D, Ali Z, Zhou J, Sun R (2021) Recent advances and challenges on removal and recycling of phosphate from wastewater using biomass-derived adsorbents. Chemosphere 278:130377. https://doi.org/10.1016/j.chemosphere.2021.130377

Kang S, Kim KH, Chung N, Li Z (2023) A new omega-3 rich euglenoid *Eutreptiella* sp. from the Korean coastal waters. Algal Res 75:103296. https://doi.org/10.1016/j.algal.2023.103296

Koyande AK, Show PL, Guo R, Tang B, Ogino C, Chang JS (2019) Bio-processing of algal bio-refinery: a review on current advances and future perspectives. Bioengineered 10(1): 574–592. https://doi.org/10.1080/21655979.2019.1679697

Kumar BR, Mathimani T, Sudhakar MP, Rajendran K, Nizami AS, Brindhadevi K, Pugazhendhi A (2021) A state of the art review on the cultivation of algae for energy and other valuable products: application, challenges, and opportunities. Renew Sust Energ Rev 138:110649. https://www.sciencedirect.com/science/article/abs/pii/S1364032120309333

Kumar MS, Buttar HS, Dalal Y, Singh RB (2022) Cardioprotective and neuroprotective effects of nutraceuticals derived from marine origin. In: Functional foods and nutraceuticals in metabolic and non-communicable diseases. Academic Press, pp 707–723. https://doi.org/10.1016/B978-0-12-819815-5.00054-9

Kumar G, Neeti K, Singh R (2024) Microalgae-based biodiesel production and its challenges and future opportunities: a review. Green Technol Sustain 2(1):100060. https://doi.org/10.1016/j.grets.2023.100060

Lisa A, Mahlia TMI, Luong NN, Hang PV, Nghiem LD (2022) Microalgae-bacteria consortium for wastewater treatment and biomass production. Sci Total Environ 838:155871. https://doi.org/10.1016/j.scitotenv.2022.155871

Magro CD, Deon MC, Rossi AD, Reinehr CO, Hemkemeier M, Colla LM (2012) Chromium (VI) biosorption and removal of chemical oxygen demand by *Spirulina platensis* from wastewater-supplemented culture medium. J Environ Sci Health Part A 47(12):1818–1824. https://doi.org/10.1080/10934529.2012.689539

Markou G, Chatzipavlidis I, Georgakakis D (2012) Cultivation of *Arthrospira (Spirulina) platensis* in olive-oil mill wastewater treated with sodium hypochlorite. Bioresour Technol 112:234–241. https://doi.org/10.1016/j.biortech.2012.02.098

Matter IA, Bui VKH, Jung M, Seo JY, Kim YE, Lee YC, Oh YK (2019) Flocculation harvesting techniques for microalgae: a review. Appl Sci 9(15):3069. https://doi.org/10.3390/app9153069

Mishra SK, Shrivastav A, Mishra S (2008) Effect of preservatives for food grade C-PC from *Spirulina platensis*. Process Biochem 43(4):339–345. https://doi.org/10.1016/j.procbio.2007.12.012

Mühling M, Belay A, Whitton BA (2005) Variation in fatty acid composition of *Arthrospira (Spirulina)* strains. J Appl Phycol 17:137–146. https://doi.org/10.1007/s10811-005-7213-9

Natasha SM, Khalid S, Murtaza B, Anwar H, Shah AH, Sardar A, Shabbir Z, Niazi NK (2023) A critical analysis of wastewater use in agriculture and associated health risks in Pakistan. Environ Geochem Health 45(8):5599–5618. https://doi.org/10.1007/s10653-020-00702-3

Nayak M, Karemore A, Sen R, Nayak M, Karemore A, Sen R (2016) Performance evaluation of microalgae for concomitant wastewater bioremediation, CO_2 biofixation and lipid biosynthesis for biodiesel application. Algal Res 16:216–223. https://doi.org/10.1016/j.algal.2016.03.020

Norton TA, Melkonian M, Andersen RA (1996) Algal biodiversity. Phycologia 35(4):308–326. https://doi.org/10.2216/i0031-8884-35-4-308.1

Otondo A, Kokabian B, Stuart-Dahl S, Gude VG (2018) Energetic evaluation of wastewater treatment using microalgae, *Chlorella vulgaris*. J Environ Chem Eng 6(2):3213–3222. https://doi.org/10.1016/j.jece.2018.04.064

Pancha I, Chokshi K, George B, Ghosh T, Paliwal C, Maurya R, Mishra S (2014) Nitrogen stress triggered biochemical and morphological changes in the microalgae *Scenedesmus* sp. CCNM 1077. Bioresour Technol 156:146–154. https://doi.org/10.1016/j.biortech.2014.01.025

Pancha I, Chokshi K, Mishra S (2019) Industrial wastewater-based microalgal biorefinery: a dual strategy to remediate waste and produce microalgal bioproducts. In: Gupta S, Bux F (eds) Application of microalgae in wastewater treatment. Vol. 2: biorefinery approaches of wastewater treatment. Springer International Publishing, Cham, pp 173–193. https://doi.org/10.1007/978-3-030-13909-4_8

Phang SM, Miah MS, Yeoh BG, Hashim MA (2000) *Spirulina* cultivation in digested sago starch factory waste water. J Appl Phycol 12(3):395–400. https://doi.org/10.1023/A:1008157731731

Prabakaran G, Moovendhan M, Arumugam A, Matharasi A, Dineshkumar R, Sampathkumar P (2019) Evaluation of chemical composition and *in-vitro* anti-inflammatory effect of marine microalgae *Chlorella vulgaris*. Waste Biomass Valori 10:3263–3270. https://doi.org/10.1007/s12649-018-0370-2

Premaratne M, Nishshanka GKSH, Anthonio RADP, Liyanaarachchi VC, Thevarajah B, Nimarshana PHV, Malik A, Ariyadasa TU (2022) Resource recovery from waste streams for production of microalgae biomass: a sustainable approach towards high-value biorefineries. Bioresour Technol Rep 18:101070. https://doi.org/10.1016/j.biteb.2022.101070

Qv M, Dai D, Liu D, Wu Q, Tang C, Li S, Zhu L (2023) Towards advanced nutrient removal by microalgae-bacteria symbiosis system for wastewater treatment. Bioresour Technol 370: 128574. https://doi.org/10.1016/j.biortech.2022.128574

Rahman KM (2020) Food and high value products from microalgae: market opportunities and challenges. In: Alam M, Xu JL, Wang Z (eds) Microalgae biotechnology for food, health and high value products. Springer, Singapore, pp 3–27. https://doi.org/10.1007/978-981-15-0169-2_1

Reynolds D, Huesemann M, Edmundson S, Sims A, Hurst B, Cady S, Beirne N, Freeman J, Berger A, Gao S (2021) Viral inhibitors derived from macroalgae, microalgae, and cyanobacteria: a review of antiviral potential throughout pathogenesis. Algal Res 57:102331. https://doi.org/10.1016/j.algal.2021.102331

Sahay S, Braganza VJ (2016) Microalgae - based biodiesel production: current and future scenario. J Exp Sci 7:31–35. https://doi.org/10.25081/jes.2016.v7.3027

Sahay S, Braganza VJ (2022) Effect of nitrogen on the biomass production and lipid accumulation of freshwater microalgae *Micractinium reisseri*—a potential strain for bio-fuel production. Biosci Biotech Res Commun 15(3):442–449. https://doi.org/10.21786/bbrc/15.3.11

Sahay S, Jain B, Solanki D, Dave S, Suresh A (2024) Flocculation of *Micractinium reisseri* for effective harvesting and potential use. Water Sci Technol 90(10):2934–2946. https://doi.org/10.2166/wst.2024.374

Su Y (2021) Revisiting carbon, nitrogen, and phosphorus metabolisms in microalgae for wastewater treatment. Sci Total Environ 762:144590. https://doi.org/10.1016/j.scitotenv.2020.144590

Sun J, Norouzi O, Mašek O (2022) A state-of-the-art review on algae pyrolysis for bioenergy and biochar production. Bioresour Technol 346:126258. https://doi.org/10.1016/j.biortech.2021.126258

Venckus P, Cicchi B, Chini ZG (2021) Effects of medium salinity on growth and biochemical composition of the green microalga *Tetraselmis suecica*. J Appl Phycol 33:3555–3563. https://doi.org/10.1007/s10811-021-02560-7

Zhang Y, Su H, Zhong Y, Zhang C, Shen Z, Sang W, Yan G, Zhou X (2012) The effect of bacterial contamination on the heterotrophic cultivation of *Chlorella pyrenoidosa* in wastewater from the production of soybean products. Water Res 46(17):5509–5516. https://doi.org/10.1016/j.watres.2012.07.025

Zhang J, Liu L, Ren Y, Chen F (2019) Characterization of exopolysaccharides produced by microalgae with antitumor activity on human colon cancer cells. Int J Biol Macromol 128:761–767. https://doi.org/10.1016/j.ijbiomac.2019.02.009

Zhang D, Sial MS, Ahmad N, Filipe AJ, Thu PA, Zia-Ud-Din M, Caleiro AB (2020) Water scarcity and sustainability in an emerging economy: a management perspective for future. Sustainability 13(1):144. https://doi.org/10.3390/su13010144

Microalgal Circular Economy: Harnessing Biomass Utilization with Low-Volume High-Value and High-Volume Low-Value Products

12

Jayshri Khadilkar and Gunjan Prakash

Abstract

Microalgae have attained significant interest in renewable energy, food production, environmental management, water purification, and synthesis, such as biofertilizers, cosmetics, and healthcare products. The rate of biomass production and biomolecule accumulation presents an opportunity to replace traditional products and foster a sustainable circular economy. Microalgal biotechnology is integral to the circular economy concept, emphasizing the production of low-volume, high-value products. This strategy enhances economic returns while utilizing waste streams and minimizing environmental impact, aligning with the circular economy, and sustainable development goals. The viability of microalgae as a renewable biofuel feedstock is highlighted alongside various commercial prospects. Microalgae yield high-value products, including polyunsaturated fatty acids (PUFA) and pigments, and other bioactive compounds for nutraceuticals, cosmetics, and industrial applications. Emphasizing high-value, low-volume outputs renders microalgal biorefineries economically sustainable, aligning with consumer preferences for eco-friendly and health-oriented products. Carotenoids such as astaxanthin, β-carotene, fucoxanthin, and lutein have been used in nutraceuticals, cosmetics, and natural coloring agents. The global carotenoid market is anticipated to surpass USD 1.5 billion and is projected to grow at a CAGR of 3.5% from 2024 to 2030, propelled by the demand for natural antioxidants and food additives. Along with pigment, EPA and DHA derived from microalgae also serve as sustainable substitutes for fish oil

J. Khadilkar · G. Prakash (✉)
Department of Biological Sciences and Biotechnology, Institute of Chemical Technology, Mumbai, Maharashtra, India
e-mail: g.prakash@ictmumbai.edu.in

K. Mohanty et al. (eds.), *Advances in Algal Research*,
https://doi.org/10.1007/978-981-95-7234-2_12

in dietary supplements and functional foods, while other bioactive compounds, such as sulfated polysaccharides and polyphenols, exhibit promising health applications in wellness industries.

Post-extraction of high-value products like pigments or PUFAs, residual biomass can be processed into biofertilizers or biostimulants rich in macronutrients for agricultural purposes, thereby ensuring effective waste management. The proteins and lipid-rich residual biomass can be utilized for aquaculture feed supplementation. Additionally, leftover wet biomass can undergo anaerobic digestion to generate biogas, with CO_2 being recycled into algal cultivation to complete the resource loop. By incorporating CO_2 sequestration into the high-value compound production process, microalgal platforms can play a role in pollution reduction and climate change mitigation. The utilization of microalgal biomass thus presents a comprehensive strategy for fostering economic growth and societal welfare while addressing environmental challenges and endorsing the principles of a circular economy.

Keywords

Algal cultivation · Biostimulants · Bioremediation · Circular bioeconomy · Microalgae · Value-added products

12.1 Introduction

Microalgae have garnered substantial attention across diverse sectors, including renewable energy, food production, environmental management, water purification, and bio-based synthesis for their versatile biomass production and bioproduct accumulation potential. Their integration into circular economy models optimizes resource efficiency and sustainability by emphasizing low-volume, high-value outputs that align with global sustainable development goals (SDGs) (Mehariya et al. 2021a). Microalgal biomass serves as a renewable feedstock for biofuels, pigments, polyunsaturated fatty acids (PUFAs), and various bioactive compounds, positioning microalgal biotechnology as a cornerstone for biocircular economy frameworks (de Souza et al. 2019). Microalgae, which are photosynthetic microorganisms capable of rapid growth, utilizing nonarable land and waste materials, and producing a broad range of bioactive compounds, have become promising feedstocks for biorefineries (de Souza et al. 2019). The microalgal circular bioeconomy is defined by (Santos et al. 2025; Ezhumalai et al. 2024; Díaz et al. 2022):

1. *Resource optimization*: Using waste CO_2 and wastewater as input.
2. *Zero-waste production*: Fractionating biomass into multiple valuable products.
3. *Environmental remediation*: Integrating bioremediation with product generation.

Key to commercialization is shifting from bulk, low-value applications (like biofuels) to low-volume, high-value biomolecules such as pigments, polyunsaturated fatty acids (PUFAs), vitamins, and specialty ingredients for health, nutrition, cosmetics, and biopolymers (Ezhumalai et al. 2024). This chapter details the multidimensional potentials of microalgae, exploring recent scientific advancements, commercial prospects, and environmental impacts, drawing on the latest literature.

12.2 Biorefinery Paradigm with Microalgal Value-Added Products

12.2.1 Low-Volume High-Value Product Driving Economic Feasibility

A plethora of microalgal species have been explored for their bioactive biomass components, in which fixed carbon is converted into food, feed, and pharmaceutical products. Through the synthesis and metabolism of bio-compounds with concentrated nutritional value and therapeutic properties, microalgae provide important raw materials for healthcare and pharmaceutical applications, including carbohydrates, proteins, lipids, nucleic acids, vitamins, and minerals (Mutanda et al. 2020). Various microalgal species have been reported to yield biomass rich in bioactive substances, and selecting suitable species or strains, along with optimized cultivation strategies, is a crucial first step in uncovering their biologically active metabolites (Fig. 12.1).

Diverse classifications of photosynthetic pigments, namely, chlorophyll, carotenoids, and phycocyanin, assist in the process of photosynthesis and serve as a protective measure against an array of stress conditions. Microalgal VAPs are considered potential alternatives to replace chemical-based commercialized products consumed by various industries. Table 12.1 lists the different value-added products from microalgal biomass and their respective applications in commercial fields. Microalgae have a significant impact on the generation of economically viable and essential nutritional substances with applications in pharmaceuticals, cosmetics, and food markets (Mutanda et al. 2020; Ibrahim et al. 2023; Mehariya et al. 2021b). This

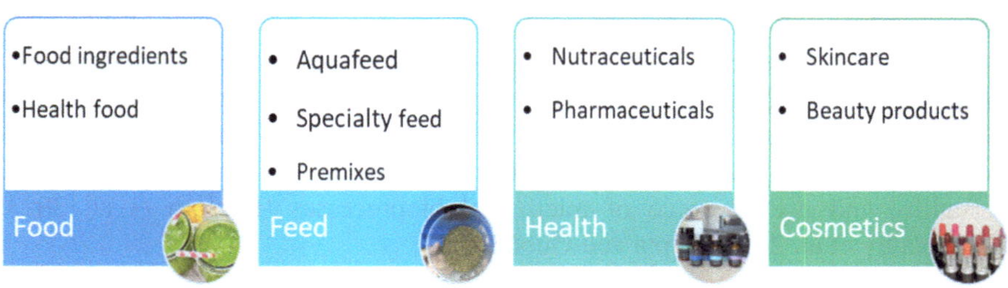

Fig. 12.1 Applications of microalgae in various fields

Table 12.1 Value-added products in microalgae and their application

VAP	Microalgae species	Application
Astaxanthin	*Haematococcus pluvialis*	Food Aquafeed Cosmetics Pharmaceuticals
	Chlorella zofingiensis	Aquafeed Cosmetics Colourant
β-Carotene	*Dunaliella salina*	Health food Cosmetics
	Dunaliella bardawil	Health food Cosmetics
Lutein	*Chlorella sorokiniana*	Health food Cosmetics
	Scenedusmus obliquus	Health food Cosmetics
Fucoxanthin	*Nitzschia laevis*	Food Aquafeed Cosmetics
	Phaeodactylum tricornutum	Food Aquafeed Pharmaceuticals
Phycocyanin	*Spirulina platensis*	Health Food Pharmaceuticals
DHA	*Schizochytrium limacinum*	Nutraceuticals Animal food Aquafeed
	Thraustochytrid spp.	Nutraceuticals Animal food Aquafeed
EPA	*Nannochloropsis* spp.	Nutraceuticals Health supplement Aquafeed
	Phaeodactylum tricornutum	Nutraceuticals Supplements Aquafeed

advances the economic viability, sustainability, safety, and efficacy of the process. This integrated utilization enhances process economics, sustainability, product safety, and functional performance. Moreover, recovering coproducts and lipids from the same biomass within a single cultivation cycle further amplifies the versatility and overall value of microalgal bioprocesses (Ibrahim et al. 2023; Nethravathy et al. 2019; Gupta and Gupta 2020).

12.2.2 Circular Economy Concept in Microalgae

The increasing energy consumption and exponential human growth are creating major environmental, social, and economic problems worldwide. Rising energy demand and rapid population growth are intensifying worldwide sustainability problems, prompting the United Nations to adopt 17 Sustainable Development Goals with 169 targets under the 2030 Agenda for Sustainable Development (Shehata et al. 2022).

These SDGs aim to secure healthy lives and a clean environment for future generations by minimizing risks and pollution. Among the SDGs, SDG 1 ("no poverty") seeks to eradicate poverty globally, while SDG 2 ("zero hunger") focuses on eliminating hunger, improving nutrition, and ensuring food security. Poverty is a major driver of hunger in developing regions, and microalgae present a promising tool to address both issues due to their adaptability and tolerance to climate stress (Shehata et al. 2022; Olabi et al. 2023; Dhahri and Omri 2020). There is a close relationship between poverty and hunger, as the former is the prime cause of hunger within developing countries. Remarkably, microalgae offer a promising and viable solution for achieving the SDGs due to their remarkable adaptability and resilience to the adverse effects of global warming. Furthermore, microalgae, along with their extracts of high-value extracts, represent an abundant source of bioactive compounds, as illustrated in Fig. 12.2 (Papadaki et al. 2024). Since microalgae produce carotenoids and other natural metabolites with strong bioactive properties, their incorporation into food, feed, and nutraceutical products can enhance health and contribute to SDG 3 (good health and well-being). These combined benefits position microalgae as a strategic component of sustainable development and circular bioeconomy initiatives (Ma et al. 2016).

Microalgae can generate several renewable energy carriers, including biodiesel, biohydrogen, bioethanol, and biogas, making them an important contributor to clean energy and directly relevant to SDG 7. They also support economic development, industrial activity, and technological innovation, aligning with SDG 8, SDG 9, and SDG 12 (Olabi et al. 2023). A key benefit of microalgae cultivation is the ability to use atmospheric carbon dioxide and nutrient-rich wastewater as inputs, providing both carbon capture and wastewater treatment functions. By doing so, microalgae help create greener communities by reducing pollution, conserving natural resources, and supporting a healthier environment, which relates to SDG 6, SDG 11, SDG 13, SDG 14, and SDG 15. Microalgal biomass can be processed to obtain clean water, diverse bioactive molecules, biofuels, and biofertilizers that can be used in agriculture and other sectors. However, progress toward these sustainability targets is hampered by factors such as the COVID-19 pandemic, economic instability, geopolitical tensions, rapid population growth, resource scarcity, weak infrastructure, inadequate policies, political constraints, and limited financing. These persistent barriers highlight the urgency of accelerating actions and investments needed to realize the Sustainable Development Goals (Fig. 12.2) (Olabi et al. 2023; Dhahri and Omri 2020; Koyande et al. 2019a). Microalgae serve as versatile biomanufacturing platforms for low-volume, high-value compounds spanning pigments, lipids, proteins, and vitamins, with expanding global markets driven by

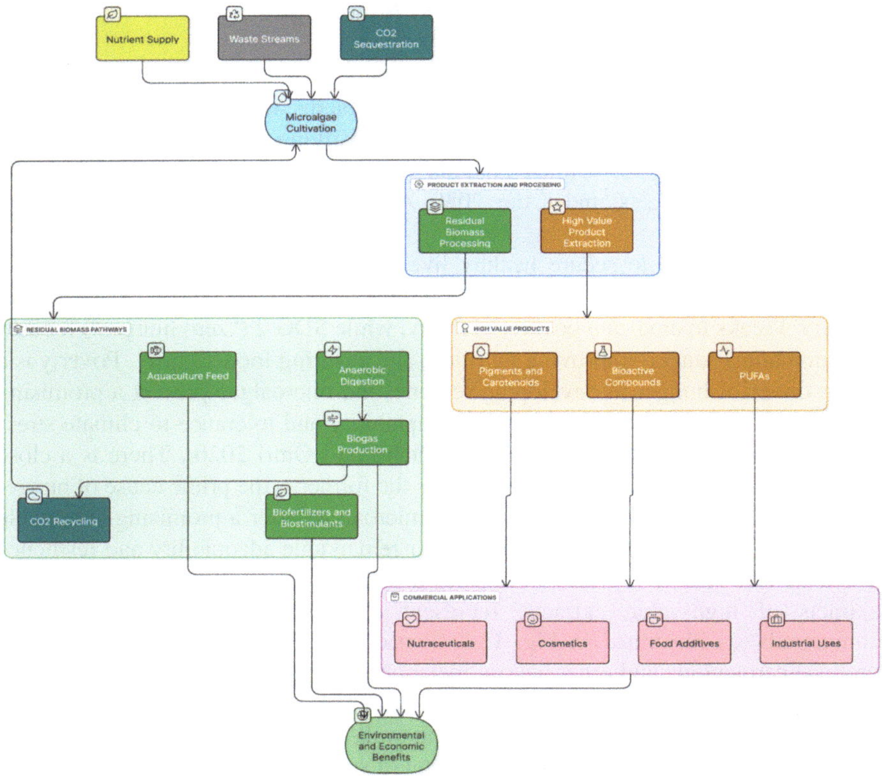

Fig. 12.2 Network visualization of microalgae applications toward circular bioeconomy

demand for natural and sustainable bioactive ingredients. These metabolites underpin diverse applications in nutraceuticals, pharmaceuticals, cosmetics, and functional foods, positioning microalgae as key contributors to the emerging bioeconomy (Olabi et al. 2023; Obaideen et al. 2021).

12.2.3 Microalgae Biomanufacturer: For Low-Volume High-Value Compounds

12.2.3.1 Carotenoids

Carotenoids are isoprenoid classes in microalgae that protect the photosynthetic apparatus and support light harvesting by quenching singlet oxygen and scavenging reactive oxygen species. These photoprotective molecules underpin applications in healthcare and pharmaceuticals as natural antioxidants (Nethravathy et al. 2019; Henríquez et al. 2016).

Commercially relevant microalgal carotenoids include astaxanthin, zeaxanthin, β-carotene, canthaxanthin, fucoxanthin, and lutein, which function as natural colorants and antioxidant ingredients in nutraceutical, cosmetic, and food formulations. The global carotenoid market is projected to surpass USD 2.63 billion by 2030 with a CAGR of about 4.6%, supported by a shift toward natural and sustainable alternatives to synthetic additives. Species such as *Haematococcus pluvialis, Dunaliella salina*, and *Dunaliella tertiolecta* are established carotenoid producers. *H. pluvialis* is recognized as the preferred industrial source of astaxanthin due to its superior accumulation capacity (Santos et al. 2025; Henríquez et al. 2016).

Additional microalgae, including *Scenedesmus species, Chromochloris zofingiensis*, and *Chlorococcum species,* also synthesize astaxanthin, while lutein can be sourced from *Chromochloris zofingiensis, Parachlorella, Chlorella strains, Scenedesmus species*, and *H. pluvialis* (Henríquez et al. 2016*;* Mularczyk et al. 2020*;* Low et al. 2020).

Although commercial lutein production from microalgae remains limited because of relatively low productivity, yields can be enhanced through strain selection, optimization of culture conditions, and targeted engineering of carotenoid biosynthetic pathways. D*unaliella salina, Dunaliella tertiolecta,* and *Chlorella species* are prominent β-carotene producers, and *D. salina* is already exploited industrially for natural β-carotene (Henríquez et al. 2016; Bogacz-Radomska and Harasym 2018).

12.2.3.2 PUFAs

Microalgae are primary producers of omega-3 and omega-6 PUFAs, including EPA and DHA, which are crucial for human health. Moreover, microalgal PUFAs are promising alternatives for conventional fish and krill-based PUFAs, eliminating the chance of ocean-borne heavy metal contamination and reducing the burden on marine diversity (Ma et al. 2016). Microalgal ω-3 fatty acids comprise α-linoleic acid (ALA C18-3n-3), arachidonic acid (ARA C20:4n-6), eicosapentaenoic acid (EPA, C20:5n-3), and docosahexaenoic acid (DHA, C22:6n-3). Among these, eicosapentaenoic acid (EPA, 20:5 ω-3) is particularly associated with cardiovascular and immune benefits (Abdelkarim et al. 2025).

The market for polyunsaturated fatty acids (PUFAs) is witnessing substantial growth and is projected to expand significantly over the coming decade. The PUFA market was estimated at around USD 6.69 billion in 2025 and is forecast to reach roughly USD 11.42 billion by 2035, corresponding to a CAGR of about 5.5%. Within this segment, the EPA and DHA market was valued at nearly USD 1.83 billion in 2024, projected to increase to about USD 1.95 billion in 2025 and USD 3.33 billion by 2033, at an anticipated CAGR of around 6.9% from 2025 to 2033 (Future Market Report 2025; Microalgae Market Size, Share, and Industry Analysis 2025).

Marine microalgae are the primary producers of ω-3 PUFAs. Trials confirm comparable bioavailability of microalgal-derived PUFAs to conventional sources. Strategic cultivation methods, including mixotrophy and metabolic engineering, augment both biomass accumulation and PUFA content. Omega-3 and omega-6 PUFAs are currently the market's most popular value-added compounds derived from microalgae. The accumulation of lipids in microalgae usually occurs in the range of 20% to 50% of total biomass. Microalgae species, including *Phaeodactylum tricornutum, Spirulina* sp., *Nannochloropsis oculata, Nannochloropsis salina, Chlorella vulgaris, Tetraselmis*, and *Scenedesmus obliquus,* have the most significant importance for both humans and animals. Microalgae species such as *Chlamydomonas reinhardtii, Dunaliella salina* and *Phaeodactylum tricornutum* are genetically engineered to improve fatty acid concentrations (Ma et al. 2016; Zakaria et al. 2025). PUFAs, majorly EPA and DHA, are reported to have a functional role in inflammatory pathologies such as Alzheimer's disease and arthritis, in which EPA is the precursor of eicosanoids and prostaglandins that are an essential class of compounds that act as cell signaling molecules (Barkia et al. 2019). Furthermore, DHA is essential for infant formulas and preparations, especially in pregnant women, as well as dietary supplements, beverages, and food products (Barkia et al. 2019; Gupta and Gupta 2020).

12.2.3.3 Phycobiliproteins

Phycobiliproteins (PBPs) are one of the colored protein groups that are predominantly found in cyanobacteria, or blue-green algae, and can sometimes be found in red algae or cryptomonads (Manirafasha et al. 2016; Tan et al. 2021). They are classified according to their structure and absorption properties into phycocyanin, phycoerythrin, and allophycocyanin, with phycocyanin and phycoerythrin being abundantly produced by *Arthrospira species* and *Porphyridium cruentum* (Manirafasha et al. 2016). Phycobiliproteins are widely utilized in food, feed, and cosmetic industries. The global phycocyanin market is expected to reach about USD 279.6 million by 2030, growing at a CAGR of approximately 28.1% in value and 33.8% in volume from 2023 to 2030. This growth is primarily fueled by the rising use of phycocyanin in nutraceutical and nutritional products, the shift away from synthetic colorants, the increasing demand for natural blue pigments, and growing venture capital investments in phycocyanin production (Manirafasha et al. 2016; Fernandes et al. 2023).

12.2.3.4 Vitamins

Microalgae offer vitamins such as vitamin B12, vitamin C, D, E, and A. Studies carried out by Koyande et al. (2019a) reported that microalgae species such as *Dunaliella tertiolecta, Chlorella* sp., and *Tetraselmis suecica* can synthesize and are excellent sources of vitamin B12 or cobalamin due to their bioavailability. The microalgae *Chlamydomonas reinhardtii, H. pluvialis, Scenedesmus* sp., *Tetraselmis tetrathele, Tetraselmis striata, B. braunii, C. vulgaris, Chlorococcum novaeangliae, Desmodesmus* sp., and *Chlorella* sp. are reported for the synthesis of vitamin E (Mudimu et al. 2017). Further, microalgae are also able to produce diverse kinds of vitamin B. *Chlorella* sp., and *D. tertiolecta* produce riboflavin (B2), biotin (B7), and cobalamin (B12) (Koyande et al. 2019a). However, further screening and development are required to unlock the full industrial potential of microalgae for vitamin production.

12.2.3.5 Protein

The essential amino acids have significant applications in the health, nutrition, cosmetic, and pharmaceutical sectors (Nehmé et al. 2017). The global algae protein market was valued at approximately USD 997.4 million in 2024 and is projected to reach around USD 1327.8 million by 2030, with a CAGR of about 4.8% between 2025 and 2030, reflecting increasing demand for environmentally friendly protein alternatives (Grand View Research 2025; Meticulous Research 2024).

Microalgal cells can synthesize amino acids, which are then converted into essential proteins. Microalgae cells such as *Porphyridium aerugineum* contain various amino acids such as tryptophan, lysine, leucine, and arginine. *Tetraselmis chuii* and *B. braunii* contain arginine, while *Nannochloropsis granulate* comprises leucine. Tibbetts et al. (2016) have reported that the *Scenedesmus* sp. AMDD WAB contains 32–44% protein in its cells, along with *Scenedesmus obliquus* (50–55%) and *Scenedesmus almeriensis* (41.7%) (Tomás-Almenar et al. 2018). The microalgae *Chlorella* sp. and *Dunaliella salina* also contain ~53.3% and 57% of crude protein in their biomass, respectively, whereas *Tetraselmis* sp. contains 59 ± 7% protein (Schwenzfeier et al. 2011). Microalgae biomass needs to be considered a rich protein source that can be used in the nutraceutical, pharmaceutical, and food or feed sectors.

12.2.4 Circular Economy for High-Volume Low-Value Products Coupled with Waste Valorization and Resource Recovery

Microalgae efficiently remove nutrients and toxic contaminants from wastewater, making them suitable for environmental management within circular systems. Recent studies highlight their cultivation on municipal, industrial, agricultural, and aquaculture effluents, achieving phycoremediation coupled with biomass production for commercial use (Fig. 12.3). Moreover, microalgae possess advantageous characteristics such as carbon fixation capabilities, higher photosynthetic efficiencies, and biomass productivity compared to terrestrial plants, as well as the absence of a need for arable land or freshwater for growth (Sathasivam et al. 2019).

Additionally, microalgae can be employed for biostimulant and biofertilizer production along with bioremediation purposes, such as the phyco-remediation of wastewater streams and the sequestration of carbon dioxide (CO_2) (Aresta and

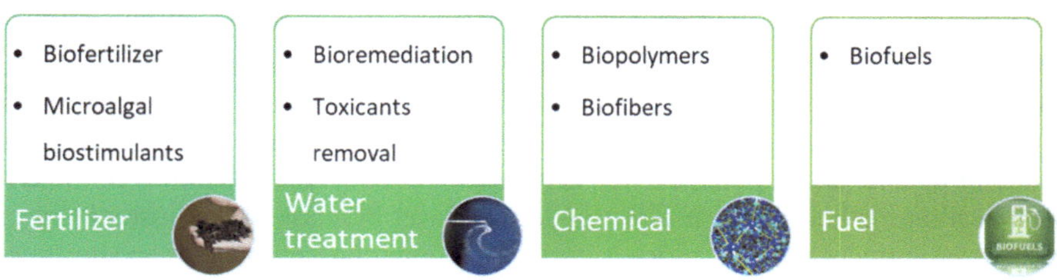

Fig. 12.3 Applications of high-volume low-value products

Dibenedetto 2019). Such strategic utilization of high-volume low-value products needs to be employed in the circular bioeconomy to support economic feasibility, along with value-added products.

12.2.4.1 Biofertilizers and Biostimulants

After the extraction of high-value compounds from microalgal biomass, the leftover material retains a rich profile of macronutrients such as proteins, lipids, and minerals. This residual biomass can be directly utilized as biofertilizers and biostimulants, promoting plant growth, enhancing soil fertility, and supporting crop resilience against abiotic stress. Microalgae-based biofertilizers incorporate live microorganisms or their derivatives, improving nutrient availability and organic matter in soil (Gupta et al. 2024).

The microalgal strains, such as *Chlamydomonas*, *Chlorella vulgaris*, and *Spirulina* were identified as the most promising in biofertilizer and biostimulant formulation studies. The production of biofertilizers derived from cyanobacteria predominantly utilizes species with high agricultural potential, including *Anabaena variabilis*, *Nostoc*, *Tolypothrix* sp., *Aulosira fertilisima*, *Calothrix* sp., and *Scytonema* sp. These microorganisms are well recognized for their ability to fix atmospheric nitrogen and enrich soil fertility, thereby serving as effective and sustainable alternatives to conventional chemical fertilizers (Díaz et al. 2022). The global market for microalgae-based biofertilizers has experienced considerable growth. In 2021, the market revenue surpassed USD 11.33 million, with projections indicating an expansion to approximately USD 22.28 million by 2028, reflecting a compound annual growth rate (CAGR) of 9.75% during the period of 2022–2028 (Díaz et al. 2022). This increasing demand is largely driven by the worldwide emphasis on sustainable and eco-friendly agricultural practices, as biofertilizers derived from cyanobacteria and microalgae provide viable solutions to the environmental and ecological challenges associated with synthetic fertilizers. Complementary market analyses for microalgae-based fertilizers further reinforce these trends, estimating the market value at USD 28.7 million in 2023, with forecasts suggesting sustained growth to USD 74.1 million by 2033, corresponding to a projected CAGR of 10% between 2023 and 2033 (Ashour et al. 2024). Collectively, these data highlight not only the agronomic relevance of cyanobacteria and microalgae in biofertilizer development but also their expanding economic value. Biostimulants in the Indian market are projected to grow at a CAGR of 16.49% through 2024, reaching a total of USD 180.95 million (Ashour et al. 2024).

Microalgal biomass, whether whole, lipid-extracted, or as digestate from bioenergy processes, serves as a biofertilizer due to its high levels of nitrogen, phosphorus, essential micronutrients, and beneficial secondary metabolites (Díaz et al. 2022; Ashour et al. 2024). This biomass improves soil fertility, improves nutrient uptake, and supplements or may replace conventional chemical fertilizers. Application methods include direct addition of wet biomass to soil, use of dry or processed extracts, and foliar sprays, each promoting robust crop yields and healthier soil microbiomes. Microalgal biomass contains phytohormones (e.g., auxins, cytokinins, and gibberellins), polysaccharides, amino acids, and signaling molecules that function as biostimulants by stimulating germination, growth, flowering, and stress tolerance in plants (Bello et al. 2021; Behera et al. 2021). Biostimulant effects can be achieved using either intact microalgal cells or cellular extracts applied to seeds or directly to crops, often resulting in increased yields and improved resistance to abiotic and biotic stressors (Behera et al. 2021). Utilization of residual biomass as a biostimulant minimizes environmental impact by recycling nutrients, reducing the carbon footprint, and offering cost-effective fertilization options for sustainable agriculture.

12.2.4.2 Aquaculture and Animal Feed

Residual microalgal biomass is particularly valuable in aquaculture as a feed supplement, effectively replacing fishmeal and fish oil. Microalgae are rich in essential amino acids, digestible proteins, valuable lipids (especially omega-3 fatty acids), vitamins, and pigments.

Polyunsaturated fatty acids, proteins, and pigments are essential components in feed for all aquatic organisms (Miranda et al. 2024). Omega-3 long-chain polyunsaturated fatty acids (n3-LC-PUFA), namely EPA and DHA, are essential fatty acids in the aquaculture industry. DHA can be produced by *Thraustochytrid*, *Schizochytrium*, *Ulkenia,* and *Crypthecodimium,* with a 20% (w/w) yield. The estimated DHA production is abo about 1000 tons per year. EPA is produced commercially using *Nannochloropsis* spp. made as a promising source of this fatty acid. This strain is excellent in aquaculture because of its high EPA content (Ma et al. 2016).

Approximately 70% of fish species raised for commercial aquaculture originate from recognized commercial stocks, with nearly 68% of these species relying on formulated fish feed in their production systems (Ronga et al. 2019). Microalgal biomass offers improved digestibility and palatability, promoting rapid fish growth and minimizing the occurrence of deformities. The nutrient profile of biomass is rich in proteins, minerals, essential fatty acids, and amino acids, which are crucial for fish health and development. It also enables a low feed conversion ratio, as most of the feed is converted directly into fish flesh, resulting in minimal waste. Fish fed with diets containing biomass tend to show improved immunity and survival rates, which makes them highly valued in the industry. These unique qualities have solidified fish meal's position as an essential component of commercial aquaculture feeds (Ronga et al. 2019; Sharma et al. 2021; Fernández et al. 2021).

Furthermore, potential microalgal applications in the feed industry are being considered for pets, horses, broilers, and cows (Ronga et al. 2019). Key challenges include mitigating heavy metal bioaccumulation and antinutrient effects. However, research and technological advances are continually improving the safety and efficacy of microalgal feeds.

12.2.4.3 Carbon Sequestration

The wet biomass remaining after extraction serves as a potent feedstock for bioenergy production through anaerobic digestion, producing biogas renewable energy source. This aligns with the biocircular economy, where every step of biomass processing is designed for maximal resource recovery and minimal waste (Miranda et al. 2024).

According to the Global CCS Institute, climate objectives by 2050 will require the deployment of roughly 2000 commercial CO_2 capture projects each year, with estimated investments ranging from \$655 billion to \$1280 billion (Nagappan et al. 2021).

The CO_2 generated during cultivation and processing can be recycled in closed photobioreactor systems, thus closing the carbon loop and lowering the net greenhouse gas emissions. Integration of CO_2 recycling with microalgal cultivation further strengthens the climate mitigation strategy within the biocircular framework (Moheimani et al. 2018). Microalgae-based biorefineries excel at valorizing waste processing it into energy, chemicals, pharmaceuticals, and food products, making them models of circular bioeconomy in practice. Key advantages include flexible product output, simple reactor designs, ambient operating conditions, and the ability to utilize various waste streams as growth media, amplifying the efficiency and sustainability of the entire value chain. For every 1 kg of algal biomass, it is reported that 1.83 kg of CO_2 is fixed through autotrophic mode (Wei et al. 2021).

Microalgae such as *Dunaliella salina, Chlorella* sp., *Teraselmis suecica, Nannochloropsis oceanica*, and *Haematococcus pluvialis* are explored for CO_2 sequestration as well as production of valuable products. Microalgae contribute over 90% of the primary production in marine ecosystems and annually sequester about 50 gigatons of atmospheric CO_2, making them pivotal in the global carbon cycle (Rasool 2021). Studies by Kadam (2001) demonstrated that a 1000-hectare open pond microalgae system can mitigate up to 50% of a 50 MW power plant's yearly CO_2 emissions, equivalent to 210,000 out of 414,000 tons of CO_2. Through photosynthesis, microalgae transform the absorbed CO_2 into nutritionally rich and economically valuable bioactive compounds and are notable for both their significant CO_2 capture potential and high productivity, with reported capture rates of 1.6–2 tons of CO_2 and biomass yields ranging from 127 to 300 tons per hectare per year (Wei et al. 2021; Rasool 2021). Species like those in the Chlorella genus exhibit especially high promise for both CO_2 fixation and conversion to bioenergy products such as biohydrogen, largely thanks to unique carbon concentrating mechanisms that enable CO_2 absorption rates 10–50 times higher than those of terrestrial plants (Rasool 2021).

Meeting the economic feasibility and cost effectiveness of the CO_2 sequestration by microalgae still needs more research. The design and operation can be focused more. Toward cost-effectiveness so that technology can compete favorably with existing technologies. Overall, the application of microalgae for CO_2 sequestration is a promising and effective technique for carbon capture, but it still has interesting areas for development for greater global environmental impact.

12.2.4.4 Microalgae-Mediated Wastewater Treatment

Microalgae play a promising and multifaceted role in wastewater treatment due to their ability to efficiently remove nutrients and organic matter while simultaneously producing valuable biomass. Industrial and urban wastewaters are typically rich in pollutants such as nitrogen and phosphorus, which must be eliminated to prevent eutrophication in natural water bodies, reduce ammonia toxicity to aquatic organisms, and minimize interference with disinfection processes such as chlorine treatment (Mehariya et al. 2021a; Ighalo et al. 2022). Microalgae are particularly effective in this regard, as they can utilize these nutrients for their own growth, achieving near-complete nutrient removal efficiencies in many cases.

Robust species like *Oscillatoria, Euglena, Chlamydomonas, Chlorella, Navicula, Nitzschia, Stigeoclonium,* and *Scenedesmus* are the eighth-most tolerant genera. *Scenedesmus* sp. and *Chlorella* sp. are widespread in freshwater habitats and widely used for industrial wastewater treatment. Studies have shown that species like *Chlorella* and *Scenedesmus* can remove up to 100% of phosphorus and COD, and around 98% of ammonium from wastewater, while also accumulating significant amounts of lipids (15–17%) and achieving robust biomass productivity (Ighalo et al. 2022; Sarwer et al. 2022). Moreover, cultivating microalgae in wastewater supports exceptionally high carbon dioxide fixation rates (1.83 kg CO_2 per kg biomass), with biomass yields surpassing those of terrestrial crops by up to 50% (Kadam 2001). However, challenges such as culture contamination, cell harvesting, and operating large-scale systems still need to be addressed to make microalgae-based wastewater treatment commercially viable. Despite these hurdles, the dual benefit of pollutant removal and resource recovery positions microalgae as a sustainable and circular solution for the treatment and valorization of wastewater streams (Sarwer et al. 2022).

12.2.5 Technoeconomic Feasibility and Challenges Toward Microalgal Circular Bioeconomy

Technoeconomic assessment (TEA) of microalgal circular bioeconomy targeting low-volume, high-value products such as pigments, omega-3 fatty acids, proteins, and specialty metabolites demonstrates that profitability depends on integrated biorefinery design, optimized scale, and effective downstream processing. Cascading valorization of multiple products from a single stream of microalgal biomass improves revenue and marketability, raising utilization rates above 90% compared to 7–28% in single-product processes (Rezvani et al. 2022; Wiatrowski et al. 2022).

Wastewater-integrated models, which enable microalgae to remediate nutrients while generating valuable biomass, suggest the cost reductions up to 96% reported for optimized nutrient recipes. However, high-value market potential is constrained by variable and limited demand, imposing a need for flexible, modular biorefinery configurations and diversified output strategies (Nagi et al. 2021).

Despite the considerable applications of microalgae for diverse biotechnological applications, the transition to large-scale cultivation and commercialization is impeded by several notable challenges. Foremost among the upstream bottlenecks are the complexities associated with selecting suitable strains tailored for specific target compounds, alongside the substantial operational costs, especially in closed photobioreactor systems. While open raceway ponds present a more economically efficient alternative, susceptible to predation and high risk of contamination, which leads to consistent biomass and results in low biomass productivity. Outdoor large-scale cultivation further contends with fluctuations in temperature, light intensity, nutrient availability, and CO_2 transfer efficiency, all of which compromise process reproducibility and stability (Nagi et al. 2021; Hoque et al. 2025; Zhou et al. 2022). Addressing these issues necessitates the deployment of sophisticated monitoring, automation, and control systems. Integrated biorefinery configurations, designed to sequentially valorise microalgal biomass into target products such as pigments, proteins, and bioactive molecules, followed by the utilization of residual fractions for fertilizer or biostimulant production, introduce substantial technical complexity. The extraction and purification of high-value molecules from robust microalgal cells is both energy and cost-intensive. Although wet processing and mild extraction protocols may reduce operational costs, they frequently do so at the expense of product purity (Ezhumalai et al. 2024; Occhipinti et al. 2024). Each step of extraction, separation, and residual conversion needs to be carefully balanced; harsh treatment for one product might reduce the yield or quality of another. Achieving effective synchronization across cultivation, harvesting, and downstream processing in the context of fluctuating effluent and energy inputs constitutes a major integration challenge, one that demands predictive modeling and the implementation of modular production systems (Ezhumalai et al. 2024; Koyande et al. 2019b).

Integrated solutions that combine wastewater remediation, carbon sequestration, and the generation of biostimulant and biofertilizer products have the potential to enhance both sustainability and economic viability. However, such approaches require advanced engineering integration, real-time process control, and ongoing optimization to achieve economic sustainability. In addition to the significant capital (CAPEX) and operational expenditures (OPEX), regulatory gaps and societal acceptance, particularly in relation to novel microalgal ingredients for food or cosmetic applications, impose further barriers to widespread market adoption.

12.3 Conclusion

Microalgal biotechnology serves as a vital nexus for aligning renewable resource production with the principles of a circular economy and sustainable development goals. Its multifunctional applications, from producing carotenoids, PUFAs, and bioactive compounds to generating biofertilizers, aquafeeds, and bioenergy, demonstrate a holistic strategy to foster economic viability, environmental responsibility, and societal benefit. Overcoming existing technical and market challenges through innovation and policy support will further entrench microalgal biorefineries as sustainable pillars of future bioeconomies, reinforcing their role in pollution reduction, climate change mitigation, and resource circularity.

References

Abdelkarim OH, Wijffels RH, Barbosa MJ (2025) Microalgal lipid production: a comparative analysis of Nannochloropsis and Microchloropsis strains. J Appl Phycol 37:15–34. https://doi.org/10.1007/s10811-024-03318-7

Aresta M, Dibenedetto A (2019) Beyond fractionation in the utilization of microalgal components. In: Bioenergy with carbon capture and storage. Elsevier, pp 173–193. https://doi.org/10.1016/B978-0-12-816229-3.00009-0

Ashour M, Mansour AT, Alkhamis YA, Elshobary M (2024) Usage of Chlorella and diverse microalgae for CO2 capture - towards a bioenergy revolution. Front Bioeng Biotechnol:12. https://doi.org/10.3389/fbioe.2024.1387519

Barkia I, Saari N, Manning SR (2019) Microalgae for high-value products towards human health and nutrition. Mar Drugs 17:304. https://doi.org/10.3390/md17050304

Behera B, Venkata Supraja K, Paramasivan B (2021) Integrated microalgal biorefinery for the production and application of biostimulants in circular bioeconomy. Bioresour Technol 339: 125588. https://doi.org/10.1016/j.biortech.2021.125588

Bello AS, Saadaoui I, Ben-Hamadou R (2021) "Beyond the Source of Bioenergy": microalgae in modern agriculture as a biostimulant, biofertilizer, and anti-abiotic stress. Agronomy 11:1610. https://doi.org/10.3390/agronomy11081610

Bogacz-Radomska L, Harasym J (2018) β-Carotene—properties and production methods. Food Qual Safe 2:69–74. https://doi.org/10.1093/fqsafe/fyy004

de Souza MP, Hoeltz M, Gressler PD, Benitez LB, Schneider RCS (2019) Potential of microalgal bioproducts: general perspectives and main challenges. Waste Biomass Valoriz 10:2139–2156. https://doi.org/10.1007/s12649-018-0253-6

Dhahri S, Omri A (2020) Foreign capital towards SDGs 1 & 2—ending poverty and hunger: the role of agricultural production. Struct Chang Econ Dyn 53:208–221. https://doi.org/10.1016/j.strueco.2020.02.004

Díaz V, Leyva-Díaz JC, Almécija MC, Poyatos JM, del Mar MM, Martín-Pascual J (2022) Microalgae bioreactor for nutrient removal and resource recovery from wastewater in the paradigm of circular economy. Bioresour Technol 363:127968. https://doi.org/10.1016/j.biortech.2022.127968

Ezhumalai G, Arun M, Manavalan A, Rajkumar R, Heese K (2024) A holistic approach to circular bioeconomy through the sustainable utilization of microalgal biomass for biofuel and other value-added products. Microb Ecol 87:61. https://doi.org/10.1007/s00248-024-02376-1

Fernandes R, Campos J, Serra M, Fidalgo J, Almeida H, Casas A et al (2023) Exploring the benefits of Phycocyanin: from Spirulina cultivation to its widespread applications. Pharmaceuticals 16: 592. https://doi.org/10.3390/ph16040592

Fernández FGA, Reis A, Wijffels RH, Barbosa M, Verdelho V, Llamas B (2021) The role of microalgae in the bioeconomy. New Biotechnol 61:99–107. https://doi.org/10.1016/j.nbt.2020.11.011

Future Market Report (2025) Polyunsaturated fatty acids market trends 2025 to 2035. https://www.futuremarketinsights.com/reports/polyunsaturated-fatty-acids-market

Grand View Research (2025) GVR report cover algae protein market (2025–2030). https://www.grandviewresearch.com/industry-analysis/algae-protein-market

Gupta J, Gupta R (2020) Nutraceutical status and scientific strategies for enhancing production of Omega-3 fatty acids from microalgae and their role in healthcare. Curr Pharm Biotechnol 21: 1616–1631. https://doi.org/10.2174/1389201021666200703201014

Gupta R, Mishra N, Singh G, Mishra S, Lodhiyal N (2024) Microalgae cultivation and value-based products from wastewater: insights and applications. Blue Biotechnol 1:20. https://doi.org/10.1186/s44315-024-00019-1

Henríquez V, Escobar C, Galarza J, Gimpel J (2016) Carotenoids in microalgae, pp 219–237. https://doi.org/10.1007/978-3-319-39126-7_8

Hoque MM, Iannelli V, Padula F, Radice RP, Saha BK, Martelli G et al (2025) Microalgae: green engines for achieving carbon sequestration, circular economy, and environmental sustainability—a review based on last ten years of research. Bioengineering 12:909. https://doi.org/10.3390/bioengineering12090909

Ibrahim TNBT, Feisal NAS, Kamaludin NH, Cheah WY, How V, Bhatnagar A et al (2023) Biological active metabolites from microalgae for healthcare and pharmaceutical industries: a comprehensive review. Bioresour Technol 372:128661. https://doi.org/10.1016/j.biortech.2023.128661

Ighalo JO, Dulta K, Kurniawan SB, Omoarukhe FO, Ewuzie U, Eshiemogie SO et al (2022) Progress in microalgae application for CO_2 sequestration. Clean Chem Eng 3:100044. https://doi.org/10.1016/j.clce.2022.100044

Kadam KL (2001) Microalgae production from power plant flue gas: environmental implications on a life cycle basis

Koyande AK, Chew KW, Rambabu K, Tao Y, Chu D-T, Show P-L (2019a) Microalgae: a potential alternative to health supplementation for humans. Food Sci Human Wellness 8:16–24. https://doi.org/10.1016/j.fshw.2019.03.001

Koyande AK, Show P-L, Guo R, Tang B, Ogino C, Chang J-S (2019b) Bio-processing of algal bio-refinery: a review on current advances and future perspectives. Bioengineered 10:574–592. https://doi.org/10.1080/21655979.2019.1679697

Low KL, Idris A, Mohd YN (2020) Novel protocol optimised for microalgae lutein used as food additives. Food Chem 307:125631. https://doi.org/10.1016/j.foodchem.2019.125631

Ma XN, Chen TP, Yang B, Liu J, Chen F (2016) Lipid production from nannochloropsis. Mar Drugs:14. https://doi.org/10.3390/md14040061

Manirafasha E, Ndikubwimana T, Zeng X, Lu Y, Jing K (2016) Phycobiliprotein: potential microalgae derived pharmaceutical and biological reagent. Biochem Eng J 109:282–296. https://doi.org/10.1016/j.bej.2016.01.025

Mehariya S, Goswami RK, Verma P, Lavecchia R, Zuorro A (2021a) Integrated approach for wastewater treatment and biofuel production in microalgae biorefineries. Energies (Basel) 14:2282. https://doi.org/10.3390/en14082282

Mehariya S, Goswami RK, Karthikeysan OP, Verma P (2021b) Microalgae for high-value products: a way towards green nutraceutical and pharmaceutical compounds. Chemosphere 280:130553. https://doi.org/10.1016/j.chemosphere.2021.130553

Meticulous Research (2024) Microalgae market size, share, forecast, & trends analysis. https://www.meticulousresearch.com/product/microalgae-market-5197

Microalgae Market Size, Share & Industry Analysis (2025) Fortune business insights. https://www.fortunebusinessinsights.com/microalgae-market-110314

Miranda AM, Hernandez-Tenorio F, Villalta F, Vargas GJ, Sáez AA (2024) Advances in the development of biofertilizers and biostimulants from microalgae. Biology (Basel) 13:199. https://doi.org/10.3390/biology13030199

Moheimani NR, Vadiveloo A, Ayre JM, Pluske JR (2018) Nutritional profile and in vitro digestibility of microalgae grown in anaerobically digested piggery effluent. Algal Res 35:362–369. https://doi.org/10.1016/j.algal.2018.09.007

Mudimu O, Koopmann IK, Rybalka N, Friedl T, Schulz R, Bilger W (2017) Screening of microalgae and cyanobacteria strains for α-tocopherol content at different growth phases and the influence of nitrate reduction on α-tocopherol production. J Appl Phycol 29:2867–2875. https://doi.org/10.1007/s10811-017-1188-1

Mularczyk M, Michalak I, Marycz K (2020) Astaxanthin and other nutrients from Haematococcus pluvialis—multifunctional applications. Mar Drugs 18:459. https://doi.org/10.3390/md18090459

Mutanda T, Naidoo D, Bwapwa JK, Anandraj A (2020) Biotechnological applications of microalgal oleaginous compounds: current trends on microalgal bioprocessing of products. Front Energy Res:8. https://doi.org/10.3389/fenrg.2020.598803

Nagappán S, Das P, AbdulQuadir M, Thaher M, Khan S, Mahata C et al (2021) Potential of microalgae as a sustainable feed ingredient for aquaculture. J Biotechnol 341:1–20. https://doi.org/10.1016/j.jbiotec.2021.09.003

Nagi GK, Minhas AK, Gaur S, Jain P, Mandal S (2021) Integration of algal biofuels with bioremediation coupled industrial commodities towards cost-effectiveness. Front Energy Res:9. https://doi.org/10.3389/fenrg.2021.735141

Nehmé R, Atieh C, Fayad S, Claude B, Chartier A, Tannoury M et al (2017) Microalgae amino acid extraction and analysis at nanomolar level using electroporation and capillary electrophoresis with laser-induced fluorescence detection. J Sep Sci 40:558–566. https://doi.org/10.1002/jssc.201601005

Nethravathy MU, Mehar JG, Mudliar SN, Shekh AY (2019) Recent advances in microalgal bioactives for food, feed, and healthcare products: commercial potential, market space, and sustainability. Compr Rev Food Sci Food Saf 18:1882–1897. https://doi.org/10.1111/1541-4337.12500

Obaideen K, Nooman AlMallahi M, Alami AH, Ramadan M, Abdelkareem MA, Shehata N et al (2021) On the contribution of solar energy to sustainable developments goals: case study on Mohammed bin Rashid Al Maktoum Solar Park. Int J Thermofluids 12:100123. https://doi.org/10.1016/j.ijft.2021.100123

Occhipinti PS, Russo N, Foti P, Zingale IM, Pino A, Romeo FV et al (2024) Current challenges of microalgae applications: exploiting the potential of non-conventional microalgae species. J Sci Food Agric 104:3823–3833. https://doi.org/10.1002/jsfa.13136

Olabi AG, Shehata N, Sayed ET, Rodriguez C, Anyanwu RC, Russell C et al (2023) Role of microalgae in achieving sustainable development goals and circular economy. Sci Total Environ 854:158689. https://doi.org/10.1016/j.scitotenv.2022.158689

Papadaki S, Tricha N, Panagiotopoulou M, Krokida M (2024) Innovative bioactive products with medicinal value from microalgae and their overall process optimization through the implementation of life cycle analysis—an overview. Mar Drugs 22:152. https://doi.org/10.3390/md22040152

Rasool D (2021) Unlocking private finance to support CCS investments. Global CCS Institute. https://www.globalccsinstitute.com/wpcontent/uploads/2025/08/Unlocking-Private-Finance-For-CCS-Thought-Leadership-Report

Rezvani S, Saadaoui I, Al Jabri H, Moheimani NR (2022) Techno-economic modelling of high-value metabolites and secondary products from microalgae cultivated in closed photobioreactors with supplementary lighting. Algal Res 65:102733. https://doi.org/10.1016/j.algal.2022.102733

Ronga D, Biazzi E, Parati K, Carminati D, Carminati E, Tava A (2019) Microalgal biostimulants and biofertilisers in crop productions. Agronomy 9:192. https://doi.org/10.3390/agronomy9040192

Santos B, Freitas F, Sobral AJFN, Encarnação T (2025) Microalgae and circular economy: unlocking waste to resource pathways for sustainable development. Int J Sustain Eng:18. https://doi.org/10.1080/19397038.2025.2501488

Sarwer A, Hamed SM, Osman AI, Jamil F, Al-Muhtaseb AH, Alhajeri NS et al (2022) Algal biomass valorization for biofuel production and carbon sequestration: a review. Environ Chem Lett 20:2797–2851. https://doi.org/10.1007/s10311-022-01458-1

Sathasivam R, Radhakrishnan R, Hashem A, Abd-Allah EF (2019) Microalgae metabolites: a rich source for food and medicine. Saudi. J Biol Sci 26:709–722. https://doi.org/10.1016/j.sjbs.2017.11.003

Schwenzfeier A, Wierenga PA, Gruppen H (2011) Isolation and characterization of soluble protein from the green microalgae Tetraselmis sp. Bioresour Technol 102:9121–9127. https://doi.org/10.1016/j.biortech.2011.07.046

Sharma GK, Khan SA, Shrivastava M, Bhattacharyya R, Sharma A, Gupta DK et al (2021) Circular economy fertilization: phycoremediated algal biomass as biofertilizers for sustainable crop production. J Environ Manag 287:112295. https://doi.org/10.1016/j.jenvman.2021.112295

Shehata N, Mohamed OA, Sayed ET, Abdelkareem MA, Olabi AG (2022) Geopolymer concrete as green building materials: recent applications, sustainable development and circular economy potentials. Sci Total Environ 836:155577. https://doi.org/10.1016/j.scitotenv.2022.155577

Tan HT, Yusoff FM, Khaw YS, Ahmad SA, Shaharuddin NA (2021) Uncovering research trends of Phycobiliproteins using bibliometric approach. Plants 10:2358. https://doi.org/10.3390/plants10112358

Tibbetts MS, Melanson JR, Park CK, Banskota HA, Stefanova R, McGinn JP (2016) Nutritional evaluation of whole and lipid-extracted biomass of the microalga Scenedesmus sp. AMDD isolated in Saskatchewan, Canada for animal feeds: proximate, amino acid, fatty acid, carotenoid and elemental composition. Curr Biotechnol 4:530–546. https://doi.org/10.2174/2211550104666150827201854

Tomás-Almenar C, Larrán AM, de Mercado E, Sanz-Calvo MA, Hernández D, Riaño B et al (2018) Scenedesmus almeriensis from an integrated system waste-nutrient, as sustainable protein source for feed to rainbow trout (Oncorhynchus mykiss). Aquaculture 497:422–430. https://doi.org/10.1016/j.aquaculture.2018.08.011

Wei M, Parrish CC, Guerra NI, Armenta RE, Colombo SM (2021) Extracted microbial oil from a novel Schizochytrium sp. (T18) as a sustainable high DHA source for Atlantic salmon feed: impacts on growth and tissue lipids. Aquaculture 534:736249. https://doi.org/10.1016/j.aquaculture.2020.736249

Wiatrowski M, Klein BC, Davis RW, Quiroz-Arita C, Tan ECD, Hunt RW et al (2022) Techno-economic assessment for the production of algal fuels and value-added products: opportunities for high-protein microalgae conversion. Biotechnol Biofuels Bioprod 15:8. https://doi.org/10.1186/s13068-021-02098-3

Zakaria MF, Mat Zaidi FNA, Ahmad Kamal AH, Aleng NA, Abd Wahid ME, Katayama T et al (2025) PUFA from microalgae: challenges, factors affecting high production and industrial application. Aquac Fish. https://doi.org/10.1016/j.aaf.2025.01.004

Zhou Y, Liu L, Li M, Hu C (2022) Algal biomass valorisation to high-value chemicals and bioproducts: recent advances, opportunities and challenges. Bioresour Technol 344:126371. https://doi.org/10.1016/j.biortech.2021.126371

Marine Macroalgae and Microbial Synergies for Carbon Neutrality

13

Vitasta Jad and Cathrine Sumathi Manohar

Abstract

Macroalgae, often referred to as seaweeds, are rightfully recognized as ecosystem engineers, given that they are the most productive marine macrophytes, thriving in intertidal coastal zones. They are increasingly acknowledged for their role in global carbon cycling, as they exhibit shorter growth cycles, higher photosynthetic rates, and high primary productivity. This enables them to sequester atmospheric CO_2 effectively, an essential function in the pursuit of carbon neutrality. Seaweed ecosystems contribute to oceanic carbon sequestration through the production of both dissolved and particulate organic carbon. Notably, seaweeds are known to produce highly recalcitrant compounds, some of which have been reported to get buried in sediments or exported to the deep sea, allowing for carbon storage on millennial timescales. In addition, macroalgae can absorb sulfur and nitrogen oxides, further contributing to the reduction of atmospheric pollutants and greenhouse gas equivalents. The full potential of macroalgae in achieving carbon neutrality is best realized when considering their intricate associations with diverse microbial assemblages. These microbiomes, comprising bacteria, archaea, and fungi, form a functionally integrated consortium with their macroalgal hosts that support essential physiological processes, including nutrient acquisition, organic matter transformation, stress tolerance, and even morphogenesis. Macroalgae are known to enrich microbial communities involved in carbon and nitrogen cycling, further amplifying their role in regulating ecosystem-level carbon dynamics. This

V. Jad · C. S. Manohar (✉)
Biological Oceanography Division, CSIR- National Institute of Oceanography, Dona Paula, Goa, India

Academy of Scientific and Innovative Research (AcSIR), Ghaziabad, Uttar Pradesh, India
e-mail: cathrine.nio@csir.res.in

K. Mohanty et al. (eds.), *Advances in Algal Research*,
https://doi.org/10.1007/978-981-95-7234-2_13

chapter examines the synergistic roles of macroalgae and their associated microorganisms in facilitating carbon capture and long-term storage. It highlights microbial contributions to algal growth optimization, carbon flux modulation through decomposition and mineralization, and long-term carbon burial through particulate organic matter. With the aid of emerging genomic and metagenomic tools, the complexity and metabolic versatility of these macroalgae-associated microbiomes are being progressively understood. Beyond their role in carbon neutrality, these biomes support ecosystem resilience and marine biodiversity, placing them at the forefront of nature-based climate mitigation strategies.

Keywords

Marine · Macroalgae · Microbes · Carbon capture · Climate change · Mitigation

13.1 Introduction

Achieving carbon neutrality by balancing human-driven carbon emissions into the atmosphere with equivalent removal remains one of the most pressing and complex challenges of the twenty-first century. Since the onset of industrialization, atmospheric carbon dioxide (CO_2) concentrations have surged to approximately 425 ppm, driven largely by deforestation and fossil fuel combustion. While the ocean buffers the worst impacts of climate change by absorbing nearly 9 billion tons of CO_2 annually, this uptake has come at a cost of rising sea surface temperatures, declining pH, and accelerating ocean acidification. Projections warn that if these emissions remain unchecked, atmospheric CO_2 levels could reach up to 700 ppm by 2100, with a global temperature increase of 0.6–2 °C and a drop in ocean pH by 0.3 units. These changes threaten climate stability, disrupt marine ecosystems, and compromise global biodiversity (Fine et al. 2017; Lindsey 2023). To achieve global carbon neutrality goals, carbon dioxide removal strategies must complement emission reductions. While terrestrial forests and soils have traditionally dominated carbon offset frameworks, the ocean, Earth's largest and most underutilized carbon sink, offers untapped potential. Among the most promising ocean-based solutions are marine macroalgae, commonly known as seaweeds, and their diverse microbial consortia. Macroalgal ecosystems are among the most expansive and productive coastal vegetated habitats on the planet, covering 10–40 times more area than other marine macrophytic landscapes, such as the mangroves and seagrasses (Duarte 2017). Despite this, they have long been overlooked in the global blue carbon assessments and recent scientific policy discourse has begun to recognize their vital role in capturing and storing atmospheric CO_2, both in the wild and in farmed systems (Duarte et al. 2025; Krause-Jensen and Duarte 2016; Macreadie et al. 2021).

The central aim of this chapter is to examine marine macroalgae and their complex, metabolically active microbial consortia as integrated biological systems capable of contributing significantly to global carbon neutrality efforts. The discussion synthesizes ecological, biochemical, and emerging biotechnological insights to outline the multiple mechanisms by which macroalgae contribute to atmospheric CO_2 drawdown. Particular emphasis is placed on the critical yet underappreciated role of macroalgae-associated microbiota, including bacteria, fungi, and viruses, in mediating carbon cycling, enhancing host resilience, and transforming labile organic matter into long-lived carbon reservoirs. The microbial dimension is positioned not merely as supportive, but as a key lever for optimizing macroalgal carbon sequestration potential in both natural and cultivated systems. Building on this foundation, this chapter introduces macroalgae as high-throughput carbon fixers, provides insights into the macroalgal productivity, export, and carbon storage capacity, and delineates the principal sequestration pathways that underpin their blue carbon potential. This chapter also provides an understanding of the restoration and cultivation frameworks, including innovations in integrated multi-trophic aquaculture that enhance scalability and ecological benefits. It also explores the macroalgal holobiont concept, where the seaweed host and its associated microbes function as an integrated ecological unit, highlighting the role of microbial symbionts in shaping physiological performance and carbon turnover, along with the emerging field of macroalgal microbiome engineering for enhanced carbon sequestration. By synthesizing the ecological understanding with emerging technological advances, this chapter positions macroalgal holobionts as critical players in the marine CO_2 removal landscape, offering sustainable pathways toward long-term carbon neutrality.

Marine macrophytes have long been recognized as natural global carbon sinks and are integral to the blue carbon framework. Among these, macroalgae (seaweeds) are gaining increasing attention for their pivotal role in the oceanic carbon cycle and their capacity to contribute significantly to carbon sequestration (Lovelock and Duarte 2019). These multicellular, photosynthetic eukaryotes typically inhabit intertidal and subtidal coastal zones, often anchoring to rocky substrates. They are classified into three major phyla, Chlorophyta (green), Rhodophyta (red), and Ochrophyta (brown), that have chemically and morphologically diverse characteristics (Filbee-Dexter et al. 2024). Beyond their taxonomic diversity, macroalgae demonstrate remarkable ecological versatility that extends their environmental impact well beyond carbon sequestration. Macroalgae hold substantial ecological significance owing to their exceptional capacity to bioaccumulate a wide range of pollutants, including heavy metals, POPs, and even microplastics (Seng et al. 2020). Through these bioaccumulative properties, macroalgae contribute effectively to natural bioremediation processes. As primary producers, they form the foundation of complex trophic networks, offering habitat, shelter, and nutritional resources to a multitude of marine organisms. Moreover, they enhance carbon sequestration and play a mitigating role in ocean acidification by absorbing excess CO_2 from surrounding waters (Gao et al. 2019). The recognition of macroalgae as significant carbon sinks has evolved considerably over the past several decades. Initially excluded from blue carbon assessments due to uncertainties around carbon tracking, measurement, and policy interventions; their role as natural CO_2 sinks has been recently acknowledged (Duarte et al. 2013; Chung et al. 2011). Early studies largely emphasized their contribution to carbon sequestration through the export of organic matter to the deep sea (Krause-Jensen and Duarte 2016; Queirós et al. 2019).

Historically, calcifying macroalgae species like *Halimeda* spp., which form carbonate deposits, were thought to contribute minimally to long-term carbon sequestration due to their lack of soft sedimentary habitats and the assumption that their carbon storage was restricted to transient above-ground biomass (Hill et al. 2015). However, this paradigm has been fundamentally challenged by growing evidence demonstrating that macroalgae produce recalcitrant organic compounds that can either be buried in sediments or exported to the deep ocean, where they contribute to long-term carbon storage (Lovelock and Duarte 2019). This paradigm shift has fundamentally altered how we conceptualize macroalgal contributions to global carbon cycling. This broader understanding has repositioned macroalgal habitats as key players within the coastal carbon cycle supporting long-term storage in the adjacent deep sea systems (Hill et al. 2015). The first global estimate of macroalgae derived carbon sequestration suggested that these underwater forests could surpass the combined sequestration potential of all other recognized blue carbon ecosystems (Krause-Jensen and Duarte 2016). As the urgency to discover scalable, ocean-based solutions to climate change intensifies, both natural and farmed macroalgae are being re-evaluated as essential, previously underestimated, components of the global blue carbon strategy (Macreadie et al. 2021). The quantitative assessment of macroalgal carbon dynamics reveals the substantial scale of their contribution to global carbon cycling. Macroalgae dominate primary production in coastal ecosystems, with an estimated carbon sequestration potential between 61 and 268 TgC annually (Krause-Jensen and Duarte 2016). However, it is not merely the rate of carbon fixation that distinguishes macroalgae, but the downstream fate of this carbon that determines its true sequestration value. Their global net primary production (NPP) has been projected to be around 1521 TgC yr^{-1}, of which, on average of about 0.4% gets buried within soft-sediments, accounting for approximately 6.2 TgC yr^{-1}. Unlike other blue carbon ecosystems, macroalgae follow a different carbon export strategy. Contrary to seagrasses and mangroves, which often retain and bury carbon in situ, macroalgae release a significant portion of their fixed carbon into the surrounding environment. On average, approximately 44% of their NPP, around 679 TgC yr^{-1} is exported as organic carbon into surrounding waters (Duarte et al. 2025; Krause-Jensen and Duarte 2016). Of this, 52% (355 TgC yr^{-1}) is in dissolved organic carbon (DOC) form, and 48% (323 TgC yr^{-1}) as particulate organic carbon (POC). An estimated 33% of the DOC (117 TgC yr^{-1}) is transported below the ocean's mixed layer, beyond the reach of rapid remineralization, positioning it as a candidate for longer-term carbon sequestration. Similarly, about 11% of POC export (35 TgC yr^{-1}) is believed to reach deep sea sediments, where low oxygen and biological activity allow for slow decomposition or burial. An additional 4.6% (14 TgC yr^{-1}) of the macroalgae derived POC contributes to continental shelf burial, with the remainder presumed to undergo remineralization in coastal waters (Ortega et al. 2019). Carbon decomposed in the upper ocean usually equilibrates with atmospheric CO_2 relatively quickly, meaning it does not contribute to long-term sequestration. However, macroalgal detritus that reaches depths greater than 1000 m is often either buried or broken down very slowly, remaining out of contact with the atmosphere for centuries or longer, hence being sequestered (Filbee-Dexter et al. 2024).

Recent global modeling efforts accounting for decomposition rates and coastal residence times suggest that macroalgal forests may export about 30 g of Cm^{-2} yr^{-1} as POC to deeper waters. Applied globally to the known extent of brown seaweed forests globally, this implies that roughly 56 TgC yr^{-1} may be transported beyond 200 m depth. Further modeling, based on how CO_2 behaves when injected into different ocean depths, indicated that of this exported carbon:15–65% of material remineralized beyond 200 m will take over 25 years to return to the surface, 6–11% (4–6Tg C yr^{-1}) will take more than 100 years, 55–96% of carbon reaching 530 m depth will remain sequestered for over 25 years, 23–45% (13–25 TgC yr^{-1}) for over a century, and up to 44 TgC yr^{-1} transported to depths beyond 1000 m could remain out of circulation for more than 100 years and account truly for long-term sequestration (Filbee-Dexter et al. 2024). Importantly, macroalgal export represents an active biological process rather than passive sedimentation. Hydrodynamic forces, fragment buoyancy, and biological interactions influence how far and fast carbon travels. For example, *Sargassum*, a floating brown alga with gas bladders, can drift thousands of kilometers before sinking, especially after colonization by epibionts increases its density. Some fragments have been tracked reaching abyssal depths of 5000 m within days, highlighting a rapid biological pump powered by natural dispersal vectors (Johnson and Richardson 1977). Thus, while macroalgae may not retain carbon where they grow, their high productivity, efficient export, and deep-sea delivery potential make them formidable players in the global carbon cycle, especially if restoration or farming efforts are designed to maximize these flows into long-lived storage pools.

13.2 Macroalgae for Carbon Sequestration: Importance, Mechanisms, and Potential

Macroalgal forests are recognized not only for their remarkable primary productivity even in nutrient-limited conditions but also for their ability to channel substantial amounts of fixed carbon into long-term storage via multiple ecological pathways (Duarte 2017). Several pathways have been recognized by means of which macroalgae sequester carbon; these include the production of recalcitrant dissolved organic carbon (RDOC, a pool of organic molecules resistant to microbial degradation and capable of persisting in the ocean for millennia), burial of POC, and vertical export of organic matter to the deep ocean. Each of these routes contributes uniquely to their blue carbon potential, either directly through sedimentary accumulation or indirectly by feeding into distant, slower-moving carbon pools (Fig. 13.1).

Among these, the production of RDOC is increasingly recognized as a key mechanism (Hansell 2013). Macroalgae are prolific sources of DOC, with average release rates surpassing those of other coastal vegetated ecosystems such as seagrasses, mangroves, and even phytoplankton (Pessarrodona et al. 2023). Although the majority of this DOC is eventually remineralized through microbial or photochemical processes, a smaller fraction resists degradation and can persist up

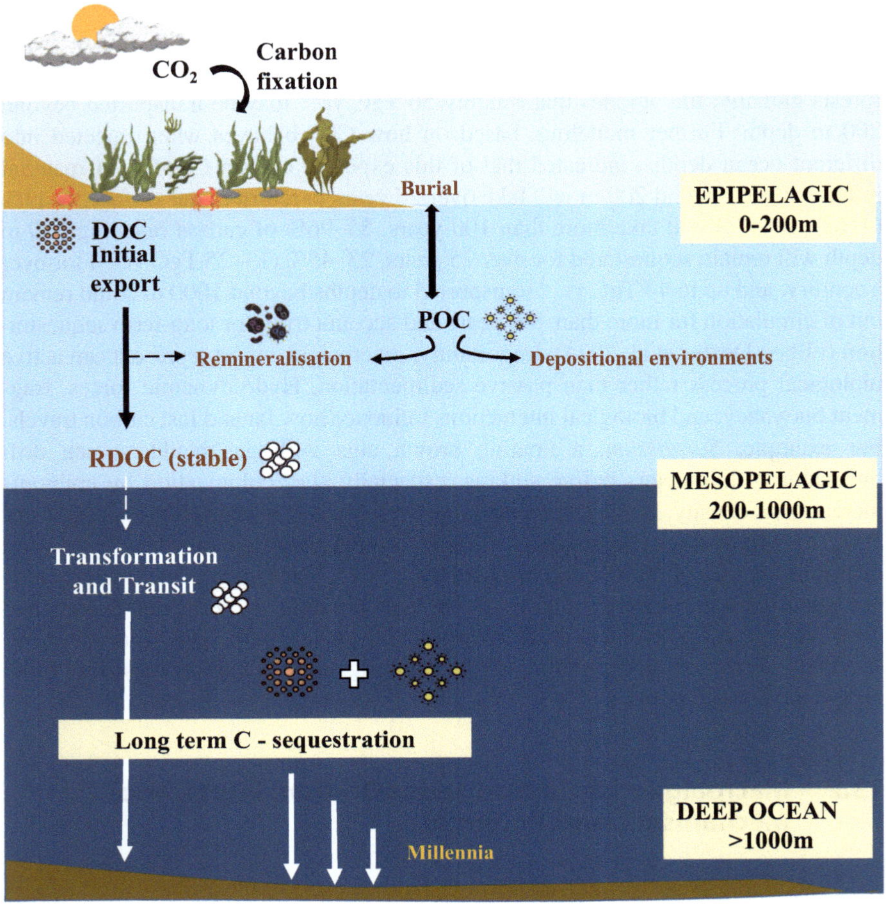

Fig. 13.1 Schematic representation of *macroalgal carbon sequestration pathways* across the ocean water column. This includes CO_2 fixation by macroalgae in the epipelagic zone, followed by the release of carbon as DOC or POC. Organic carbon can be buried in sediments or exported to depositional environments, with partial remineralization during transit. Fixed carbon may undergo remineralization or persist as refractory organic matter, which resists microbial degradation and contributes to long-term storage over centennial to millennial timescales

to centennial timescales in the ocean, contributing to long-term carbon storage (Hansell 2013). The degree to which DOC becomes refractory depends on multiple factors, including its concentration, biochemical composition, environmental conditions, and interactions with surrounding microbial communities (Baltar et al. 2021; Jiao et al. 2014). Notably, macroalgal DOC is often enriched with phenolic and humic-like compounds, molecular structures that are more resistant to microbial breakdown and thus decompose more slowly, enhancing their potential for long-term persistence in the marine environment (Krause-Jensen and Duarte 2016). Burial of macroalgal POC in the marine sediments represents another significant sequestration pathway. Unlike rooted seagrasses and saltmarshes, macroalgae predominantly colonize rocky substrates where in situ burial is limited, the POC has to be exported to depositional environments. Mean release rates are comparable to those from tidal marshes and exceed those of seagrasses and phytoplankton. POC production is highly variable and influenced by seasonal cycles, hydrodynamics, and biotic interactions, including biofouling (Pessarrodona et al. 2023). The efficiency of this process hinges on multiple physical and biological factors, including fragment buoyancy, detrital degradation rates, current dynamics, and seabed morphology. Submarine canyons, fjords, and depositional basins act as natural traps, enhancing burial efficiency (Krause-Jensen and Duarte 2016; Filbee-Dexter et al. 2024). However, the degree of burial depends on detritus quality, local decomposition rates, and sediment characteristics. Remineralization processes, especially in the mixed surface layer, can limit long-term sequestration unless POC is buried below active zones. In deeper sediments, particularly in the deep sea, macroalgal carbon may persist much longer, even with available oxygen, due to low biological activity. Empirical evidence supports long-term burial in Nordic sediments for over 120 years, but it will, however, only contribute to carbon sequestration if it remains buried for periods of time actually relevant to climate mitigation (Frigstad et al. 2021).

The third, and potentially most impactful, involves the export of organic carbon, both POC and DOC, to the deep ocean interior. This mechanism is increasingly recognized as a critical component of macroalgal carbon dynamics, particularly when carbon is transported beyond depths of 1000 to 1500 m, where it may be effectively sequestered for centuries or longer (Krause-Jensen and Duarte 2016; Ortega et al. 2019). However, the permanence of this sequestration depends on how deep the carbon travels and where it ends up in the ocean. If it sinks below 1000–1500 m into regions with slow water mixing (low ventilation), it can remain isolated from the atmosphere for centuries. However, in areas with faster water circulation, the carbon may return to the surface and be released as CO_2 much sooner (Baker et al. 2022). The North Atlantic basin shows relatively rapid turnover, with nearly a third of carbon exported to 1000 m returning to the atmosphere within 100 years. In contrast, the Indian and Pacific Oceans offer considerably longer storage due to slower ventilation. Furthermore, the biogeochemical signature of exported macroalgal material, its C:N ratio, particulate size, and degradability, can influence its fate during transit. Fine particles often remain suspended longer, increasing their odds of reaching deeper basins, while larger or more labile fragments tend to settle closer to the source and face rapid microbial decomposition. Nevertheless, evidence from observational studies and particle tracking suggests that macroalgal export represents a meaningful addition to the long-term carbon budget of the ocean, particularly when linked to hydrodynamic pathways that facilitate offshore delivery (Johnson and Richardson 1977). Thus, macroalgae contribute to oceanic carbon sequestration through a multipronged pathway system that blends chemical resistance to degradation with physical transport processes and ecological interactions, making macroalgae valuable allies in long-term carbon storage. Recognizing and optimizing these pathways, through restoration, cultivation, or system design, could transform these often-overlooked primary producers into central agents of marine-based climate mitigation. Restoration and sustainable cultivation of macroalgae stands out for its scalability and ecological co-benefits. Unlike land-based carbon solutions that often compete with agriculture or forest conservation, seaweed systems harness vast marine space to deliver climate benefits without compromising on food or land security (Duarte et al. 2021). Historically, macroalgal restoration was pursued primarily to revive coastal habitats and rebuild fisheries. Iconic systems like kelp forests and nearshore algal beds were valued as biodiversity hotspots and nursery grounds for marine life. However, with climate change intensifying, these ecosystems are now being revisited for their critical role in strengthening marine biodiversity and expanding blue carbon sinks. Unfortunately, many of these natural habitats including kelp forests and nearshore seaweed beds are in steep decline driven by ocean warming, eutrophication, overgrazing, and coastal modifications. Since the mid-twentieth century, macroalgal forests have been vanishing at an average rate of 1.8% per year (Krumhansl et al. 2016), with over 1.5 million hectares lost in the current century alone (Filbee-Dexter et al. 2022). This erosion has diminished not only their carbon donor capacity but also a suite of associated ecosystem services. Restoration holds the potential to reverse these losses, reviving local biodiversity, rebuilding fisheries, and reactivating the biological carbon pump. Estimates indicate that such recovery could significantly boost global carbon standing stocks by several tens of Tg, while also increasing carbon sequestration rates by multiple Tg yr^{-1} (Filbee-Dexter et al. 2022). Beyond restoration, sustainable seaweed cultivation, whether offshore, nearshore, or within integrated aquaculture systems, extends this potential by engineering productive carbon sinks in regions previously unoccupied by macroalgae. A particularly promising innovation is integrated multi-trophic aquaculture, which co-cultivates seaweeds with animal species such as shellfish or finfish. By harnessing nutrient flows from animal waste as natural fertilizer for seaweed, this approach supports a closed-loop system that enhances water quality, nutrient recycling, and carbon capture. Empirical comparisons illustrate this synergy in mariculture systems, which showed that while kelp monoculture sequestered 4387.5 kgC hm^{-2} $yr^{-1}$and abalone monoculture 1808.3 kgC hm^{-2} yr^{-1}, the integrated multi-trophic aquaculture model combining both achieved an impressive 12,311.9 kgC hm^{-2} yr^{-1} (Tang 2014). This integrated design amplifies CO_2 uptake across trophic levels and strengthens the biological pump, making it a powerful, nature-based solution for marine carbon management. Despite this potential, seaweed-based sustainable mariculture remains underdeveloped in the global carbon capture, utilization, and storage framework. Its broader contributions to long-term carbon sequestration, environmental bioremediation, and potential to underpin a sustainable marine bioeconomy warrant greater research and policy attention (Wang et al. 2025).

13.3 Ecological and Biotechnological Significance of Macroalgal Microbiomes

While environmental parameters such as light, nutrients, and temperature are well-known regulators of macroalgal productivity, a critical yet underappreciated component of macroalgal health and resilience lies in its dynamic association with diverse microbial partners, collectively referred to as the macroalgal holobiont. These microbial consortia, comprising epiphytic and endophytic bacteria, archaea, fungi, protists, and viruses, colonize the algal surface and tissues, actively participating in host metabolism and ecosystem functioning (Barbato et al. 2022). Dominated by groups such as Proteobacteria, Actinobacteria, Verrucomicrobia, and Planctomycetes, macroalgal-associated microbial communities play integral roles in nutrient cycling, particularly nitrogen turnover and organic matter degradation. Studies reveal a metabolically rich microbial community, underscoring vital roles in maintaining nutrient and energy dynamics, and overall holobiont function, essential for maintaining productivity and ecological stability in seaweed ecosystems (Akrong et al. 2023; Burke et al. 2011; Parab et al. 2025). These associated microbes, including bacteria and fungi, are known to secrete extracellular enzymes such as alginate lyases, cellulases, and agarases, which hydrolyze complex macroalgal polysaccharides into simpler, bioavailable forms. This not only supports microbial nutrition but also facilitates internal nutrient recycling, benefiting the host (Jagtap and Manohar 2021; Singh and Reddy 2014). Additionally, by regulating the density and composition of the biofilm on algal surfaces, they prevent microbial overgrowth, which can otherwise impede light penetration, photosynthetic efficiency, and gas exchange. Importantly, macroalgal microbes contribute to long-term carbon sequestration by catalyzing the conversion of labile DOC into RDOC, a stable and biologically resistant carbon pool that can persist in ocean waters for centuries. This transformation is influenced by microbial metabolic traits, including oxidative stress tolerance and the capacity to produce aromatic, humic-like compounds that resist rapid degradation (Jiao et al. 2014). Their contributions toward host resilience and stress tolerance are also significant. The brown alga *Ascophyllum nodosum* hosts the endophytic fungus *Stigmidium ascophylli*, whose synchronized life cycle and physiological interactions have been linked to enhanced desiccation resistance during tidal exposure (Bonthond et al. 2022). Other macroalgal–microbial associations enhance host tolerance to salinity fluctuations, oxidative stress, and thermal extremes by modulating antioxidant pathways and producing protective secondary metabolites (Singh and Reddy 2014). Some of these microbes also produce bioactive metabolites with antimicrobial properties, contributing to host defense and microbiome stability (Asharaf and Chakraborty 2023). Viruses, often overlooked, are emerging as key regulators of the macroalgal holobiont function. Beyond their roles in microbial turnover through lytic cycles, many marine viruses encode auxiliary metabolic genes that modulate microbial metabolism and biogeochemical fluxes. In macroalgal systems, auxiliary metabolic genes have been implicated in DOC turnover, sulfur metabolism, and nitrogen assimilation, suggesting viral influence extends to reshaping microbial function and potentially

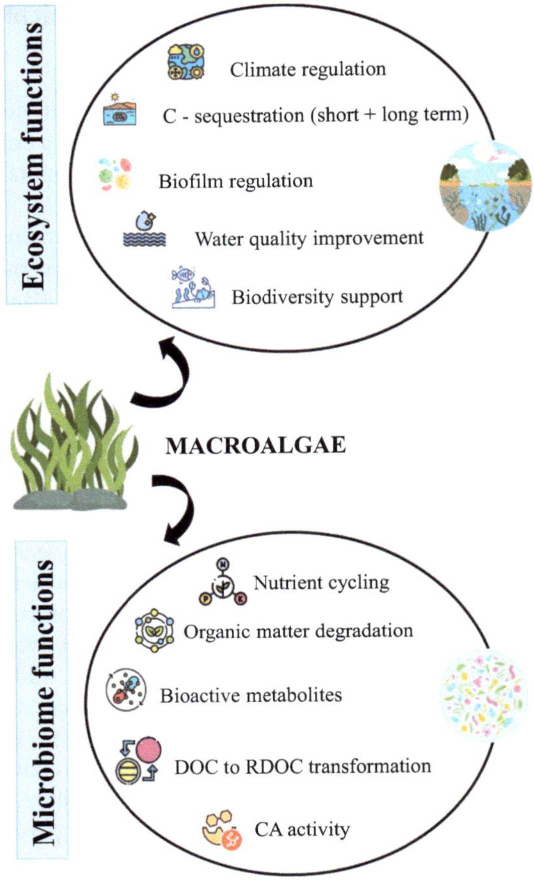

Fig. 13.2 Complementary *ecosystem* and *microbiome functions* within macroalgal holobionts. Macroalgae contribute to ecosystem services such as climate regulation and short- and long-term carbon sequestration. The associated microbiomes further reinforce these benefits through nutrient cycling, organic matter degradation, production of bioactive metabolites, transformation of DOC into RDOC, and CA-mediated inorganic carbon processing. Together, these coupled functions highlight the *holobiont* as a *synergistic unit* with enhanced potential for carbon retention and climate mitigation

enhancing carbon retention (Zhao et al. 2024). The functional contributions of these microbial communities to both macroalgal health and ecosystem services illustrates the synergy between microbiome activities and host-mediated carbon retention, nutrient cycling, and biodiversity support (Fig. 13.2).

Despite the recognized importance of macroalgae as fast-growing carbon sinks, the functional role of their associated microbiomes in carbon fixation and long-term sequestration remains insufficiently explored. While the macroalgal host contributes significantly to coastal blue carbon stocks through rapid biomass accumulation and detrital export, its microbial symbionts may also influence carbon dynamics, not only by transforming organic matter but potentially by mediating inorganic carbon uptake pathways. In contrast, microalgal systems have been extensively studied for CO_2 bio-sequestration, particularly in the context of engineered cultivation platforms and carbon capture strategies. The role of microbial enzymes like carbonic anhydrase (CA) has been primarily highlighted in facilitating inorganic carbon acquisition and enhancing photosynthetic efficiency in microalgal cultures (Mondal et al. 2016). Carbonic anhydrase (CA) is a zinc metalloenzyme widely distributed across all domains of life. It catalyzes the interconversion of CO_2 and bicarbonate (HCO_3^-), playing a central role in carbon concentrating mechanisms. Its activity helps maintain CO_2 availability around RuBisCO, the primary enzyme involved in carbon fixation (Bose and Satyanarayana 2017). Unlike in microalgae and cyanobacteria, where the contribution of CA to carbon-concentrating mechanisms and photosynthetic enhancement is well established, the novelty in macroalgal holobionts lies in the microbial mediation of carbon fluxes external to host chloroplasts. Associated microbes, including bacteria, archaea, and fungi, may generate auxiliary CA activity at the host–microbe interface, influencing both the supply of inorganic carbon to the alga and the stabilization of escaped CO_2 within the boundary layer. This distinction moves the conversation beyond intracellular algal physiology into a multi-partner holobiont framework, expanding carbon cycling from a host-centric to a community-integrated process. Preliminary genomic analyses suggest that macroalgal-associated bacteria, archaea, and fungi, commonly encode CA genes, often with multiple copies per genome, implying adaptive capabilities for inorganic carbon concentration (Franco et al. 2025).

Investigating microbial CA activity within macroalgal holobionts may reveal powerful mechanisms for enhancing carbon sequestration. By accelerating the conversion of CO_2 to HCO_3^-, CA maintained by epiphytic or endophytic microbes could boost local bicarbonate availability, which macroalgal carbon concentrating mechanisms preferentially use, especially under fluctuating CO_2 conditions. This microbial assistance may improve overall carbon assimilation efficiency, effectively enhancing photosynthetic productivity. Moreover, CA activity situated at the host–microbe interface could play a protective role against carbon loss. By rapidly rehydrating escaped CO_2 back into HCO_3^-, microbes may help stabilize inorganic carbon pools near algal surfaces, reducing carbon leakage during diel light–dark cycles and improving net CO_2 uptake. These increased carbon assimilation rates may translate into greater biomass production, which corresponds with elevated DOC exudation. Thus, microbial CA might indirectly reinforce blue carbon retention by supporting downstream microbial pathways that stabilize organic carbon. Moreover, recent advances suggest that microbial CA may facilitate microbially induced carbonate precipitation, a biomineralization process where bicarbonate combines with calcium to form stable calcium carbonate ($CaCO_3$). In terrestrial and engineered systems, CA-mediated microbially induced carbonate precipitation is being explored as a biocementation strategy for carbon capture and construction material stabilization. Localized CA activity could modulate pH within the algal biofilm, triggering $CaCO_3$ formation and reinforcing both structural stability and carbon permanence (Faridi and Satyanarayana 2015; Nair et al. 2025; Ruan et al. 2024). Integrating this process into macroalgal holobionts presents an exciting avenue for dual-function carbon retention, enhancing organic sequestration while enabling inorganic carbon deposition. While empirical studies on CA in macroalgal ecosystems are still lacking, integrating CA active microbes into seaweed cultivation frameworks offers a promising strategy for improving biomass yield and enhancing carbon retention, marking a significant step toward high-efficiency blue carbon farming.

Crucially, the success and long-term viability of macroalgal cultivation systems are increasingly understood to be inseparable from the microbial communities that inhabit and interact with them. Seaweeds act as ideal substrates for microbial colonization, supporting structurally complex and metabolically dynamic microbial assemblages comprising bacteria, archaea, fungi, protists, and viruses that exert a profound influence on ecosystem functioning (Barbato et al. 2022). These microbial consortia are not passive residents but active participants in shaping and regulating macroalgal physiology, productivity, buffering environmental stressors, and, importantly, facilitating the conversion of photosynthetically fixed carbon into recalcitrant forms that support long-term oceanic carbon sequestration (Barbato et al. 2022; Singh and Reddy 2014). Such microbial functions not only improve the robustness and consistency of biomass yields under variable environmental conditions but also enhance the resilience and sustainability of large-scale macroalgal cultivation systems. Offshore waters chosen for seaweed farming, often oligotrophic, rely heavily on and benefit from these microbial partnerships. Nitrogen-fixing bacteria and ammonia-oxidizing archaea are especially vital, as they transform atmospheric nitrogen and ammonia into bioavailable forms. These microbial pathways reduce reliance on external nutrient inputs, enabling seaweed cultivation in nutrient—poor offshore waters and decreasing dependence on eutrophic coastal zones. Studies on species such as *Gigartina lemaneiformis* (red algae) grown in offshore farms confirm that its surface microbiome is enriched with epiphytic diazotrophs, bacteria actively expressing the nifH gene, at notably higher levels than found in surrounding seawater. This highlights their essential role in augmenting nitrogen availability to the algal host (Pei et al. 2024). Similarly, kelp-associated microbiomes like those of *Macrocystis pyrifera* (giant kelp) harbor functional genes for nitrate and nitrite reduction, ammonification, and other nutrient cycling pathways that enable sustained productivity in open ocean farms (Hochroth and Pfister 2024). As our understanding of these associations deepens, incorporating microbial functionality into restoration and cultivation designs may become central in maximizing both biomass yield and carbon sequestration potential. Despite these advances, two key challenges temper enthusiasm for large-scale macroalgal farming as a climate solution. First, the permanence of sequestration remains contested, while detrital export and DOC release may deliver carbon to the deep ocean, remineralization during transit or microbial degradation in mid-waters can substantially reduce long-term storage efficiency. Second, offshore aquaculture faces ecological and economic constraints such as exposure to storms, high operational costs, and uncertainties around ecological impacts (altered nutrient dynamics, habitat shading, or biodiversity displacement). These trade-offs underscore the importance of embedding microbial innovations such as CA-mediated carbon retention or nutrient cycling consortia within a realistic understanding of both biogeochemical permanence and practical feasibility.

Traditional macroalgal cultivation strategies, while promising for carbon sequestration, are increasingly vulnerable to environmental stressors such as ocean warming, acidification, and pollution. These challenges undermine both productivity and carbon retention, highlighting the need for next-generation interventions that go beyond host-level solutions. Among the most promising frontiers is microbiome engineering, the deliberate manipulation of seaweed-microbiome to enhance macroalgal health, resilience, and carbon capture efficiency (Kaur et al. 2022). Natural macroalgal microbiomes, though well-adapted to their hosts, can become unstable under farming pressures or shifting ocean conditions. These communities may break down, lose beneficial members, or become dominated by harmful microbes, making it difficult to maintain healthy and productive seaweed crops over time. Therefore, instead of relying on naturally assembling microbiomes, researchers are exploring the use of probiotic or synthetic microbial consortia tailored to specific macroalgal species (Ruan et al. 2024). Inspired by advances in terrestrial agriculture and aquaculture (Jin et al. 2024; Kaur et al. 2022), microbiome engineering seeks to overcome these limitations by designing synthetic or probiotic consortia tailored to the physiological and ecological needs of specific macroalgal species. Experimental manipulations have confirmed that microbiome optimization can boost algal biomass, improve defense against disease, and reduce oxidative stress. Targeted inoculation with beneficial bacterial strains such as *Pseudoalteromonas*, *Bacillus*, and *Alteromonas* has demonstrated promising outcomes. *Pseudoalteromonas porphyrae*, isolated from *Laminaria japonica*, has been reported to produce catalase and other bio-stimulatory compounds that enhance plant growth and stress tolerance (Dimitrieva et al. 2006). Similarly, probiotic applications of *Pseudoalteromonas* sp. and *Phaeobacter* sp. have reduced pathogen-induced bleaching in red macroalgae like *Delisea pulchra* and *Agarophyton vermiculophyllum* (Li et al. 2022). These engineered associations are typically host-specific, shaped by coevolution and chemical signaling, suggesting that a "one-microbe-fits-all" strategy is unlikely to succeed. Instead, species-specific microbial formulations offer the highest potential for stable colonization and function under field conditions. This approach not only enhances productivity but also expands the potential for long-term carbon storage under variable climatic conditions. Beyond improving growth and resilience, microbiomes also serve as bioindicators of ecosystem health. Shifts in microbial composition or functional competencies, such as increased virulence or reactive oxygen species sensitivity, can signal nutrient imbalances, stress, or disease onset, enabling predictive management in intensive mariculture systems. Integrating microbial strategies into climate-resilient cultivation shows great promise, such as enriching farms with thermotolerant, acidification-resistant, or oxidative stress-mitigating taxa that can extend the viability across seasonal extremes and less hospitable environments. These tailored microbiomes might also enhance yield and carbon sequestration in next-generation cultivation systems.

13.4 Conclusion

This chapter highlights the pivotal role of marine macroalgae and their associated microbial communities as a powerful, nature-based strategy for carbon neutrality. Despite being historically underrepresented in blue carbon discourse, macroalgae are now recognized for their exceptional carbon fixation capacity. Through diverse pathways, including the production of RDOC, POC burial, and deep-sea carbon export, they offer scalable and long-term carbon storage potential. These benefits are deeply interlinked with the macroalgal microbiome. These microbial inhabitants, encompassing bacteria, archaea, fungi, protists, and viruses, form the holobiont, actively driving nutrient acquisition, stress tolerance, and the transformation of organic carbon into stable forms. Emerging evidence suggests that microbial enzymes such as CA, from macroalgae-associated microbes, may enhance inorganic carbon assimilation and reduce CO_2 loss, representing a promising mechanism for boosting carbon retention. Tailored microbial consortia, aligned with specific macroalgal hosts, offer solutions to overcome the instability of natural communities and enhance productivity, disease resistance, and carbon storage, especially under climate stress.

To unlock the full sequestration potential of macroalgal systems, the next frontier can be realized by:

- Conserving seaweed species and their genetic reservoirs to ensure ecosystem resilience, stable holobiont interactions, and sustained carbon sequestration.
- Advancing sustainable farming systems that integrate microbial support to enhance productivity and long-term carbon storage.
- Elucidating the mechanisms, particularly microbial CA-mediated carbon cycling, underpinning carbon fixation and storage using multi-omics tools.
- Designing and validating species-specific microbial consortia for large-scale cultivation.
- Conducting field-scale trials to quantify microbiome-enhanced carbon sequestration.
- Ensuring engineered microbiomes remain stable and effective under diverse and dynamic marine conditions.

Reimagining macroalgal ecosystems as holobiont systems, where host and microbiome are co-engineered for performance, can transform seaweed aquaculture into a cornerstone of resilient blue carbon solutions. Hence, the integration of macroalgal holobionts into policy and market frameworks is essential. If microbial contributions to carbon fixation and permanence can be quantified and standardized, they could strengthen the case for including macroalgal systems within blue carbon accounting, carbon credit schemes, and climate mitigation protocols. Embedding microbial functionality into restoration, conservation, and aquaculture policies would not only enhance sequestration reliability but also align seaweed-based solutions with global climate targets. In this way, holobiont-driven macroalgal systems move from an ecological innovation to a recognized pillar of blue carbon governance.

References

Akrong MO, Anning AK, Addico GND, Hogarh JN, Adu-Gyamfi A, deGraft-Johnson KAA et al (2023) Variations in seaweed-associated and planktonic bacterial communities along the coast of Ghana. Mar Biol Res 19(4–5):219–233

Asharaf S, Chakraborty K (2023) Seaweed-associated heterotrophic *Bacillus altitudinis* MTCC13046: a promising marine bacterium for use against human hepatocellular adenocarcinoma. Arch Microbiol 205(1):10

Baker CA, Martin AP, Yool A, Popova E (2022) Biological carbon pump sequestration efficiency in the North Atlantic: a leaky or a long-term sink? Glob Biogeochem Cycles 36(6): e2021GB007286

Baltar F, Alvarez-Salgado XA, Arístegui J, Benner R, Hansell DA, Herndl GJ et al (2021) What is refractory organic matter in the ocean? Front Mar Sci 8:642637

Barbato M, Vacchini V, Engelen AH, Patania G, Mapelli F, Borin S et al (2022) What lies on macroalgal surface: diversity of polysaccharide degraders in culturable epiphytic bacteria. AMB Express 12(1):98

Bonthond G, Barilo A, Allen RJ, Cunliffe M, Krueger-Hadfield SA (2022) Fungal endophytes vary by species, tissue type, and life cycle stage in intertidal macroalgae. Lane C, editor. J Phycol 58(2):330–342

Bose H, Satyanarayana T (2017) Microbial carbonic anhydrases in biomimetic carbon sequestration for mitigating global warming: prospects and perspectives. Front Microbiol 8:1615

Burke C, Thomas T, Lewis M, Steinberg P, Kjelleberg S (2011) Composition, uniqueness and variability of the epiphytic bacterial community of the green alga *Ulva australis*. ISME J 5(4): 590–600

Chung IK, Beardall J, Mehta S, Sahoo D, Stojkovic S (2011) Using marine macroalgae for carbon sequestration: a critical appraisal. J Appl Phycol 23(5):877–886

Dimitrieva GY, Crawford RL, Yuksel GU (2006) The nature of plant growth-promoting effects of a pseudoalteromonad associated with the marine algae *Laminaria japonica* and linked to catalase excretion. J Appl Microbiol 100(5):1159–1169

Duarte CM (2017) Reviews and syntheses: hidden forests, the role of vegetated coastal habitats in the ocean carbon budget. Biogeosciences 14(2):301–310

Duarte CM, Losada IJ, Hendriks IE, Mazarrasa I, Marbà N (2013) The role of coastal plant communities for climate change mitigation and adaptation. Nat Clim Chang 3(11):961–968

Duarte CM, Bruhn A, Krause-Jensen D (2021) A seaweed aquaculture imperative to meet global sustainability targets. Nat Sustain 5(3):185–193

Duarte CM, Delgado-Huertas A, Marti E, Gasser B, Martin IS, Cousteau A et al (2025) Carbon burial in sediments below seaweed farms matches that of Blue Carbon habitats. Nat Clim Chang 15(2):180–187

Faridi S, Satyanarayana T (2015) Applicability of carbonic Anhydrases in mitigating global warming and development of useful products from CO_2. Clim Change Environ Sustain 3(2):77

Filbee-Dexter K, Wernberg T, Barreiro R, Coleman MA, De Bettignies T, Feehan CJ et al (2022) Leveraging the blue economy to transform marine forest restoration. Edwards M, editor. J Phycol 58(2):198–207

Filbee-Dexter K, Pessarrodona A, Pedersen MF, Wernberg T, Duarte CM, Assis J et al (2024) Carbon export from seaweed forests to deep ocean sinks. Nat Geosci 17(6):552–559

Fine RA, Willey DA, Millero FJ (2017) Global variability and changes in ocean total alkalinity from Aquarius satellite data. Geophys Res Lett 44(1):261–267

Franco MEE, Singer E, Roux S, Meredith LK, U'Ren JM (2025) Genomic and metagenomic survey of microbial carbonic anhydrase genes reveals novel clades, high diversity, and biome-specificity. Cold Spring Harbor Laboratory

Frigstad H, Gundersen H, Andersen A, Borgersen G, Kvile KØ, Krause-Jensen D et al (2021) Blue Carbon–climate adaptation, CO_2 uptake and sequestration of carbon in Nordic blue forests: results from the Nordic Blue Carbon Project. Nordic Council of Ministers

Gao K, Beardall J, Häder DP, Hall-Spencer JM, Gao G, Hutchins DA (2019) Effects of ocean acidification on marine photosynthetic organisms under the concurrent influences of warming, UV radiation, and deoxygenation. Front Mar Sci 6:322

Hansell DA (2013) Recalcitrant dissolved organic carbon fractions. Annu Rev Mar Sci 5(1): 421–445

Hill R, Bellgrove A, Macreadie PI, Petrou K, Beardall J, Steven A et al (2015) Can macroalgae contribute to blue carbon? An A ustralian perspective. Limnol Oceanogr 60(5):1689–1706

Hochroth A, Pfister CA (2024) Ammonification by kelp associated microbes increases ammonium availability. Melzner F, editor. PLoS One 19(3):e0296622

Jagtap AS, Manohar CS (2021) Overview on microbial enzymatic production of algal Oligosaccharides for nutraceutical applications. Mar Biotechnol 23(2):159–176

Jiao N, Robinson C, Azam F, Thomas H, Baltar F, Dang H et al (2014) Mechanisms of microbial carbon sequestration in the ocean – future research directions. Biogeosciences 11(19): 5285–5306

Jin Y, Chen C, Ai C, Song W, Zhao J (2024) Development of probiotic complex based on the synthetic microbial community and probiotic effects in farming Yunlong grouper. Aquaculture 586:740708

Johnson DL, Richardson PL (1977) On the wind-induced sinking of Sargassum. J Exp Mar Biol Ecol 28(3):255–267

Kaur S, Egidi E, Qiu Z, Macdonald CA, Verma JP, Trivedi P et al (2022) Synthetic community improves crop performance and alters rhizosphere microbial communities. J Sustain Agric Environ 1(2):118–131

Krause-Jensen D, Duarte CM (2016) Substantial role of macroalgae in marine carbon sequestration. Nat Geosci 9(10):737–742

Krumhansl KA, Okamoto DK, Rassweiler A, Novak M, Bolton JJ, Cavanaugh KC et al (2016) Global patterns of kelp forest change over the past half-century. Proc Natl Acad Sci 113(48): 13785–13790

Li J, Weinberger F, Saha M, Majzoub ME, Egan S (2022) Cross-host protection of marine bacteria against macroalgal disease. Microb Ecol 84(4):1288–1293

Lindsey R (2023) Climate.gov. NOAA

Lovelock CE, Duarte CM (2019) Dimensions of Blue Carbon and emerging perspectives. Biol Lett 15(3):20180781

Macreadie PI, Costa MDP, Atwood TB, Friess DA, Kelleway JJ, Kennedy H et al (2021) Blue carbon as a natural climate solution. Nat Rev Earth Environ 2(12):826–839

Mondal M, Khanra S, Tiwari ON, Gayen K, Halder GN (2016) Role of carbonic anhydrase on the way to biological carbon capture through microalgae-a mini review. Environ Prog Sustain Energy 35(6):1605–1615

Nair S, Zhang Z, Wang X, Zhang B, Jiao N, Zhang Y (2025) Engineering microbiomes to enhance macroalgal health, biomass yield, and carbon sequestration. Green Carbon 3(1):63–73

Ortega A, Geraldi NR, Alam I, Kamau AA, Acinas SG, Logares R et al (2019) Important contribution of macroalgae to oceanic carbon sequestration. Nat Geosci 12(9):748–754

Parab AS, Ghose M, Jad V, Phakatkar SS, Jayan AK, Manohar CS (2025) Comparative Metabarcoding analysis of bacterial communities and functional profiles in the seaweed and mangrove blue carbon ecosystems of Goa, India. Mar Ecol 46(3):e70020

Pei P, Aslam M, Yang C, Ye P, Ke X, Liang Z et al (2024) Temporal variations of biological nitrogen fixation and diazotrophic communities associated with artificial seaweed farms. Front Mar Sci 11:1408958

Pessarrodona A, Franco-Santos RM, Wright LS, Vanderklift MA, Howard J, Pidgeon E et al (2023) Carbon sequestration and climate change mitigation using macroalgae: a state of knowledge review. Biol Rev 98(6):1945–1971

Queirós AM, Stephens N, Widdicombe S, Tait K, McCoy SJ, Ingels J et al (2019) Connected macroalgal-sediment systems: blue carbon and food webs in the deep coastal ocean. Ecol Monogr 89(3):e01366

Ruan Z, Chen K, Cao W, Meng L, Yang B, Xu M et al (2024) Engineering natural microbiomes toward enhanced bioremediation by microbiome modeling. Nat Commun [Internet] 15(1):4694

Seng N, Lai S, Fong J, Saleh MF, Cheng C, Cheok ZY et al (2020) Early evidence of microplastics on seagrass and macroalgae. Mar Freshw Res 71(8):922

Singh RP, Reddy CRK (2014) Seaweed-microbial interactions: key functions of seaweed-associated bacteria. FEMS Microbiol Ecol 88(2):213–230

Tang Q (2014) Management strategies of marine food resources under multiple stressors with particular reference of the Yellow Sea large marine ecosystem. Front Agric Sci Eng 1(1):85

Wang Q, Sun X, Lin S, Dong Y, Shen H, He Z et al (2025) Large-scale seaweed cultivation as a nature solution for carbon-negative economy and restorative environmental stewardship: lessons from China. Renew Sust Energ Rev 207:114954

Zhao J, Nair S, Zhang Z, Wang Z, Jiao N, Zhang Y (2024) Macroalgal virosphere assists with host–microbiome equilibrium regulation and affects prokaryotes in surrounding marine environments. ISME J 18(1):wrae083

Part III

Environmental Applications and Emerging Technologies

Engineered Microalgae: A Sustainable Solution for the Remediation of Organic Contaminants

14

Ankita Roy, Boda Ravi Kiran, and Krishna Mohan Poluri

Abstract

The growing concern of persistent organic pollutants in the environment causes hazardous effects on the ecosystem and living biota. Conventional strategies for cleaning organic pollutants, such as adsorption, advanced oxidation processes, membrane filtration, and photocatalysis, are expensive, inefficient, and generate secondary pollutants. Bioremediation is a cost-effective and eco-friendly means for environmental remediation. Microalgae are ubiquitous photosynthetic microorganisms offering sustainable environmental remediation, concomitantly producing high-value-added products and mitigating carbon emissions in a circular biorefinery framework. Integrating synthetic biology in microalgae-mediated bioremediation improves the efficacy of the process and minimizes the time. Molecular and genetic engineering advances have made it possible to rewire algal metabolic pathways to detoxify and mitigate the organic pollutants within the cellular systems. Further, the performance of natural communities can be tailored and enhanced using omics approaches and AI/ML tools, aiding in in situ remediation. This chapter explores the potential of different microalgal species for

A. Roy · B. R. Kiran
Department of Biosciences and Bioengineering, Indian Institute of Technology (IIT) Roorkee, Roorkee, Uttarakhand, India

K. M. Poluri (✉)
Department of Biosciences and Bioengineering, Indian Institute of Technology (IIT) Roorkee, Roorkee, Uttarakhand, India

Centre for Transportation Systems, Indian Institute of Technology (IIT) Roorkee, Roorkee, Uttarakhand, India
e-mail: krishna.poluri@bt.iitr.ac.in

K. Mohanty et al. (eds.), *Advances in Algal Research*,
https://doi.org/10.1007/978-981-95-7234-2_14

mitigating organic pollutants. Omics-based approaches for identifying target genes involved in the degradation mechanism and synthetic biology tools employed for enhancing bioremediation performance will be discussed.

Keywords

Organic pollutants · Microalgae · Synthetic biology · Omics · AI/ML techniques · Phycoremediation

14.1 Introduction

The growing surge in chemical and agrochemical industries has released tenacious and persistent organic pollutants (POPs) into the environment since the 1900s. According to the European chemicals legislation REACH (https://eur-lex.europa.eu/; accessed on 25 August 2025), these elements are categorized as PBT (persistent, bioaccumulating, and toxic) or vPvB (very persistent and very bioaccumulating). POPs are chemicals of anthropogenic origin found in various environmental matrices released from pesticides, pharmaceuticals, endocrine disruptors, steroid hormones, personal care products, brominated flame retardants, industrial chemicals such as chlorinated compounds, polycyclic aromatic hydrocarbons (PAHs), and per- and polyfluoroalkyl substances (PFAS) (Negrete-Bolagay et al. 2021). Among them, twelve are listed as the "dirty twelve" in Stockholm Convention, which include nine organochloride pesticides (dichlorodiphenyltrichloroethane (DTT), heptachlor, aldrin, toxaphene, chlordane, dieldrin, endrin, mirex, and hexachlorobenzene (HCB)) and three byproducts of polychlorinated biphenyls (PCBs), dibenzo-p-dioxins, and dibenzofurans (https://chm.pops.int/theconvention/thepops/the12initialpops; accessed on 25 August 2025). Former studies reported POPs classification based on QSPR models utilizing structural and physicochemical information. Recently, machine learning techniques have enabled the classification of POPs into four major classes: (1) perfluorinated compounds (PFOS, PFDA, and PFOA); (2) pesticides (acetochlor, carbaryl, and dimetachlor); (3) polycyclic aromatic hydrocarbons (PAHs) and other organic hydrocarbons; and (4) mixed class (PCBs, PBDEs, and DDT) (Nagar et al. 2023) (Fig. 14.1). These highly carcinogenic pollutants eventually enter the food web, affecting the living biota (Meng et al. 2017). In humans and animals, exposure to POPs has been linked to cancer, sterility, immune dysfunction, cardiovascular diseases, and various neurological, hormonal, and metabolic disorders (Martínez 2024). Prominently, POPs have bioaccumulation potential, adversely affecting fauna within aquatic ecosystems, leading to metabolic disorders and further increasing the risk of biomagnification (Zhou et al. 2020).

Despite their toxicity and environmental threat, very little attention has been given to POPs until recently. Among the POPs, volatile chlorinated hydrocarbons (vinyl chloride, chloroform, etc.), chlorinated aromatic compounds (polychlorinated biphenyls, chlorophenols, and hexachlorobenzene), chlorinated cyclic compounds (organochlorine pesticides), and other halogenated compounds, such as Perfluoro compounds, are extremely difficult to treat (Chen et al. 2024a).

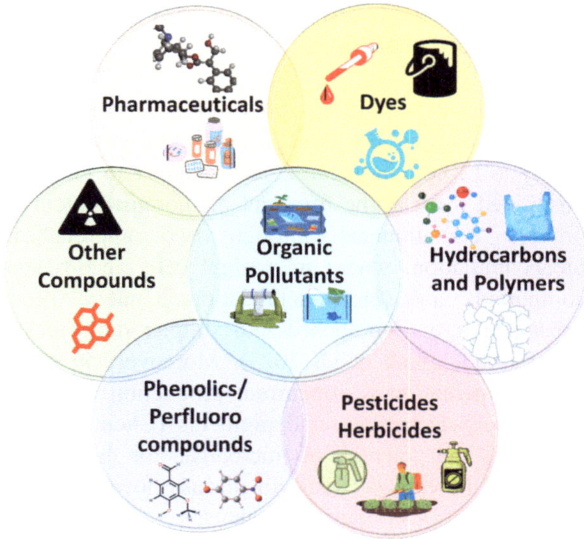

Fig. 14.1 Different types of organic pollutants

The most widely used remediation approaches for these compounds include the following: (1) physical methods such as screening, flocculation, sedimentation, flotation, membrane filtration, thermal treatment, and adsorption; and (2) chemical methods involving precipitation or chemical transformation, oxidation–reduction processes (including advanced oxidation processes, AOPs), and chemical dosing (Chen et al. 2024a; Hanafi and Sapawe 2020). Conventional methods are inefficient due to the instability and high propensity of organic contaminants to volatilize or convert into toxic intermediates, which cause secondary pollution. Biological approaches have gained considerable attention because of their lower environmental footprint and more effective and sustainable processes (Reynolds 2023). Bioremediation by these organisms involves aerobic, anaerobic, consortia-based, phycoremediation, phytoremediation, and enzyme-mediated approaches (Muhammad et al. 2024). While bacteria and fungi are often employed for degrading organic pollutants, a key limitation is that complete mineralization of pollutants releases carbon into the atmosphere, contributing to global warming and secondary environmental impacts. Microalgae are photosynthetic organisms that have shown potential for assimilation and biodegradation of many POPs, such as pharmaceuticals, antibiotics, endocrine disruptors, and hydrocarbons (Touliabah et al. 2022). Over the years, most microalgal-based bioremediation studies have centered on heavy metals, wastewater treatment, and nutrient removal. However, sustainable remediation strategies are urgently needed with the rising prevalence of persistent organic pollutants contaminating soil, air, and water. Recognizing this, this chapter summarizes microalgae-based remedial approaches for removing organic pollutants and technological advancements such as omics technologies, genetic engineering techniques, and AI/ML innovations in treating organic pollutants.

14.2 Algae-Mediated Removal of Persistent Organic Contaminants

Microalgae are ubiquitous photosynthetic organisms that are gaining significant attention for their potential in mitigating environmental pollutants. Their unique ability to be carbon-negative and adaptive to varied environmental conditions has positioned them at the forefront of bioremediation research. Microalgae clean up organic contaminants through two approaches: (1) Bioadsorption and bioaccumulation, where microalgal cells absorb specific organic and inorganic compounds, and (2) Biodegradation, a natural process where complex organic substances are decomposed into simpler molecules such as carbon dioxide and water (Abdelfattah et al. 2023). Moreover, an added advantage of microalgae-based remediation is concurrent production of high-value byproducts, including lipids, polysaccharides, and pigments (Choudhary and Poluri 2024). Table 14.1 highlights the potential of microalgae for the remediation of organic pollutants, and the following section will discuss how microalgal processes are involved in removing various organic pollutants.

14.2.1 Microalgae-Based Dye Removal

The widespread use of complex dyes, particularly azo dyes, in sectors like paints, textiles, and the automobile industries poses significant environmental hazards. 60–70% of the dyes used in the food and textile sectors are azo dyes, which can decompose into carcinogenic chemicals and harm aquatic life (Berradi et al. 2019). Microalgae mitigate ecosystems' eutrophication by breaking down complex dye molecules into monomers, carbon and nitrogen, and utilizing them as nutrients for their growth, thereby facilitating dye removal from water. The efficacy of this process varies based on the molecular structure of the dye, the microalgal strain, and the azo-reductase enzyme activity. Microalgae employ numerous methods to remove dyes, of which bioadsorption, bioconversion, and biodegradation play a pivotal role (Kundu et al. 2024). Microalgal species such as *Chlorella* sp., *Scenedesmus* sp., *Micractinium* sp., *Desmodesmus* sp., and *Chlamydomonas* sp. have been reported to break down dye molecules (Touliabah et al. 2022).

Chlorella vulgaris has been extensively studied to degrade azo dyes up to 80–90% because of its robustness and adaptability (Ishchi and Sibi 2019). Similarly, Moradi et al. (2024) reported the breakdown of the synthetic dye Direct Green 6 (DG6) using *C. vulgaris*. The study revealed that the azoreductase enzyme first breaks down the dye's azo bond and then peroxidases and laccases decompose it. The degradation was highly proactive in acidic conditions ($pH < 7$) and temperatures above 40 °C (Moradi et al. 2024). A study by Aragaw & Asmare (2018) further confirmed that azo linkages in dyes may be cleaved and produce aromatic amines using FTIR analysis (Aragaw and Asmare 2018). In addition, microbial consortia are often more potent at bioremediating toxic pollutants. *Chlorella* sp., in conglomerate with *Aspergillus* sp. XJ-2, efficiently decolorizes the

Table 14.1 Microalgal-based bioremediation of organic contaminants

Pollutant type	Specification	Concentration	Microalgal species	Mechanism	Removal % or degradation %	References
PAHs	Pyrene	50–250 μM	*Chlorella* sp. MM3	Biodegradation	30–70	Subashchandrabose et al. (2017a)
	Petroleum hydrocarbons	–	*Chlorella vulgaris* BS1	Biodegradation	98	Das and Deka (2019)
	PAHs in produced water (PW)	–	*Nannochloropsis oculate*	Bioadsorption	89–99	Marques et al. (2021)
	Phenanthrene and Anthracene	5 mg/L	*Gonium pectorale*	Biodegradation	98	Hoque et al. (2023)
	Fluorene	2–50 mg/L	*C. vulgaris*	Biodegradation	–	Asghari et al. (2020)
	Phenanthrene	0.1%	*Scenedesmus* sp. ISC 94, *Chlorella* sp. ISC 23, *Nostoc calcicola* ISC 89, *Anabaena* sp. ISC 88, *and Leptolyngbya fragilis* ISC 108	–	46–86	Ghanbarzadeh et al. (2022)
	Fluoranthene	25 μM	*C. vulgaris*	Biodegradation	90–94	Tomar and Jajoo (2021)
	Benzo[a]pyrene (BaP)		*S. capricornutum* and *Scenedesmus acutus*	Bioadsorption and Biodegradation	95–99	García de Llasera et al. (2016)
	Benzo[a]pyrene (BaP)	2–50 mg/L	*C. vulgaris*	Biodegradation	52.07–74.13	Chen et al. (2024b)
Dye	Azo dye (Reactive black 5; Direct Blue 71; Disperse red 1)	200–300 mg/L	*C. vulgaris*	Biodegradation	78–84	Ishchi and Sibi (2019)
	Azo dye (DG6)	20–100 mg/L	*C. vulgaris*	Biosorption	32–78	Moradi et al. (2024)
	Azo dye (Methylene blue)	100 mg/L	*C. vulgaris*	Bioadsorption	83.04	Chin et al. (2020)

(continued)

Table 14.1 (continued)

Pollutant type	Specification	Concentration	Microalgal species	Mechanism	Removal % or degradation %	References
	Cationic dye (Safranin & Malachite green)	5 mg/L	*Oscillatoria* sp.	Biodegradation	52–93	Gelebo et al. (2020)
	Anionic dye (Aniline blue)	25–100 mg/L	*C. vulgaris*	Bioadsorption	58	Carbajo Arteaga et al. (2018)
	Cationic dye (Malachite green)	100 mg/L	*Haematococcus pluvialis*	Bioadsorption	66–95	Liu et al. (2018)
	Textile wastewater containing azo dyes	–	Co-culture containing *Scenedesmus* sp., *Chlorella* sp., *Synedra* sp., *Achnanthidium* sp.	Biodegradation	Upto 82.6	Aragaw and Asmare (2018)
	Disperse red 3B	0.1–0.5 g/L	*Chlorella sorokiniana* XJK	Biodegradation	50	Tang et al. (2019)
	Disperse red 3B	0.1–0.5 g/L	*C. sorokiniana* XJK consortium with *Aspergillus* sp. XJ-2	Biodegradation	97	Tang et al. (2019)
	Red HE8B, reactive green 27, and acid blue 29,	5 mg/L	*Chlamydomonas mexicana*	Biotransformation	39–64	Mustafa et al. (2024)
	Reactive yellow 2, reactive red 195, reactive blue 19, and reactive black 5	10–30 mg/L	*Ulva fasciata* and *Pterocladia capillacea*	Bioadsorption	72–100	El-Agawany et al. (2023)
Pharmaceuticals	Antibiotics (sulfamethoxazole; metronidazole; trimethoprim; metronidazole; ofloxacin; Ciprofloxacin)	2.9–5 ng/L	*Scenedesmus almeriensis*	–	44–99	Morillas-España et al. (2025)
	Paracetamol; ibuprofen; olanzapine	100 μg/ml	*Nannochlorpsis* sp.	Bioadsorption	55–67	Encarnação et al. (2020)

	Atenolol, carbamazepine, ibuprofen, and naproxen	0.5 mg/L	*Navicula* sp.	Biodegradation	Greater than 90	Ding et al. (2020)
	Estradiol, diclofenac, and triclosan	–	Microalgal consortia	Biodegradation	74–91	Bano et al. (2021)
	Analgesic (tradmol)	50 mg/L	*Scenedesmus obliquus*	Bioadsorption	91	Ali et al. (2018)
	Antibiotic (tetracycline)	40–50 mg/L	*Chlorella* sp.	Bioadsorption	68–76	Suárez-Martínez et al. (2022)
	Metronidazole	5 μM	*C. vulgaris*	Bioadsorption	100	Hena et al. (2020)
	Diclofenac	16 μg/mL	*N. oculata*	Bioadsorption	68.8	Sánchez-Sandoval et al. (2021)
	Ciprofloxacin and sulfadiazine	1–10 mg/L	*Chlamydomonas* sp. Tai-03	Biodegradation	65 and 17	Xie et al. (2020)
	Florfenicol	46–159 mg/L	*Chlorella* sp. (UTEX1602 and L38)	Biodegradation	74–97	Song et al. (2019)
	Caffeine, fluconazole, trimetoprim, ofloxacin, codeine, carbamazepin, oxazepam, hydroxyzine, tramadol, mirtazapine, diphenhydramine, memantine, orphenadrine, bupropion, biperiden, flecainide, trihexyphenidyl, Clomipramine, and amitriptyline	1 μg/L	*Chlorella* sp., *Coelastrella* sp., *Desmodesmus* sp., *Scenedesmus* sp.	–	More than 70	Gojkovic et al. (2019)
	Sulfamerazine, sulfamethoxazole,	20–100 μg/L	*Haematococcus pluvialis*, *S. capricornutum*,	Biodegradation	9–99	Kiki et al. (2020)

(continued)

Table 14.1 (continued)

Pollutant type	Specification	Concentration	Microalgal species	Mechanism	Removal % or degradation %	References
	sulfamonomethoxine, trimethoprim, clarithromycin, azithromycin, roxithromycin, lomefloxacin, levofloxacin, and flumequine		*S. quadricauda* and *C. vulgaris*			
	Cephalexin	482.92 mg/L	*Chlorella* sp.	Bioadsorption	71–82	Angulo et al. (2018)
Pesticides and Insecticides	Thiamethoxam	60 mg/L	*Scenedesmus* sp. TXH	Bioadsorption, bioaccumulation, and biodegradation	97.5	Quan et al. (2023)
	Trichlorofon (TCF)	100 mg/L	*Chlamydomonas reinhardtii*	Biodegradation	100	Wan et al. (2020)
	Diazinon	20 mg/L	*C. vulgaris*	Biodegradation	94	Kurade et al. (2016)
	Malathion	25–100 mg/L	*Chlorella sorokiniana*	Biodegradation	90	Nanda et al. (2019)
	Chlorpyrifos, cypermethrin, and oxadiazon	1 mg/L	*Chlorella* sp. and *Scenedesmus* sp.	Biodasorption	33–62	Avila et al. (2021)
	Fipronil	0.1 mg/L	*S. quaclricauda* and *C. vulgaris*	Biotransformation and biodegradation	60–70	Ou et al. (2022)
		various	*Skeletonema costatum*	Biodegradation	66–76	Zhang et al. (2023)

	Chlorpyrifos, acetochlor, and dicofol					
	Metalaxyl, pyrimethanil, fenhexamid, iprodione, and triclopyr	4 mg/L	*S. quaclricauda* and *C. vulgaris*	Biodegradation	80–90	Baglieri et al. (2016)
	Chlorpyrifos	1.5 mg/L	*Chlorella, Scenedesmus* sp., *and Hapalosyphon* sp.	–	50–100	Castellanos-Estupiñan et al. (2022)
	Imidacloprid, fenamiphos, and oxamyl	50–450 mg/L	*C. vulgaris and Ulva lactuca*	Bioadsorption	20–66	Reyad et al. (2022)
Others	Phthalic acid esters	0.1 mg/L	*Chaetoceros muelleri, Cylindrotheca Closterium, and Dunaliella salina*	Biodegradation	95–97	Chi et al. (2019)
	Di-n-butyl phthalate	0.1 mg/L	*C. muelleri, C. Closterium, and D. salina*	Biodegradation	22–91	Chi et al. (2019)
	Phenol	0.8 g/L	*C. pyrenoidosa*	Biodegradation	97	Dayana Priyadharshini and Bakthavatsalam (2016)
	Phenol	100–450 mg/L	*Chlorella* sp., *Pseudochlorococcum* sp., *and Chlamydomonas* sp.	Biodegradation	40–90	Al-Zuhair (2016)
	Phenol	25–200 mg/L	*C. pyrenoidosa*	Biodegradation	30–95	Das et al. (2015)
	PFOS and PFOA	–	*Synechocystis* sp. PCC6803	–	88 and 37	Marchetto et al. (2021)
	PFOA	1 mg/L	*C. pyrenoidosa*	Bioadsorption and Bioaccumulation	Less than 10	Hu et al. (2020)
	PFOA	5–20 mg/L	*C. reinhardtii* and *S. obliquus*	Bioadsorption and Bioaccumulation	5–18	Hu et al. (2014)

anthraquinone dye Disperse Red 3B. The enzymes lignin peroxidase and manganese peroxidase achieved activity levels of 86.7 U/L and 122.5 U/L, respectively, driving the degradation process (Tang et al. 2019). Similarly, *C. mexicana* alone could decolorize the textile azo dyes such as Red HE8B, Reactive Green 27, and Acid Blue 29. At the same time, in consortium with activated sludge, the decolorization percentage was found to be increased by 75–79% (Mustafa et al. 2024). Moreover, dye removal generally depends on environmental factors such as pH, salinity, dye concentration, and exposure period. It was observed that marine algae *Ulva fasciata* and *Pterocladia capillacea* exhibited enhanced dye removal efficiency in saline water compared to freshwater, with an increase in contact time (El-Agawany et al. 2023).

14.2.2 Microalgae-Based Removal of Pesticides/Herbicides

Chemical pesticides/herbicides are commonly used to eradicate pests and increase crop yields. When applied extensively, it results in ecological damage and curtails soil fertility (Webster et al. 2024). Biosorption, bio-uptake, or bioaccumulation followed by biodegradation are the processes used in microalgae-based pesticide remediation. Pesticide removal by microalgae generally begins with bioadsorption, a rapid and passive binding of pesticide molecules to functional groups on the algal cell wall, which can occur in live and dead cells. This is followed by bioaccumulation, where pesticides are engulfed by the cell through surface proteins or ion transport mechanisms. Subsequently, biodegradation occurs, covering enzymatic transformation of pesticides into simpler organic matter, facilitated by enzymes such as hydrolases, esterases, and transferases, with concurrent oxygen release (Wang et al. 2019).

Several microalgal strains exhibited pesticide removals greater than 90% for atrazine, trichlorfon, dimethoate, pyriproxyfen, metolachlor, molinate, carbofuran, simazine, and cypermethrin (Fayaz et al. 2024). *C. reinhardtii* attained nearly 100% removal for trichlorfon at 100 mg/L concentration (Wan et al. 2020), while *C. sorokiniana* degraded 90% of the pesticide malathion (Nanda et al. 2019). *Hapalosyphon* sp., *Scenedesmus* sp., and *Chlorella* sp. exhibited chlorpyrifos removal efficiency ranging from 50% to 100% (Castellanos-Estupiñan et al. 2022). Further, *C. vulgaris* and *Ulva lactuca* exhibited 40.76% to 28.44% removal for imidacloprid, 66.2% to 61.91% removal for fenamiphos, and 70% for oxamyl removal (Reyad et al. 2022). Conjugation of two or more species showed enhanced removal of pesticides/herbicides. Microalgal consortia have demonstrated 100% removal of the insecticide lindane in BG-11 medium and 85–97% removal of pesticides chloropyrifos, oxadiazon, and cypermethrin (Avila et al. 2021). In addition, microalgae can utilize cyanide and insecticides as nutrient sources, aiding in their further removal and breakdown (Sutherland and Ralph 2019).

14.2.3 Microalgae-Based Removal of Organic Hydrocarbons

Organic hydrocarbons primarily originated from crude oil and, as byproducts of industrial processes, include petroleum products and polyaromatic hydrocarbons (PAHs). PAHs comprise carbon and hydrogen bonds that are linked to form multiple aromatic rings. They exhibit mutagenic and carcinogenic properties, and are classified into low molecular and high molecular weight PAHs (Barathi et al. 2023). Microalgae play a crucial role in the remediation of crude oils by fragmenting complex hydrocarbons into simpler molecules (Das and Deka 2019). *C. vulgaris, S. quadricauda,* and *Selenastrum capricornutum* microalgae cultures degrade high molecular weight PAH benzo[a]pyrene (BaP) into sulfate ester and glucoside conjugates using the dioxygenase pathway (García de Llasera et al. 2016; Chen et al. 2024b). Similarly, *Chlorella* sp. has the potential to degrade and remediate pyrene-contaminated soils through dihydrolipoamide acetyltransferase activity (Subashchandrabose et al. 2017a). The degradation competence varies across species and is largely determined by algal biomass concentration.

14.2.4 Microalgae-Based Removal of Phenolics and Perfluoro Compounds

Phenolic compounds are the most carcinogenic pollutants in marine systems, with concentrations in India reported as high as 1.79×10^8 in surface waters. Structurally, they entail one or more aromatic rings bonded to hydroxyl groups. Pentachlorophenol, tetrabromobisphenol, phenolphthalein, polychlorophenols, bisphenol A (BPA), octylphenols, and nonylphenols are some of the toxic phenolic compounds (Ladeia Ramos et al. 2024). *C. pyrenoidosa* exhibited phenol degradation up to 97–100% (Dayana Priyadharshini and Bakthavatsalam 2016). Further, Laccases from *Chlamydomonas moewusii* and *Tetracystis aeria* play a crucial role in phenolic oxidation (Otto and Schlosser 2014). Species like *Pseudochlorococcum* sp., *Chlorella* sp., and *Chlamydomonas* sp. exhibited their potential in treating phenol-contaminated wastewaters (Al-Zuhair 2016; Das et al. 2015). Per- and polyfluoroalkyl substances (PFAS) are considered emerging contaminants of concern. Microalgae such as *C. reinhardtii* and *S. obliquus* have shown the potential to adsorb PFOA, although the removal efficiency was low, around 5.5 to 7.5% (Hu et al. 2014). The limitation of nitrogen and phosphate in *C. pyrenoidosa* demonstrated a greater bioadsorption capability (Hu et al. 2020). Most notably, the cyanobacterium *Synechocystis* sp. PCC6803 demonstrated an 88% removal efficiency of PFOS and a 37% removal efficiency of PFOA (Marchetto et al. 2021).

14.2.5 Microalgae-Based Removal of Antibiotics/Pharmaceuticals

The prevalent use of antibiotics in human medicine, animal husbandry, and aquaculture is the major source of environmental contamination. These antibiotics and antibiotic-resistant genes (ARGs) accumulate in soil, water, and air and concomitantly transfer to the trophic food-web (Barathe et al. 2024). The hysterical use of anti-inflammatory drugs, analgesics, disinfectants, and endocrine disruptors adversely affects humans and nontarget organisms (Eapen et al. 2024). Microalgae are candidate species for treating antibiotics and pharmaceutical compounds through bioadsorption, bioaccumulation, and biodegradation. The process normally takes place in three stages: (1) passive adsorption driven by physicochemical interactions between pollutants and the surface functional groups, (2) transport of molecules across the cell membrane, and (3) subsequent bioaccumulation and/or biodegradation within the cell (Hena et al. 2021). Adsorption primarily occurs through direct binding to functional groups on the cell surface or is facilitated by extracellular polysaccharides (EPS) (Lu et al. 2021). *Haematococcus pluvialis, Selenastrum capricornutum, Scenedesmus quadricauda,* and *Chlorella vulgaris* have shown removal competencies ranging from 92% to 99% for lomefloxacin, 39% to 94% for sulfamonomethoxine, 30% to 99% for sulfamerazine, 40% to 70% for flumequine, 68% to 93% for azithromycin, 25% to 67% for trimethoprim, and 32% to 94% for sulfamethoxazole (Gojkovic et al. 2019; Kiki et al. 2020). *Chlorella* sp. removed 68–76% of tetracycline (Suárez-Martínez et al. 2022) while *Chlamydomonas* sp. Tai-03 achieved 54.5% removal of sulfadiazine and 65% removal of ciprofloxacin (Xie et al. 2020). In addition, *C. sorokiniana, S. obliquus* RISE (UTEX 417), *Desmodesmus* sp. RUC2, *Coelastrella* sp., and *C. vulgaris* achieved 65–89% removal of ofloxacin (Gojkovic et al. 2019). Remarkably, *C. vulgaris* was also capable of completely removing metronidazole (Hena et al. 2020). Furthermore, *Chlorella* sp. standalone removed ~90% of florfenicol and 71% of cephalexin (Song et al. 2019; Angulo et al. 2018).

14.3 Synthetic Biology for Enhanced Removal of Organic Pollutants

Biotechnological approaches can enhance microalgal efficiency in both nutrient and contaminant removal and biomass and bio-product accumulation. Such modifications target mitochondrial, chloroplast, and especially nuclear genomes, which are crucial since they encode enzymes liable for secondary metabolite synthesis (Arora et al. 2019a). Synthetic biology approaches used for enhanced bioremediation of organic pollutants are depicted in Fig. 14.2. Genetic engineering involves transformation methods such as glass-bead agitation, electroporation, particle bombardment, and agrobacterium-mediated transfer, followed by selection of desired traits. Advanced genome-editing tools enable precise site-directed mutagenesis, including zinc-finger nucleases (ZFNs), transcription activator-like effectors (TALENs), and CRISPR-Cas9. Meanwhile, gene editing and interference remain the most effective strategies for targeted modification (Bashir et al. 2016). ZFN technology was first induced in microalgae *Chlamydomonas reinhardtii* in 2013, targeting the COP3 gene (Sizova et al. 2013). TALENs were subsequently introduced in *Phaeodactylum tricornutum* and *Thalassiosira pseudonana* in 2014, marking the beginning of their broader use in microalgal genome editing (Hao et al.

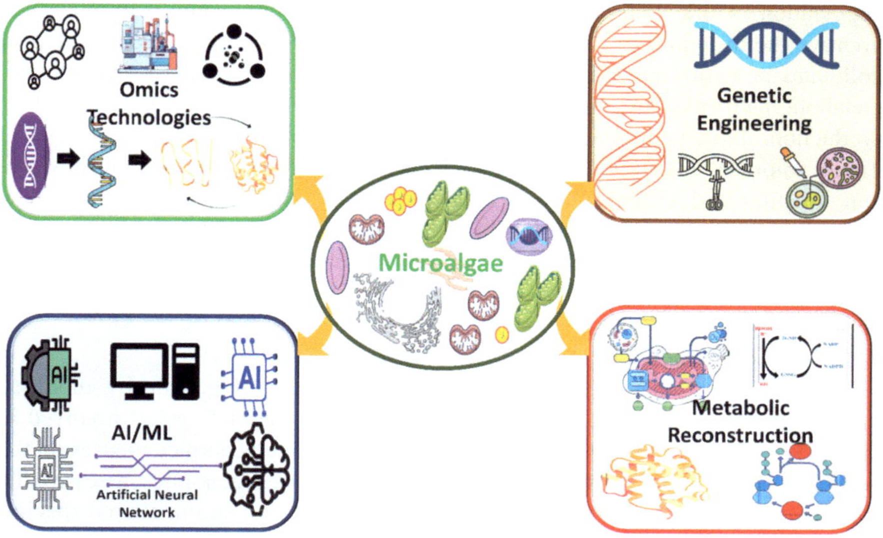

Fig. 14.2 Schematic view of synthetic biology approaches for enhanced bioremediation of persistent organic contaminants

2018; Daboussi et al. 2014). However, their large size poses significant challenges in efficient cellular delivery. CRISPR/Cas9 gene editing has gained considerable interest due to its precision, adaptability, and ease of use across various organisms (Zhu 2022). Long-term gene expression and integration of gene cassettes into the host genome are prerequisites for genetically modified algae in industrial applications (Patel et al. 2019). CRISPR-dCas9, microRNA (miRNA), and silent RNA (siRNA) are some of the gene-interfering tools that regulate gene expression, either by triggering or silencing specific targets, thereby achieving desired microalgal phenotypes (Ng et al. 2017). To meet the hovering demands, several researchers have been trying to create universal cloning toolkits that include selection markers, tags, untranslated regions (UTRs), promoters, terminators, reporters, and introns for effective assembly of transgenic expression strains. *P. tricornutum, C. vulgaris,* and *S. elongatus* have the potential to be utilized as workhorses for this purpose; however, currently *C. reinhardtii* is the widely used algal species for pollutant removal. The toolkits specific to microalgal species are limited, with notable examples being the *C. reinhardtii* MoClo toolkit (Crozet et al. 2018) and CyanoGate (Vasudevan et al. 2019).

14.3.1 Omics Technologies for Removal of Organic Pollutants

Omics approaches such as genomics, transcriptomics, proteomics, and metabolomics provide critical insights into algal physiology and metabolism, enabling strain selection and identification of genetic targets for enhanced pollutant degradation (Arora et al. 2019b). Combining molecular phylogeny with omics can improve the biosynthesis of high-value compounds or enhance the degradation of pollutants in wastewater (Mishra et al. 2019). “Algomics” reveals protein synthesis, metabolic pathways, and environmental responses, supporting advances in energy production, bioremediation, and industrial biotechnology (Kadri et al. 2023). Transcriptomics uncovers gene expression patterns, proteomics accomplishes protein function and trafficking, and metabolomics tracks metabolites to optimize pathways (Arora et al. 2018a). Overall, Pathway deletion, flux enhancement, and tolerance improvements drive metabolic optimization processes for efficient biodegradation of persistent pollutants.

14.3.1.1 Genomics and Functional Metagenomics

Whole genome studies help in genome-wide screening of genetically modified microalgae encoding stress tolerance and prospective candidate genes involved in the biodegradation of pollutants (Kadri et al. 2023). While genomics focuses on the genetic makeup of individual organisms, functional metagenomics assesses the collective genetic makeup of a microbial population, including uncultured species. Functional metagenomics integrates whole-genome sequencing, gene assembly, gene prediction, and functional annotations with databases like UniProtKB, InterPro database, and KEGG (Chen et al. 2022). This approach provides a compilation of microbial genetic information from a population, aiding in identifying novel genes, new enzymes, complex pathways that are involved in nutrient recycling (nitrogen and phosphorus), contaminant transformation, antibiotic resistance, and other important processes related to wastewater treatment (Rodríguez et al. 2020). For instance, metagenomic analysis revealed that antioxidant-producing microbes in activated sludge reduced the mutation of antibiotic resistance genes (ARG) (Ren et al. 2021). By offering insights into microbial community structure, diversity, and function at specific locales, metagenomics aids in creating effective in situ bioremediation strategies (Mariano et al. 2024). Metatranscriptomics and metaproteomics enhance this understanding by capturing gene expression changes and protein profiles in microbial communities. Additionally, metagenomic libraries identify catabolic genes for the degradation of pesticides, polychlorinated bisphenol (PCB), and other aromatic compounds (Singh et al. 2018). Similarly, metagenomic analysis revealed the adaptability of the microbial community toward sulfamethoxazole (Fan et al. 2023). Thus, efficient bioremediation of recalcitrant pollutants requires combining analytical techniques with multi-omics strategies, greatly enhancing the potential of microalgal and microbial systems for environmental cleanup.

14.3.1.2 Transcriptomics

Transcriptomic profiling has been widely used in understanding how microorganisms accumulate, transform, and biodegrade pollutants like heavy metals, hydrocarbons, antibiotics, and pharmaceuticals (Eregie et al. 2025). *C. sorokiniana*, through transcriptome analysis, unveiled key regulatory genes in azo dye degradation and antibiotic ciprofloxacin (CIP) removal (Li et al. 2022; Tarbajova et al. 2023). Genes associated with aldehyde dehydrogenase, ABC transporters, decarboxylase, oxidoreductase, hydrolase, O-acetyltransferase, and lyase were upregulated in the microalgae cultures during the assimilation of drugs like sulfamethoxazole, clarithromycin, and sulfacetamide (Peng et al. 2021; Zhang et al. 2022). A study on exposure of microalgae species to antibiotics and nanoparticles revealed that antibiotic exposure upregulated 186 differentially expressed genes (DEGs) linked to membrane components. In contrast, nanoparticle exposure upregulated 24 DEGs (Cao et al. 2023). These approaches highlight the molecular mechanisms and gene regulatory networks of pollutant biotransformation in aquatic ecosystems for environmental cleanup.

14.3.1.3 Proteomics

Proteomics approaches have been employed to annotate the degradation mechanism of organic pollutants such as PAHs (Guo et al. 2022). However, only a limited number of studies have focused on the role of microalgae in organic pollutant removal using a proteomics approach. One major study delved into the proteome of *Thalassiosira* sp. OUC2 during the breakdown of p-xylene. The research highlighted the role of notable enzymes such as monooxygenase, alcohol dehydrogenase, benzaldehyde dehydrogenase, catechol 2,3-dioxygenase, and benzoate 1,2-dioxygenase, whose activity was found to be increased prominently with exposure period (Duan et al. 2020). Likewise, the protein expression profiles of *Phaeodactylum tricornatum* grown in dairy manure wastewater (DMW) provided information about its physiological response (Burch et al. 2021). Further, a proteome analysis of *Microcystis aeruginosa* subjected to varying nitrogen and phosphorus concentrations, along with the antibiotic spiramycin, revealed that spiramycin influenced the microcystin production at the proteome level. Microcystin synthases (mcyB and mcyC), enolase, superoxide dismutase, protein GrpE, DNA-directed RNA polymerase, and serine proteases were the key targets upregulated during the analysis (Chen et al. 2017). Similarly, norflurazon resistance in *C. reinharditii* influenced an upsurge in photosynthesis and enhancement of photoprotection processes (Leetanasaksakul et al. 2025).

14.3.1.4 Metabolomics

Metabolomics provides a dynamic overview of the cell's physiological state. Changes in gene expression or protein synthesis profiles lead to substantial shifts in metabolome, providing a distinctive edge over other omics fields (Arora et al. 2018b). For example, researchers used GC–MS-based metabolomics to evaluate microalgal-based wastewater remediation containing diclofenac sodium (DFS). The study revealed 32 metabolites linked to lipid, amino acids, and osmolytes, which were upregulated compared to the control (Sharma et al. 2023). MS-based untargeted metabolomics of *S. obliquus* cultures exposed to the antibiotic ciprofloxacin demonstrated 160 upregulated and 151 downregulated metabolites. Pathway enrichment analysis revealed ABC transporters, sugar metabolism (starch, glucose, galactose, and maltose), sorbitol, and arachidonic acid-related pathways contributing toward increased extracellular polymeric substance (EPS) production or improved resistance to ciprofloxacin toxicity (Wang et al. 2023). Recent findings highlight the effect of pollutant and nutrient interactions on the algal metabolome. For instance, *M. aeruginosa*, upon exposure to bisphenol A (BPA), exhibited increased synthesis of microcystins mediated primarily by amino acid and lipid metabolism. The metabolic network of 3-PGA, glucose 6-phosphate, UDP-α-D galactose, and UDP-GalNAc primarily controlled cell growth, energy production, and photosynthesis (Yang et al. 2024). These studies reveal that metabolomics is an effective tool to elucidate the impacts of pollutants on the central carbon flux of microalgae. Moreover, metabolite alterations influence biotic interactions between algae and bacteria or algae and pollutants, ultimately providing insights into possible routes or targets for enhancing bioremediation, biomass production, and bio-product generation.

14.3.2 Genetic Engineering for POPs Degradation

Microalgae have shown potential for degrading various pollutants in Sect. 14.2; however, bioaugmentation techniques can greatly enhance efficacy and applicability. One major technique is genetic and metabolic engineering of microalgae to develop robust strains, displaying greater tolerance and enhanced degradative capability. Through targeted genetic modifications, it is possible to optimize enzyme activity by overexpressing surface receptors to promote greater adsorption and internalization of dye molecules, or by enhancing the production of key enzymes such as azoreductases and peroxidases that facilitate detoxification (Maqsood et al. 2023). In parallel, metabolic reprogramming aimed at increasing the extracellular secretion of these enzymes can also be explored to boost on-site dye breakdown efficiency. Genetic modification of microalgae, achieved either by introducing a target gene into the host cell or by deleting a specific DNA fragment, represents a sophisticated approach that enables direct expression of desired traits and enhancement of cellular metabolic activity, thereby reducing the need for prolonged screening and acclimation (Sreenikethanam et al. 2022). *C. reinhardtii* is a known model organism for genetic engineering in phycoremediation studies. Sobieh et al. (2022) developed a transgenic strain via Agrobacterium-mediated transformation by introducing the cyanobacterial cyanase gene into *C. reinhardtii* (Sobieh et al. 2022). The strain exhibited tolerance to 150 ppm potassium cyanide, underscoring its inherent capacity for treating cyanide-polluted industrial wastewaters. Similarly, Ismaiel et al. (2019) engineered *C. reinhardtii* with glutathione-S-transferase enzyme, enhancing its catalytic activity against herbicides. The modified strain removed 93.6% of the rice herbicide penoxsulam, a noticeable improvement relative to the 52% removal by the wild type (Ismaiel et al. 2019). Beyond *C. reinhardtii*, genetic engineering of cyanobacteria has also demonstrated promising outcomes in pollutant degradation. *Anabaena* sp. has been reported to biodegrade the persistent organochlorine pesticide lindane. Overexpression of the linA2 transgene facilitated the complete removal of 10 ppm lindane within 6–10 days (Chaurasia et al. 2013). Similarly, studies on *Synechococcus elongatus* have highlighted its utility in dye bioremediation. Wild-type *S. elongatus* eliminated 99.5% of the carcinogenic dye malachite green, and heterologous expression of the *katmr* gene (encoding triphenylmethane reductase) further enhanced degradation efficiency to 99.8% (Han et al. 2020). More recently, engineering advances in *C. reinhardtii* have also focused on pesticide degradation. Overexpression of a modified Cytochrome P450 enzyme increased the degradation of the herbicide Diuron from only 6% in wild-type strains to 52% in genetically engineered strains (Helvig et al. 2025).

Modern analytical techniques have improved bioremediation efficiency by enhancing the microbial degradation of PAHs. This can be further accelerated by increasing key enzyme activity in PAH-degrading pathways (Imam et al. 2022). Advances in molecular biology, including gene conversion, duplication, and plasmid- or transposon-mediated delivery, provide powerful tools for boosting microbial catabolic capacity (Rafeeq et al. 2023). However, the activity of genetically engineered microorganisms (GEMs) depends on recombinant plasmid stability, and ensuring their genetic stability is critical before field application. Additionally, enzyme-based approaches offer sustainable alternatives, where extracellular enzymes such as oxidases, monooxygenases, esterases, and phenol oxidases play key roles in pollutant breakdown (Rafeeq et al. 2023). Plastic degradation has largely focused on the PETase enzyme since its discovery in *I. sakaiensis* (Sevilla et al. 2023). Although algae cannot naturally degrade PET, transferring PETase and MHETase genes into algal cells offers significant degradation. Kim et al. (2020) expressed wild-type PETase in *C. reinhardtii* using the Sh-Ble-2A fusion expression system, but activity was limited due to intracellular retention and slow degradation (Kim et al. 2020). In contrast, Moog et al. (2019) expressed a mutant IsPETaseR280A in *P. tricornutum* with a secretion tag, successfully releasing active enzyme into the medium and degrading PET surfaces (Moog et al. 2019). PETase, a cutinase-like hydrolase, cleaves PET into MHET, which MHETase further processes into ethylene glycol and terephthalate, usable as carbon and energy sources. While MHETase remains less studied, a chimeric PETase–MHETase construct has shown enhanced activity compared to the individual enzymes (Knott et al. 2020).

Genetic engineering enables efficient degradation of phenol and related compounds. Chiaiese et al. (2011) cloned the poxA1b laccase gene from *Pleurotus ostreatus* into a plant expression vector with a secretion signal under the CaMV 35S promoter. Using Agrobacterium-mediated transformation, recombinant laccase was expressed and secreted by *Chlamydomonas, Chlorella*, and *Ankistrodesmus* sp., removing up to 40% of phenol from the medium. Overexpression of phenol oxidases offers further potential for broad-spectrum phenolic degradation (Chiaiese et al. 2011). Microalgae, however, employ multiple mechanisms for drug removal, including surface adsorption and metabolic transformation, making them promising candidates for bioremediation (Zhou et al. 2022). Despite this potential, their ability to degrade recalcitrant pharmaceuticals is limited by the low expression of key enzymes or the absence of certain metabolic pathways. Genetic engineering can be applied to overcome these challenges, for example, by incorporating cytochrome P450 genes into microalgal genomes to expand their degradative capacity (Kariyawasam et al. 2024). Similarly, laccase genes have been shown to enhance oxidoreductase stability and improve contaminant degradation in algal strains

(Subashchandrabose et al. 2013). Such approaches collectively offer a viable strategy to strengthen the efficiency of pharmaceutical and antibiotic removal. Microalgae offer a low-cost, sustainable platform for heterologous expression, thriving in simple media with sunlight as the main energy source. Beyond remediation, the harvested biomass can produce carbon-neutral fuels, photosynthetic pigments, feed/food, biofertilizers, and bioplastics supporting a circular economy (Choudhary et al. 2025).

Random mutagenesis is also an effective technique for manipulating the characteristics of microalgae, having lesser prerequisite of skilled labor and lower cost as compared to targeted genetic engineering. Several techniques, including chemical methods, plasma, and UV irradiation, produce mutants through random mutagenesis (Arora and Philippidis 2021). The process essentially relies on selection pressure to screen for mutants of interest. This has proven a reliable method for generating strains with higher metabolite production rates, better productivity, improved resilience to stress and environmental contaminants (Arora et al. 2020). Such strains are more robust at bioremediating pollutants of a persistent nature. Developing a strategy for creating adaptive strains involves proactively exposing the culture or strain to low levels of pollutants and gradually increasing the dosage. For example, after three growth cycles, the *C. sorokiniana* cultures showed a significant increment in their maximum specific growth rate and enhanced stress resistance (Khalid et al. 2018). While GMOs have been explored for heavy metal detoxification, studies on organic pollutant remediation remain limited, leaving microalgae underutilized despite advances in genetic engineering. This field of study offers promise as robust technology for sustainable environmental remediation; therefore, studies in this regard must be performed. Table 14.2 summarizes the recent developments in microalgal genetic engineering for the remediation of organic pollutants.

14.3.3 AI/ML Technologies

Microalgae have emerged as potential candidates for biotechnological solutions to mitigate environmental pollution. This potential is largely attributed to their ease of cultivation, rapid growth, cost-effectiveness, and ability to proliferate efficiently. The biodegradation rate of refractory pollutants can be further enhanced by optimizing cultivation conditions and improving growth strategies. However, there is a pressing need for innovative approaches that enable real-time monitoring, automated regulation, and accurate prediction of microalgal productivity. Internet of things (IoT) technologies offer significant opportunities in microalgae-based biorefineries (Lim et al. 2022). IoT applications can support automated cultivation processes, continuous monitoring and adjustment of key parameters, productivity enhancement, detection of toxic algal strains, screening and classification of desired microalgal species, and viability assessment of microalgal cells. In aquatic environments, algae play a pivotal role in biomonitoring and the remediation of organic pollutants (Touliabah et al. 2022). High-rate algal ponds (HRAPs) are large-

Table 14.2 Genetic engineering applications in microalgal bioremediation of organic contaminants

S. no	Microalgal /cyanobacterial species	Gene	Tool used	Target pollutant	% removal	References
1	*Synechococcus elongatus* PCC 7942	*katmr*	Plasmid transformation	Malachite green dye	99.5%	Han et al. (2020)
2.	*C. reinhardtii*	*GST*	Agrobacterium transformation	Pesticides	93.6% and 54%	Ismaiel et al. (2019)
3.	*C. reinhardtii*	*PETase*	Plasmid electroporation	PET plastic	35%	Kim et al. (2020)
4.	*C. reinhardtii* CC-124	*CYN*	Agrobacterium transformation	Cyanide	up to 100%	Sobieh et al. (2022)
5	*Phaodactylum tricornutum*	*IsPETase*	Biolistic transformation	PET plastic	–	Moog et al. (2019)
6.	*Anabaena* sp.	*LinA2*	Plasmid electroporation	Lindane pesticide	More than 98%	Chaurasia et al. (2013)
7	*S. elongatus* PCC 7942	*Cu/Zn SOD (sodC)*	Overexpression	Azo dye	6.2%	Mohandass et al. (2021)
8	*Chlamydomonas* sp.	*P450 BM3*	Heterologous overexpression	Herbicide Diuron	52%	Helvig et al. (2025)
9.	*C. reinhardtii*	*PHL7*	Heterologous expression and secretion	PET plastic	–	Dutra Molino et al. (2025)
10.	*Chlamydomonas* sp., *Chlorella* sp., *and Ankistrodesmus* sp.	*poxA1b*	Heterologous expression and secretion	Mono- and poly-aromatic compounds	40% enhancement	Chiaiese et al. (2011)
11.	*Synechococcus* sp. PCC 7002	*cyp1a1*	Heterologous expression	Atrazine	–	Berepiki et al. (2016)

scale cultivation systems widely used to remove emerging organic contaminants (EOCs) from urban wastewaters. Compounds such as acetaminophen, caffeine, hydro-cinnamic Acid, ibuprofen, and methyl dihydro-jasmonate can be effectively eliminated using HRAPs. HRAPs removal efficiencies range from 60 to 90% for galaxolide, naproxen, octylphenol, oxybenzone, tonalide, tributyl phosphate, triclosan, and bisphenol A; 40–60% for atrazine, benzotriazole, cashmeran, diazinon, celestolide, diclofenac, OH-benzothiozole, and triphenyl phosphate; and about 30% for 2,4-D, benzothiazole, carbamazepine, methylparaben, and tris(2-chloroethyl) phosphate (Matamoros et al. 2015). Notably, ecotoxicological risk assessment indicated that the treated effluents posed no acute toxicity risk, as hazard quotient indices for influent wastewater contaminants were reduced up to 90%. Artificial intelligence (AI) and machine learning (ML) serve as powerful tools for detecting, predicting, and managing risks (Olawade et al. 2024). Methods such as artificial neural networks (ANN), support vector machines (SVM), genetic algorithms (GA), decision trees (DT), deep learning (DL), and random forest algorithms (RFA) have demonstrated substantial potential in enhancing process optimization (Teng et al. 2020). Despite significant advances, operational monitoring and optimization applications remain limited. Ansari et al. (2021) demonstrated that nutrient-enriched secondary effluents supplemented with BG11 significantly boosted algal biomass, with ANN-based modeling precisely predicting lipid, protein, and carbohydrate yields ($R^2 = 0.983$) using six key variables (temperature, pH, dissolved oxygen, conductivity, nitrate, and phosphate) (Ansari et al. 2021). The approach improved nutrient removal, biochemical composition, and economic feasibility, with a payback period shorter than the project lifespan. In recent times, Sayed et al. (2024) combined ANN with forensic-based investigation (FBI) algorithms to optimize microbial fuel cell-algal reactors, considering wastewater and yeast concentrations as decision factors. The hybrid ANN-FBI model enhanced COD removal and power density, outperforming conventional methods by 2.24% (Sayed et al. 2024).

AI/ML techniques are used exuberantly to select candidate strains and optimize conditions for microalgal pollutant treatments. Algorithms such as genetic algorithms (GA), decision trees (DT), and particle swarm optimization have been employed to determine optimal nutrient concentrations and dosing regimens, thereby enhancing nutrient removal efficiencies while reducing resource consumption (Singh and Mishra 2022). Under optimized conditions (CO_2 2.5% v/v, 25 °C, pH 7, light 200 μmol m^{-2} s^{-1}, N/P 16), *Tetradesmus arenicola* achieved a biomass yield of 4.45 g/L in wastewater (Singh et al. 2023). AI models have also contributed to robust strain classification. For instance, ANN-based image analysis successfully differentiated *Scenedesmus almeriensis* and *C. vulgaris* (Otálora et al. 2021). Similarly, SVM facilitated precise identification of six microalgal species via fluorescence imaging flow cytometry (Harmon et al. 2020). BP–GA models combined with fluorescence spectroscopy enabled monitoring *C. reinhardtii* growth (Liu et al. 2020). Further, SVM with hyperspectral imaging has been used to study interspecies interactions under differing pH levels (Bi et al. 2019). The toxicity of polycyclic aromatic hydrocarbons (PAHs) in microalgae can be effectively assessed using an artificial neural network (ANN) and genetic algorithms (GA), integrating ANN and GA with a 2^4 factorial design (Imamoglu 2024). Using this framework, the researchers have assessed the combined effect of PAHs (benzo[a]pyrene and phenanthrene) and heavy metals (cadmium and lead) on microalgal systems (Subashchandrabose et al. 2017b). Predictive modeling can help select the most efficient strain for the remediation of the target compound.

The European Union's BIOSYSMO and EDAPHOS projects aim to develop an integrated bioremediation platform that integrates biosensors, Internet of things (IoT), and artificial intelligence (AI) to clean up contaminated soils and groundwaters (https://www.biosysmo.eu, https://edaphos.eu/about; accessed on 30 August 2025). The IoT-enabled sensors monitor environmental factors such as pH, temperature, moisture, and pollutant levels, while biosensors detect microbial activity to assess treatment performance. ML-based algorithms analyze sensor data in real time to properly manage and improve bioremediation processes. These and other projects, such as GREENER and Symbiorem, develop state-of-the-art sustainable in situ bioremediation techniques. Compared to traditional methods, these methods harbor the potential to achieve better efficiency with lesser investments (https://www.greener-h2020.eu/en/, https://symbiorem.eu/about/; accessed on 30 August 2025). The EU-funded MIBIREM project also aims to develop a toolbox involving microbiomes and biotechnological advancements for efficient bioremediation (https://mibirem.eu/about/; accessed on 30 August 2025). Likewise, the EU and China jointly funded the project EiCLaR, which aims to develop technological advancements for in situ treatments of various pollutants such as heavy metals and hydrocarbons (https://eiclar.eu/; accessed on 30 August 2025). Integrating IoT sensors (soil moisture, temperature, oxygen levels, pollutant concentrations, etc.) with AI-driven controls enables automated aeration, irrigation, and nutrient dosing to maximize microbial activity. When combined with real-time biosensor data on microbial populations and metabolism, these technologies can efficiently guide process adjustments and aid in developing resilient and efficient strategies for in situ bioremediation. Pilot-scale applications of an AI-based decision forecast plume behavior and optimize in situ bioremediation strategies, achieving significant reductions in solvent concentrations and restoration of groundwater quality, highlighting the potential of this system for remote monitoring, and control of remediation processes (Kraiem et al. 2024). Collectively, these initiatives demonstrate how IoT, AI, and biosensors can transform bioremediation by providing real-time, site-specific insights, reducing investments, and enabling the development of adaptive, contaminant-targeted treatment models.

14.4 Integration of Algal Biotechnology and Synthetic Biology for Advanced Remediation, Biomonitoring, and Biosensors

Biomonitoring encompasses biological responses elicited by organisms for the qualitative and quantitative monitoring of environmental change, which aid in quality control assessments (Omar 2010). Microalgae, due to their longevity, ubiquitous occurrence, and high capacity to accumulate metals and pollutants, are employed as bioindicators and predictive tools for assessing pollutants in the aquatic environments (Zhou et al. 2008). For example, mosses are useful for mapping the spread of POPs in remote areas such as Antarctica (Wu et al. 2014). Biosensors work with an amalgamation of biological knowledge and technological advancements like nanotechnology, material sciences, microfluidics, and electronics (Mishra et al. 2025). Nano-biotechnology has been pivotal in enhancing bioremediation applications using nanoparticles to catalyze breakdown. Nanoparticles act as carriers that enhance the efficiency of nutrients or electron acceptors for the microbial species, thereby improving the metabolic activity and pollutant breakdown. Correspondingly, the inherent properties of these systems enable direct interactions with pollutants, increasing bioavailability and the target pollutants' susceptibility toward enzymatic degradation (Hidangmayum et al. 2023; Das et al. 2025). Most often, biological recognition molecules such as antibodies, enzymes, receptors, or nucleic acids are used in these sensors. Since they are highly sensitive, affordable, and less susceptible to physicochemical changes than other bioreceptors, algae-based biosensors are a great tool for continuously monitoring environmental fluctuations and detecting insecticides and herbicides in food and water in real time (Samal et al. 2023).

Algal biosensing systems, first developed in the 1990s, used simple algal bioassays to detect xenobiotics through changes in growth and photosynthetic activity (Wlodkowic and Karpiński 2021). Later studies linked xenobiotic interactions with PSII, where herbicides block electron transfer from QA to QB via the D1 protein, leading to cell death. This principle advanced second-generation biosensors, including whole-cell algae immobilized on Clark electrodes to monitor oxygen evolution, and optical sensors exploiting chlorophyll a fluorescence changes. Despite rapid detection and simple preparation, issues of stability and selectivity limited commercialization. Recently, alginate-encapsulated *Scenedesmus acutus* and *Pseudokirchneriella subcapitata* demonstrated reliable detection of copper (7.5 ppm), glyphosate (10–15 ppm), atrazine (50 ppb), and chromium (250 ppb), offering a user-friendly and precise sensing approach (Prudkin-Silva et al. 2021). Fluorescence-based algal biosensors detect toxicants through shifts in emission signals (Wong et al. 2013). For example, *Selenastrum capricornutum* biosensor was optimized under varying light and oxygen conditions (Cho et al. 2008), while *C. vulgaris* immobilized on screen-printed carbon electrodes was applied for amperometric oxygen measurement (Shitanda et al. 2009). Other approaches include *Spirulina subsalsa* coupled with gas-diffusion electrodes for estuarine toxicity (Campanella et al. 2001), herbicide-sensitive *Chlorella vulgaris* biosensors integrated with Teflon-covered electrodes (Pandard and Rawson 1993), and the AlgaTox device, which employs Clark electrodes to measure oxygen evolution under 680 nm irradiation (Buckova et al. 2017). Additionally, microbial fuel cells (MFCs) function as bioelectrochemical systems that utilize biological catalysts to harness electricity or valuable bioproducts.

14.5 Conclusion and Future Perspectives

The current global challenges of rapid population growth, escalating environmental pollution, and rising fuel crisis necessitate developing economically viable and environmentally sustainable remediation strategies. In this context, microalgal-based technologies offer promising solutions by serving as green remediation systems and enabling the production of bioenergy and high-value bioproducts within a closed-loop approach. This chapter explores the potential of microalgae for bioremediation of persistent organic pollutants and discusses state-of-the-art strategies for enhancing the remediation efficiency. This chapter highlights recent advancements in microalgal biotechnology, including the application of omics technologies, genetic engineering, and artificial intelligence/machine learning (AI/ML)-driven approaches aimed at strain development and optimization. High-throughput screening platforms and molecular biology tools, coupled with progress in DNA synthesis and genome editing, have significantly advanced the engineering of complex microalgal systems. Further, future research should prioritize the development of specific-purpose bioengineered strains capable of degrading a diverse array of co-existing environmental contaminants, as pollutants rarely occur in isolation in natural ecosystems. Additionally, bioaugmentation of native microalgal populations or consortia can enhance in situ remediation efficiency. Furthermore, integrating synthetic biology with extraction techniques and AI/ML-guided process optimization can significantly enhance the competitiveness of microalgal bioremediation systems. Nonetheless, stringent regulatory frameworks and comprehensive monitoring are essential to assess and mitigate potential environmental and human health risks, especially when deploying genetically modified strains in outdoor cultivation settings. In conclusion, with appropriate safety assessments and regulatory oversight, the convergence of synthetic biology and microalgal bioremediation is poised to revolutionize sustainable pollution mitigation and bioresource recovery.

Acknowledgments AR acknowledges the MHRD fellowship from IIT-Roorkee. **KMP** acknowledges the support received through Grants BT/PR45943/BCE/8/1785/2023 from DBT (Department of Biotechnology), and SERB-STAR fellowship (STR/2022/000008) from ANRF/SERB (Science and Engineering Research Board) Government of India.

References

Abdelfattah A et al (2023) Microalgae-based wastewater treatment: Mechanisms, challenges, recent advances, and future prospects. https://doi.org/10.1016/j.ese.2022.100205

Ali M, Abd El-Aty AM, Badawy MI, Ali RK (2018) Removal of pharmaceutical pollutants from synthetic wastewater using chemically modified biomass of green alga Scenedesmus obliquus. Ecotoxicol Environ Saf 151:144–152. https://doi.org/10.1016/J.ECOENV.2018.01.012

Al-Zuhair S (2016) High concentration phenol removal using freshwater microalgae. Int J Biotechnol Wellness Ind 5:39–45. https://doi.org/10.6000/1927-3037.2016.05.02.2

Angulo E, Bula L, Mercado I, Montaño A, Cubillán N (2018) Bioremediation of Cephalexin with non-living Chlorella sp., biomass after lipid extraction. Bioresour Technol 257:17–22. https://doi.org/10.1016/J.BIORTECH.2018.02.079

Ansari FA, Nasr M, Rawat I, Bux F (2021) Artificial neural network and techno-economic estimation with algae-based tertiary wastewater treatment. J Water Process Eng 40:101761. https://doi.org/10.1016/J.JWPE.2020.101761

Aragaw TA, Asmare AM (2018) Phycoremediation of textile wastewater using indigenous microalgae. Water Pract Technol 13:274–284. https://doi.org/10.2166/wpt.2018.037

Arora N, Philippidis GP (2021) Microalgae strain improvement strategies: random mutagenesis and adaptive laboratory evolution. Trends Plant Sci 26:1199–1200. https://doi.org/10.1016/j.tplants.2021.06.005

Arora N, Dubey D, Sharma M, Patel A, Guleria A, Pruthi PA, Kumar D, Pruthi V, Poluri KM (2018a) NMR-based metabolomic approach to elucidate the differential cellular responses during mitigation of Arsenic(III, V) in a green microalga. ACS Omega 3:11847–11856. https://doi.org/10.1021/acsomega.8b01692

Arora N, Pienkos PT, Pruthi V, Poluri KM, Guarnieri MT (2018b) Leveraging algal omics to reveal potential targets for augmenting TAG accumulation. Biotechnol Adv 36:1274–1292. https://doi.org/10.1016/J.BIOTECHADV.2018.04.005

Arora N, Tripathi S, Poluri KM, Pruthi V (2019a) Advanced gene technology and synthetic biology approaches to custom design microalgae for biodiesel production. In: Microalgae biotechnology for development of biofuel and wastewater treatment. Springer Singapore, Singapore, pp 147–175. https://doi.org/10.1007/978-981-13-2264-8_8

Arora N, Kumari P, Kumar A, Gangwar R, Gulati K, Pruthi PA, Prasad R, Kumar D, Pruthi V, Poluri KM (2019b) Delineating the molecular responses of a halotolerant microalga using integrated omics approach to identify genetic engineering targets for enhanced TAG production. Biotechnol Biofuels 12:2. https://doi.org/10.1186/s13068-018-1343-1

Arora N, Jaiswal D, Sengupta S, Wangikar PP (2020) Metabolic engineering of cyanobacteria for production of platform chemicals: a synthetic biology approach. In: Handbook of algal science, technology and medicine, pp 127–145. https://doi.org/10.1016/B978-0-12-818305-2.00008-5

Asghari S, Rajabi F, Tarrahi R, Salehi-Lisar SY, Asnaashari S, Omidi Y, Movafeghi A (2020) Potential of the green microalga Chlorella vulgaris to fight against fluorene contamination: evaluation of antioxidant systems and identification of intermediate biodegradation compounds. J Appl Phycol 32:411–419. https://doi.org/10.1007/s10811-019-01921-7

Avila R, Peris A, Eljarrat E, Vicent T, Blánquez P (2021) Biodegradation of hydrophobic pesticides by microalgae: transformation products and impact on algae biochemical methane potential. Sci Total Environ 754:142114. https://doi.org/10.1016/j.scitotenv.2020.142114

Baglieri A, Sidella S, Barone V, Fragalà F, Silkina A, Nègre M, Gennari M (2016) Cultivating Chlorella vulgaris and Scenedesmus quadricauda microalgae to degrade inorganic compounds and pesticides in water. Environ Sci Pollut Res 23:18165–18174. https://doi.org/10.1007/s11356-016-6996-3

Bano F, Malik A, Ahammad SZ (2021) Removal of Estradiol, Diclofenac, and Triclosan by naturally occurring microalgal consortium obtained from wastewater. Sustainability 13:7690. https://doi.org/10.3390/su13147690

Barathe P, Kaur K, Reddy S, Shriram V, Kumar V (2024) Antibiotic pollution and associated antimicrobial resistance in the environment. J Hazard Mater Lett 5:100105. https://doi.org/10.1016/J.HAZL.2024.100105

Barathi SJG, Rathinasamy G, Sabapathi N, Aruljothi KN, Lee J, Kandasamy S (2023) Recent trends in polycyclic aromatic hydrocarbons pollution distribution and counteracting bio-remediation strategies. Chemosphere 337:139396. https://doi.org/10.1016/J.CHEMOSPHERE.2023.139396

Bashir KMI, Kim M-S, Stahl U, Cho M-G (2016) Microalgae engineering toolbox: selectable and screenable markers. Biotechnol Bioprocess Eng 21:224–235. https://doi.org/10.1007/s12257-015-0386-4

Berepiki A et al (2016) Tapping the unused potential of photosynthesis with a heterologous electron sink. https://doi.org/10.1021/acssynbio.6b00100

Berradi M, Hsissou R, Khudhair M, Assouag M, Cherkaoui O, El Bachiri A, El Harfi A (2019) Textile finishing dyes and their impact on aquatic environs. Heliyon 5:e02711. https://doi.org/10.1016/j.heliyon.2019.e02711

Bi X, Lin S, Zhu S, Yin H, Li Z, Chen Z (2019) Species identification and survival competition analysis of microalgae via hyperspectral microscopic images. Optik (Stuttg) 176:191–197. https://doi.org/10.1016/J.ALGAL.2019.101739

Buckova M, Licbinsky R, Jandova V, Krejci J, Pospichalova J, Huzlik J (2017) Fast ecotoxicity detection using biosensors. Water Air Soil Pollut 228:166. https://doi.org/10.1007/s11270-017-3341-5

Burch AR, Yothers CW, Salemi MR, Phinney BS, Pandey P, Franz AK (2021) Quantitative label-free proteomics and biochemical analysis of Phaeodactylum tricornutum cultivation on dairy manure wastewater. J Appl Phycol 33:2105–2121. https://doi.org/10.1007/s10811-021-02483-3

Campanella L, Cubadda F, Sammartino MP, Saoncella A (2001) An algal biosensor for the monitoring of water toxicity in estuarine environments. Water Res 35:69–76. https://doi.org/10.1016/S0043-1354(00)00223-2

Cao M, Wang F, Zhou B et al (2023) Nanoparticles and antibiotics stress proliferated antibiotic resistance genes in microalgae-bacteria symbiotic systems. J Hazard Mater 443:130201. https://doi.org/10.1016/j.jhazmat.2022.130201

Carbajo Arteaga L, Ponce Zavaleta M, Moreno Eustaquio W, Mendoza Bobadilla J (2018) Removal of aniline blue dye using live microalgae Chlorella vulgaris. J Energy Environ Sci 2:6. https://doi.org/10.32829/eesj.v2i1.51

Castellanos-Estupiñan M, Carrillo-Botello A, Rozo-Granados L, Becerra-Moreno D, García-Martínez J, Urbina-Suarez N, López-Barrera G, Barajas-Solano A, Bryan S, Zuorro A (2022) Removal of nutrients and pesticides from agricultural runoff using microalgae and cyanobacteria. Water (Basel) 14:558. https://doi.org/10.3390/w14040558

Chaurasia AK, Adhya TK, Apte SK (2013) Engineering bacteria for bioremediation of persistent organochlorine pesticide lindane (γ-hexachlorocyclohexane). Bioresour Technol 149:439–445. https://doi.org/10.1016/J.BIORTECH.2013.09.084

Chen S, Liu Y, Zhang J, Gao B (2017) iTRAQ-based quantitative proteomic analysis of Microcystis aeruginosa exposed to spiramycin at different nutrient levels. Aquat Toxicol 185:193–200. https://doi.org/10.1016/J.AQUATOX.2017.02.015

Chen G, Bai R, Zhang Y, Zhao B, Xiao Y (2022) Application of metagenomics to biological wastewater treatment. Sci Total Environ 807:150737. https://doi.org/10.1016/J.SCITOTENV.2021.150737

Chen Z, Chen Z, Li Y, Zhang R, Liu Y, Hui A, Cao W, Liu J, Bai H, Song J (2024a) A review on remediation of chlorinated organic contaminants in soils by thermal desorption. J Ind Eng Chem 133:112–121. https://doi.org/10.1016/J.JIEC.2023.12.022

Chen Q, Li Z, Li Y, Liu M, Wu Y, Chen Z, Zhu B (2024b) Biodegradation of benzo[a]pyrene by a marine Chlorella vulgaris LH-1 with heterotrophic ability. Mar Pollut Bull 198:115848. https://doi.org/10.1016/J.MARPOLBUL.2023.115848

Chi J, Li Y, Gao J (2019) Interaction between three marine microalgae and two phthalate acid esters. Ecotoxicol Environ Saf 170:407–411. https://doi.org/10.1016/J.ECOENV.2018.12.012

Chiaiese P, Palomba F, Tatino F, Lanzillo C, Pinto G, Pollio A, Filippone E (2011) Engineered tobacco and microalgae secreting the fungal laccase POXA1b reduce phenol content in olive oil mill wastewater. Enzym Microb Technol 49:540–546. https://doi.org/10.1016/J.ENZMICTEC.2011.06.002

Chin JY, Chng LM, Leong SS, Yeap SP, Yasin NHM, Toh PY (2020) Removal of synthetic dye by Chlorella vulgaris microalgae as natural adsorbent. Arab J Sci Eng 45:7385–7395. https://doi.org/10.1007/s13369-020-04557-9

Cho C-W, Pham TPT, Jeon Y-C, Min J, Jung HY, Lee DS, Yun Y-S (2008) Microalgal photosynthetic activity measurement system for rapid toxicity assessment. Ecotoxicology 17:455–463. https://doi.org/10.1007/s10646-008-0197-x

Choudhary S, Poluri KM (2024) Microalgal-based biorefinery approaches toward a sustainable future. In: Industrial microbiology and biotechnology. Springer Nature Singapore, Singapore, pp 229–275. https://doi.org/10.1007/978-981-97-1912-9_10

Choudhary S, Venkatrayalu G, Ravi Kiran B, Poluri KM (2025) Advances and perspectives of microalgal biorefineries for bioenergy. Energy Fuel 39:12367–12406. https://doi.org/10.1021/acs.energyfuels.5c01676

Crozet P, Navarro FJ, Willmund F et al (2018) Birth of a photosynthetic Chassis: a MoClo toolkit enabling synthetic biology in the microalga *Chlamydomonas reinhardtii*. ACS Synth Biol 7:2074–2086. https://doi.org/10.1021/acssynbio.8b00251

Daboussi F, Leduc S, Maréchal A et al (2014) Genome engineering empowers the diatom Phaeodactylum tricornutum for biotechnology. Nat Commun 5:3831. https://doi.org/10.1038/ncomms4831

Das B, Deka S (2019) A cost-effective and environmentally sustainable process for phycoremediation of oil field formation water for its safe disposal and reuse. Sci Rep 9:15232. https://doi.org/10.1038/s41598-019-51806-5

Das B, Mandal TK, Patra S (2015) A comprehensive study on Chlorella pyrenoidosa for phenol degradation and its potential applicability as biodiesel feedstock and animal feed. Appl Biochem Biotechnol 176:1382–1401. https://doi.org/10.1007/s12010-015-1652-9

Das J, Yadav E, Poluri KM (2025) A review on the role of nanotechnological interventions in sequestration, mitigation and value-added product conversion of micro−/nanoplastics. Environ Sci Nano 12:189–218. https://doi.org/10.1039/D4EN00267A

Dayana Priyadharshini S, Bakthavatsalam AK (2016) Optimization of phenol degradation by the microalga Chlorella pyrenoidosa using Plackett–Burman design and response surface methodology. Bioresour Technol 207:150–156. https://doi.org/10.1016/J.BIORTECH.2016.01.138

Ding T, Wang S, Yang B, Li J (2020) Biological removal of pharmaceuticals by Navicula sp. and biotransformation of bezafibrate. Chemosphere 240:124949. https://doi.org/10.1016/J.CHEMOSPHERE.2019.124949

Duan W, Du S, Meng F, Peng X, Peng L, Lin Y, Wang G, Wu J (2020) The pathways by which the marine diatom Thalassiosira sp. OUC2 biodegrades p-xylene, combined with a mechanistic analysis at the proteomic level. Ecotoxicol Environ Saf 198:110687. https://doi.org/10.1016/J.ECOENV.2020.110687

Dutra Molino JV, Saucedo B, Kang K, Walsh C, Diaz CJ, Tessman M, Simkovsky R, Mayfield S (2025) Efficient secretion of a plastic degrading enzyme from the green algae Chlamydomonas reinhardtii. Sci Rep 15:24690. https://doi.org/10.1038/s41598-025-09100-0

Eapen JV, Thomas S, Antony S, George P, Antony J (2024) A review of the effects of pharmaceutical pollutants on humans and aquatic ecosystem. Explor Drug Sci:484–507. https://doi.org/10.37349/eds.2024.00058

El-Agawany NI, Abdel-Razik SAR, Abdel-Kareem MS, Kaamoush MIA (2023) Environmental performance of two marine algae Ulva fasciata and Pterocladia capillacea in the biological treatment of four reactive dyes from aqueous solutions based on fresh and dried biomass. Appl Water Sci 13:82. https://doi.org/10.1007/s13201-023-01893-7

Encarnação T, Palito C, Pais AACC, Valente AJM, Burrows HD (2020) Removal of pharmaceuticals from water by free and imobilised microalgae. Molecules 25:3639. https://doi.org/10.3390/molecules25163639

Eregie SB, Sanusi IA, Olaniran AO (2025) Bibliometric review of transcriptomic microalgae-based biodegradation of lubricant oil waste hydrocarbon: current research outlook. Bioresour Technol Rep 29:102074. https://doi.org/10.1016/J.BITEB.2025.102074

Fan Y, Wang H, Sun C, Wang Y, Jin C, Zhao Y, Gao M, Guo L (2023) Metagenomic and transcriptome elucidating the sulfamethoxazole degradation and metabolism pathways in

fermentative bacteria and microalgae coupling system for mariculture wastewater treatment. Chem Eng J 474:145560. https://doi.org/10.1016/J.CEJ.2023.145560
Fayaz T, Rana SS, Goyal E, Ratha SK, Renuka N (2024) Harnessing the potential of microalgae-based systems for mitigating pesticide pollution and its impact on their metabolism. J Environ Manag 357:120723. https://doi.org/10.1016/J.JENVMAN.2024.120723
García de Llasera MP, de Olmos-Espejel JJ, Díaz-Flores G, Montaño-Montiel A (2016) Biodegradation of benzo(a)pyrene by two freshwater microalgae Selenastrum capricornutum and Scenedesmus acutus: a comparative study useful for bioremediation. Environ Sci Pollut Res 23:3365–3375. https://doi.org/10.1007/s11356-015-5576-2
Gelebo GG, Tessema LH, Kehshin KT, Gebremariam HH, Gebremikal ET, Motuma MT, Ayele A, Getachew D, Benor S, Suresh A (2020) Phycoremediation of synthetic dyes in an aqueous solution using an indigenous *Oscillatoria* sp., from Ethiopia. Ethiop J Sci Sustain Dev 7(2): 14–20. https://doi.org/10.20372/ejssdastu:v7.i2.2020.186
Ghanbarzadeh M, Niknam V, Soltani N, Ebrahimzadeh H, Shahavi MH (2022) Removal of Phenanthrene by some microalga species and study of antioxidative compounds in Nostoc calcicola ISC 89. J Soils Sediments 22:109–119. https://doi.org/10.1007/s11368-021-03065-z
Gojkovic Z, Lindberg RH, Tysklind M, Funk C (2019) Northern green algae have the capacity to remove active pharmaceutical ingredients. Ecotoxicol Environ Saf 170:644–656. https://doi.org/10.1016/J.ECOENV.2018.12.032
Guo P, Jin L, Cui W, Yang Y, Cheng Y, Wang D, Pan Y, Jin J (2022) Proteomics analysis of pyrene biodegradation using strain *Mycobacterium* sp. 16F. Environ Pollut Bioavailab 34:236–245. https://doi.org/10.1080/26395940.2022.2081613
Han S, Han W, Chen J, Sun Y, Dai M, Zhao G (2020) Bioremediation of malachite green by cyanobacterium Synechococcus elongatus PCC 7942 engineered with a triphenylmethane reductase gene. Appl Microbiol Biotechnol 104:3193–3204. https://doi.org/10.1007/s00253-020-10438-w
Hanafi MF, Sapawe N (2020) A review on the current techniques and technologies of organic pollutants removal from water/wastewater. Mater Today Proc 31:A158–A165. https://doi.org/10.1016/j.matpr.2021.01.265
Hao X, Luo L, Jouhet J et al (2018) Enhanced triacylglycerol production in the diatom Phaeodactylum tricornutum by inactivation of a Hotdog-fold thioesterase gene using TALEN-based targeted mutagenesis. Biotechnol Biofuels 11:312. https://doi.org/10.1186/s13068-018-1309-3
Harmon J, Mikami H, Kanno H, Ito T, Goda K (2020) Accurate classification of microalgae by intelligent frequency-division-multiplexed fluorescence imaging flow cytometry. OSA Contin 3:430. https://doi.org/10.1364/OSAC.387523
Helvig C, Kariyawasam T, Vriens B, Petkovich M (2025) Genetically engineered bacteria and microalgae expressing a mutant of cytochrome P450 BM3 for efficient Diuron degradation in wastewater treatment. Microbiol Spectr 13:e0290524. https://doi.org/10.1128/spectrum.02905-24
Hena S, Gutierrez L, Croué JP (2020) Removal of metronidazole from aqueous media by C. vulgaris. J Hazard Mater 384:121400. https://doi.org/10.1016/J.JHAZMAT.2019.121400
Hena S, Gutierrez L, Croué JP (2021) Removal of pharmaceutical and personal care products (PPCPs) from wastewater using microalgae: a review. J Hazard Mater 403:124041. https://doi.org/10.1016/J.JHAZMAT.2020.124041
Hidangmayum A, Debnath A, Guru A, Singh BN, Upadhyay SK, Dwivedi P (2023) Mechanistic and recent updates in nano-bioremediation for developing green technology to alleviate agricultural contaminants. Int J Environ Sci Technol 20:11693–11718. https://doi.org/10.1007/s13762-022-04560-7
Hoque MZ, Alqahtani A, Sankaran S, Anand D, Musa MM, Nzila A, Guerriero G, Siddiqui KS, Ahmad I (2023) Enhanced biodegradation of phenanthrene and anthracene using a microalgal-bacterial consortium. Front Microbiol. https://doi.org/10.3389/fmicb.2023.1227210

Hu C, Luo Q, Huang Q (2014) Ecotoxicological effects of perfluorooctanoic acid on freshwater microalgae *Chlamydomonas reinhardtii* and *Scenedesmus obliquus*. Environ Toxicol Chem 33: 1129–1134. https://doi.org/10.1002/etc.2532

Hu Y, Meng F-L, Hu Y-Y, Habibul N, Sheng G-P (2020) Concentration- and nutrient-dependent cellular responses of microalgae Chlorella pyrenoidosa to perfluorooctanoic acid. Water Res 185:116248. https://doi.org/10.1016/j.watres.2020.116248

Imam A, Kumar Suman S, Kanaujia PK, Ray A (2022) Biological machinery for polycyclic aromatic hydrocarbons degradation: a review. Bioresour Technol 343:126121. https://doi.org/10.1016/J.BIORTECH.2021.126121

Imamoglu E (2024) Artificial intelligence and/or machine learning algorithms in microalgae bioprocesses. Bioengineering (Basel). https://doi.org/10.3390/bioengineering11111143

Ishchi T, Sibi G (2019) Azo dye degradation by Chlorella vulgaris: optimization and kinetics. Int J Biol Chem 14:1–7. https://doi.org/10.3923/ijbc.2020.1.7

Ismaiel M, El-Ayouty Y, Al-Badwy AH, El-Ayouty YM (2019) Biosorption analysis and Penoxsulam Herbicide removal efficiency by transgenic Chlamydomonas reinhardtii overexpression the Cyanobacterial Enzyme Glutathione-s-transferase. https://www.researchgate.net/publication/342720713

Kadri MS, Singhania RR, Haldar D, Patel AK, Bhatia SK, Saratale G, Parameswaran B, Chang JS (2023) Advances in Algomics technology: application in wastewater treatment and biofuel production. Bioresour Technol 387:129636. https://doi.org/10.1016/J.BIORTECH.2023.129636

Kariyawasam T, Helvig C, Petkovich M, Vriens B (2024) Pharmaceutical removal from wastewater by introducing cytochrome P450s into microalgae. Microb Biotechnol 17:e14515. https://doi.org/10.1111/1751-7915.14515

Khalid AAH, Yaakob Z, Abdullah SRS, Takriff MS (2018) Growth improvement and metabolic profiling of native and commercial Chlorella sorokiniana strains acclimatized in recycled agricultural wastewater. Bioresour Technol 247:930–939. https://doi.org/10.1016/J.BIORTECH.2017.09.195

Kiki C, Rashid A, Wang Y, Li Y, Zeng Q, Yu CP, Sun Q (2020) Dissipation of antibiotics by microalgae: kinetics, identification of transformation products and pathways. J Hazard Mater 387:121985. https://doi.org/10.1016/J.JHAZMAT.2019.121985

Kim JW, Park S-B, Tran Q-G, Cho D-H, Choi D-Y, Lee YJ, Kim H-S (2020) Functional expression of polyethylene terephthalate-degrading enzyme (PETase) in green microalgae. Microb Cell Factories 19:97. https://doi.org/10.1186/s12934-020-01355-8

Knott BC, Erickson E, Allen MD et al (2020) Characterization and engineering of a two-enzyme system for plastics depolymerization. Proc Natl Acad Sci 117:25476–25485. https://doi.org/10.1073/pnas.2006753117

Kraiem Z, Zouari K, Trabelsi R (2024) Harnessing machine learning tools for water quality assessment in the Kebili shallow aquifers, Southwestern Tunisia. Acta Geochimica 43:1065–1086. https://doi.org/10.1007/s11631-024-00689-z

Kundu P, Dutta N, Bhattacharya S (2024) Application of microalgae in wastewater treatment with special reference to emerging contaminants: a step towards sustainability. Front Anal Sci. https://doi.org/10.3389/frans.2024.1513153

Kurade MB, Kim JR, Govindwar SP, Jeon BH (2016) Insights into microalgae mediated biodegradation of diazinon by Chlorella vulgaris: microalgal tolerance to xenobiotic pollutants and metabolism. Algal Res 20:126–134. https://doi.org/10.1016/J.ALGAL.2016.10.003

Ladeia Ramos R, Rezende Moreira V, Santos Amaral MC (2024) Phenolic compounds in water: review of occurrence, risk, and retention by membrane technology. J Environ Manag 351: 119772. https://doi.org/10.1016/J.JENVMAN.2023.119772

Leetanasaksakul K, Intha T, Roytrakul S, Phaonakrop N, Thaisakun S, Sirikhachornkit A (2025) Proteomic analysis of differential responses to norflurazon herbicide in the model green alga Chlamydomonas reinhardtii. Sci Rep 15:31601. https://doi.org/10.1038/s41598-025-17119-6

Li Z, Li S, Li T, Gao X, Zhu L (2022) Physiological and transcriptomic responses of Chlorella sorokiniana to ciprofloxacin reveal molecular mechanisms for antibiotic removal. iScience 25: 104638. https://doi.org/10.1016/J.ISCI.2022.104638

Lim HR, Khoo KS, Chia WY, Chew KW, Ho SH, Show PL (2022) Smart microalgae farming with internet-of-things for sustainable agriculture. Biotechnol Adv 57:107931. https://doi.org/10.1016/J.BIOTECHADV.2022.107931

Liu JH, Zhang L, Zha DC, Chen LQ, Chen XX, Qi ZM (2018) Biosorption of malachite green onto *Haematococcus pluvialis* observed through synchrotron Fourier-transform infrared microspectroscopy. Lett Appl Microbiol 67:348–353. https://doi.org/10.1111/lam.13043

Liu JY, Zeng LH, Ren ZH, Du TM, Liu X (2020) Rapid in situ measurements of algal cell concentrations using an artificial neural network and single-excitation fluorescence spectrometry. Algal Res 45:101739. https://doi.org/10.1016/j.algal.2019.101739

Lu X, Xu W, Liu C, Zhao Q, Ye Z (2021) Insight into the role of extracellular polymeric substances in denitrifying biofilms under nitrobenzene exposure. Ecotoxicol Environ Saf 222:112539. https://doi.org/10.1016/J.ECOENV.2021.112539

Maqsood Q, Sumrin A, Waseem R, Hussain M, Imtiaz M, Hussain N (2023) Bioengineered microbial strains for detoxification of toxic environmental pollutants. Environ Res 227: 115665. https://doi.org/10.1016/j.envres.2023.115665

Marchetto F, Roverso M, Righetti D, Bogialli S, Filippini F, Bergantino E, Sforza E (2021) Bioremediation of per- and Poly-Fluoroalkyl Substances (PFAS) by Synechocystis sp. PCC 6803: a Chassis for a synthetic biology approach. Life 11:1300. https://doi.org/10.3390/life11121300

Mariano DCO, Dias GM, Castro MR, Tschoeke DA, de Oliveira FJS, Sérvulo EFC, Neves BC (2024) Exploring the diversity and functional profile of microbial communities of Brazilian soils with high salinity and oil contamination. Heliyon 10:e34336. https://doi.org/10.1016/J.HELIYON.2024.E34336

Marques IM, Oliveira ACV, de Oliveira OMC, Sales EA, Moreira ÍTA (2021) A photobioreactor using Nannochloropsis oculata marine microalgae for removal of polycyclic aromatic hydrocarbons and sorption of metals in produced water. Chemosphere 281:130775. https://doi.org/10.1016/J.CHEMOSPHERE.2021.130775

Martínez A (2024) Toxicity of persistent organic pollutants: a theoretical study. J Mol Model 30:97. https://doi.org/10.1007/s00894-024-05890-8

Matamoros V, Gutiérrez R, Ferrer I, García J, Bayona JM (2015) Capability of microalgae-based wastewater treatment systems to remove emerging organic contaminants: a pilot-scale study. J Hazard Mater 288:34–42. https://doi.org/10.1016/J.JHAZMAT.2015.02.002

Meng J, Hong S, Wang T, Li Q, Yoon SJ, Lu Y, Giesy JP, Khim JS (2017) Traditional and new POPs in environments along the Bohai and Yellow Seas: an overview of China and South Korea. Chemosphere 169:503–515. https://doi.org/10.1016/J.CHEMOSPHERE.2016.11.108

Mishra A, Medhi K, Malaviya P, Thakur IS (2019) Omics approaches for microalgal applications: prospects and challenges. Bioresour Technol 291:121890. https://doi.org/10.1016/J.BIORTECH.2019.121890

Mishra R, Minocha S, Goel R, Gaur PK, Lata K, Singh R (2025) Bioconvergence: advancing biosensors with nanotechnology for real-time health monitoring. Bull Natl Res Cent 49:14. https://doi.org/10.1186/s42269-025-01308-4

Mohandass S, Ragavan M, Gnanasekaran D, Lakshmanan U, Dharmar P, Saha SK (2021) Overexpression of Cu/Zn superoxide dismutase (Cu/Zn SOD) in Synechococcus elongatus PCC 7942 for enhanced Azo Dye removal through hydrogen peroxide accumulation. Biology (Basel). https://doi.org/10.3390/biology10121313

Moog D, Schmitt J, Senger J, Zarzycki J, Rexer K-H, Linne U, Erb T, Maier UG (2019) Using a marine microalga as a chassis for polyethylene terephthalate (PET) degradation. Microb Cell Factories 18:171. https://doi.org/10.1186/s12934-019-1220-z

Moradi Z, Haghjou MM, Zarei M, Sharifan H (2024) Harnessing Chlorella vulgaris for the phycoremediation of azo dye: a comprehensive analysis of metabolic responses and antioxidant system. Algal Res 82:103660. https://doi.org/10.1016/J.ALGAL.2024.103660

Morillas-España A, López-Serna R, Rodríguez Chikri LY, Jiménez JJ, Lafarga T, Uggetti E, Acién G, González-López CV (2025) Microalgae wastewater treatment: pharmaceutical removal and biomass valorization. J Environ Manag 380:124942. https://doi.org/10.1016/J.JENVMAN.2025.124942

Muhammad M, Batool S, Hivare V, Li W-J, Waheed A, Sinha D (2024) Bioremediation techniques—classification, principles, advantages, limitations, and prospects. In: Microbiome-assisted bioremediation. Elsevier, pp 1–23. https://doi.org/10.1016/B978-0-443-21911-5.00003-9

Mustafa G, Zahid MT, Kurade MB, Alvi A, Ullah F, Yadav N, Park HK, Khan MA, Jeon BH (2024) Microalgal and activated sludge processing for biodegradation of textile dyes. Environ Pollut 349:123902. https://doi.org/10.1016/J.ENVPOL.2024.123902

Nagar N, Saxena H, Pathak A, Mishra A, Poluri KM (2023) A review on structural mechanisms of protein-persistent organic pollutant (POP) interactions. Chemosphere 332:138877. https://doi.org/10.1016/J.CHEMOSPHERE.2023.138877

Nanda M, Kumar V, Fatima N, Pruthi V, Verma M, Chauhan PK, Vlaskin MS, Grigorenko AV (2019) Detoxification mechanism of organophosphorus pesticide via carboxylestrase pathway that triggers de novo TAG biosynthesis in oleaginous microalgae. Aquat Toxicol 209:49–55. https://doi.org/10.1016/J.AQUATOX.2019.01.019

Negrete-Bolagay D, Zamora-Ledezma C, Chuya-Sumba C, De Sousa FB, Whitehead D, Alexis F, Guerrero VH (2021) Persistent organic pollutants: the trade-off between potential risks and sustainable remediation methods. J Environ Manag 300:113737. https://doi.org/10.1016/J.JENVMAN.2021.113737

Ng I, Tan S, Kao P, Chang Y, Chang J (2017) Recent developments on genetic engineering of microalgae for biofuels and bio-based chemicals. Biotechnol J. https://doi.org/10.1002/biot.201600644

Olawade DB, Wada OZ, Ige AO, Egbewole BI, Olojo A, Oladapo BI (2024) Artificial intelligence in environmental monitoring: advancements, challenges, and future directions. Hygiene Environ Health Adv 12:100114. https://doi.org/10.1016/J.HEHA.2024.100114

Omar WM (2010) Perspectives on the use of algae as biological indicators for monitoring and protecting aquatic environments, with special reference to Malaysian freshwater ecosystems. Trop Life Sci Res 21(2):51. PMID: 24575199

Otálora P, Guzmán JL, Acién FG, Berenguel M, Reul A (2021) Microalgae classification based on machine learning techniques. Algal Res 55:102256. https://doi.org/10.1016/J.ALGAL.2021.102256

Otto B, Schlosser D (2014) First laccase in green algae: purification and characterization of an extracellular phenol oxidase from Tetracystis aeria. Planta 240:1225–1236. https://doi.org/10.1007/s00425-014-2144-9

Ou Y, Yan Z, Shi G, Yu Z, Cai Y, Ma R (2022) Enantioselective toxicity, degradation and transformation of the chiral insecticide fipronil in two algae culture. Ecotoxicol Environ Saf 235:113424. https://doi.org/10.1016/J.ECOENV.2022.113424

Pandard P, Rawson DM (1993) An amperometric algal biosensor for herbicide detection employing a carbon cathode oxygen electrode. Environ Toxicol Water Qual 8:323–333. https://doi.org/10.1002/tox.2530080309

Patel VK, Soni N, Prasad V, Sapre A, Dasgupta S, Bhadra B (2019) CRISPR–Cas9 system for genome engineering of photosynthetic microalgae. Mol Biotechnol 61:541–561. https://doi.org/10.1007/s12033-019-00185-3

Peng J, Guo J, Lei Y, Mo J, Sun H, Song J (2021) Integrative analyses of transcriptomics and metabolomics in Raphidocelis subcapitata treated with clarithromycin. Chemosphere 266:128933. https://doi.org/10.1016/J.CHEMOSPHERE.2020.128933

Prudkin-Silva C, Lanzarotti E, Álvarez L et al (2021) A cost-effective algae-based biosensor for water quality analysis: development and testing in collaboration with peasant communities. Environ Technol Innov 22:101479. https://doi.org/10.1016/J.ETI.2021.101479

Quan L, Cheng Y, Wang J, Chen Y, Li D, Wang S, Li B, Zhang Z, Yang L, Wu L (2023) Efficient removal of thiamethoxam by freshwater microalgae Scenedesmus sp. TXH: removal mechanism, metabolic degradation and application. J Environ Manag 332:117388. https://doi.org/10.1016/J.JENVMAN.2023.117388

Rafeeq H, Afsheen N, Rafique S, Arshad A, Intisar M, Hussain A, Bilal M, Iqbal HMN (2023) Genetically engineered microorganisms for environmental remediation. Chemosphere 310:136751. https://doi.org/10.1016/J.CHEMOSPHERE.2022.136751

Ren C-Y, Wu E-L, Hartmann EM, Zhao H-P (2021) Biological mitigation of antibiotic resistance gene dissemination by antioxidant-producing microorganisms in activated sludge systems. Environ Sci Technol 55:15831–15842. https://doi.org/10.1021/acs.est.1c04641

Reyad A, Abbassy M, Kh I, Marei G, Rabea E, Bedawy M (2022) Potential removal of some insecticides from water using microalgae and their determination by a validated UV-vis spectrophotometric method. Alex Sci Exch J 43:431–450. https://doi.org/10.21608/asejaiqjsae.2022.254546

Reynolds N (2023) Bioremediation vs. traditional remediation methods: a comparative analysis. Pre QC No P. https://doi.org/10.37421/2376-0214.2023.9.60

Rodríguez A, Castrejón-Godínez ML, Salazar-Bustamante E, Gama-Martínez Y, Sánchez-Salinas E, Mussali-Galante P, Tovar-Sánchez E, Ortiz-Hernández ML (2020) Omics approaches to pesticide biodegradation. Curr Microbiol 77:545–563. https://doi.org/10.1007/s00284-020-01916-5

Samal S, Mohanty RP, Mohanty PS, Giri MK, Pati S, Das B (2023) Implications of biosensors and nanobiosensors for the eco-friendly detection of public health and agro-based insecticides: a comprehensive review. Heliyon 9:e15848. https://doi.org/10.1016/J.HELIYON.2023.E15848

Sánchez-Sandoval DS, González-Ortega O, Navarro-Martínez MF, Castro-Tapia JM, de la Fernando García Cruz R, Soria-Guerra RE (2021) Photodegradation and removal of Diclofenac by the Green Alga Nannochloropsis oculata. Phyton-Int J Exp Bot 90(5):1519–1533. https://doi.org/10.32604/phyton.2021.015087

Sayed ET, Rezk H, Abdelkareem MA, Olabi AG (2024) Artificial neural network based modelling and optimization of microalgae microbial fuel cell. Int J Hydrog Energy 52:1015–1025. https://doi.org/10.1016/J.IJHYDENE.2022.12.081

Sevilla ME, Garcia MD, Perez-Castillo Y, Armijos-Jaramillo V, Casado S, Vizuete K, Debut A, Cerda-Mejía L (2023) Degradation of PET bottles by an engineered Ideonella sakaiensis PETase. Polymers (Basel) 15:1779. https://doi.org/10.3390/polym15071779

Sharma J, Mariam I, Suresh Kareya M, Pavan Jutur P, Joshi M, Harish BA, Chaurasia AK, Nigam S (2023) Metabolomic response of microalgae towards diclofenac sodium during its removal from water and concomitant recovery of pigments and lipids. Bioresour Technol 371:128617. https://doi.org/10.1016/J.BIORTECH.2023.128617

Shitanda I, Takamatsu S, Watanabe K, Itagaki M (2009) Amperometric screen-printed algal biosensor with flow injection analysis system for detection of environmental toxic compounds. Electrochim Acta 54:4933–4936. https://doi.org/10.1016/j.electacta.2009.04.005

Singh V, Mishra V (2022) Evaluation of the effects of input variables on the growth of two microalgae classes during wastewater treatment. Water Res 213:118165. https://doi.org/10.1016/j.watres.2022.118165

Singh DP, Prabha R, Gupta VK, Verma MK (2018) Metatranscriptome analysis Deciphers multifunctional genes and enzymes linked with the degradation of aromatic compounds and pesticides in the wheat rhizosphere. Front Microbiol. https://doi.org/10.3389/fmicb.2018.01331

Singh V, Verma M, Chivate MS, Mishra V (2023) Machine learning-based optimisation of microalgae biomass production by using wastewater. J Environ Chem Eng 11:111387. https://doi.org/10.1016/J.JECE.2023.111387

Sizova I, Greiner A, Awasthi M, Kateriya S, Hegemann P (2013) Nuclear gene targeting in Chlamydomonas using engineered zinc-finger nucleases. Plant J 73(5):873–882. https://doi.org/10.1111/tpj.12066

Sobieh SS, Abed El-Gammal R, El-Kheir WSA, El-Sheimy AA, Said AA, El-Ayouty YM (2022) Heterologous expression of cyanobacterial Cyanase gene (CYN) in microalga Chlamydomonas reinhardtii for bioremediation of cyanide pollution. Biology (Basel). https://doi.org/10.3390/biology11101420

Song C, Wei Y, Qiu Y, Qi Y, Li Y, Kitamura Y (2019) Biodegradability and mechanism of florfenicol via Chlorella sp. UTEX1602 and L38: experimental study. Bioresour Technol 272: 529–534. https://doi.org/10.1016/J.BIORTECH.2018.10.080

Sreenikethanam A, Raj S, JRB, Gugulothu P, Bajhaiya AK (2022) Genetic engineering of microalgae for secondary metabolite production: recent developments, challenges, and future prospects. Front Bioeng Biotechnol. https://doi.org/10.3389/fbioe.2022.836056

Suárez-Martínez D, Angulo-Mercado E, Mercado-Martínez I, Vacca-Jimeno V, Tapia-Larios C, Cubillán N (2022) Enhanced tetracycline removal from highly concentrated aqueous media by lipid-free *Chlorella* sp. Biomass ACS Omega 7(14128–14137):14137. https://doi.org/10.1021/acsomega.2c00696

Subashchandrabose SR, Ramakrishnan B, Megharaj M, Venkateswarlu K, Naidu R (2013) Mixotrophic cyanobacteria and microalgae as distinctive biological agents for organic pollutant degradation. Environ Int 51:59–72. https://doi.org/10.1016/j.envint.2012.10.007

Subashchandrabose SR, Logeshwaran P, Venkateswarlu K, Naidu R, Megharaj M (2017a) Pyrene degradation by Chlorella sp. MM3 in liquid medium and soil slurry: possible role of dihydrolipoamide acetyltransferase in pyrene biodegradation. Algal Res 23:223–232. https://doi.org/10.1016/J.ALGAL.2017.02.010

Subashchandrabose SR, Wang L, Venkateswarlu K, Naidu R, Megharaj M (2017b) Interactive effects of PAHs and heavy metal mixtures on oxidative stress in Chlorella sp. MM3 as determined by artificial neural network and genetic algorithm. Algal Res 21:203–212. https://doi.org/10.1016/J.ALGAL.2016.11.018

Sutherland DL, Ralph PJ (2019) Microalgal bioremediation of emerging contaminants - opportunities and challenges. Water Res 164:114921. https://doi.org/10.1016/J.WATRES.2019.114921

Tang W, Xu X, Ye B-C, Cao P, Ali A (2019) Decolorization and degradation analysis of Disperse Red 3B by a consortium of the fungus *Aspergillus* sp. XJ-2 and the microalgae *Chlorella sorokiniana* XJK. RSC Adv 9:14558–14566. https://doi.org/10.1039/C9RA01169B

Tarbajova V, Kolackova M, Chaloupsky P et al (2023) Physiological and transcriptome profiling of Chlorella sorokiniana: a study on azo dye wastewater decolorization. J Hazard Mater 460: 132450. https://doi.org/10.1016/J.JHAZMAT.2023.132450

Teng SY, Yew GY, Sukačová K, Show PL, Máša V, Chang JS (2020) Microalgae with artificial intelligence: a digitalized perspective on genetics, systems and products. Biotechnol Adv 44: 107631. https://doi.org/10.1016/J.BIOTECHADV.2020.107631

Tomar RS, Jajoo A (2021) Enzymatic pathway involved in the degradation of fluoranthene by microalgae Chlorella vulgaris. Ecotoxicology 30:268–276. https://doi.org/10.1007/s10646-020-02334-w

Touliabah HE-S, El-Sheekh MM, Ismail MM, El-Kassas H (2022) A review of microalgae- and cyanobacteria-based biodegradation of organic pollutants. Molecules 27:1141. https://doi.org/10.3390/molecules27031141

Vasudevan R, Gale GAR, Schiavon AA et al (2019) CyanoGate: a modular cloning suite for engineering cyanobacteria based on the plant MoClo syntax. Plant Physiol 180:39–55. https://doi.org/10.1104/pp.18.01401

Wan L, Wu Y, Ding H, Zhang W (2020) Toxicity, biodegradation, and metabolic fate of organophosphorus pesticide Trichlorfon on the freshwater algae *Chlamydomonas reinhardtii*. J Agric Food Chem 68:1645–1653. https://doi.org/10.1021/acs.jafc.9b05765

Wang L, Xiao H, He N, Sun D, Duan S (2019) Biosorption and biodegradation of the environmental hormone nonylphenol by four marine microalgae. Sci Rep 9:5277. https://doi.org/10.1038/s41598-019-41808-8

Wang Y, Ning W, Li S, Gao C, Cui R, Guo W, Chang JS, Ho SH (2023) Metabonomics analysis of microalga Scenedesmus obliquus under ciprofloxacin stress. Environ Res 237:116974. https://doi.org/10.1016/J.ENVRES.2023.116974

Webster LJ, Villa-Gomez D, Brown R, Clarke W, Schenk PM (2024) A synthetic biology approach for the treatment of pollutants with microalgae. Front Bioeng Biotechnol. https://doi.org/10.3389/fbioe.2024.1379301

Wlodkowic D, Karpiński TM (2021) Live-cell systems in real-time biomonitoring of water pollution: practical considerations and future perspectives. Sensors 21:7028. https://doi.org/10.3390/s21217028

Wong LS, Lee YH, Surif S (2013) Whole cell biosensor using *Anabaena torulosa* with optical transduction for environmental toxicity evaluation. J Sens 2013:1–8. https://doi.org/10.1155/2013/567272

Wu Q, Wang X, Zhou Q (2014) Biomonitoring persistent organic pollutants in the atmosphere with mosses: performance and application. Environ Int 66:28–37. https://doi.org/10.1016/J.ENVINT.2013.12.021

Xie P, Chen C, Zhang C, Su G, Ren N, Ho SH (2020) Revealing the role of adsorption in ciprofloxacin and sulfadiazine elimination routes in microalgae. Water Res 172:115475. https://doi.org/10.1016/J.WATRES.2020.115475

Yang M, Du D, Zhu F, Wang X (2024) Metabolomic analysis reveals the toxicity mechanisms of bisphenol A on the Microcystis aeruginosa under different phosphorus levels. Environ Pollut 342:123022. https://doi.org/10.1016/J.ENVPOL.2023.123022

Zhang Y, He D, Bu Z, Li Y, Guo J, Li Q (2022) The transcriptomic analysis revealed sulfamethoxazole stress at environmentally relevant concentration on the mechanisms of toxicity of cyanobacteria Synechococcus sp. J Environ Chem Eng 10:107637. https://doi.org/10.1016/J.JECE.2022.107637

Zhang Z, Chen Q, Chen B, Dong T, Chen M (2023) Toxic effects of pesticides on the marine microalga Skeletonema costatum and their biological degradation. Sci China Earth Sci 66:663–674. https://doi.org/10.1007/s11430-022-1064-7

Zhou Q, Zhang J, Fu J, Shi J, Jiang G (2008) Biomonitoring: an appealing tool for assessment of metal pollution in the aquatic ecosystem. Anal Chim Acta 606:135–150. https://doi.org/10.1016/J.ACA.2007.11.018

Zhou H, Chen CM, Diao X (2020) The metabolic responses of aquatic animal exposed to POPs. Environmental metabolomics: applications in field and laboratory studies to understand from exposome to metabolome, pp 121–161. https://doi.org/10.1016/B978-0-12-818196-6.00005-4

Zhou J-L, Yang L, Huang K-X, Chen D-Z, Gao F (2022) Mechanisms and application of microalgae on removing emerging contaminants from wastewater: a review. Bioresour Technol 364:128049. https://doi.org/10.1016/j.biortech.2022.128049

Zhu Y (2022) Advances in CRISPR/Cas9. Biomed Res Int 2022:9978571. https://doi.org/10.1155/2022/9978571

Harnessing Macroalgae for Environmental Remediation

15

Meera Bai Santhakumari, Gopika Ajit, Chinchu Elezebeth, Aanandhalakshmi Raghuraman, Prathyusha Bogguru, and Vanavil Balakrishnan

Abstract

Macroalgae, or seaweeds, are increasingly recognized as valuable resources in the field of environmental remediation due to their high productivity, adaptability, and rich biochemical composition. Native macroalgae play a direct role in remediating aquatic ecosystems through mechanisms such as biosorption, bioaccumulation, and nutrient uptake. Additionally, processed macroalgal products such as alginates, carrageenans, biochar, and polysaccharide-based materials exhibit enhanced capabilities for removing heavy metals, dyes, pharmaceuticals, and excess nutrients from contaminated water and soils. This chapter reviews recent advances in practical applications and the dual potential of macroalgae and their derived products in wastewater treatment, soil remediation, and integrated bioremediation systems to remove pollutants. Environmental benefits, scalability, economic feasibility, and sustainability aspects will also be discussed, positioning macroalgae as promising agents and material sources for next-generation green remediation technologies.

M. B. Santhakumari
Department of Biotechnology, Kalasalingam Academy of Research and Education, Krishnankoil, Tamil Nadu, India

Department of Biotechnology & Biochemical Engineering, Sree Buddha College of Engineering, Alappuzha, Kerala, India

G. Ajit · C. Elezebeth
Department of Biotechnology & Biochemical Engineering, Sree Buddha College of Engineering, Alappuzha, Kerala, India

A. Raghuraman · P. Bogguru · V. Balakrishnan (✉)
Department of Biotechnology, Kalasalingam Academy of Research and Education, Krishnankoil, Tamil Nadu, India
e-mail: b.vanavil@klu.ac.in

K. Mohanty et al. (eds.), *Advances in Algal Research*,
https://doi.org/10.1007/978-981-95-7234-2_15

Keywords

Seaweeds · Marine macroalgae · Environmental remediation · Biosorption · Bioenergy · Wastewater treatment

15.1 Introduction

Seaweed, or macroalgae, are macrophytic aquatic organisms commonly found in the intertidal zone, where they anchor to solid substrates like rocks and pebbles along the shoreline (Sharma et al. 2025). Unlike terrestrial crops, seaweed exhibits rapid growth and does not require arable land or freshwater for cultivation, making it a highly sustainable resource (Ozogul et al. 2024; Matos et al. 2021). Rich in essential minerals and nutrients, seaweed extracts are widely utilized as biofertilizers and soil conditioners, contributing to improved soil health and enhanced crop productivity (Akbar et al. 2025; Holdt and Kraan 2011). Seaweed plays a pivotal role in the blue bioeconomy, driving sustainability and innovation across diverse sectors. However, nearly 30% of the overall production cost is attributed to processing and handling (Sharma et al. 2025; Matos et al. 2021). In recent years, seaweed farming has gained attention for its environmental benefits, including carbon sequestration, nutrient uptake, and the restoration of marine habitats, contributing to healthier ecosystems and mitigating the impacts of climate change.

With growing interest in sustainable alternatives to conventional packaging, seaweed-based materials have emerged as eco-friendly options (Chung et al. 2011; Aitken et al. 2014). Furthermore, seaweed holds promise as a bioresource for the production of biofuels and bioplastics, aligning with global efforts toward renewable energy and sustainable material development. Its applications also extend to agriculture as a plant biostimulant and to animal husbandry as a nutritious feed supplement (Matos et al. 2021; Trif et al. 2024). This chapter highlights the environmental remediation potential of seaweeds. Macroalgae are well-suited for phytoremediation, especially of heavy metal pollution, due to their ability to bioaccumulate metals through their polysaccharide-rich cell walls. Functional groups like hydroxyl, carboxyl, and sulfate in alginates, carrageenans, and agar enable effective metal ion binding and pollutant removal from contaminated water (Akbar et al. 2025).

Marine biorefineries have the potential to significantly enhance the sustainability of ocean-based industries and support the transition to a blue economy (Akbar et al. 2025). However, achieving this requires coordinated efforts across economic and social sectors to ensure ocean health. Given the increasing challenges of resource scarcity and climate change, governments and industries should actively support the growth of marine biorefineries, which serve as vital components of green and blue economy strategies (Veríssimo et al. 2021).

15.2 Role of Native Macroalgae in the Remediation of Aquatic Ecosystems

Pollution from humans and industry harmed aquatic environments. Sustainable, green alternatives take over as conventional approaches fail (Prasher et al. 2021). Extensive research shows that macroalgae can significantly aid in restoring aquatic ecosystems by producing oxygen and removing toxic contaminants through biosorption and bioaccumulation. They also help prevent eutrophication by absorbing excess nutrients from the water (Henriques et al. 2015). Biosorption requires no cellular metabolism and uses nonliving or pre-treated algal biomass to clean wastewater cheaply. Light, temperature, pH, and nutrition levels affect bioaccumulation, allowing trace metals and minerals to be slowly absorbed. Macroalgae may absorb toxins and improve oxygenation and nutritional balance, making them a possible marine pollution solution. The complete potential of macroalgae in ecosystem restoration and environmental sustainability may be further unlocked by future research and development into genetically modified strains, integrated wastewater treatment systems, and enhanced production techniques (Bilal et al. 2018).

15.3 Algal Biomass in Environmental Remediation

Recent studies showed that algal biomass can remediate water and soil while producing usable biomass. Unlike costly conventional treatments that may cause secondary pollution, macroalgae both remove contaminants and yield bioproducts such as fuels and fertilizers (Gurau et al. 2025). Several studies have investigated different types of algae for their ability to remove various pollutants (Bădescu et al. 2017). Macroalgae can either absorb or capture pollutants via biosorption or bioaccumulation, thus diminishing toxicity. Studies show that some algae can degrade synthetic dyes and pharmaceuticals, and their conversion to biochar can enhance soil fertility and support sustainable farming, making macroalgae a promising natural option for further study (Omar et al. 2018). A study on *Nizamuddinia zanardinii* (brown algae) found HCl to be the most effective eluent for methylene blue desorption, achieving 64% efficiency (Daneshvar et al. 2017). Alkaline-modified *Scenedesmus obliquus* removed 91% of tramadol in 45 min via chemisorption, with adsorption capacities of 42 mg/g (tramadol), 68 mg/g (cephalexin), and 58 mg/g (paracetamol) (Zhou et al. 2014). Periphyton nutrient removal systems (PNRS) use algae to quickly absorb nitrogen and phosphorus, producing biomass for food, fertilizer, or biofuel. They are low-cost, fast to build, require about half the land of wetlands for similar nitrogen removal, work in varied ecosystems, and need further optimization (Sutherland and Craggs 2017). Adding increasing amounts of *Ulva* sp. and *Gelidium* sp. to contaminated soil reduced heavy metal uptake in lettuce, indicating potential for broader plant growth improvement in polluted soils (Abd-Elhady 2015). The biosorption of basic yellow dye onto the green macroalgae *Caulerpa scalpelliformis* through dynamic batch experiments recorded a maximum uptake of 27 mg/g, with external mass transfer identified as the rate-limiting step and kinetics conforming to a pseudo-second-order model. Equilibrium data followed the Freundlich isotherm, indicating the presence of heterogeneous adsorption sites (Aravindhan et al. 2007). Integrated multi-trophic aquaculture was used to study *Ulva*'s uptake of the antibiotic enrofloxacin (ENR). At 7.5 μg/L and 15 μg/L, uptake reached 7.76 ng/g WW and 14.51 ng/g WW within 30 and 15 min, respectively (Khan et al. 2025). A macroalgae–chitosan–pectin (Ma–Ch–Pe) biocomposite (75.4% macroalgae, 19.8% chitosan, and 4.8% pectin) was used to remove low-molecular soluble aromatic hydrocarbons from water. It achieved removal capacities of 16.64 mg/g for toluene, 6.13 mg/g for benzene, and 58.68 mg/g for naphthalene (Flores-Chaparro et al. 2018).

Gracilaria birdiae grew in shrimp pond effluents, achieving an average relative growth rate of 2.6% d^{-1} and reducing PO_4^{3-} by 93.5%, NH_4^+ by 34%, and NO_3^- by 100%. The species proved effective as a biofilter (Sharma et al. 2025). *Ulva rotundata*, *Enteromorpha intestinalis*, and *Gracilaria gracilis* were tested as biofilters for phosphate removal from aquaculture effluents. *U. rotundata* showed the highest uptake (2.86 μmol PO_4 g^{-1} h^{-1}), and all species removed over 60% at low flow rates, improving effluent quality (Hernández et al. 2002). Table 15.1 shows the algal species used for the removal of different pollutants and their efficiency.

The efficacy of pollution removal varies among macroalgal species, mostly because of differences in their ecological adaptations, surface functional groups, and polysaccharide composition. *Fucus vesiculosus* is one of the brown macroalgae that are very good at binding hazardous metals like lead and mercury, with removal

Table 15.1 Algal species used for the removal of different pollutants and their efficiency

Algal species	Target pollutants	Removal efficiency/ accumulation potential	References
Fucus vesiculosus	Hg, Pb, Cd	Hg: 96–99% Pb: up to 86%, Cd: <20%	Henriques et al. (2015)
Gracilaria verrucosa	Cr(Cr (VI),2,4-D Herbicide	Cr(VI): 113.2 mg/g 2,4-D: 22.3 mg/g	Ata et al. (2012)
Fucus spiralis	Gd	Gd: 85%;	Ferreira et al. (2020)
Gracilaria birdiae	PO_4^{-3}, NH_4^+, NO^{3-}	PO_4^{-3}: 93.5%, NH_4^+: 34%, NO^{3-}: 100%	Marinho-Soriano et al. (2009)
Ulva lactuca	Hg, Cd, Pb	Hg:93–99%, Cd: 57–96% Pd: 34–97%	Henriques et al. (2017)
Fucus spiralis	Hg, Pb	Hg: 90–99%, Pb: 80%	Costa et al. (2020)
Gracilaria sp.	Al, Zn, Cr	Al: 10–72%, Zn: 36.5–91.7% Cr: 52.5–83.4%	Kang and Sui (2010)
Ulva rotundata	NH_4^+	61%.	Hernández et al. (2002)
Ascophyllum nodosum	Cu (II)	Cu (II): >99%	Katiyar et al. (2021)
Sargassum fusiforme	Zn	86%	Poo et al. (2018)
Enteromorpha linza	Ni (II)	90%	Yalçın (2014)

efficiencies of up to 86% and 99%, respectively (Henriques et al. 2015). A similar high affinity for zinc has been shown by *Sargassum fusiforme*, which has an 86% removal efficiency (Poo et al. 2018). Based on these findings, fucoidan and alginate are useful candidates for heavy metal remediation because they improve the ion-exchange and chelation processes in brown algae.

Gracilaria verrucosa and *Gracilaria* sp. are examples of red macroalgae that are noteworthy for their adaptability in absorbing pollutants. Adsorption capabilities of 113.2 mg/g for Cr(VI) and 22.3 mg/g for herbicide 2,4-D were demonstrated by *Gracilaria verrucosa*, whereas *Gracilaria* sp. demonstrated significant removal of Al, Zn, and Cr with efficiencies ranging from 36.5 to 91.7% (Ata et al. 2012; Kang and Sui 2010). These results expand the potential uses of red algae for wastewater treatment by indicating that they are efficient at removing organic pollutants in addition to heavy metals. While *Ulva rotundata* had the highest phosphate uptake rates (2.86 μmol PO_4 g^{-1} h^{-1}) among tested species, green macroalgae, especially *Ulva lactuca*, are known for their broad-spectrum uptake, removing Hg, Cd, and Pb at efficiencies above 90%. This highlights their significance in nutrient-rich aquaculture effluents (Hernández et al. 2002; Henriques et al. 2017).

Some of the species that were analyzed have shown remarkable performance in applied remediation systems. In addition to surviving shrimp pond effluents, *Gracilaria birdiae* enhanced water quality by lowering phosphate, ammonium, and nitrate by 93.5%, 34%, and 100%, respectively, while sustaining a 2.6% d^{-1} relative growth rate (Sharma et al. 2025; Marinho-Soriano et al. 2009). Likewise, species of *Ulva* sp. exhibit excellent nutrient intake and rapid growth, which makes them perfect for integrated multi-trophic aquaculture systems (Hernández et al. 2002). While all macroalgae have the potential to be used for remediation, species like *Ulva* sp. and *Gracilaria birdiae* stand out as effective, scalable choices for practical environmental remediation applications.

15.4 Macroalgal Products in Environmental Remediation

Macroalgal-derived compounds contain valuable constituents like carbohydrates, proteins, minerals, phytochemicals, and pigments such as phycobiliproteins, chlorophylls, and carotenoids (Ortega et al. 2019). Macroalgal polysaccharides such as alginate, carrageenan, laminarin, ulvan, and macroalgae-derived biochar have emerged as versatile materials for environmental remediation applications due to their unique physicochemical properties and sustainability. Figure 15.1 shows the environmental pathway of macroalgae from harvest to sustainable products.

Alginates from *Chlorella vulgaris*, *Phormidium* sp., and *Prototheca zopfii* are widely used in wastewater treatment as beads or sheets in reactors. They help remove nitrogen, phosphorus, heavy metals, and toxins, lowering the organic load. Alginates also serve as biosensors and act as coagulants, especially with calcium or sodium ions, forming gels that precipitate pollutants (Yarkent et al. 2022). The fabrication of beads and hydrogels, utilizing alginate, increases surface area and enhances

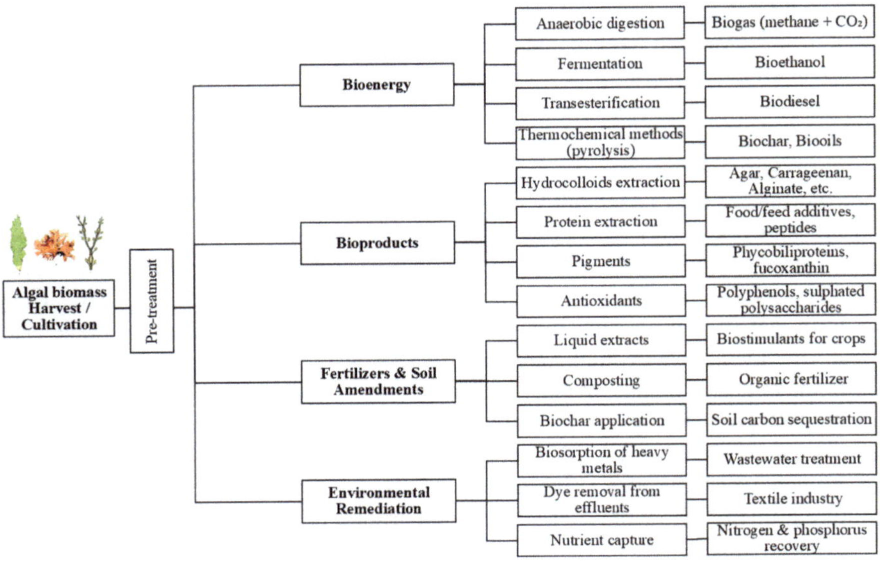

Fig. 15.1 Environmental pathway of macroalgae from harvest to sustainable products

adsorption kinetics for wastewater treatment (Chen et al. 2020). The ionotropic gelation method, utilizing alginate with a high guluronic acid content, has been used to evaluate the continuous removal of lead, copper, nickel, cadmium, and zinc ions from aqueous solutions (Demey et al. 2018). Alginate containing abundant carboxyl groups enables effective chelation and ion exchange with toxic metal ions such as Pb^{2+}, Cd^{2+}, and Hg^{2+}, promoting strong biosorption in aqueous environments (Osman et al. 2022). Moreover, immobilization of pollutant-degrading bacteria in alginate matrices enhances biodegradation of organic contaminants (Xiang et al. 2022).

Agar from red algae is valued in food and biotechnology for its gelling and stabilizing properties. Its porous gel supports adsorbents and microbes, aiding in heavy metal and dye removal. Chemical modifications enhance its binding ability. Likewise, alginate and cellulose, rich in carboxyl and hydroxyl groups, effectively adsorb pollutants. Modified cellulose removes Cr (VI), while alginate forms pollutant-absorbing beads and hydrogels. These biodegradable polysaccharides also support bioremediation by trapping microbes or nanoparticles (Pal et al. 2014). Fucoidan, a sulfated polysaccharide from brown algae, helps sequester carbon and pollutants. It also contributes to oceanic dissolved organic carbon, aiding long-term carbon storage and climate regulation beyond its role as a biosorbent (Chudasama et al. 2021). Fucoidan's anionic structure enables strong binding of heavy metals, dyes, and other pollutants, making it effective for wastewater treatment. Its antioxidant and anti-inflammatory properties support bioremediation, while combining it with biochar or other algal polysaccharides enhances adsorption and mechanical strength for multifunctional remediation (Martins et al. 2023).

Ulvan, another sulfated polysaccharide from *Ulva* species, contains sulfate esters, rhamnose, glucuronic acid, and iduronic acids, which provide an anionic nature that is favourable for chelating a variety of metal ions and radionuclides. Its gelling, emulsifying, and film-forming abilities make it suitable for food coatings, edible films, biotherapeutics, cancer support, and health supplements (Pradhan et al. 2023). *Ulvan's* complex structure enables hydrogen bonding and electrostatic adsorption, effectively removing heavy metals, radionuclides, and organic wastes. Ulvan-based hydrogels and membranes combine biodegradability, strong pollutant sequestration, and stable cross-linked networks for advanced water treatment (Roy et al. 2021). Chemical modifications and gamma irradiation improve laminarin's binding capacity, solubility, and adsorption ability, making it ideal for composite biosorbents, solution-phase remediation, and supporting microbial activity in bioremediation (Bird et al. 2011).

Red algae such as *Hypnea, Kappaphycus*, and *Laurencia* are the primary source of carrageenan, modified carrageenan-based hydrogels, and composites have demonstrated efficient adsorption of a broad spectrum of organic pollutants, including dyes and pharmaceutical residues, offering hydrophilicity and tunability to target specific contaminants. Cross-linking and grafting strategies not only improve adsorption capacity but also enhance chemical resilience, which is essential for repeated remediation cycles (Jha et al. 2023). Its performance in wastewater treatment is enhanced by its modified forms, such as hydrogels, magnetic composites, and nanoparticles. It enhances flocculation by forming aggregates, reducing turbidity. Sulfate groups of the carrageenan bind heavy metals and support microbial biofilms, enhancing water treatment. As a bio-adsorbent, it removes dye contaminants via electrostatic and hydrophobic interactions, improves sludge dewaterability, and reduces membrane fouling (Jabeen et al. 2025).

Preparing hydrogel from algal polysaccharides typically entails cross-linking after the polymer has been dissolved in water at concentrations ranging from 1% to 4% w/v. In order to create Ca-alginate beads with high mechanical stability and adsorption effectiveness, sodium alginate solutions 2–3% w/v are commonly gelled by dropwise addition into 2–5% (w/v) calcium chloride (Chen et al. 2020; Demey et al. 2018). Similar to this, 2% w/v agar is heated in water until completely dissolved, and then it is cooled to create a porous gel that can be used for microbial immobilization or pollutant adsorption (Pal et al. 2014). Solution casting is frequently used to create *Ulvan* hydrogels at a concentration of 1%–2% w/v, and cross-linking agents such as glutaraldehyde or divalent cations are used to enhance their mechanical strength and pollutant binding (Roy et al. 2021).

Formulations for biocomposite materials combine polysaccharides or macroalgal biomass with additional polymers. A well-researched example is the macroalgae–chitosan–pectin (Ma–Ch–Pe) biocomposite, which has removal capabilities of 16.64 mg/g for toluene and 58.68 mg/g for naphthalene (Flores-Chaparro et al. 2018). It is composed of 75.4% macroalgae, 19.8% chitosan, and 4.8% pectin. Typically, 1–2% carrageenan is combined with reinforcing polymers like chitosan or polyvinyl alcohol (PVA) to create modified carrageenan-based hydrogels. To improve gel stability and adsorption effectiveness, cross-linking agents like KCl or $CaCl_2$ are added at a rate of 1% to 5% (w/v) (Jha et al. 2023; Jabeen et al. 2025).

15.5 Algal-Based Hydrochars and Biochars for Pollutant Removal

Algal residues are increasingly converted into nitrogen- and phosphorus-rich biochar via pyrolysis, offering high surface area, porosity, and functional groups (Chen et al. 2015). Algal biochar provides a sustainable, multipurpose solution for environmental remediation, removing pollutants while boosting soil structure, nutrient retention, and microbial activity (Wang et al. 2024). Due to its high content of nitrogen, phosphorus, and other nutrients compared to many terrestrial biomass sources, biochar produced from algal feedstocks is assumed to possess attractive carbon sequestration and soil remediation properties (Bird et al. 2011). Marine macroalgae biochars had better heavy metal removal capacities than woody biochar (Poo et al. 2018). Biochar facilitates the degradation of polycyclic aromatic hydrocarbons (PAHs) and pesticides through adsorption processes (Beljin et al. 2024). Integration of biochar with microbes and plants further enhances remediation efficacy in complex soil systems by providing microbial habitat and promoting pollutant degradation (Chen et al. 2020).

Biochar derived from *Caulerpa scalpelliformis* algae effectively removed 81% of remazol dyes, with NaOH being the best elutant for desorption (Gokulan et al. 2019a). *Eucheuma cottonii* seaweed was pyrolyzed in a furnace and activated with sulfuric acid, producing biochar that efficiently adsorbed methylene blue from wastewater (Saeed et al. 2020). Biochar derived from green seaweed (*Ulva lactuca*) effectively removed over 77.5% of remazol dyes from complex wastewaters, with high decolorization rates and removal efficiency within 240 min (Gokulan et al. 2019b). Biochar derived from *Gracilaria rhodophyta* removed substantial amounts of silver and brilliant green dye from contaminated water (Naga Babu et al. 2023). One-pot co-carbonization with $FeCl_3$ and $ZnCl_2$ produced macroalgal biochars (399 m^2 g^{-1}, 6.14% N, high hydrophobicity) that removed up to 90 mg g^{-1} of polycyclic aromatic hydrocarbon via multilayer sorption and were readily regenerated by thermal desorption (Cheng et al. 2020). Biochar derived from *Eucheuma spinosum* and bioelectrochemical system with *Shewanella* effectively removed and biodegraded toxic azo dyes, offering a complete solution for removing noxious dyes (Gurav et al. 2021). Biochar derived from macroalgae residue at 800 °C showed a high adsorption capacity for malachite green, crystal violet, and congo red (Chen et al. 2018). Microwave-assisted pyrolysis of brown seaweed *Turbinaria conoides* biochar effectively removed remazol brilliant blue R (RBB) from wastewater, with NaOH performing better in releasing the dye (Vijayaraghavan and Ashokkumar 2019). *Ulva prolifera*-derived biochar effectively removed 85.7% of reactive red 120 dye molecules in aqueous solutions, with optimum results achieved at a pH of 2, sorbent dose of 2 g/L, and temperature of 35 °C (Mahendran et al. 2023). Biochar from *Keppaphycus alvarezii* macroalgae effectively removed 90.3% of eosin yellow dye from textile wastewater, with a pseudo-second order kinetics process (Elayappan et al. 2024).

Biochar hydrophilicity, surface charge, and reactivity are influenced by inorganic elements (K, Ca, Mg, N, and P), pH, and cation exchange capacity. These factors affect its use in soil enhancement, water purification, and pollutant adsorption (Beljin et al. 2024). Surface functional groups like (-OH, -COOH), sulfur (-SO_3H, -SH), and nitrogen (-NH_2, pyridinic N, pyrrolic N), are key to algal biochar's reactivity (Li et al. 2025; Tan et al. 2023). Biochar and its modified variants eliminate water contaminants via several physicochemical pathways like electrostatic attraction, pore filling, partitioning, hydrophobic forces, and π–π EDA interactions, with electrostatic forces playing the leading role for ionic species (Tan et al. 2023). It allows charged groups on the biochar surface to attract contaminants, depending on pH and ionic strength. Electrostatic forces have been proven by experiments on attapulgite biochar composites for norfloxacin and montmorillonite-modified biochar for phosphate removal (Murtaza et al. 2022). Pore-filling, where contaminants are physically absorbed into biochar, boosts adsorption through increased pore volume and surface area. It works best with low-volatile biochar and low pollutant levels. Pollutants enter biochar's amorphous zones, interacting via partitioning and hydrophobic effects, while graphene-like biochars mainly show π–π EDA interactions (Murtaza et al. 2022). Hydrogen bonding, van der Waals forces, and surface functional groups, including hydroxyl, carbonyl, carboxyl, and amine groups, enhance these adsorption methods (Beljin et al. 2024).

15.6 Algal-Based Resources for Carbon Capture and Bioenergy

Macroalgae play a significant role in organic carbon storage and carbon cycling. Their ability to capture inorganic carbon and convert it into biomass supports both carbon sequestration and CO_2 reduction. Integrating carbon capture and storage (CCS) with bioenergy production offers a promising strategy to lower CO_2 emissions while generating renewable energy. This combined approach holds strong potential for long-term carbon storage and plays a crucial role in climate change mitigation (Costa et al. 2020). Algae's photosynthetic ability and improved cultivation systems help reduce atmospheric CO_2 and support sustainable bioenergy production. They store inorganic carbon via a microbially induced calcium carbonate precipitation pathway. Algae convert CO_2 into biomass, which can be processed into biofuels and value-added products, aiding carbon capture and storage (Li and Yao 2024). Both macro- and microalgae produce low-emission biofuels (Zhao et al. 2022). Seaweed biomass plays a key role in sustainable fuel and byproduct production. Its ability to grow with minimal nutrients makes it highly suitable for carbon capture and bioenergy generation. Seaweed aquaculture beds contribute to CO_2 reduction, with the harvested biomass converted into biogas, as well as solid and liquid biofuels. For efficient ethanol production, enzymatic hydrolysis of polysaccharides is essential to maximize yield (Yanagisawa et al. 2013). Alkaline thermal treatment of brown seaweeds using NaOH and a Ni/ZrO_2 catalyst offers a sustainable method for bioenergy production and carbon capture. It converts energy into H_2 while capturing carbon as stable solid carbonates, enabling long-term carbon storage from seaweed biomass (Zhang et al. 2020). Optimizing carbon content in seaweeds like *Sargassum*, *Gracilaria*, and *Ulva* enhances carbon capture in marine agronomy. These seaweeds efficiently sequester carbon and can meet global energy needs through bioethanol and high-value chemical production, yielding up to 8 trillion liters of bioethanol (Johnston et al. 2023). The red macroalga *Sarcodia suae* captures carbon effectively in subtropical waters and removes excess nutrients from aquaculture wastewater (Weerakkody et al. 2023). The seaweed *Chaetomorpha* sp. was a potential raw material for biodiesel production through the transesterification process. The yield of biodiesel was 32.83 ml/100 g dry biomass, and it was also eco-friendly and promoted both environmental sustainability and energy security (Nayak and Behera 2025). The hydrolysed biomass of *Ulva lactuca* has a high concentration of fermentable sugars. By the cascading approach, this biomass was used as a substrate for the production of chemical 1,2-propanediol and biofuels like acetone, butanol, and ethanol (Bikker et al. 2016). The spent biomass obtained from the *Gracilaria corticate* after the extraction of phycobiliproteins was an alternative substrate for the production of bioethanol. Acid pretreatment followed by fermentation using baker's or MTCC yeast yielded up to 4.85% w/w ethanol (Sudhakar et al. 2020). The polysaccharides obtained from the enzyme hydrolysis of seaweeds such as *Gracilaria verrucosa* and *Saccharina japonica* were used for the sustainable feedstocks for the production of bioethanol along with *Saccharomyces cerevisiae, Clostridium,* and *Trichoderma* as the fermentation microbes (Immanuel Suresh and Divyeswari 2024). The biodiesel production was 32.3 mg g^{-1} dw from the seaweed *Ulva intestinalis.* This can be used as an alternative feedstock for the bioenergy recovery through sequential biodiesel and bioethanol production (Osman et al. 2020). Using biomasses of *Dilsea carnosa, Ulva Lactuca,* and *Laminaria digitata***,** the ethanol yield of 5.4 g L^{-1}, 7.8 g L^{-1}, and 3.2 g L^{-1}, respectively, was obtained (Kostas et al. 2019). The seaweed biomass of *Gelidium pusillum* produced an ethanol concentration of 5.7 ± 0.19 g/L using microwave-alkali treatment (Anbuselvan et al. 2024). The seaweed hydrolysate was used in the microbial fuel cells as a potential feedstock for the production of bioelectricity. This seaweed, along with the anode electrodes Fe_2O_3@AuNPs/PANI nanocomposites, showed enhanced power density, current density, and voltage of 222.78 mW/cm^3, 269 mA/cm^2, and 944 mV, respectively (Alalawy et al. 2024). The thermochemical characterization of seaweeds revealed that seaweeds had high volatile matter, carbon fixation rates and low ash content. So, seaweeds can be used as an excellent resource for biofuel production (Cassani et al. 2022).

15.7 Seaweed Extract as Fertilizers

Seaweed extracts from *Polysiphonia, Ulva, and Cladophora* have the potential to be employed as nontoxic biostimulants for plant growth (Godlewska et al. 2016). These seaweed species help to attain the germination of plants by providing essential nutrients. Seaweed extracts improve plant production, stress response, and plant growth using the extracts as foliar and soil-based applications in modern agricultural practices rather than synthetic fertilizers or any other insecticides. Seaweed-based stimulants have been shown to enhance crop yield, protein content, and overall nutritional value, particularly in marginal lands affected by drought and salinity (Deolu-Ajayi et al. 2022). Zinc-enriched seaweed biomass enhances the nutrient content of *Lepidium sativum* more effectively than inorganic salts or chelates. Its zinc biosorption ability offers a natural alternative to chemical fertilizers (Tuhy et al. 2015). Zinc-enriched seaweed biomass boosts *Lepidium sativum* nutrition and serves as an eco-friendly alternative to chemical fertilizers (Prasedya et al. 2022). Seaweed fertilizer (SF) boosts the expression of functional genes and protein domains involved in plant metabolism. A 1:1 mix of seaweed and pruning waste enhances grape yield while supporting soil fertility and vine nutrient needs. Additionally, SF increases soil nitrogen and carbon content, promotes biological crust development, and supports the growth of *Psammophyte* species (Yuan et al. 2019). Using seaweed fertilizer increased crop yield by 15.17% and boosted the sugar-acid ratio by 38.32%, enhancing crop sugar content (Pei et al. 2024). Tomato plants treated with compost from fish waste and drift seaweed yielded 5.56 kg/plant, outperforming control and mineral treatments. This compost serves as an effective organic fertilizer for boosting horticultural crop yields (Illera-Vives et al. 2015). Fermented seaweed fertilizer from *Ascophyllum nodosum* enhanced maize growth, while combining seaweed extracts with palm or rice leaf residues (6 mL/L) improved broccoli yield, growth, and quality (Manea and Abbas 2018). Seaweed fertilizer reduced replanting stress in *Malus hupehensis* seedlings. Brown seaweeds like *Sargassum crassifolium*, *S. polycystum*, and *S. cristaefolium* improve soil fertility and microbial activity, supporting green agriculture. Fertilizer from *Ulva lactuca* boosted protein content, growth, and photosynthetic pigments in *Spirulina platensis* (Prasedya et al. 2022). Foliar extracts from *Ascophyllum nodosum* increased berry and vine yields in Merlot grapes. *Sargassum* sp. was effective as a soil amendment for tomatoes in alkaline soils, while liquid fertilizer from *Gracilaria* waste boosted growth in *Capsicum frutescens* L (Pudoli et al. 2024). Seaweed-based conditioners like *Sargassum* sp. improved chickpea biomass and growth, while *Gelidium sesquipedale* residues enhanced corn leaf and strawberry fruit quality. Biofertilizers from *Sargassum muticum* and *Ascophyllum nodosum* boosted growth, yield, and germination in *Lactuca sativa* and *Oryza sativa*. In stressed maize, *Kappaphycus alvarezii* extract improved yield and soil health (Levarity and Gustave 2024).

15.8 Applications in Pollution Detection and Control

Since the 1950s, macroalgae have been used to detect environmental pollution due to their ability to accumulate heavy metals and dyes (García-Seoane et al. 2018). Their wide distribution and high metal accumulation make them effective bioindicators for monitoring metal pollution (Chalkley et al. 2019). Macroalgae act as natural buffers, helping reduce metal pollution in coastal waters and serving as reliable bioindicators, as their metal content often reflects seawater conditions. Variations in their pigments, enzymes, and morphology indicate pollutant bioavailability, highlighting their potential in biosorption applications (Conti and Cecchetti 2001). Table 15.2 shows the different algae used for metal detection.

Table 15.2 Algae used for metal detection

Macroalgae	Application in metal detection	References
C. glomerata	Cu and Pb	Akbar et al. (2025)
Fucus vesiculosus	Trace metal	Chalkley et al. (2019)
Ascophyllum nodosum	Cu	Chalkley et al. (2019)
Ulva sp.	Cu	Chakraborty et al. (2014)
Rhodophyceae, Phaeophyceae, Chlorophyceae	Fe, Mn, Cu, Zn, and Pb	Chakraborty et al. (2014)
Enteromorpha sp.	Hg, Pb, and Zn	Akbar et al. (2025)
Cystoseira sp.	Cd, Cr, and Fe	Chalkley et al. (2019)

Table 15.3 Algae in biosensor applications

Algae	Biosensor application detection	References
Gracilaria verruca	Phenoxyalkanoic acid herbicide 2,4-D and Cr	Ata et al. (2012)
Fucus serratus and Fucus versiculosus	Water quality monitoring	Moro et al. (2018)
Cladophoropsis membranacea	Organophosphate insecticides	Nunes et al. (2014)
Chlamydomonas reinhardtii	Atrazine herbicide	Czechowski et al. (1979)

Algal biosensors are effective, low-cost tools for detecting pesticides and metal ions, using macroalgae as a support for sensitive pollutant monitoring (Akbar et al. 2025; Nunes et al. 2014). The cell wall structure of algae exhibits an affinity toward metal ions, making it a promising source for a sustainable bioreceptor in biosensor development (Amaral et al. 2024). The functional groups present in biopolymers derived from macroalgae enable metal ion binding through adsorption, ion exchange, or covalent binding (Kalyani et al. 2004). Even though the challenge is to attain high sensitivity, specificity, stability, and selectivity in the electrode response. Algal-based electrochemical biosensors, fluorescent biosensors, and optical biosensors are promising areas of research (Moro et al. 2018; Thenarasu et al. 2024) . Table 15.3 shows the applications of algae as biosensors for the detection of various compounds.

15.9 Macroalgae-Based Green Materials: Applications in Packaging and Construction

Plastic pollution and greenhouse gas emissions from construction have driven the pursuit of sustainable materials (Chung et al. 2011). Alginate, carrageenan, and agar from macroalgae are utilized for the production of eco-friendly materials. Biopolymers having film-forming, gelling, and thickening capabilities are ideal for packaging and construction (Czechowski et al. 1979). For example, brown algae like *Laminaria digitata* are rich in alginate, which forms stable gels when cross-linked with calcium ions, can be processed into biodegradable films, commonly used in food packaging (Pereira 2018). Red algae such as *Kappaphycus alvarezii* provide carrageenan, known for its excellent film-forming and stabilizing properties. Agar derived from *Gracilaria* is widely used in microbiology and is now being explored in bioplastic development due to its transparency and strong gel structure.

The process of developing seaweed-based films often involves solvent casting, where the biopolymer is dissolved in water, mixed with plasticizers (like glycerol), and dried into thin, flexible films (Zinina et al. 2023). However, these films are often hydrophilic and lack tensile strength. To improve durability and barrier properties, they are frequently blended with nanoparticles (such as silver or clay), essential oils, or reinforced with other biopolymers like chitosan and PLA (Senturk Parreidt et al. 2018). Beyond packaging, macroalgae are used in green construction as natural binders or reinforcements in biocomposites for insulation, bricks, and panels. Countries like Denmark and Germany use seaweed mats for insulation, while calcified red algae like *Corallina officinalis* help make low-emission biobricks (Arfat et al. 2017). Alginate improves the strength of bio-based construction materials by forming cross-linked networks in lime or clay matrices, balancing flexibility and support. However, seaweed-based materials are hydrophilic and less mechanically strong than plastics unless reinforced (Senturk Parreidt et al. 2018). Scalability is limited by seasonal seaweed availability, high extraction costs, and insufficient processing infrastructure (Johnston et al. 2023). To address these challenges, researchers promote seaweed biorefineries that convert entire biomass into diverse products like food additives, fertilizers, biofuels, and construction materials, enhancing viability and reducing waste. Automation, cold storage, and local processing units can ease seasonal supply issues and cut costs. Public-private partnerships, consumer education, eco-labelling, and cross-disciplinary collaboration can further drive commercialization and product optimization. Expanding the seaweed industry supports green jobs and coastal economies, especially in developing nations, due to its low-cost, low-tech nature. Advances in synthetic biology now enable high-yield, genetically optimized macroalgae with tailored polymers for specific material uses. The future also lies in blending macroalgae biopolymers with other biodegradable materials to enhance strength, water resistance, and scalability. For the construction sector, ongoing research into hybrid composites, natural resins, and bio-cementation techniques will further enable seaweed's role in sustainable building.

15.10 Scalability and Economics, Sustainability

Sustainable, zero-waste strategies like cascade processing can boost the economic viability of seaweed industries while reducing environmental impact, supporting the United Nations Sustainable Development Goals. Advancing research on extraction, scalability, biomass availability, and environmental effects is key to realizing seaweed's full commercial potential (Matos et al. 2021). Compared to traditional terrestrial crops, macroalgae offer distinct advantages for biorefinery, which has

Table 15.4 Comparative overview of macroalgal remediation materials in terms of preparation inputs, efficiency, and cost implications

Material/ approach	Preparation method	Pollutant removal efficiency	Reusability potential	Economic consideration	References
Raw biomass (*Ulva*, *Gracilaria*)	Direct harvest, washing, drying, and grinding	50–90%	Limited—single use	Lowest upfront cost, but frequent replacement is needed	Abd-Elhady (2015). Flores-Chaparro et al. (2018)
Alginate hydrogel/ bead	2–3% alginate crosslinked with 2–5% $CaCl_2$	99%	3–5 treatment cycles	Moderate cost; more economical over repeated use	Chen et al. (2020), Demey et al. (2018)
Agar hydrogel	2% agar heated and cooled into gel form	60–80%	2–3 cycles	Affordable and simple to prepare	Pal et al. (2014)
Ulvan hydrogel/ membrane	1–2% ulvan with ionic or chemical crosslinking	70–95%	4–5 cycles	Higher preparation input, but stable and efficient	Roy et al. (2021)
Carrageenan composites	1–2% carrageenan blended with polymers (e.g., PVA and chitosan)	80–98%	More than 5 cycles	Higher initial cost, but excellent long-term value	Roy et al. (2021), Jabeen et al. (2025)

sparked growing interest in macroalgae biorefineries as a promising solution to the environmental and economic challenges of land-based biomass systems (Trif et al. 2024). To provide a clearer understanding of the trade-offs between cost, pollutant removal efficiency, and long-term reusability of different macroalgal remediation approaches, a comparative overview is presented in Table 15.4. This summary highlights the balance between initial preparation costs and the benefits of material stability and regeneration potential.

Seaweed farming is growing due to demand for biomass supply and to reduce pressure on wild stocks. Seaweed cultivation is practiced through various methods, including offshore, deep-sea, and integrated multi-trophic aquaculture (IMTA) systems. Offshore farming involves growing seaweed in open ocean areas beyond the coastal zone, nutrient-rich and less polluted waters, enhancing yield and quality. Deep-sea farming takes place in high-pressure and low-light zones (typically below 100 m). IMTA mimics natural ecosystems by co-cultivating species from different trophic levels. Nutrients from fed species like fish or shrimp support seaweed and algae growth, while filter feeders (e.g., mussels and oysters) clean the water, and detritivores (e.g., sea cucumbers and polychaetes) maintain seabed health (Sharma et al. 2025).

Seaweed farm monitoring and analysis tools, which include monitoring seaweed growth with drones and sensors for weather, water quality, and nutrient levels, were developed to optimize farming operations with real-time data. These tools can be integrated with artificial intelligence and machine learning to construct algorithms and models to analyze data, improve yields, detect diseases, and predict optimal harvest dates, then automate seaweed planting and harvesting systems to minimize labor costs and increase efficiency (Sharma et al. 2025).

Feedstock processing and handling account for about 30% of seaweed production costs, but this can be reduced through technological improvements. Optimizing processes like separation, pre-treatment, shredding, and storage can enhance yields and lower costs (Sharma et al. 2025). Since pricing influences consumer acceptance, reducing costs supports commercialization. Seaweed processing also generates various by-products, which can be promoted to improve profitability. Producing these by-products individually is costly, but integrated production offers a cost-effective alternative. Stand-alone facilities may not be viable without supportive policies and government initiatives to drive large-scale, practical seaweed-based production (Akbar et al. 2025; Veríssimo et al. 2021). Coastal regions have the potential to emerge as centers for macroalgae cultivation, generating employment in farming, harvesting, and processing. Additionally, improved water quality supports healthier fisheries, boosts tourism, and enhances overall ecosystem health-benefiting coastal communities both economically and ecologically (Sharma et al. 2025; Akbar et al. 2025). A key challenge lies in the technical difficulty of scaling up macroalgae cultivation for large-scale remediation. Moreover, the long-term ecological effects of extensive farming on natural ecosystems remain insufficiently understood. Assessing the economic viability of large-scale macroalgae operations is also crucial to their broader adoption (Sharma et al. 2025; Akbar et al. 2025).

15.11 Conclusion

Seaweed has enormous potential as a sustainable resource for a variety of environmental applications. In wastewater treatment, they are highly effective as they can remove nutrients like phosphorus and nitrogen, lower COD and BOD levels, and adsorb heavy metals. In addition, they provide a natural remedy for mitigating climate change through their innate ability to fix carbon, which greatly lowers atmospheric CO_2. In addition to their application in pollution control, macroalgae are employed as biofertilizers that improve soil productivity and health, and their sensitivity to environmental changes makes them useful as bioindicators in ecological monitoring. Macroalgae's ability to detect pollutants with great specificity and sensitivity is further demonstrated by their implementation into biosensors. Macroalgae are a perfect fit for integrated environmental management systems because of their quick development, ecological compatibility, and regenerative nature. Overall, cleaner technology and more sustainable ecosystems can be supported by utilizing the multipurpose qualities of macroalgae.

References

Abd-Elhady ES (2015) Evaluation of algae dry biomass as a biochemical soil remediation for polluted soil. Int J Environ 4(4):309–314

Aitken D, Bulboa C, Godoy-Faundez A, Turrion-Gomez JL, Antizar-Ladislao B (2014) Life cycle assessment of macroalgae cultivation and processing for biofuel production. J Clean Prod 75: 45–56. https://doi.org/10.1016/j.jclepro.2014.03.080

Akbar SA, Lestari AN, Fazli RR, Gunawan G (2025) Harnessing macroalgae for heavy metal phytoremediation: a sustainable approach to aquatic pollution control. In: BIO web of conferences 2025, vol 156. EDP Sciences, p 02013. https://doi.org/10.1051/bioconf/202515602013

Alalawy AI, Zidan NS, Sakran M, Hazazi AY, Salama ES, Alotaibi MA (2024) Enhancing bioelectricity generation in seaweed-derived microbial fuel cells using modified anodes with Fe2O3@ AuNPs/PANI nanocomposites. Biomass Bioenergy 182:107104. https://doi.org/10.1016/j.biombioe.2024.107104

Amaral DS, Tholozan LV, Bonemann DH, Jansen-Alves C, Boschetti W, Novo DL, Carreno NL, Pereira CM (2024) Algal biosensors for detection of potentially toxic pollutants and validation by advanced methods: a brief review. Chem 12(11):235. https://doi.org/10.3390/chemosensors12110235

Anbuselvan VA, Krishnamurthy V, Muthuvelu KS (2024) Comparative assessment of pre-treatment techniques and process economics on marine seaweed, *Gelidium pusillum*, for enhanced bioethanol production. Biofuels 15(10):1259–1269. https://doi.org/10.1080/17597269.2024.2358563

Aravindhan R, Rao JR, Nair BU (2007) Removal of basic yellow dye from aqueous solution by sorption on green alga *Caulerpa scalpelliformis*. J Hazard Mater 142(1–2):68–76. https://doi.org/10.1016/j.jhazmat.2006.07.058

Arfat YA, Ahmed J, Jacob H (2017) Preparation and characterization of agar-based nanocomposite films reinforced with bimetallic (Ag-Cu) alloy nanoparticles. Carbohydr Polym 155:382–390. https://doi.org/10.1016/j.carbpol.2016.08.097

Ata A, Nalcaci OO, Ovez B (2012) Macro algae *Gracilaria verrucosa* as a biosorbent: a study of sorption mechanisms. Algal Res 1(2):194–204. https://doi.org/10.1016/j.algal.2012.07.001

Bădescu IS, Bulgariu D, Bulgariu L (2017) Alternative utilization of algal biomass (*Ulva* sp.) loaded with Zn (II) ions for improving of soil quality. J Appl Phycol 29(2):1069–1079. https://doi.org/10.1007/s10811-016-0997-y

Beljin J, Đukanović N, Anojčić J, Simetić T, Apostolović T, Mutić S, Maletić S (2024) Biochar in the remediation of organic pollutants in water: a review of polycyclic aromatic hydrocarbon and pesticide removal. Nano 15(1):26. https://doi.org/10.3390/nano15010026

Bikker P, van Krimpen MM, van Wikselaar P, Houweling-Tan B, Scaccia N, van Hal JW, Huijgen WJ, Cone JW, López-Contreras AM (2016) Biorefinery of the green seaweed *Ulva lactuca* to produce animal feed, chemicals and biofuels. J Appl Phycol 28(6):3511–3525. https://doi.org/10.1007/s10811-016-0842-3

Bilal M, Rasheed T, Sosa-Hernández JE, Raza A, Nabeel F, Iqbal HM (2018) Biosorption: an interplay between marine algae and potentially toxic elements—a review. Mar Drugs 16(2):65. https://doi.org/10.3390/md16020065

Bird MI, Wurster CM, de Paula Silva PH, Bass AM, De Nys R (2011) Algal biochar–production and properties. Bioresour Technol 102(2):1886–1891. https://doi.org/10.1016/j.biortech.2010.07.106

Cassani L, Lourenço-Lopes C, Barral-Martinez M, Chamorro F, Garcia-Perez P, Simal-Gandara J, Prieto MA (2022) Thermochemical characterization of eight seaweed species and evaluation of their potential use as an alternative for biofuel production and source of bioactive compounds. Int J Mol Sci 23(4):2355. https://doi.org/10.3390/ijms23042355

Chakraborty S, Bhattacharya T, Singh G, Maity JP (2014) Benthic macroalgae as biological indicators of heavy metal pollution in the marine environments: a biomonitoring approach for pollution assessment. Ecotoxicol Environ Saf 100:61–68. https://doi.org/10.1016/j.ecoenv.2013.12.003

Chalkley R, Child F, Al-Thaqafi K, Dean AP, White KN, Pittman JK (2019) Macroalgae as spatial and temporal bioindicators of coastal metal pollution following remediation and diversion of acid mine drainage. Ecotoxicol Environ Saf 182:109458. https://doi.org/10.1016/j.ecoenv.2019.109458

Chen H, Zhou D, Luo G, Zhang S, Chen J (2015) Macroalgae for biofuels production: Progress and perspectives. Renew Sust Energ Rev 47:427–437. https://doi.org/10.1016/j.rser.2015.03.086

Chen YD, Lin YC, Ho SH, Zhou Y, Ren NQ (2018) Highly efficient adsorption of dyes by biochar derived from pigments-extracted macroalgae pyrolyzed at different temperature. Bioresour Technol 259:104–110. https://doi.org/10.1016/j.biortech.2018.02.094

Chen Y, Li J, Huang Z, Su G, Li X, Sun Z, Qin Y (2020) Impact of short-term application of seaweed fertilizer on bacterial diversity and community structure, soil nitrogen contents, and plant growth in maize rhizosphere soil. Folia Microbiol 65(3):591–603. https://doi.org/10.1007/s12223-019-00766-4

Cheng H, Ji R, Bian Y, Jiang X, Song Y (2020) From macroalgae to porous graphitized nitrogen-doped biochars–using aquatic biota to treat polycyclic aromatic hydrocarbons-contaminated water. Bioresour Technol 303:122947. https://doi.org/10.1016/j.biortech.2020.122947

Chudasama NA, Sequeira RA, Moradiya K, Prasad K (2021) Seaweed polysaccharide based products and materials: an assessment on their production from a sustainability point of view. Molecules 26(9):2608. https://doi.org/10.3390/molecules26092608

Chung IK, Beardall J, Mehta S, Sahoo D, Stojkovic S (2011) Using marine macroalgae for carbon sequestration: a critical appraisal. J Appl Phycol 23(5):877–886. https://doi.org/10.1007/s10811-010-9604-9

Conti ME, Cecchetti G (2001) Biological monitoring: lichens as bioindicators of air pollution assessment—a review. Environ Pollut 114(3):471–492. https://doi.org/10.1016/S0269-7491(00)00224-4

Costa M, Henriques B, Pinto J, Fabre E, Viana T, Ferreira N, Amaral J, Vale C, Pinheiro-Torres J, Pereira E (2020) Influence of salinity and rare earth elements on simultaneous removal of Cd, Cr, Cu, Hg, Ni and Pb from contaminated waters by living macroalgae. Environ Pollut 266:115374. https://doi.org/10.1016/j.envpol.2020.115374

Czechowski F, Jankowska A, Marsh H, Siemieniewska T, Tomkow K (1979) The development of porosity in group components of brown coals on carbonisation and activation. In: Gregg SJ, Sing KSW, Stoeckli HF (eds) Characterization of porous solid. Soc Chem Ind, London, pp 67–70

Daneshvar E, Vazirzadeh A, Niazi A, Kousha M, Naushad M, Bhatnagar A (2017) Desorption of methylene blue dye from brown macroalga: effects of operating parameters, isotherm study and kinetic modeling. J Clean Prod 152:443–453. https://doi.org/10.1016/j.jclepro.2017.03.119

Demey H, Vincent T, Guibal E (2018) A novel algal-based sorbent for heavy metal removal. Chem Eng J 332:582–595. https://doi.org/10.1016/j.cej.2017.09.083

Deolu-Ajayi AO, van der Meer IM, Van der Werf A, Karlova R (2022) The power of seaweeds as plant biostimulants to boost crop production under abiotic stress. Plant Cell Environ 45(9):2537–2553. https://doi.org/10.1111/pce.14391

Elayappan T, Jayanarayanan B, Daniel AP (2024) Production of biochar from *Kappaphycus alvarezii* (macroalgae) for the removal of eosin yellow: desorption, kinetic, and isotherm studies. Biomass Convers Biorefinery 14(21):27697–27708. https://doi.org/10.1007/s13399-022-03424-x

Ferreira N, Ferreira A, Viana T, Lopes CB, Costa M, Pinto J, Soares J, Pinheiro-Torres J, Henriques B, Pereira E (2020) Assessment of marine macroalgae potential for gadolinium removal from contaminated aquatic systems. Sci Total Environ 749:141488. https://doi.org/10.1016/j.scitotenv.2020.141488

Flores-Chaparro CE, Rodriguez-Hernandez MC, Chazaro-Ruiz LF, Alfaro-De la Torre MC, Huerta-Diaz MA, Rangel-Mendez JR (2018) Chitosan-macroalgae biocomposites as potential adsorbents of water-soluble hydrocarbons: organic matter and ionic strength effects. J Clean Prod 197:633–642. https://doi.org/10.1016/j.jclepro.2018.06.200

García-Seoane R, Fernández JA, Villares R, Aboal JR (2018) Use of macroalgae to biomonitor pollutants in coastal waters: optimization of the methodology. Ecol Indic 84:710–726. https://doi.org/10.1016/j.ecolind.2017.09.015

Godlewska K, Michalak I, Tuhy Ł, Chojnacka K (2016) Plant growth biostimulants based on different methods of seaweed extraction with water. Biomed Res Int 2016(1):5973760. https://doi.org/10.1155/2016/5973760

Gokulan R, Avinash A, Prabhu GG, Jegan J (2019a) Remediation of remazol dyes by biochar derived from *Caulerpa scalpelliformis*—an eco-friendly approach. J Environ Chem Eng 7(5): 103297. https://doi.org/10.1016/j.jece.2019.103297

Gokulan R, Prabhu GG, Jegan J (2019b) Remediation of complex remazol effluent using biochar derived from green seaweed biomass. Int J Phytoremediation 21(12):1179–1189. https://doi.org/10.1080/15226514.2019.1612845

Gurau S, Imran M, Ray RL (2025) Algae: a cutting-edge solution for enhancing soil health and accelerating carbon sequestration–a review. Environ Technol Innov 37:103980. https://doi.org/10.1016/j.eti.2024.103980

Gurav R, Bhatia SK, Choi TR, Choi YK, Kim HJ, Song HS, Lee SM, Park SL, Lee HS, Koh J, Jeon JM (2021) Application of macroalgal biomass derived biochar and bioelectrochemical system with *Shewanella* for the adsorptive removal and biodegradation of toxic azo dye. Chemosphere 264:128539. https://doi.org/10.1016/j.chemosphere.2020.128539

Henriques B, Rocha LS, Lopes CB, Figueira P, Monteiro RJ, Duarte AD, Pardal MA, Pereira E (2015) Study on bioaccumulation and biosorption of mercury by living marine macroalgae: prospecting for a new remediation biotechnology applied to saline waters. J Chem Eng 281: 759–770. https://doi.org/10.1016/j.cej.2015.07.013

Henriques B, Rocha LS, Lopes CB, Figueira P, Duarte AC, Vale C, Pardal MA, Pereira E (2017) A macroalgae-based biotechnology for water remediation: simultaneous removal of Cd, Pb and Hg by living *Ulva lactuca*. J Environ Manag 191:275–289. https://doi.org/10.1016/j.jenvman.2017.01.035

Hernández I, Martínez-Aragón JF, Tovar A, Pérez-Lloréns JL, Vergara JJ (2002) Biofiltering efficiency in the removal of dissolved nutrients by three species of estuarine macroalgae cultivated with sea bass (*Dicentrarchus labrax*) waste waters 2. Ammonium. J Appl Phycol 14(5):375–384. https://doi.org/10.1023/A:1022178417203

Holdt SL, Kraan S (2011) Bioactive compounds in seaweed: functional food applications and legislation. J Appl Phycol 23(3):543–597. https://doi.org/10.1007/s10811-010-9632-5

Illera-Vives M, Labandeira SS, Brito LM, López-Fabal A, López-Mosquera ME (2015) Evaluation of compost from seaweed and fish waste as a fertilizer for horticultural use. Sci Hortic 186:101–107. https://doi.org/10.1016/j.scienta.2015.02.008

Immanuel Suresh J, Divyeswari S (2024 Aug 28) Seaweeds are potential source for production of sustainable bioethanol for the imminent future. In: Multidisciplinary applications of marine resources: a step towards green and sustainable future. Springer Nature Singapore, Singapore, pp 141–160. https://doi.org/10.1007/978-981-97-5057-3_7

Jabeen F, Ahmad R, Mir S, Awwad NS, Ibrahium HA (2025) Carrageenan: structure, properties and applications with special emphasis on food science. RSC Adv 15(27):22035–22062. https://doi.org/10.1039/D5RA03296B

Jha S, Gaur R, Shahabuddin S, Tyagi I (2023) Biochar as sustainable alternative and green adsorbent for the remediation of noxious pollutants: a comprehensive review. Toxics 11(2): 117. https://doi.org/10.3390/toxics11020117

Johnston KG, Abomohra A, French CE, Zaky AS (2023) Recent advances in seaweed biorefineries and assessment of their potential for carbon capture and storage. Sustainability 15(17):13193. https://doi.org/10.3390/su151713193

Kalyani S, Rao PS, Krishnaiah A (2004) Removal of nickel (II) from aqueous solutions using marine macroalgae as the sorbing biomass. Chemosphere 57(9):1225–1229. https://doi.org/10.1016/j.chemosphere.2004.08.057

Kang KH, Sui Z (2010) Removal of eutrophication factors and heavy metal from a closed cultivation system using the macroalgae, *Gracilaria* sp. (Rhodophyta). Chin J Oceanol Limnol 28(6):1127–1130. https://doi.org/10.1007/s00343-010-9902-8

Katiyar R, Patel AK, Nguyen TB, Singhania RR, Chen CW, Dong CD (2021) Adsorption of copper (II) in aqueous solution using biochars derived from *Ascophyllum nodosum* seaweed. Bioresour Technol 328:124829. https://doi.org/10.1016/j.biortech.2021.124829

Khan AA, Naqvi SR, Ali I, Farooq W, Anjum MW, Almohamadi H, Lam SS, Verma M, Ng HS, Liew RK (2025) Algal biochar: a natural solution for the removal of Congo red dye from textile wastewater. J Taiwan Inst Chem Eng 166:105312. https://doi.org/10.1016/j.jtice.2023.105312

Kostas ET, Williams OS, Duran-Jimenez G, Tapper AJ, Cooper M, Meehan R, Robinson JP (2019) Microwave pyrolysis of *Laminaria digitata* to produce unique seaweed-derived bio-oils. Biomass Bioenergy 125:41–49. https://doi.org/10.1016/j.biombioe.2019.04.006

Levarity TB, Gustave W (2024) The use of seaweed *Sargassum sp.* extract as a bio-fertilizer. Int J Bahamian Stud 30(1):33–42. https://doi.org/10.15362/ijbs.v30i1.565

Li G, Yao J (2024) A review of algae-based carbon capture, utilization, and storage (Algae-Based CCUS). Gases 4(4):468–503. https://doi.org/10.3390/gases4040024

Li L, Wang J, Sun W, Peng X, Qi X (2025) Production of algae-derived biochar and its application in pollutants adsorption—a mini review. Separations 12(2):32. https://doi.org/10.3390/separations12020032

Mahendran S, Gokulan R, Aravindan A, Rao HJ, Kalyani G, Praveen S, Pushpa TB, Senthil KM (2023) Production of *Ulva prolifera* derived biochar and evaluation of adsorptive removal of Reactive Red 120: batch, isotherm, kinetic, thermodynamic and regeneration studies. Biomass Convers Biorefinery 13(6):5379–5390. https://doi.org/10.1007/s13399-021-01483-0

Manea AI, Abbas KA (2018) Influence of seaweed extract, organic and inorganic fertilizer on growth and yield broccoli. Int J Veg Sci 24(6):550–556. https://doi.org/10.1080/19315260.2018.1446205

Marinho-Soriano E, Nunes SO, Carneiro MA, Pereira DC (2009) Nutrients' removal from aquaculture wastewater using the macroalgae *Gracilaria birdiae*. Biomass Bioenergy 33(2):327–331. https://doi.org/10.1016/j.biombioe.2008.07.002

Martins A, Alves C, Silva J, Pinteus S, Gaspar H, Pedrosa R (2023) Sulfated polysaccharides from macroalgae—a simple roadmap for chemical characterization. Polymers 15(2):399. https://doi.org/10.3390/polym15020399

Matos GS, Pereira SG, Genisheva ZA, Gomes AM, Teixeira JA, Rocha CM (2021) Advances in extraction methods to recover added-value compounds from seaweeds: sustainability and functionality. Foods 10(3):516. https://doi.org/10.3390/foods10030516

Moro L, Pezzotti G, Turemis M, Sanchís J, Farré M, Denaro R, Giacobbe MG, Crisafi F, Giardi MT (2018) Fast pesticide pre-screening in marine environment using a green microalgae-based optical bioassay. Mar Pollut Bull 129(1):212–221. https://doi.org/10.1016/j.marpolbul.2018.02.036

Murtaza G, Ahmed Z, Dai DQ, Iqbal R, Bawazeer S, Usman M, Rizwan M, Iqbal J, Akram MI, Althubiani AS, Tariq A (2022) A review of mechanism and adsorption capacities of biochar-based engineered composites for removing aquatic pollutants from contaminated water. Front Environ Sci 10:1035865. https://doi.org/10.3389/fenvs.2022.1035865

Naga Babu A, Srinivasa Reddy D, Krishna Mohan GV, Suresh Kumar G, Dora TK (2023) Mathematical investigation into the sequential adsorption of silver ions and brilliant green dye using biochar derived from *Gracilaria* Rhodophyta algae. Biomass Convers Biorefinery 13(11): 10065–10084. https://doi.org/10.1007/s13399-021-01897-w

Nayak J, Behera SK (2025) Valorization of the Seaweed biomass from the Odisha coast of India by converting the Seaweed lipid into biodiesel. Biomass Bioenergy 202:108191. https://doi.org/10.1016/j.biombioe.2025.108191

Nunes GS, Lins JA, Silva FG, Araujo LC, Silva FE, Mendonça CD, Badea M, Hayat A, Marty JL (2014) Design of a macroalgae amperometric biosensor; application to the rapid monitoring of organophosphate insecticides in an agroecosystem. Chemosphere 111:623–630. https://doi.org/10.1016/j.chemosphere.2014.05.028

Omar H, El-Gendy A, Al-Ahmary K (2018) Bioremoval of toxic dye by using different marine macroalgae. Turk J Bot 42(1):15–27. https://doi.org/10.3906/bot-1703-4

Ortega A, Geraldi NR, Alam I, Kamau AA, Acinas SG, Logares R, Gasol JM, Massana R, Krause-Jensen D, Duarte CM (2019) Important contribution of macroalgae to oceanic carbon sequestration. Nat Geosci 12(9):748–754. https://doi.org/10.1038/s41561-019-0421-8

Osman ME, Abo-Shady AM, Elshobary ME, Abd El-Ghafar MO, Abomohra AE (2020) Screening of seaweeds for sustainable biofuel recovery through sequential biodiesel and bioethanol production. Environ Sci Pollut Res 27(26):32481–32493. https://doi.org/10.1007/s11356-020-09534-1

Osman AI, Fawzy S, Farghali M, El-Azazy M, Elgarahy AM, Fahim RA, Maksoud MA, Ajlan AA, Yousry M, Saleem Y, Rooney DW (2022) Biochar for agronomy, animal farming, anaerobic digestion, composting, water treatment, soil remediation, construction, energy storage, and carbon sequestration: a review. Environ Chem Lett 20(4):2385–2485. https://doi.org/10.1007/s10311-022-01424-x

Ozogul F, Trif M, Rusu A (eds) (2024) Seaweeds and seaweed-derived compounds: meeting the growing need for healthy biologically active compounds. Springer International Publishing: Cham, Switzerland, pp 459–487

Pal A, Kamthania MC, Kumar A (2014) Bioactive compounds and properties of seaweeds—a review. OAlib 1(4):1–7. https://doi.org/10.4236/oalib.1100752

Pei B, Zhang Y, Liu T, Cao J, Ji H, Hu Z, Wu X, Wang F, Lu Y, Chen N, Zhou J (2024) Effects of seaweed fertilizer application on crops' yield and quality in field conditions in China-a meta-analysis. PLoS One 19(7):e0307517. https://doi.org/10.1371/journal.pone.0307517

Pereira L (2018) Seaweeds as source of bioactive substances and skin care therapy—cosmeceuticals, algotheraphy, and thalassotherapy. Cosmetics 5(4):68. https://doi.org/10.3390/cosmetics5040068

Poo KM, Son EB, Chang JS, Ren X, Choi YJ, Chae KJ (2018) Biochars derived from wasted marine macro-algae (*Saccharina japonica* and *Sargassum fusiforme*) and their potential for heavy metal removal in aqueous solution. J Environ Manag 206:364–372. https://doi.org/10.1016/j.jenvman.2017.10.056

Pradhan B, Bhuyan PP, Ki JS (2023) Immunomodulatory, antioxidant, anticancer, and pharmacokinetic activity of Ulvan, a seaweed-derived sulfated polysaccharide: an updated comprehensive review. Mar Drugs 21(5):300. https://doi.org/10.3390/md21050300

Prasedya ES, Kurniawan NS, Kirana IA, Ardiana N, Abidin AS, Ilhami BT, Jupri A, Widyastuti S, Sunarpi H, Nikmatullah A (2022) Seaweed fertilizer prepared by EM-fermentation increases abundance of beneficial soil microbiome in paddy (*Oryza sativa* L.) during vegetative stage. Fermentation 8(2):46. https://doi.org/10.3390/fermentation8020046

Prasher P, Mudila H, Sharma M (2021 Jan 19) Biosorption and bioaccumulation of pollutants for environmental remediation. In: Microbial rejuvenation of polluted environment: volume 2. Springer Singapore, Singapore, pp 379–405. https://doi.org/10.1007/978-981-15-7455-9_15

Pudoli I, Prasetyati SB, Fadhlullah M (2024) Valorization of seaweed Gracilaria sp. biomass waste into liquid organic fertilizer: assessment on cayenne pepper *Capsicum frutescens* L. growth. In: AIP conference proceedings 2024 Mar 7 (Vol. 3080, No. 1, p. 100001). AIP Publishing LLC. https://doi.org/10.1063/5.0198556

Roy A, Sharma A, Yadav S, Jule LT, Krishnaraj R (2021) Nanomaterials for remediation of environmental pollutants. Bioinorg Chem Appl 2021(1):1764647. https://doi.org/10.1155/2021/1764647

Saeed AA, Harun NY, Sufian S, Siyal AA, Zulfiqar M, Bilad MR, Vagananthan A, Al-Fakih A, Ghaleb AA, Almahbashi N (2020) *Eucheuma cottonii* seaweed-based biochar for adsorption of methylene blue dye. Sustainability 12(24):10318. https://doi.org/10.3390/su122410318

Senturk Parreidt T, Müller K, Schmid M (2018) Alginate-based edible films and coatings for food packaging applications. Foods 7(10):170. https://doi.org/10.3390/foods7100170

Sharma A, Dubey S, Singh K, Mittal R, Quille P, Rajauria G (2025) Innovative processing and industrial applications of seaweed. Phycology 5(1):10. https://doi.org/10.3390/phycology5010010

Sudhakar MP, Arunkumar K, Perumal K (2020) Pretreatment and process optimization of spent seaweed biomass (SSB) for bioethanol production using yeast (*Saccharomyces cerevisiae*). Renew Energy 153:456–471. https://doi.org/10.1016/j.renene.2020.02.032

Sutherland DL, Craggs RJ (2017) Utilising periphytic algae as nutrient removal systems for the treatment of diffuse nutrient pollution in waterways. Algal Res 25:496–506. https://doi.org/10.1016/j.algal.2017.05.023

Tan X, Zhang F, Wang H, Ho SH (2023) The magic of algae-based biochar: advantages, preparation, and applications. Bioengineered 14(1):2252157. https://doi.org/10.1080/21655979.2023.2252157

Thenarasu A, Chai MK, Tan YH, Wong LS, Rajamani R, Djearamane S (2024) Study of Chlorella vulgaris from different growth phases as biosensor for detection of titanium and silver nanoparticles in water. Nat Environ Pollut Technol 23(2):959–969. https://doi.org/10.46488/NEPT.2024.v23i02.030

Trif M, Rusu AV, Bellahcen TO, Cherifi O, El Bakali M (2024 Oct 22) Sustainable and cost-effective management of residual aquatic seaweed biomass. Business opportunity for seaweeds biorifineries. In: Seaweeds and seaweed-derived compounds: meeting the growing need for healthy biologically active compounds. Springer International Publishing, Cham, pp 367–396. https://doi.org/10.1007/978-3-031-65529-6_12

Tuhy Ł, Samoraj M, Basladynska S, Chojnacka K (2015) New micronutrient fertilizer biocomponents based on seaweed biomass. Pol J Environ Stud 24(5):2213–2221. https://doi.org/10.15244/pjoes/39552

Veríssimo NV, Mussagy CU, Oshiro AA, Mendonça CM, de Carvalho S-EV, Pessoa A, de Souza Oliveira RP, Pereira JF (2021) From green to blue economy: marine biorefineries for a sustainable ocean-based economy. Green Chem 23(23):9377–9400. https://doi.org/10.1039/D1GC03191K

Vijayaraghavan K, Ashokkumar T (2019) Characterization and evaluation of reactive dye adsorption onto Biochar Derived from *Turbinaria conoides* Biomass. Environ Prog Sustain 38(4):13143. https://doi.org/10.1002/ep.13143

Wang C, Lin X, Zhang X, Show PL (2024) Research advances on production and application of algal biochar in environmental remediation. Environ Pollut 348:123860. https://doi.org/10.1016/j.envpol.2024.123860

Weerakkody WS, Ling KH, Hsieh HH, Abedneko VG, Shyu JF, Lee TM, Shih YY, Ranatunga RR, Santschi PH, Hung CC (2023) Carbon capture by macroalgae *Sarcodia suae* using aquaculture wastewater and solar energy for cooling in subtropical regions. Sci Total Environ 855:158850. https://doi.org/10.1016/j.scitotenv.2022.158850

Xiang L, Harindintwali JD, Wang F, Redmile-Gordon M, Chang SX, Fu Y, He C, Muhoza B, Brahushi F, Bolan N, Jiang X (2022) Integrating biochar, bacteria, and plants for sustainable remediation of soils contaminated with organic pollutants. Environ Sci Technol 56(23):16546–16566. https://doi.org/10.1021/acs.est.2c02976

Yalçın S (2014) The mechanism of heavy metal biosorption on green marine macroalga *Enteromorpha linza*. Clean–Soil Air Water 42(3):251–259. https://doi.org/10.1002/clen.201200500

Yanagisawa M, Kawai S, Murata K (2013) Strategies for the production of high concentrations of bioethanol from seaweeds: production of high concentrations of bioethanol from seaweeds. Bioengineered 4(4):224–235. https://doi.org/10.4161/bioe.23396

Yarkent C, Guler BA, Gurlek C, Sahin Y, Kose A, Oncel SS, Imamoglu E (2022) Algal alginate in biotechnology:biosynthesis and applications. In: Properties and applications of alginates, 1st ed.; Deniz, I., Imamoglu, E., Keskin–Gundogdu, T., Eds.; IntechOpen Print: London, UK, 2022; ISBN 978-1-83969-500-1

Yuan M, Xiao H, Wang R, Duan Y, Cao Q (2019) Effects of changes in precipitation pattern and of seaweed fertilizer addition on plant traits and biological soil crusts. J Appl Phycol 31(6): 3791–3802. https://doi.org/10.1007/s10811-019-01838-1

Zhang K, Kim WJ, Park AH (2020) Alkaline thermal treatment of seaweed for high-purity hydrogen production with carbon capture and storage potential. Nat Commun 11(1):3783. https://doi.org/10.1038/s41467-020-17627-1

Zhao Y, Bourgougnon N, Lanoisellé JL, Lendormi T (2022) Biofuel production from seaweeds: a comprehensive review. Energies 15(24):9395. https://doi.org/10.3390/en15249395

Zhou GJ, Ying GG, Liu S, Zhou LJ, Chen ZF, Peng FQ (2014) Simultaneous removal of inorganic and organic compounds in wastewater by freshwater green microalgae. Environ Sci Process Impacts 16(8):2018–2027. https://doi.org/10.1039/C4EM00094C

Zinina O, Merenkova S, Galimov D (2023) Development of biodegradable alginate-based films with bioactive properties and optimal structural characteristics with incorporation of protein hydrolysates. Sustainability 15(20):15086. https://doi.org/10.3390/su152015086

Algae and Synthetic Biology for Environmental Bioremediation

16

Chandrima Roy, Pramita Sen, Sushma Lavudya, and Himabindu Vurimindi

Abstract

Algae culturing utilizing sewage/industrial wastewater is one of the cost-effective and sustainable approaches toward environmental bioremediation. Standalone algae have been proven to successfully remove pollutants from a variety of wastewater, like sewage, household/kitchen wastewater, and dairy wastewater. Studies show efficient removal (up to 90%) of nutrients by algae alone, whereas the removal of chemical oxygen demand (COD) is restricted to 75%–80%. Researchers overcome this limitation by algal–bacterial co-culturing, where symbiotic bacteria help algal growth and enhance biomass yield, as well as COD removal up to 95% can be achieved. Other than COD/nutrients depollution, several microalgal species, namely, *Chlorella vulgaris*, *Phaeodactylum tricornatum*, *Arthrospira platensis*, and *Chlamydomonas reinhardtii*, have been successfully employed to eliminate heavy metals, salts, plastics, and pesticides from wastewater/surface water.

Synthetic biology holds huge potential in enhancing microalgae-based bioremediation of wastewater. A large number of genes are engineered for producing target-driven microalgal workhorses capable of bioremediation of heavy metals. This includes genes that drive sequestration of metal ions into the vacuole and those that facilitate intracellular metal transfer. The utilization of synthetic biology in environmental bioremediation is greatly acknowledged by the ability of bioengineered bacteria/algae to remove recalcitrant pollutants in high temperature

C. Roy · S. Lavudya · H. Vurimindi (✉)
Centre for Environment, UCEST, JNTUH, Hyderabad, Telangana, India
e-mail: drvhimabindu@jntuh.ac.in

P. Sen
Department of Chemical Engineering, Heritage Institute of Technology, Kolkata, West Bengal, India

K. Mohanty et al. (eds.), *Advances in Algal Research*,
https://doi.org/10.1007/978-981-95-7234-2_16

and harsh chemical conditions. Emerging pollutants like persistent organic pollutants (POPs) and pharmaceutical and personal care products (PPCPs) are difficult to remove by conventional biological routes, employing bioengineered algae as a feasible and sustainable alternative to remove these pollutants from water. Synthetic biology, however, faces major limitations in bioremediation, including biosafety and biosecurity risks, ethical concerns, regulatory challenges, and risk of secondary contamination. By combining rigorous research, careful risk assessment, public awareness, and strong regulatory frameworks, synthetic biology can be harnessed to develop more effective and sustainable solutions for environmental cleanup.

Keywords

Algae · Environmental bioremediation · COD and nutrients removal · Algal–bacterial synergy · Industrial flue gas sequestration · Synthetic biology for bioremediation

16.1 Introduction

16.1.1 Algae in Environmental Bioremediation

Microalgae are gaining interest in recent years for their excellent potential of environmental bioremediation in terms of organic and nutrient removal and as a source of various value-based products like pigments, proteins, bioplastics, and biofuels. Algae are preferred biological means of environmental bioremediation over bacteria for their ease of photosynthetic growth, utilizing wastewater organic load and nutrient source. Planned mass cultivation of algae also leads to higher biomass yield followed by value-based recovery of by-products, which make the process sustainable and economically feasible, aiming for an algal biorefinery. Certain algal–bacterial co-culture systems pertaining to symbiotic correlation, where algae utilize the CO_2 released by bacteria and bacteria help in enhanced removal of organic load from wastewater/industrial effluent, are also at the center of researchers' attraction nowadays, which also leads to higher biomass yield apart from improved bioremediation.

16.1.2 Algae Cultivation Modes: Autotrophic Versus Heterotrophic/Mixotrophic Growth of Algae and Co-culture

Algae can grow in different media utilizing various sources of carbon and nutrients. When they use atmospheric CO_2 (inorganic carbon) and light for their growth, the mode is called autotrophic. As autotrophic growth mimics natural photosynthesis, it is sustainable and ecologically benign. But its limitation is lower biomass productivity because of the uncertainty of proper availability of light and the changing pattern of light scattering and absorption in algal cells. Continuous CO_2 or flue gas supply and its mass transfer limitations in the cells are also bottlenecks to autotrophic cultivation. In heterotrophic growth, they use organic carbon compounds present in water/wastewater, and in mixotrophic more common in nature where they use both. Wastewater treatment in outdoor ponds/photobioreactors mostly maintains mixotrophic growth conditions. Industrial flue gas introduction to algal cultivation in wastewater also belongs to the mixotrophic category.

Experiments showed that sewage and food industry wastewater behaved almost the same way to heterotrophic and mixotrophic cultivation; however, sewage water is the most common growing media for mixotrophic growth as it is readily available, abundant, and easier to process. Kurniawan et al. (2025) compared the autotrophic and heterotrophic growth modes for algae cultivated in wastewater or organic-rich effluent. Heterotrophic growth conditions showed better COD/nutrients removal from wastewater, whereas the removal of phosphorus did not increase considerably. Biomass yield and nutrient assimilation for algae and co-culture both increase with hetero−/mixotrophic growth rather than autotrophy, but the main challenges in these types are the continuous supply of carbon source, which can be fulfilled by adding organic-rich effluent to the growth media. However, autotrophic cultivation seems attractive for its low cost, operational simplicity, and better CO_2/flue gas sequestration, leading to climate mitigation. Mixotrophic cultivation of microalgae in textile wastewater increased the biomass of *C. vulgaris*, increased the fatty acid methyl ester content (FAME), and decreased the chemical oxygen demand (COD) compared to the cultivation in photoautotrophic and heterotrophic conditions (Javed et al. 2022).

16.1.3 Application of Synthetic Biology for Enhanced Bioremediation

Synthetic biology in microalgal species refers to the engineering of the genetic and regulatory systems of the microorganism to expand the natural capacity of the algal species for environmental applications, especially the removal, transformation, or monitoring of pollutants. It is an interdisciplinary field that applies engineering principles to design, modify, and reconstruct novel genetic circuits, metabolic pathways, and regulatory networks within algal cells, for incorporating essential traits such as pollutant degradation, nutrient recovery, biosensing, and stress tolerance for environmental bioremediation. By expanding the natural capacities of algal systems, synthetic biology offers a powerful toolkit to overcome the limitations of conventional bioremediation strategies and develop robust, scalable, and targeted solutions for environmental cleanup. It may be mentioned that algae are spontaneously capable of removing pollutants from wastewater and contaminated environments through processes such as nutrient uptake, biosorption of heavy metals and micropollutants, and biodegradation of organic compounds. However, the efficiency of these natural mechanisms is often challenged by the slow growth rates, sensitivity to environmental stress, or the absence of specific degradation pathways. The success of synthetic biology lies in overcoming these challenges and creating microbial workhorses capable of bioremediation in stressful conditions. The role played by genetic engineering in improving algal bioremediation is discussed in Sect. 2.

16.2 Algae in Wastewater/Industrial-Effluent Treatment

16.2.1 Growing Algae in Sewage/Sewage-Contaminated Lake Water: A Nature-Based Approach to Treat Wastewater

Untreated sewage from domestic or commercial sources are rich sources of nutrients and when disposed of in water bodies, causing natural overgrowth of algae. For many decades, algae have been used in wastewater treatment ponds in India, South Africa, the United States, and Europe. One recent project in Montetema (Schoeman 2025) South Africa, introduced microalgae in wastewater treatment ponds for removing nutrients and pathogens from domestic sewage, and this rural-based project effectively cleans the water without any addition of chemicals or electricity. Algae naturally bloom in sewage-contaminated lake water, causing eutrophication, reducing the DO level of the water significantly. Researchers isolated algal strains from this contaminated lake water and grow them under controlled conditions in the lab and later used them for phytoremediation of wastewater. Thus, the grown biomass is harvested, dried, and undergone extraction and recovery of value-based products like lipid, proteins, bio-oil, and biopolymer. Table 16.1 depicts a few recent works on bioremediation of sewage/urban wastewater utilizing standalone algae and co-culture (Verma et al. 2020; Mantovani et al. 2020; Alazaiza et al. 2023; Bai and Saravana 2023; Aparicio et al. 2024). Most of the studies reveal significant nutrient removal with algae alone, whereas co-culturing leads to organic removal too. Some studies also claimed nitrate removal is mainly due to nitrifying-denitrifying bacterial action, whereas phosphorus uptake is solely by algal cells. Work reported from our lab (Verma et al. 2020) on algae/co-culture cultivation in batch lab flaks of 500 ml revealed around 70% removal of COD by co-culture, whereas in similar environment research group from China and Malaysia (Alazaiza et al. 2023) achieved over 80% COD removal, which can be resulted from their low initial COD of the collected sewage (93 g/L) than the sewage contaminated lake water (400 g/L) in Hyderabad. In later studies, we observed a higher COD removal when mode of cultivation changed from batch flasks to open pond or photobioreactors; this fact is also justified by recent study by Aparico et al. (Aparicio et al. 2024), who attained 96% organic matter removal from urban wastewater in a continuous membrane high-rate algal pond (MHRAP) as a pilot plant with HRT 2–6 days. However, enhancement in overall biomass yield in the case of co-culturing proves the symbiotic correlation of naturally existing algae–bacterial consortia, which eventually helps in effective bioremediation of the wastewater.

Table 16.1 Performance of algal and algal–bacterial systems in the treatment of sewage/sewage-contaminated lake water and urban wastewater

Sl. no.	Microalgal strain	Year and research group	Type of wastewater and cultivation	Outcome	Remarks/challenges	Reference
1	*Chlorella* sp.	2020, Verma et al., Hyderabad, India	Sewage contaminated lake water/500 ml batch flasks	Removal of nitrates (NO_3^-) and phosphates (PO_4^{3-}) was 93% and 99% removal of COD 73% by co-culture	Not much effective in COD removal	Verma et al. (2020)
2	Microalgae–bacteria consortia (*chlorella* sp.; *Scenedesmus* sp.; *Chlamydomonas*)	Mantovani et al. (2020), Italy	Raceway pond (1.2 m^3 capacity) supernatant from Blackwater dewatering (centrate) of urban WWTP	Removal efficiencies were 96% and 71% for NH4-N and P-PO_4, respectively	COD removal was not reported	Mantovani et al. (2020)
3	*Chlorella vulgaris*	2023, Alazaiza et al., Oman, China, and Malaysia	Raw wastewater from domestic sewage plant facility, 500 ml batch flasks	84% COD removal, 95% ammoniacal nitrogen, 97% phosphorus	More insights required for mass-scale cultivation	Alazaiza et al. (2023)
4	*Spirulina* sp.	Mercy Bai et al., 2023, Coimbatore, India	Wastewater from sewage treatment plant	Effective in nutrient and pollutant removal Blue green algae is a promising source of proteins, lipids, pigments, etc.	Contamination of *Spirulina* with pathogens and toxins may limit its application as feed supplements	Bai and Saravana (2023)
5	Microalgae–bacteria co-culture	Aparicio et al., 2024, Spain	Urban wastewater, membrane high-rate algal pond (MHRAP) pilot plant, continuous mode, HRT 2–6 days	91% Nitrogen removal, 71% phosphorus removal, 96% organic matter removal	Not significant phosphorus removal	Aparicio et al. (2024)

16.2.2 Algae Cultivation in Food Industry Effluent

Dairy industry effluent is one of the most voluminous wastewaters generated from food industries. Dairy/food industry effluent is a potential media for algal/co-culture cultivation for its high organic carbon content (COD typically ranges from 600 to 1500 g/L) and presence of micro-/macro-nutrients. The physicochemical characteristics of dairy wastewater vary depending on the change in milk processing steps, collection tanks, change in seasons, etc. Primary or raw collection tanks have wastewater with a high organic load, which reduced after oil and grease removal, followed by the primary clarifier. Algal growth is better in the effluent with oil and grease removed than in the raw effluent. There is a threshold COD value of dairy wastewater above which algal growth hinders. A high amount of floating oil causes hindrance to light penetration, and photosynthetic activity of algae stops. However, a high amount of fatty acids in dairy/food industry wastewater supplies an excellent carbon source for algae, leading to the accumulation of good amount of lipids in algal cells, which in turn can be utilized as a good source of algal oil and converted to biodiesel after transesterification. Table 16.2 lists the recent significant research works in this direction. Reported works of algae cultivation in food industry wastewater show satisfactory biomass growth and bioremediation in terms of COD/nutrients removal for dairy wastewater; however, other food wastes like fruit/vegetable waste, brewery wastewater, and starch waste require certain pretreatment to remove suspended particles and oil/grease, which impart turbidity to water, hence hindering algal photosynthetic activities. Studies by Gramegna et al. (2020) in 2020 show moderate COD removal efficiency, which is different from our studies with dairy wastewater in 2023. This fact is justified as Italian researchers collected dairy effluent after RO, as Retentate D1 (COD 19,000 mg/l) and Permeate D2 (COD 630 mg/L). Retentate D1 is concentrated and rich in organics and microorganisms, which may hinder the microalgal activities even after diluting to achieve COD removal significantly. Whereas most of the dairy effluent collected in the city of Hyderabad have an initial COD ranging from 800 to 2000 mg/L with a higher turbidity. After careful removal of turbidity by coagulation/flocculation or dilution, this water serves as a perfect growth media for lake waterborne algae and results in enhanced COD and nutrients removal.

Apart from dairy wastewater, effluents derived from starch, meat, brewery, oil, and fruit/vegetable processing industries gained importance as nutrient-rich media for algae/co-culture cultivation. There are some special benefits of cultivating microalgae in the food industry or food processing wastewater. Algae cultivation is always a cost-effective, sustainable approach contributing to climate change mitigation; moreover, food waste has all the essential nutrients (nitrogen and phosphorus) in abundance to serve as growth media for algae. The biomass, being rich in lipids, is a potential source of biofuels and other value-based products; the leftover biomass after the nutrient/products recovery can be directly used as biofertilizer, as the biomass is free from heavy metals or other toxic compounds. However, they need special attention, such as pre-treatment, dilution, stagewise cultivation, water-biomass ratio etc., to enhance the algal biomass recovery from food waste

Table 16.2 Significant research works on Algae cultivation utilizing dairy/food industry wastewater: performance analysis and challenges faced

Sl. no.	Microalgal strain	Year and research group	Type of wastewater and cultivation	Outcome	Remarks/challenges	Reference
1	*Chlorella species and recombinant Chlamydomonas reinhardtii strain*	Gramegna et al., 2020, Italy	Heat-treated dairy wastewater with nutrient supplement in batch flasks	COD reduction 65%, phosphorus removal 77%, total nitrogen removal 43%	Scaling up and economic feasibility of the process depend on high-capacity algal cultivation in real-time dairy effluent	Gramegna et al. (2020)
2	*Chlorella vulgaris*	Handayani et al. 2020	Dairy industry wastewater, 10 m^3 raceway bioreactor type	74% N-ammonia, 77% total nitrogen (TN), 78% total phosphate (TP), 92% chemical oxygen demand (COD), and 77% biological oxygen demand (BOD) within 10 days of the process	Nutrient recycling is needed to sustain the microalgae continuous growth	Handayani et al. (2020)
3	*Chlorella vulgaris*	Roy et al., 2023, Hyderabad, India	Dairy industry effluent, experiments were carried out in shake flasks with microalgae and algae–bacteria co-culture	COD removal 80% with standalone algae and 85% with algae–bacteria co-culture Nitrate removal 70% and 75% Phosphate removal percentages were 75% and 80% for standalone algae and co-culture, respectively	Significant decrease in COD and nutrients level in lab flasks by microalgae growth model predicted the biomass yield	Roy et al. (2023)
4	*Chlorella fusca and Spirulina* sp.	Cruz et al., 2023, Brazil	Diluted dairy effluent with BG11 as supplement media	BOD removal efficiency of 98.5% and 99.1%, COD 96.5%	Dairy effluent is an excellent alternate media for microalgae	Cruz et al. (2023)

(continued)

Table 16.2 (continued)

Sl. no.	Microalgal strain	Year and research group	Type of wastewater and cultivation	Outcome	Remarks/challenges	Reference
				and 97.7%, total phosphorus 98.8% and 85.3%, and ammonia nitrogen 98% and 99% for *Chlorella fusca* and *Spirulina* sp., respectively Biomass production 1.03 and 1.98 g L^{-1}	cultivation with treated effluent having a potential for application in agriculture	
5	*Chlorella vulgaris Scenedesmus Filamentous microalgae* sp. (*Tribonema (T.) and Lyngbya (L.)*)	Suvidha Gupta and Jorge M. Marchetti, 2024, Norway	Dairy wastewater	86.7% of chemical oxygen demand (COD), >88.7% of NO_3^--N and >98.5% of PO_4^{3-}-P biomass productivity of 84.25 mg L^{-1} d^{-1}	Co-culturing different microalgae species in a consortium significantly increases biomass production	Gupta and Marchetti (2024)
6	*Chlorella vulgaris*, *Nannochloropsis oculata*, *and* Scenedesmus *sp*	Anagnostopoulou et al., 2024, Greece	Food industry wastewater (mixture comprising of brewery wastewater, cheese-whey, and expired orange juice)	COD, TOC, and nutrient removal efficiencies are within 80%–90%	Significant bioremediation and waste valorization from food industry wastewater	Anagnostopoulou et al. (2024)
7	*Chlorella sp*, *spirulina sp*, *Tribonema* sp. *and Synechocystissp.*, *Arthrospira sp*, *Scenedesmus*, *Chlamydomonas*, etc.	Gupta et al., 2024, India	Soyabean, brewery, tannery, textile, dairy, molasses, starch, slaughter,etc.	% Removal of COD and nutrients vary depending on the type of wastewater and algal strains used	In the future, more focus on refining cultivation techniques and optimizing biorefinery processes is required	Gupta et al. (2024)

(Handayani et al. 2020; Roy et al. 2023; Cruz et al. 2023; Gupta and Marchetti 2024). This study also reveals that the two most significant algal strains that removed most of the COD, total organic carbon, and nutrients from food wastewater are *Chlorella vulgaris* and *Nanochloropsis occulate*. A detailed study on microalgae cultivation in diverse wastewater by Anagnostopoulou et al. and Gupta et al in 2024 illustrated recent advancements in the field of microalgae-based wastewater treatment and biorefinery processes focusing on challenges and an integrated approach to algal bioremediation (Anagnostopoulou et al. 2024; Gupta et al. 2024). The section concludes that food industry wastewater can serve as a promising source for microalgae cultivation; however, more research is required in the direction of pre-treatment of the effluent, optimizing growth parameters, and economic feasibility in large-scale cultivation. Moreover, industry -academia collaboration and investments in R&D will pave the way to microalgae-based wastewater treatment on a larger scale.

16.2.3 Growing Algae in Aquaculture Wastewater

Aquaculture industries can employ algae to grow in water after culturing to remediate the water by naturally removing excess nutrients. This is a sustainable approach to treat the water, promotes nutrient recycling, and even nutritious aquafeed can also be produced from the harvested algal biomass. It is also known that certain algae can destroy harmful microorganisms, thus presenting a healthier and cleaner environment for aqua-cultured species. The treated water can be recycled, and the algae-based food supplement thus obtained can decrease the dependency on costly aquafeed, thus promoting the circular economy within the system. Reported works show over 90% removal of organic matter and nutrients by standalone and co-culturing algae in aquaculture wastewater, making it a promising growth media; however, optimization of system performance and exploring economic feasibility of the process is needed for large-scale application (Table 16.3).

16.2.4 Growing Algae in Treated Landfill Leachate

Landfill leachate (LL) is a complex wastewater that is generated from municipal solid waste dumpsite. Its composition is variable and characterized by high organic matter, ammoniacal nitrogen, salts, and recalcitrant compounds. Conventional physicochemical and biological treatment systems are usually employed in treating the LL to degrade the organic matter, reduce the toxicity, and improve overall treatability of LL. In recent times, the cultivation of microalgae in treated LL has gained focus as it is a sustainable strategy that integrates pollutant removal with simultaneous resource recovery. This approach has gained a lot of attention as LL contains nutrients like dissolved inorganic nitrogen and phosphates which are essential for algal growth. However, pretreatment of LL is required for cultivation of microalgae to create a suitable condition for the subsequent algae-based treatments.

Table 16.3 Few recent works on algae/co-culture cultivation in aquaculture wastewater, mode of cultivation, major outcome, and challenges

Sl. no.	Microalgal strain	Year and research group	Type of wastewater and cultivation	Outcome	Remarks/challenges	References
1	*Chlorella minutissima*	Hawrot-Paw et al., 2019, Poland	Wastewater resulting from saline aquaculture 14 L tubular photobioreactors (PBRs)	N and P removal 88% and over 99%, respectively. N-NO_3 was reduced by 88.6%, N-NO_2 by 74.3%, and dissolved orthophosphates (V) by 99%	An integrated system for the purification of saline post culture water and the simultaneous production of microalgal biomass is feasible	Hawrot-Paw et al. (2020)
3	*Tetraselmis* sp., *Pavlova* sp., *Chlorella* sp., *Isochrysis* sp., *Chaetoceros* sp., *Phaeodactylum* sp., *Skeletonema* sp., *Thalassiosira* sp., *and Nannochloropsis sp*	Vijayaram et al., 2024, Norway and Iran	Aquaculture water	94% COD removal, over 90% nitrate and phosphate removal	Aquaculture scientists must aware of the specific regulations in different region and country	Vijayaram et al. (2024)
4	Microalgae consortium composed of *Chlorella* sp., *Scenedesmus* sp. *and Phormidium* sp.	Christodoulou et al., 2025, Switzerland	Aquaculture wastewater, biofilm photobioreactor (PBR)	Nutrient removal efficiencies with 92% reduction of PO P (removal rate: 0.07 mg/L d) and a 62% reduction of NO_3^--N (removal rate: 1.1 mg/L d)	Requirement of optimization of system performance and exploring the economic potential of large-scale applications	Christodoulou et al. (2025)

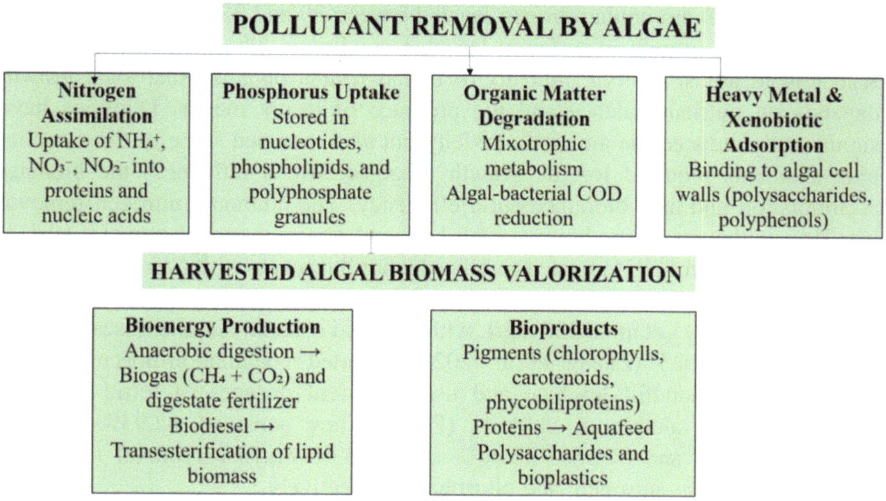

Fig. 16.1 Schematic representation of pollutant removal mechanisms by algae and subsequent valorization pathways of harvested algal biomass. The process involves nutrient assimilation (nitrogen and phosphorus uptake), organic matter degradation, and adsorption of heavy metals and xenobiotics, followed by conversion of the algal biomass into bioenergy and high-value by-products

Pretreatments encompass a range of physical, chemical, biological, and advanced treatment systems that work synergistically to degrade or transform the pollutants. Treated landfill leachate (TLL) exhibits less color, turbidity, heavy metals, and organic fraction when compared to raw landfill leachate (RLL). The viability and yield of algae in the TLL depend on optimization of the growth conditions, nutrient balance, and degradation of organic content (Fig. 16.1).

In 2016, Kumari et al. (2016) co-cultivated *Scenedesmus* sp. and *Paenibacillus* sp. in 20% LL concentrations and reported significant heavy metal and organic content removal with reduced cytotoxicity and genotoxicity. In a study by El ouaer et al. (2017), LL was treated using *Chlorella* sp. The impact of varying time and concentration of LL was investigated on algae growth and COD/ammonia removal. The results indicated COD and ammonia removal were 60% and 90% in 13 days. They also reported 10% leachate concentration was the optimum for maximum algal growth and also bioremediation. The same year Nuygen et al. (Nguyen et al. 2017) utilized microalgae in microbial fuel cell (MFC) for bioremediation and energy production using 5%–40% LL concentrations. Application of microalgae in MFC can be advantageous as oxygen is evolved using the photosynthesis mechanism and generates electricity. With the initial concentration of ammonia at 300 mg/L, 100% removal was observed. Likewise, COD removal of 96.8% was observed in the anode chamber and 52.9% in the cathode chamber. Phosphorus was reduced by 61% in 4 days. In 2018, Paskuliakova et al. (2018) examined the performance of *Chlamydomonas* sp. in six different landfill leachate samples, which were collected from separate landfill sites. The LL concentrations ranged from 10% to 100%, maintaining total ammonia nitrogen levels between 30 and 220 mg L^{-1}. The algal strain exhibited growth in different leachate conditions. Phosphate was added as K2HPO4 to adjust the N:P ratio to 16:1, and results obtained that algal growth depends on nutrient balance and the presence of heavy metals. Dilutions have significantly reduced the ammonia toxicity but also removed some of the essential trace elements required for the growth of algae, which influences the biomass accumulation and also bioremediation efficiency. The ammonia nitrogen removal was 70%, with a maximum biomass of 1.2 g/L and a specific growth rate of 0.19 d^{-1}. In 2025, Liew et al. (2025) treated stabilized landfill leachate with *Chlorella vulgaris* and achieved 97% ammonia–nitrogen and 80% phosphate removal. The maximum biomass yield they got up to 220 mg/L with 8% lipid content. Another research work from Asia in 2023 (Zhang et al. 2023) reported coagulation-flocculation as pretreatment of landfill leachate and used different dosages of ferric chloride, alum, and poly aluminum chloride (PAC). They achieved 62.93%–72.43%, 74.93%–75.55%, and 87.58%–93.40% and 73.63%–86.73% removal for COD, BOD5, ammonium–nitrogen, and phosphate, respectively.

In conclusion, it can be stated that cultivation of algae in landfill leachate is challenging due to the presence of residual recalcitrant compounds, xenobiotic, and trace heavy metals, which limit the algal growth. Seasonal light intensity fluctuations, temperature also influence the algal growth, especially in open pond systems. Integrating advanced systems like photobioreactors, immobilized algal systems, or genetically modified strains in a conventional wastewater system will help overcome these limitations.

16.3 Nutrient Removal Challenges

16.3.1 Nitrate Removal

Algae cultivation offers a sustainable route to remove nutrients—chiefly nitrogen and phosphorus—and reduce chemical oxygen demand (COD) from landfill leachate. Untreated leachate discharged to surface waters drives eutrophication, oxygen depletion, and long-term ecological harm. In algal systems, ammoniacal nitrogen (NH_4^+/NH_3) is most problematic: at high pH and temperature, free NH_3 disrupts photosynthesis, inhibits nitrate reductase, and can lyse cells. Most microalgae tolerate only moderate NH_4^+ levels, often below 200–300 mg L^{-1}, so dilution or pretreatment is commonly required. In contrast, nitrate and nitrite are more readily assimilated but demand energy-intensive reduction steps, which become rate-limiting under low light.

16.3.2 Phosphate Removal

Phosphorus uptake similarly depends on soluble phosphate; much phosphorus in leachate is bound in organic complexes or precipitated with Ca^{2+}, Mg^{2+}, and Fe^{3+}, so bioavailability varies across leachates and algal species. Phosphorus removal depends on the presence of bioavailable phosphate. Photosynthetic activity in algal cells increases the pH level, which leads to loss of ammonia and enhances nitrate removal and phosphate precipitation. Co-cultivating algae with bacteria or fungi can improve nutrient uptake, accelerate COD reduction, and increase resistance to inhibitory compounds in LL. Beyond bioremediation, algal systems are also used for nutrient recovery. Nutrients in algal biomass can be used as a biofertilizer, which reduces the synthetic fertilizers, conversion to biogas through anaerobic digestion, and biofuel production, which improves the economic feasibility. Following circular economy principles, algae-based approaches have the potential to turn landfill leachate from a disposal problem into a source of recoverable resources.

16.4 Industrial Flue Gas/CO_2 Sequestration in Algal Cultivation: Benefits and Challenges

Mass cultivation of algae is a promising means of CO_2 sequestration already existing in the atmosphere or emitted from industries. CO_2/flue gas sequestration technology by algae is gaining importance nowadays due to the adverse effects of CO2 and industrial emissions. Algae use CO_2/industrial flue gas in their photoautotrophic growth, fixing it in their cells. Harmful gases in industrial flue gas like CO_2, NO_x, and SO_x can be captured in the algal cells, which act like a biological sink providing a sustainable, cost-effective alternative to conventional carbon capture technologies. This process helps in climate change and generating carbon credit from algae in terms of biofuels, biopolymers, and other value-added products. However, major challenges in this area are the presence of toxic gases in industrial emissions, and high-temperature gases can hinder the algal growth. Large-scale flue gas sequestration demands bioengineered algal strains and efficient bioreactor designs. Some recent works in this direction are discussed below.

In 2024, Udaypal et al. (2024) reviewed the current and future status of microalgae-based carbon sequestration technologies and explained how to optimize microalgal CO_2 sequestration, integrate valuable product generation, and explore novel techniques like genetic manipulations, phytohormones, quantum dots, and AI tools to enhance the efficiency of CO_2 sequestration. A study on direct flue gas capture and subsequent valorization (Singh et al. 2025) reported life cycle assessment and techno-economic analysis of a pilot-scale photobioreactor that uses flue gas directly from natural gas or biogas combustion at 3%–5% CO_2 concentration. They pointed out the suitability of photobioreactors over open ponds for large-scale microalgae cultivation, utilizing CO2/industrial flue gases as open ponds may lead to CO_2 loss and associated high land water requirement. Research work conducted earlier (Kandimalla et al. 2016) reported a mixotrophic cultivation of *Scenedesmus quadricauda*, *Chlorella vulgaris*, and *Botryococcus braunii* 6% CO_2 concentration was optimum for maximum algal growth.

16.5 Synthetic Biology Approach to Algal Bioremediation

Synthetic biology in microalgal species refers to the engineering of the genetic and regulatory systems of the microorganism to expand the natural capacity of the algal species for environmental applications, especially the removal, transformation, or monitoring of pollutants. It is an interdisciplinary field that applies engineering principles to design, modify, and reconstruct novel genetic circuits, metabolic pathways, and regulatory networks within algal cells, for incorporating essential traits such as pollutant degradation, nutrient recovery, biosensing, and stress tolerance for environmental bioremediation. By expanding the natural capacities of algal systems, synthetic biology offers a powerful toolkit to overcome the limitations of conventional bioremediation strategies and develop robust, scalable, and targeted solutions for environmental cleanup. It may be mentioned that algae are spontaneously capable of removing pollutants from wastewater and contaminated environments through processes such as nutrient uptake, biosorption of heavy metals and micropollutants, and biodegradation of organic compounds. However, the efficiency of these natural mechanisms is often challenged by the slow growth rates, sensitivity to environmental stress, or the absence of specific degradation pathways. The success of synthetic biology lies in overcoming these challenges and creating microbial workhorses capable of bioremediation in stressful conditions. The natural algal systems serve as promising alternatives to bioremediation, as discussed in the following subsection.

16.5.1 Need for Synthetic Biology in Algae-Based Bioremediation

Although algae possess excellent innate capabilities for pollutant uptake, nutrient removal, and detoxification, their natural performance is often insufficient for industrial scale or targeted bioremediation. Several advantages that highlight the necessity of interventions by synthetic biology include the introduction of tailored degradation pathways through the application of synthetic biology for overcoming the challenges of low pollutant specificity of natural algae, engineering stress-resilient strains through synthetic networks, redirecting the metabolic flux toward bioremediation, and designing genetic biosensors for better response to specific pollutants.

16.5.1.1 Low Pollutant Specificity

Many algal species lack the metabolic pathways necessary to degrade complex or synthetic contaminants (e.g., pharmaceuticals, pesticides, plastics). Synthetic biology enables the introduction of tailored degradation pathways from bacteria or fungi into algae.

16.5.1.2 Limited Stress Tolerance

High pollutant concentrations, extreme pH, salinity levels, or heavy metal toxicity often inhibit algal growth. Engineering stress-resilient strains through synthetic circuits and regulatory elements allows algae to remain effective under harsh environmental conditions.

16.5.1.3 Trade-off between Growth and Remediation

Natural algae prioritize growth and biomass accumulation rather than pollutant degradation. Synthetic biology can redirect metabolic flux toward bioremediation functions without compromising viability.

16.5.1.4 Scalability Challenges

In large-scale wastewater or industrial effluent systems, maintaining consistent performance is difficult. Synthetic biology offers tools for stability (e.g., genetic safeguards, kill-switches) and scalability in engineered photobioreactors.

16.5.1.5 Resource Recovery Optimization

Although algal biomass can be used for biofuels or fertilizers, its productivity during remediation is variable. Synthetic engineering can couple pollutant degradation with enhanced biomass or metabolite production, improving economic viability.

In summary, synthetic biology is needed to transform algae from naturally capable yet inconsistent bio-remediators into robust, programmable, and application-specific platforms for environmental cleanup.

16.5.2 Synthetic Biology Tools and Strategies

Synthetic biology can be implemented in organisms with reverse genetic tools and methodologies like genetic transformation, selection markers, plasmid vectors, promoters, and the advent of advanced molecular biology techniques like ZFN, TALEs, and CRISPR. Therefore, synthetic biology in algae is limited to a handful of established model organisms like *Chlamydomonas reinhardtti*, *Cyanidioschyzon merolae*, *Ostreococcus tauri*, *Phaeodactylum tricornutum*, *Thalassiosira pseudonana*, and *Nanochloropsis* sp. (Hlavová et al. 2016). Several genetic engineering strategies are devised for generating engineered microalgal strains suitable for the removal of pollutants. The classical genetic engineering involves the introduction and expression of foreign DNA in algal cells without precise control over the integration site. The methods include transformation methods like electroporation, biolistic, PEG, and agrobacterium-mediated transformation mutagenesis-based approaches. The targeted genetic engineering provides site-specific modification of genes or regulatory components using programmable nucleases. These systems allow gene knockout, knock-in, or precise base editing greatly improving control over metabolic pathways. The conventional and target-based genetic techniques are discussed in the following subsections.

16.5.2.1 Classical Genetic Engineering Techniques

The application of random mutagenesis has established itself as a robust and effective tool for generating desirable traits within microalgae strains, and microalgae can be rapidly cultured, mated, and selected for strains exhibiting higher bioproduct yields within a short span of time (Trovão et al. 2022). In physical mutagenesis, the use of laser mutagenesis could increase the population of two variants of Chlorella strains, FACHB 9 and FACHB 31 (Xing et al. 2021). Similarly, in a chemical mutagenesis example, researchers (Nayak et al. 2022) utilized ethyl methane sulfonate (EMS) mutagenesis coupled with fluorescence-activated cell sorting (FACS)-based screening to generate a mutant variety of *Chlorella* sp. *HS2* mutant with higher lipid content and productivity compared to the non-mutagenized and wild strains. Random mutagenesis technology has significantly improved the removal efficiency of bacteria and microalgae for the removal of pollutants like CO2, petroleum hydrocarbons, and nutrients (Wang and Hong 2025). In one of the studies, the wastewater containing cheese whey, tequila vinasses was subjected to UV mutagenesis with ~69% removal of total nitrogen (TN), ~32% total phosphate (TPO4), and 75% COD from the tequila vinasses-containing wastewater (Najar-Almanzor et al. 2025). Copper-tolerant *Chlamydomonas reinhardtii* strains with disrupted genes AIDA and CYP720B1 were identified from an insertional mutant library, resulting in increased biomass growth (Li et al. 2025). While a mutant strain of *Scenedesmus* sp. could thrive in undiluted cellulosic ethanol wastewater with substantial removal of COD (33.5%), total ammoniacal nitrogen (71.2%), and total phosphorus (46.4%), a mutant Chlorella vulgaris was able to grow in the same medium and effectively remove COD, TAN, and TP, with removal rates of 54.8%, 97.6%, and 84.2%, respectively (Calatrava et al. 2024).

Agrobacterium-mediated gene transfer possesses the ability to transfer low copies of transgene and integrate foreign DNA into transcriptionally active genomic regions (Sharma et al. 2021). Genetically engineered *C. sorokiniana* expressing purple acid phosphatase from *Phaeodactylum tricornutum* exhibited complete biodegradation of organophosphate pesticide glyphosate at concentration levels below 10 ppm in 4–6 days (Liu et al. 2025).

16.5.2.2 Targeted Genetic Engineering Tools

Synthetic biology leverages advanced genetic engineering tools for modifying algal genomes to enhance pollutant uptake, degradation, and tolerance. The most recent genetic engineering tools applicable in microalgal systems include the targeted genetic modification techniques, namely, zinc finger nucleases (ZFN), transcription activator-like effector nucleases (TALENs), and clustered regularly interspaced short palindromic repeats (CRISPR/Cas9).

ZFN is capable of introducing a number of genetic modifications by allowing DNA double-strand breaks that simulate error at specific genomic sites (Hassanien et al. 2023). TALENs use transcription activator-like effector (TALE) proteins derived from Xanthomonas bacteria and are distinguished by their ability to recognize and bind to single base pairs of a DNA sequence. Studies have reported the ZFN-mediated modification of COP3 genes in *Chlamydomonas reinhardtii* (Sizova et al. 2013), the generation of enhanced lipid-producing mutant strains of *Phaeodactylum ticornatum* using TALE nucleases; however, very few studies have reported the ZFN and TALEN-mediated improvement in bioremediation abilities of microalgal species (Greiner et al. 2017; Jeong et al. 2023).

Though limited studies have reported the application of genome editing using CRISPR-Cas9 in water environments or wastewater treatment, applications of CRISPR-Cas technology have increased remarkably in recent years and can be expanded to bioremediation, be it through significant utilization of microorganisms for elimination of xenobiotics, conversion of toxic compounds, degradation of pesticide, or through wastewater surveillance by pathogen detection and ARG/AMR control (Mamatha Bhanu et al. 2024). The CRISPR-Cas platform has been effectively employed for microalgal strain improvement in industrial-scale production of chemicals, biofuels, and food (Patel et al. 2019). To enhance microalgae-mediated wastewater treatment through genetic engineering, two strategies can be applied: (a) improving strain tolerance and degradability of certain pollutants, such as hydrocarbons or heavy metals, or (b) expressing and producing degradation-aiding molecules, such as surfactants or antifouling ingredients. In both cases, the first step is the identification of target genes for transformation into selected microalgae strains.

Dunaliella salina and *Porphyridium cruentum* are two marine microalgal species producing extracellular polymeric substances that can be used as emulsifiers to metabolize oil hydrocarbons. Also, *Arthrospira* sp. and other cyanobacteria and microalgae can produce biosurfactant as a by-product in the presence of certain organic carbon, which can also help in hydrocarbon removal. In this context, genetic engineering has been suggested as a useful tool to increase the potential of microalgae to secrete biosurfactant. Moreover, using wastewater as a substrate for *Pseudozyma tsukubaensis* made it possible to successfully increase the production of mannosylerythritol lipids (MELs), which are a class of biosurfactant. Using gene-editing technologies like CRISPR further increases the feasibility of using these microalgae for wastewater treatment. (Hassanien et al. 2023). In a separate study, CRISPR-Cas was used to introduce about 61.5% bi-allelic mutation through knock-out of the urease gene of *Thalassiosira pseudonana*. The results indicated that knockout of the urease gene did not prevent the nitrogen uptake and growth of the microalgal strain but slowed down the growth rate (Hopes et al. 2016).

In one of the studies, a synthetic biology workflow was proposed based on design–build–test–learn cycle for enhancing the bioremoval of heavy metals from wastewater using microalgae. For instance, expression of mercuric reductase from *Bacillus megaterium* in microalgae *Chlorella* sp. promotes the bioremoval of mercury, and the study proposes that the construction of biotransformation pathways can be an effective option in enhancing the mercury removal ability of the microalgae (Sattayawat et al. 2021).

A synthetic biology-based Modular Cloning toolkit (MoClo) for the green microalga *Chlamydomonas reinhardtii* was designed for generating five antibiotic-resistant microalgal strains. While one of the promoter genes, PNIT1 controls the expression of ble-GFP gene responsible for strong resistance to zeocin antibiotic when ammonium is used as the nitrogen source, the PPSAD promoter confers strong resistance to both ammonium and nitrate sources (Crozet et al. 2018). In another study, the effectiveness of resistance markers in genetically engineered strains of *Nanochloropsis oceanica* toward the respective antibiotics was assessed. Four of the markers, namely, hygromycin B-phosphotransferase, bleomycin-resistant protein, aminoglycoside 3′-phosphotransferase, and nourseothricin acetyltransferase, showed greater than 85% survival rates of the mutant strains (Poliner et al. 2019).

Plastic pollutants, including PET, pose a menace to the environment. The bacterial species *Ideonella sakaiensis* isolated from a PET bottle recycling site was employed for the extraction of two essential PET-degrading enzymes, namely, PET hydrolase, which converts PET to mono (2-hydroxyethyl) terephthalic acid (MHET), and MHET hydrolase for converts MHET to terephthalic acid (TPA). For bioremediation in aquatic environments, extracellular MHET activity has been transferred to the marine microalga *Phaeodactylum tricornutum* for an eco-friendly biological degradation of synthetic PET polymers into reusable monomers like TPA. TPA and MHET were efficiently produced from shredded PET substrates when incubated with the mutants, whereas the wild-type culture did not produce any PET degradation products in control experiments (Rylott and Bruce 2020).

16.6 Chapter Summary

The present chapter provides a detailed discussion on the application of microalgae, both as standalone cultures and in co-culture systems, for the bioremediation of diverse wastewater streams, including sewage, food industry effluents, aquaculture wastewater, and landfill leachate. It elucidates the mechanisms underlying the nutrient and chemical oxygen demand (COD) removal, highlighting processes such as bio-adsorption, bioaccumulation, and metabolic assimilation. The chapter also emphasizes the integration of biomass valorization within a circular economy framework, where the treated biomass can be further utilized for the production of biofuels, biofertilizers, and other high-value bioproducts, thereby promoting resource recovery and sustainability.

Furthermore, the chapter explores the potential of synthetic biology in enhancing microalgal performance for bioremediation applications. It discusses the deployment of advanced genetic engineering tools, including CRISPR-Cas systems, TALENs, and ZFNs, along with traditional mutagenesis and adaptive evolution strategies, to develop engineered strains with improved pollutant degradation efficiency, enhanced stress tolerance, and optimized metabolic pathways. The integration of omics-based approaches and computational modeling is also highlighted as a means to design next-generation microalgae with superior bioremediation capabilities. Collectively, the chapter presents a holistic perspective on coupling environmental remediation with biomass valorization through the synergistic use of biotechnology and synthetic biology.

References

Alazaiza MYD, He S, Su D, Abu Amr SS, Toh PY, Bashir MJK (2023) Sewage water treatment using *Chlorella vulgaris* microalgae for simultaneous nutrient separation and biomass production. Separations 10:229

Anagnostopoulou C, Papachristou I, Kontogiannopoulos KN, Mourtzinos I, Panagiotis G, Kougias PG (2024) Optimization of microalgae cultivation in food industry wastewater using microplates. Sustain Chem Pharm 39:101510

Aparicio S, Alejandro R-M, Gallardo-Mejías JP, Robles A, Borrás L (2024) Microalgae-bacteria consortia dynamics in a long term operated membrane-coupled high-rate algal pond (MHRAP). J Environ Manag 371:123186

Bai M, Saravana K (2023) Recent analysis of sewage treatment plan (STP) using blue-green algae. Int J Adv Res 11(05):22–31

Calatrava V, Gonzalez-Ballester D, Dubini A (2024) Microalgae for bioremediation: advances, challenges, and public perception on genetic engineering. BMC Plant Biol 24(1):1261

Christodoulou X, M'ahmed C, Zili F, Bessadok B, Sadok S, Monney I, Rothlisberger R, Bagnoud M (2025) Design and development of pilot photobioreactor for simultaneous microalgae cultivation and aquaculture wastewater treatment. Process Biochem 153:182–190

Crozet P, Navarro FJ, Willmund F, Mehrshahi P, Bakowski K, Lauersen KJ et al (2018) Birth of a photosynthetic chassis: a MoClo Toolkit enabling synthetic biology in the microalga *Chlamydomonas reinhardtii*. ACS Synth Biol 7:2074–2086

Cruz CG, Centeno da Rosa AP, Strentzle BR, Vieira Costa JA (2023) Microalgae-based dairy effluent treatment coupled with the production of agricultural biostimulant. J Appl Phycol 35: 2881–2890

El Ouaer M, Kallel A, Kasmi M, Hassen A, Trabelsi I (2017) Tunisian landfill leachate treatment using Chlorella sp.: effective factors and microalgae strain performance. Arab J Geosci 10(19): 457

Gramegna G, Scortica A, Scafati V, Ferella F, Gurrieri L, Giovannoni M, Bassi R, Sparla F, Mattei B, Benedetti M (2020) Exploring the potential of microalgae in the recycling of dairy wastes. Bioresour Technol Rep 12:100604

Greiner A, Kelterborn S, Evers H, Kreimer G, Sizova I, Hegemann P (2017) Targeting of photoreceptor genes in *Chlamydomonas reinhardtii* via zinc-finger nucleases and CRISPR/Cas9. Plant Cell 29:2498–2518

Gupta S, Marchetti JM (2024) Co-cultivation of high-value microalgae species with filamentous microalgae for dairy wastewater treatment. npj Clean Water 7:119

Gupta R, Niranjan Mishra N, Singh G, Mishra S, Lodhiyal N (2024) Microalgae cultivation andvalue-based products from wastewater: insights and applications. Blue Biotechnol 1:20

Handayani T, Mulyanto A, Priyanto FE, Nugroho R (2020) Utilization of dairy industry wastewater for nutrition of microalgae *Chlorella vulgaris*. J Phys Conf Ser 1655:012123

Hassanien A, Saadaoui I, Schipper K, Al-Marri S, Dalgamouni T, Aouida M et al (2023) Genetic engineering to enhance microalgal-based produced water treatment with emphasis on CRISPR/ Cas9: a review. Front Bioeng Biotechnol 10:1104914

Hawrot-Paw M, Koniuszy A, Gałczyńska M, Zając G, Szyszlak-Bargłowicz J (2020) Production of microalgal biomass using aquaculture wastewater as growth medium. Water 12:106

Hlavová M, Vítová M, Bišová K (2016) Synchronization of green algae by light and dark regimes for cell cycle and cell division studies. In: Plant cell division: methods and protocols. Springer, New York, pp 3–16

Hopes A, Nekrasov V, Kamoun S, Mock T (2016) Editing of the urease gene by CRISPR-Cas in the diatom *Thalassiosira pseudonana*. Plant Methods 12:49

Javed F, Rehman F, Khan AU, Fazal T, Hafeez A, Rashid N (2022) Real textile industrial wastewater treatment and biodiesel production using microalgae. Biomass Bioenergy 165: 106559

Jeong B, Jang J, Jin ES (2023) Genome engineering via gene editing technologies in microalgae. Bioresour Technol 373:128701

Kandimalla P, Desi S, Vurimindi H (2016) Mixotrophic cultivation of microalgae using industrial flue gases for biodiesel production. Environ Sci Pollut Res 23:9345–9354

Kumari M, Ghosh P, Thakur IS (2016) Landfill leachate treatment using bacto-algal co-culture: an integrated approach using chemical analyses and toxicological assessment. Ecotoxicol Environ Saf 128:44–51

Kurniawan SB, Čížková M, Ahmad A, Wibowo YG, Said NSM, Jusoh HHW, Ismail A, Imron MF (2025) Autotrophic vs. heterotrophic microalgae: juxtaposition of performances in treating organic-rich effluent. Desalin Water Treat 322:101159

Li Y, Jiang L, Xu C, Wang S, Qian Y, Wu Y, Miao C, Dong Z, Wang L (2025) Insertional mutagenesis of AIDA or CYP720B1 in the green alga *Chlamydomonas reinhardtii* confers copper(II) tolerance and increased biomass. J Hazard Mater 486:137026

Liew LW, Bashir JKM, Yi Toh P, Alazaiza MYD, Abu Amr SS, Khoo KS, Raksasat R, Lim JW (2025) Microalgae cultivation in stabilized landfill leachate for simultaneous treatment and biomass production. J Taiwan Inst Chem Eng 166(Part 1):105068

Liu S-F, Yi Z-C, Huang Z-Q, Yuan Z-D, Yang Y-C, Zhao Y, He Q-Y, Yang W-D, Li H-Y, Lin CSK, Wang X (2025) Enhanced biodegradation of glyphosate by *Chlorella sorokiniana* engineered with exogenous purple acid phosphatase. Water Res 268:122737

Mamatha Bhanu LS, Kataki S, Chatterjee S (2024) CRISPR: new promising biotechnological tool in wastewater treatment. J Microbiol Methods 227:107066

Mantovani M, Marazzi F, Fornaroli R, Bellucci M, Ficara E, Mezzanotte V (2020) Outdoor pilot-scale raceway as a microalgae-bacteria side stream treatment in a WWTP. Sci Total Environ 710:135583. ISSN 0048-9697

Najar-Almanzor CE, González-Díaz RL, García-Cayuela T, Carrillo-Nieves D (2025) Adaptation and bioremediation efficiency of UV-mutagenized microalgae in undiluted agro-industrial effluents from Mexico. Environments 12(9):307

Nayak M, Suh WI, Oh YT, Ryu AJ, Jeong KJ, Kim M, Mohapatra RK, Lee B, Chang YK (2022) Directed evolution of Chlorella sp. HS2 towards enhanced lipid accumulation by ethyl methanesulfonate mutagenesis in conjunction with fluorescence-activated cell sorting based screening. Fuel 316:123410

Nguyen HTH, Kakarla R, Min B (2017) Algae cathode microbial fuel cells for electricity generation and nutrient removal from landfill leachate wastewater. Int J Hydrog Energy 42(46): 29433–29442

Paskuliakova A, McGowan T, Tonry S, Touzet N (2018) Microalgal bioremediation of nitrogenous compounds in landfill leachate—the importance of micronutrient balance in the treatment of leachates of variable composition. Algal Res 32:162–171

Patel VK, Soni N, Prasad V, Sapre A, Dasgupta S, Bhadra B (2019) CRISPR–Cas9 system for genome engineering of photosynthetic microalgae. Mol Biotechnol 61:541–561

Poliner E, Clark E, Cummings C, Benning C, Farre EM (2019) A high-capacity gene stacking toolkit for the oleaginous microalga, *Nannochloropsis oceanica*. Algal Res 45:101664

Roy C, Sen P, Vurimindi H (2023) Kinetic modeling and experiments on removal of COD/nutrients from dairy ffluent using chlorella and co-culture. Bioprocess Biosyst Eng 46(8):1099–1110

Rylott EL, Bruce NC (2020) How synthetic biology can help bioremediation. Curr Opin Chem Biol 58:86–95

Sattayawat P, Yunus IS, Noirungsee N, Mukjang N, Pathom-Aree W, Pekkoh J et al (2021) Synthetic biology-based approaches for microalgal bio-removal of heavy metals from wastewater effluents. Front Environ Sci 9:778260

Schoeman Y (2025) Algae can clean sewage without electricity or chemicals—we put it to the test in South Africa. University of the Free State. https://theconversation.com/algae-can-clean-sewage-without-electricity-or-chemicals-we-put-it-to-the-test-in-south-africa-256339

Sharma PK, Goud VV, Yamamoto Y, Sahoo L (2021) Efficient *Agrobacterium tumefaciens*-mediated stable genetic transformation of green microalgae, *Chlorella sorokiniana*. 3 Biotech 11(4):196

Singh U, Naaz F, Hawkins TR, Weinberg E, Banerjee S, Mroz R, Ohri N, Zhang J, Lee Y-Y, Chen F, Hill R, Lid Y (2025) Direct flue gas capture for algae cultivation and subsequent valorization: evaluating life cycle emissions and costs. Sustainable Energy Fuels 9:4392

Sizova I, Greiner A, Awasthi M, Kateriya S, Hegemann P (2013) Nuclear gene targeting in Chlamydomonas using engineered zinc-finger nucleases. Plant J 73:873–882

Trovão M, Schüler LM, Machado A, Bombo G, Navalho S, Barros A, Pereira H, Silva J, Freitas F, Varela J (2022) Random mutagenesis as a promising tool for microalgal strain improvement towards industrial production. Mar Drugs 20(7):440

Udaypal GRK, Mehariya S, Verma P (2024) Advances in microalgae-based carbon sequestration: current status and future perspectives. Environ Res 249:118397

Verma K, Kumar PK, Krishna SV et al (2020) Phycoremediation of sewage-contaminated lake water using microalgae–bacteria co-culture. Water Air Soil Pollut 231:299

Vijayaram S, Ringø E, Ghafarifarsani H, Hoseinifar SH, Ahani S, Chou C-C (2024) Use of algae in aquaculture: a review. Fishes 9:63

Wang Z, Hong Y (2025) Enhancing the pollutant removal performance of bacteria and microalgae through random mutagenesis technology: a comprehensive review of effect and mechanism. Curr Pollut Rep 11(1):28

Xing W, Zhang R, Shao Q, Meng C, Wang X, Wei Z, Sun F, Wang C, Cao K, Zhu B, Gao Z (2021) Effects of laser mutagenesis on microalgae production and lipid accumulation in two economically important fresh Chlorella strains under heterotrophic conditions. Agronomy 11(5):961

Zhang X, Han H, Chai H, Gao P, Wei W, Li M, Lv Q, Guo X, Meng Q, Liu X, Gao P (2023) Pre-treatment of landfill leachate via coagulation-flocculation: optimization of process parameters using response surface methodology (RSM). J Water Process Eng 54:103937

Microalgae as Green Tools for Microplastic Remediation 17

Rayanee Chaudhuri and Paramasivan Balasubramanian

Abstract

Microplastics (MPs) are widespread in aquatic environments due to their persistence, toxicity, and bioaccumulation. Detected in oceans, rivers, sediments, and even potable water, MPs threaten ecosystem food and human health. As global plastic production exceeds 367 million tons annually—only ~15% of which is recycled due to mishandling and poor waste infrastructure—a large fraction enters natural waterbodies, worsening the environmental crisis. In response, microalgae have attracted increasing scientific interest as a sustainable and biologically driven solution for MP remediation. Owing to their high adaptability and biomass productivity, microalgae offer a promising avenue for MP bioremediation via surface biosorption, entrapment by extracellular polymeric substances, enzymatic degradation, and bioflocculation—collectively providing an eco-friendly additional approach alongside the mainstreaming treatment processes. Despite this growing interest, the present literature still lacks an intuitive comprehension of the mechanisms underpinning the interactions between microalgae as well as the factors affecting these processes. This chapter fills that gap by providing a comprehensive overview of the mechanisms involved in MP remediation, evaluating the benefits and limitations of these approaches. Finally, it outlines future research directions with strategic insights for developing effective, scalable solutions to mitigate MP pollution.

Keywords

MPs · Microalgal bioremediation · Algal-MP interactions · Bioflocculation

R. Chaudhuri · P. Balasubramanian (✉)
BioEnergy and Environmental (BEE) Lab, Department of Biotechnology and Medical Engineering, National Institute of Technology Rourkela, Rourkela, Odisha, India
e-mail: biobala@nitrkl.ac.in

K. Mohanty et al. (eds.), *Advances in Algal Research*,
https://doi.org/10.1007/978-981-95-7234-2_17

17.1 What Is MP Pollution, and Why Is it Our Concern?

Plastics have been widely used in developed societies since the early twentieth century, with a major industrial expansion in the 1930s driven by their low production cost and durability (Iqbal et al. 2022). At present, these plastics are among the most widely produced synthetic materials and have become an essential part of modern life. As of 2020, global plastic production had already surpassed 367 million tons annually, and if current trends in consumption and manufacturing persist, this figure is expected to rise to approximately 590 million tons by 2050 (Hoang 2022). However, even with such high levels of production, global average recycling rates remain at just 15%, while an estimated 40% of plastic waste—almost 90 million tons annually—is mismanaged through littering, inadequate waste handling, and is dispersed into the environment by wind, rainfall, and surface runoff, resulting in widespread plastic pollution (Ragossnig and Agamuthu 2021). Among the various forms of plastic waste, microplastics (MPs)—defined as plastic particles smaller than 5 mm—have emerged as a particularly concerning pollutant due to their persistence, ubiquity, and biological impact (especially on marine life and humans) (Yu and Singh 2023).

MPs are now found in nearly every environment—marine surfaces, deep-sea sediments, soils, mountain peaks, polar snow, and freshwater. In aquatic systems, they are consumed by plankton, fish, and seabirds, causing gut blockage, oxidative damage, and cellular stress, which disrupts food webs and reduces biodiversity (Swain et al. 2025). Humans are exposed through seafood, water, and inhalation, raising toxicological concerns due to additives like bisphenol A (BPA) and phthalates (Manosuthi 2024). These polymers are highly resistant to degradation and progressively fragment into smaller, more reactive particles, complicating removal and remediation efforts. Their ubiquity and persistence mean that virtually no environment remains untouched by contamination. Notably, even remote ecosystems such as Antarctica show evidence of MPs in sediments, organisms, and snowfall, further underscoring their global dispersion (Pellegrino et al. 2025). Beyond environmental damage, MPs present multifaceted threats to food safety, public health, ecological stability, and critical economic sectors, including agriculture and fisheries.

Given these far-reaching implications, the following sections will explore various types and sources of MPs, their impact on the environment, human health and the emerging role of microalgae in their remediation. This chapter will also examine the underlying mechanisms and influencing factors of microalgae-based MP removal, evaluate current methodological feasibility and constraints, and propose future research directions.

17.2 Categories, Sources, and Environmental Impact of MPs

MP pollutants, plastic particles up to 5 mm in diameter, originate either from manufacturing processes or the degradation of larger plastic debris. They are widely used across sectors like textiles, medicine, and personal care products. Based on their origin, MPs are classified into two categories: (1) primary MPs, which are intentionally produced in small sizes for industrial or commercial applications—such as resin pellets, cosmetic microbeads, exfoliants, and abrasives; and (2) secondary MPs, which form from the breakdown of larger plastic items (e.g., plastic bottles, bags) through environmental weathering processes such as sunlight exposure, UV radiation, mechanical abrasion, and biodegradation, producing fragments that can persist in the environment for a really longtime (Yu and Singh 2023) (see Fig. 17.1). These plastics also vary widely in chemical composition and density. Denser types like polyamide (PA), polyvinyl chloride (PVC), and polyethylene terephthalate (PET) tend to sink and accumulate in sediments, whereas less dense or lighter ones like polystyrene (PS), polylactic acid (PLA), high- and low-density polyethylene (HDPE/LDPE), polypropylene (PP), and polyurethane (PUR) remain buoyant, float longer, and can undergo greater horizontal transport across aquatic surfaces (Yang et al. 2021; Vercauteren et al. 2024). Stride et al. further reported a density threshold around 1.32–1.4 g/cm^3, above which MPs tend to deposit unless flow velocities exceed ~0.101–0.169 m/s (Stride et al. 2024). Beyond density, aging processes increase surface roughness, promoting biofilm, and algal attachment that enhance biofouling, which can cause originally buoyant plastics to sink, altering their vertical transport over time (though this might vary depending on their type, size, and surface texture) (Vercauteren et al. 2024; Coyle et al. 2020). Additionally, MP morphology and size also influence mobility, with fibers exhibiting higher

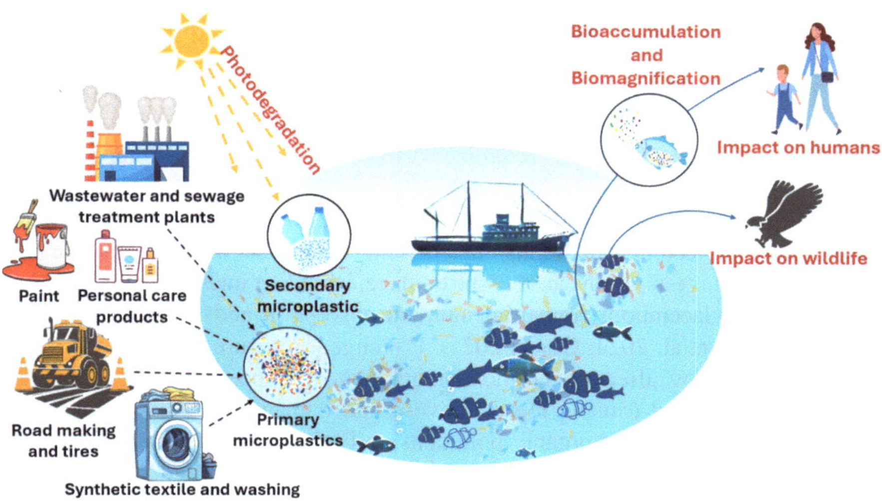

Fig. 17.1 MP pathways from production through environmental transport to biological exposure

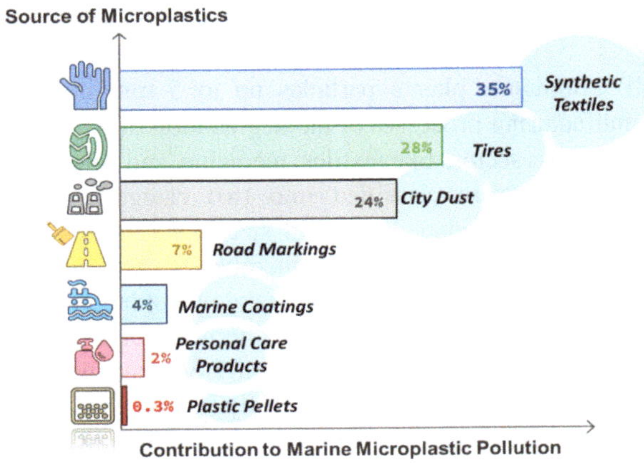

Fig. 17.2 Different sources of primary MP pollution in marine environments

vertical transport than flakes or films, particularly in terrestrial systems under rainfall or infiltration, while larger particles (500–1000 μm) migrate deeper than smaller ones (Park et al. 2023).

Wang et al. reported that in 2015, mainland China generated approximately 7×10^8 kg of primary MP waste, with one-sixth entering the hydrosphere (Wang et al. 2019). *Among the various sources*, tire dust and synthetic fibers accounted for 53.91% and 28.77% of the total primary MP waste, respectively, *making them the dominant contributors*. Similarly, Boucher and Friot reported that globally, around 1.5 million tons of primary MPs enter the oceans each year (Boucher and Friot 2017). The authors further mentioned that synthetic textiles (~35%), tire wear (~28%), and urban dust (~24%) as the major sources, with smaller contributions from road markings (~7%), marine coatings (~4%), personal care products (~2%), and plastic pellets (~0.3%) (see Fig. 17.2) (Boucher and Friot 2017).

In contrast, secondary MPs make up most environmental MP pollution, with models estimating that over 70% in marine and freshwater systems originate from them (An et al. 2020). These particles vary in shape and polymer type, often appearing as fragments or films from PE, PP, and PS. Under business-as-usual scenarios, secondary MP input is projected to double by 2040 (Thompson et al. 2024).

Once in the environment, MPs persist and move through multiple pathways and pose significant ecological, human health, and economic risks. In marine and freshwater life, ingestion can cause tissue damage, including intestinal inflammation, blockages, and lacerations, impairing nutrient absorption and potentially causing death (Ofoegbu et al. 2024; Ghosh 2025). Prolonged exposure also disrupts reproductive systems by affecting hormone regulation and sperm quality, ultimately reducing fertility and causing reproductive organ damage. MPs can also adsorb and concentrate persistent organic pollutants or POPs (such as dichloro diphenyl trichloroethane, polychlorinated biphenyls), pharmaceuticals, and heavy metals (e.g., lead, cadmium), increasing their mobility and persistence (Iqbal et al. 2022; Arienzo and Donadio 2023). These interactions depend on polymer type, aging, and environmental conditions (Arienzo and Donadio 2023). Arienzo and Donadio noted that due to their surface properties, MPs can carry both hydrophobic and hydrophilic drugs, alter their toxicity, and promote bioaccumulation in aquatic life (Arienzo and Donadio 2023). This interaction is influenced by pH, salinity, temperature, organic matter, and biofilms, which modify toxicity and intensify ecological effects.

MPs accumulate in fish tissues—muscle, liver, and gills—posing a human exposure risk, especially when whole fish or shellfish are consumed. A review by Danopoulos et al. found that seafood consumption can lead to up to 55,000 MP particles ingested per person annually, depending on diet and region. Highest contamination is reported in Asia, particularly China, due to high seafood intake and pollution (Danopoulos et al. 2020). MPs act as vectors for antibiotics, polycyclic aromatic hydrocarbons (PAHs), and heavy metals, raising concerns about antibiotic resistance and carcinogenicity (Sui et al. 2024). Once ingested, they can translocate to organs, triggering inflammation, oxidative stress, immune suppression, and endocrine disruption (Smith et al. 2018).

MPs introduced through sludge and compost disrupt soil physical and chemical properties, including bulk density, water retention, pH, and nutrient cycling. These changes negatively affect soil fertility and microbial communities, lowering crop productivity and ecosystem functioning (Hanif et al. 2024). For instance, de Souza Machado et al. reported that polyester fibers and polyamide beads significantly altered soil structure and water dynamics (de Souza Machado et al. 2019). Due to their shape and surface area, these MPs impair soil aggregate formation, reduce water retention and porosity, and limit nutrient availability, thereby impacting microbial activity. They can also affect plant biomass, elemental composition, and root development, reducing agricultural productivity. In another recent study, Sakali et al. investigated three agricultural sites where compost derived from wastewater sludge (rich in MPs, primarily fibers, followed by fragments and films) was applied as fertilizer (Sakali et al. 2024). The findings revealed that ingested MPs caused gut blockage, oxidative stress, and lower reproduction in earthworms, raising concerns about terrestrial food web biomagnification. According to the FAO report, around 20% of the global population relies on fish as their main source of animal protein, and beyond the ecological risks, MPs also threaten the blue economy by lowering fishery yields, damaging marine infrastructure, and harming tourism, leading to significant economic losses for coastal communities and national economies (Gupta et al. 2024; FAO 2012).

17.3 What Are the Different MP Remediation Techniques, and Why Should We Look More into Microalgal Remediation?

MP remediation is a multifaceted challenge, with several conventional and emerging strategies showing varying degrees of effectiveness. Physical or chemical techniques such as chemical coagulation (or electrochemical coagulation) and membrane filtration have achieved high removal efficiencies, often exceeding 90% (Zhang et al.

2020; Teran et al. 2023; Park et al. 2021). However, membrane-based filtration systems are typically size-specific, prone to membrane fouling, and have high operational costs, which limit their scalability and long-term application (Amparán et al. 2025). Similarly, adsorption-based methods, while useful, are highly dependent on MP size and surface characteristics, which constrain their broad applicability (Amparán et al. 2025). Additionally, chemical methods like coagulation and flocculation, though effective, often produce sludge loaded with pollutants, necessitating secondary treatment and disposal. In addition, more recent chemical approaches, such as photocatalytic degradation and ozonation have demonstrated only partial effectiveness. Specifically, photocatalytic degradation, though capable of breaking down MPs, suffers from high operational costs, while ozonation alone is often insufficient for complete degradation and may promote the formation of harmful by-products (Mecha and Chollom 2020; Sacco et al. 2023). Methods like advanced oxidation processes (AOPs), combining ozonation with hydrogen peroxide, generate hydroxyl radicals (•OH) that chemically degrade MPs like polyethylene (PE) (Hu et al. 2023). Despite their effectiveness, AOPs pose concerns regarding by-product formation. Shen et al. reported that prolonged exposure of PE to AOP led to the formation of smaller plastic fragments and potentially toxic leachates, raising concerns about nanoplastic generation and their ecological toxicity (Shen et al. 2022). Du et al. further highlighted that ozonation combined with hydrogen peroxide caused significant release of dissolved organic matter from plastic additives, which may contribute to disinfection by-product formation in treated water (Du et al. 2023).

Collectively, these methods are size-selective, pose secondary pollution risks, and require energy-intensive, high-end infrastructure, making them less viable for sustainable large-scale application. Given these trade-offs, biologically based strategies have gained attention for their environmental compatibility. Among them, microalgae have emerged as a promising option, mainly due to their ability to secrete extracellular polymeric substances (EPS), which facilitate the heteroaggregation of MPs (however, this process is species-specific and depends on the type and quantity of EPS produced). Thus, microalgae offer a low-energy, eco-friendly solution for MP remediation, with the added benefit of biomass production. The following sections explore their potential, limitations, and strategies for improvement.

17.4 Mechanisms of MP Interaction with Microalgae

Microalgae have shown significant potential in MP remediation through a variety of physicochemical and biological mechanisms. Among these, the production of EPS is particularly important. EPS form a protective layer around microalgal cells and are noted for their diverse functional properties, including applications in environmental pollutant remediation (Kausar et al. 2021; Moreira et al. 2022; Chaudhuri and Balasubramanian 2023, 2025). In the context of MP exposure, this layer acts as a defensive shield, reducing cell damage, protecting photosynthetic activity, and lowering oxidative stress. Microalgal species such as *Chlorella vulgaris*, *Dunaliella salina*, and *Scenedesmus abundans* are well-known for their EPS production, which

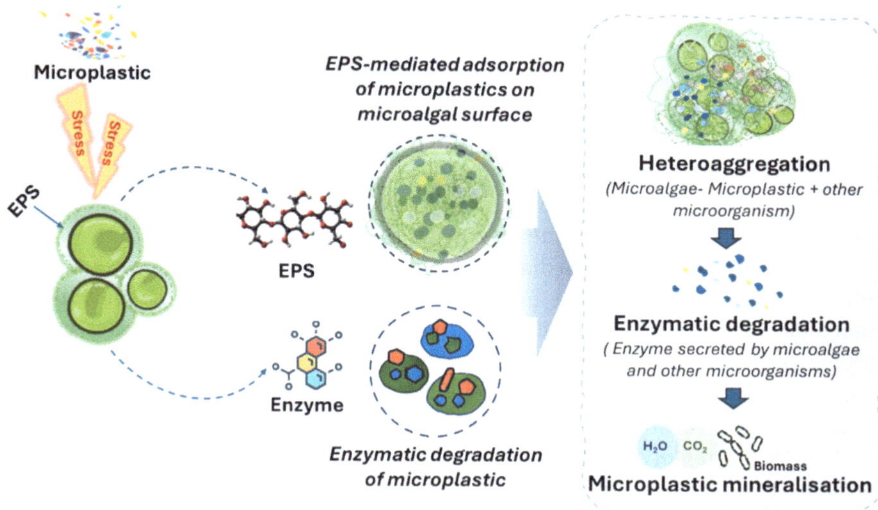

Fig. 17.3 Possible pathway of microalgae-based MP remediation pathway

not only helps the cells endure stress but also plays a crucial role in facilitating MP removal (Hadiyanto et al. 2022; Gowthami et al. 2025; Song et al. 2020). The sticky, anionic nature of EPS enhances MP adsorption onto the algal cell surface (as shown in Fig. 17.3). When secreted in a loosely bound form into the surrounding medium, EPS promotes MP heteroaggregation, leading to the formation of stable flocs and biofouling structures that aid in immobilizing and removing MPs from aquatic systems (Su et al. 2023). In this regard, Sousa et al. also emphasized that surface charge interactions are a key mechanism for microalgae to aggregate MPs in water and wastewater systems, reinforcing the role of electrostatic attraction in EPS-mediated flocculation (Sousa et al. 2025). Further elaborating on EPS-mediated pathways, Hadiyanto et al. mentioned that the microalga itself does not directly degrade MPs nor use them as a carbon source (Hadiyanto et al. 2022). Instead, microalgae such as *D. salina* release EPSs, which have a sticky and adhesive nature that encourages the formation of a biofilm on MP surfaces. This biofilm facilitates the attachment and colonization of other microorganisms, such as bacteria and fungi, which act as the primary agents of MP biodegradation. Thus, microalgal EPS create a microenvironment that boosts microbial metabolic activity, ultimately leading to the breakdown of MPs into simpler compounds like methane and hydrocarbons. This indicates that in realistic environments, where native bacterial consortia coexist with microalgae, MP remediation is not an isolated algal process but occurs through close algae–bacteria interactions, leading to more effective removal than algal monocultures alone.

In a controlled photobioreactor experiment using *Dunaliella salina* with oxium and oxidized HDPE MPs, Hadiyanto et al. reported that *D. salina* supported the effective biodegradation of both types (Hadiyanto et al. 2022). Among the two, oxium degraded more rapidly due to the presence of pro-oxidant additives, while oxidized HDPE showed notable structural changes, particularly a significant crystallinity loss. The study also found that pre-treating HDPE with hydrogen peroxide (H_2O_2) enhanced its degradability by chemically breaking some polymer bonds, introducing carbonyl and hydroxyl groups that increased hydrophilicity and facilitated microbial attachment. Overall, these results suggest that integrating chemical pre-treatment with microalgal EPS pathways can enhance the biodegradation of persistent plastic pollutants in aquatic environments.

The efficiency of EPS-mediated hetero-aggregation in MP remediation is influenced by several factors, including the microalgal strain, the quantity of microalgal EPS produced, and the type of MP involved. Different species exhibit varying aggregation abilities; as demonstrated by Su et al. that among several tested strains, *Scenedesmus obliquus* showed the highest aggregation with MPs, followed sequentially by *Chlorella vulgaris*, *Tetraselmis subcordiformis*, *Chaetoceros müelleri*, and *Streptococcus westermani* (Su et al. 2023). This variation is largely attributed to differences in EPS production and surface characteristics among species. In a related study, where *Scenedesmus abundans* was tested against PS, PMMA, and PLA, it showed ~98% removal of PMMA, while removal rates for PS and PLA exceeded 70% only after pre-exposure, indicating that prior contact may enhance EPS-mediated aggregation or cellular adaptation (Cheng and Wang 2022). Further reinforcing this mechanism, Cunha et al. reported that *Cyanothece* sp. significantly increased EPS production in response to exposure to 10 mg L^{-1} of polystyrene nanoplastics (Cunha et al. 2020). This overproduction, interpreted as a self-defense response, facilitated aggregation through electrostatic and hydrogen bonding interactions between EPS and MP surfaces, forming stable, distinct aggregates. Moreover, other microorganisms with strong flocculating capacity can boost removal via bioflocculation. For instance, Hadiyanto et al. reported that while *Chlorella vulgaris* alone achieved 55.7% removal of PP and 95.0% of PET, the addition of *Aspergillus niger* pellets (3.5% w/v) boosted removal to 90 $\pm$ 0.031% for PP and 94 $\pm$ 0.015% for PET, confirming the synergistic potential of bioflocculation in MP remediation outcomes (Hadiyanto et al. 2025).

In addition to hetero-aggregation, microalgae can also contribute to MP degradation through enzymatic mechanisms (see Fig. 17.3). Gowthami et al. demonstrated that marine microalgae, particularly *Synechocystis* sp., possess the ability to degrade PS MPs by forming biofilms and secreting specific enzymes like laccase (Gowthami et al. 2025). Upon interaction with PS MPs, *Synechocystis* sp. facilitates plastic adhesion to its cell surface, where biofilm formation creates a localized microenvironment conducive to sustained enzymatic activity. Within this biofilm matrix, laccase enzymes actively oxidize the aromatic benzyl rings of polystyrene, introducing functional groups such as hydroxyl and carbonyl groups. These chemical modifications compromise the structural integrity of the polymer chains, leading to fragmentation into smaller compounds like alkanes, alcohols, and acids. These degradation products can subsequently enter microbial metabolic pathways such as the Krebs cycle, enabling further breakdown and assimilation into cellular biomass. This enzymatic degradation pathway highlights a biologically active role of microalgae in MP remediation, going beyond passive physical entrapment. Another similar observation was made by Molino et al., who emphasized the active role of *Chlamydomonas reinhardtii* in plastic degradation (Molino et al. 2024). The authors mentioned that when recombinant *Chlamydomonas reinhardtii* was cultured along with PET, the microalgae secreted the PET-degrading enzyme PHL7, and this degradation of PET was validated by the detection of terephthalic acid (TPA), a breakdown product of PET, through mass spectrometry analysis. However, these intermediates accumulated in the medium and were not assimilated, with the authors noting that further engineering is required to enable their metabolic assimilation by microalgae. Overall, these findings show that microalgae act not only as passive filters through EPS-mediated flocculation but also as enzymatically active biocatalysts, reinforcing their dual functionality in MP remediation (Table 17.1).

17.5 Factors Influencing Microalgae–MP Interactions

The interactions between microalgae and MPs are governed by a complex interplay of physicochemical, biological, and environmental factors that ultimately determine the efficiency of MP aggregation, removal, or degradation. A thorough understanding of these factors is essential for optimizing microalgae-based remediation systems. The interactions between microalgae and MPs are mainly shaped by three interconnected aspects: (1) the properties of MPs (such as their size and degree of aging), (2) the biological characteristics of the microalgae (including the quality and quantity of their EPSs and enzymatic activity), and (3) the culture environment (including operational conditions and the presence of co-pollutants). Together, these factors govern key mechanisms such as adsorption, heteroaggregation, and biodegradation. Each of these aspects will be discussed in detail in the following sections.

17.5.1 MP Properties

Physical and chemical characteristics of the MPs determine their interaction with the microalgae. Properties like particle size, surface morphology, as well as degree of weathering influence adsorption, biofilm production and MP–microalgal interaction during treatment. The following sections discuss all these aspects in detail.

17.5.1.1 Effect of MP Size

Particle sizes and shapes are important factors affecting the interactions of MPs with microalgae. Smaller particles, particularly those in the nano- to micro-scale range, possess a higher surface area-to-volume ratio, offering more contact sites and hence increasing their capability to adhere to algal cells as well as aggregate. In a study by Su et al., where the heteroaggregation behavior of PS and PLA MPs with different microalgae was examined across varying particle sizes, it was found that smaller MP particles (74 μm) aggregated significantly faster than larger ones (613 μm), likely due to their greater surface area, which provided more interaction sites and facilitated

Table 17.1 Microalgal-based MP remediation

Microalgal species	MP type	Remediation technique	Process efficiency and other observations	Reference
Chlorella sp. L38 and *Phaeodactylum tricornutum* MASCC-0025	PP, PE, PET, and PVC	Heteroaggregation via EPSs	Microalgae secrete EPS—mainly polysaccharides and proteins—which enhance the adsorption and aggregation of MP particles on algal cell surfaces EPS-mediated hetero-aggregation facilitates physical entrapment and flocculation of MPs	Song et al. (2020)
Synechocystis sp.	PS	Enzymatic breakdown by laccase enzyme	45 days, *Synechocystis* sp. reduced the PS MPs weight by 23.2 ± 0.21% Laccase enzyme plays a central role in PS degradation via oxidative pathways	Gowthami et al. (2025)
Chlamydomonas reinhardtii	PET	Enzymatic breakdown by PHL7 enzyme	Microalgae secrete PHL7 enzyme, which degrades PET plastics into monomers like terephthalic acid	Molino et al. (2024)
Dunaliella salina	Oxidised HDPE and Oxium MPs	Heteroaggregation via EPSs	Microalgae secrete biofilm around the MPs, which host bacteria and fungi, which secrete enzymes/acids. Together, these EPS + microbes cause heteroaggregation and physical damage	Hadiyanto et al. (2022)
Chlorella vulgaris	PS and PLA (613 μm)	Heteroaggregation via EPSs	MPs-microalgae heteroaggregates adsorbed more co-existing pollutants, such as Cu2+, than pristine (clean, unaggregated) MPs. This also negatively impacted the microalgal growth due to enhanced toxicity	Su et al. (2025)
Chlorella vulgaris	PP and PET	Heteroaggregation via EPSs	*Chlorella vulgaris* can enhance MP removal via EPS-mediated flocculation PET MPs led to higher EPS production and better removal than PP	Joelyna et al. (2024)

Cyanothece sp.	PS nano- and MPs	Heteroaggregation and flocculation via EPSs	Even at low concentrations, EPS from *Cyanothece* sp. showed strong bioflocculant properties, effectively aggregating both nano- and micro-sized PS particles EPS reduced turbidity by up to 60% in flocculation assays, showing strong MP aggregation MP exposure reduced microalgal growth by up to 47%	Cunha et al. (2020)
Scenedesmus abundans	Polymethyl methacrylate, PLA and PS	Heteroaggregation via EPSs	MPs presence increased both bound and soluble EPS production and EPS mediated heteroaggregation contributed to more than 70% of the MP removal Removal efficiency by type: 98% for polymethyl methacrylate, 70% for PLA, and 70% for PS	Cheng and Wang (2022)
Scenedesmus obliquus, *Chlorella vulgaris*, *Tetraselmis subcordiformis*, *Chaetoceros müelleri*, and *Streptococcus westermani*	PS and PLA	Heterogeneous aggregation	Heterogeneous aggregation efficiency: *Scenedesmus obliquus* > *Chlorella vulgaris* > *Tetraselmis subcordiformis* > *Chaetoceros müelleri* > *Streptococcus westermani*	Su et al. (2023)
C. vulgaris	PP and PET	Biofloculation	Removal efficiencies of different MPs: 55.7% for PP and 95.0% for PET	Hadiyanto et al. (2025)
C. vulgaris + *Aspergillus niger* pellets			Removal efficiencies of different MPs: 90 ± 0.031% for PP and 94 ± 0.015% for PET	
Natural green and filamentous algae with bacteriaBacteria (CyanobacteriaCyanobacteria,	PS of different sizes (0.065, 0.5, and 5 μm)	Heterogeneous aggregation via algal–bacterial protein-rich biofilm	PS MPs of size 0.5 and 5 μm, accumulated and adhered to the biofilm surface	Gong et al. (2023)

(continued)

Table 17.1 (continued)

Microalgal species	MP type	Remediation technique	Process efficiency and other observations	Reference
Proteobacteria Firmicutes, Actinobacteria)				
Desmodesmus sp. + algae-associated microbial communities	PVC, PET, PE, PS	Heterogeneous aggregation via algal–bacterial	Algal–bacterial consortia built around *Desmodesmus* sp. are capable of degrading MPs through EPS release, biofilm formation, and chemical alteration of MPs	Fan et al., (2024)

quicker attachment and aggregation with microalgal cells (Su et al. 2023). Song et al. further mentioned that the size of the MP also influences what type and how much EPS the algae produce (Song et al. 2023). When PS MPs of various sizes (specifically 20 nm, 100 nm, and 1 μm) came in contact with *Microcystis aeruginosa*, it produced distinct EPS profiles: for 100 nm PS MPs, the algae produced more aromatic proteins and fulvic acids, whereas 20 nm PS MPs led to increased secretion of polysaccharides and humic acids.

17.5.1.2 Effect of MP Aging

Apart from this, the age of the MPs plays a crucial role in this process. Aged MPs, those that have undergone environmental weathering, showed higher surface roughness and the occurrence of hydroxyl and carbonyl groups due to oxidation. These changes make aged MPs more reactive and likely to bind with biofilms or microalgal EPS (Yu et al. 2023). It was also reported that aged MPs have a higher capacity to adsorb co-contaminants compared to pristine particles, further intensifying their toxic impact. Su et al. confirmed that aged MPs combined with heavy metals such as Cu^{2+} were markedly more detrimental to microalgal growth than new MPs mixed with the same contaminants (Su et al. 2025).

17.5.2 Microalgal Properties

The unique biological traits of microalgae—including their capacity to produce EPS and secrete specific enzymes—shape their ability to capture, aggregate, or degrade MPs. These points are discussed in detail in the following sections.

17.5.2.1 Quality and Quantity of the Microalgal EPS

The quantity and quality of microalgal EPS are central to their interactions with microplastics, as these sticky biopolymers drive the adhesion of plastic particles and the formation of stable heteroaggregates. The quantity and biochemical composition of EPSExopolysaccharides (EPSs) vary widely across microalgal strains (Cunha et al. 2019) and under different environmental conditions (presence of any environmental stressor like salinity or nutrient-limited conditions) (Chaudhuri et al. 2025), ultimately influencing the heteroaggregation dynamics and the MP removal efficiency of the whole system. For instance, Cunha et al. observed that strains such as *Microcystis panniformis* produce relatively small EPS molecules (10–110 μm) that are prone to disaggregation, thereby limiting aggregate formation (Cunha et al. 2019). In contrast, strains like *Tetraselmis* sp. secrete larger EPS molecules with greater aggregation potential. Authors further noted that, among the different microalgal strains tested, *Gloeocapsa* sp. produced EPS in abundance (~2.15 mg mL^{-1}) that was dense, thick, and viscous, demonstrating the strongest aggregation capacity by adhering to, incorporating, and colonizing diverse types of microplastics. In another study, Cunha et al. mentioned that although the precise roles of different biomolecules' (such as carbohydrate and protein) fractions in determining flocculation potential are yet to be understood fully, existing evidence suggests that carbohydrate-rich EPS generally enhance flocculation capacity (Cunha et al. 2020). Together, these findings highlight that both the quantity of EPS produced and its biochemical composition (molecular size, viscosity, and carbohydrate-to-protein balance) act in concert to regulate heteroaggregation dynamics and ultimately define the remediation potential of microalgae.

17.5.2.2 Enzymatic Activity

Some microalgae have demonstrated promising enzymatic potential for MP biodegradation through the production of oxidative enzymes. For instance, *Synechocystis* sp. has been shown to produce laccases that can break down polymer structures in persistent plastics like PS; this enzymatic activity has resulted in weight loss of up to approximately 23% over 45 days, indicating real biodegradation capacity for such resistant materials (Gowthami et al. 2025). Similarly, *Arthrospira platensis* (commonly known as *Spirulina*), a widely cultivated protein-rich microalga, can produce enzymes such as laccase and catalase, which in silico docking studies have been shown to have strong binding affinities for bioplastic polymers, suggesting its potential applicability for MP degradation as well (Sharma and Vuppu 2023). In addition, recombinant *Chlamydomonas reinhardtii* has been successfully engineered to secrete the PHL7 enzyme, which degrades PET plastics into monomers like terephthalic acid, highlighting the potential of genetically modified microalgae as a sustainable platform for breaking down persistent polyesters (Molino et al. 2024). However, it is important to note that the effectiveness of such enzymatic degradation processes depends strongly on the metabolic capability and genetic characteristics of the microalgal strain used.

17.5.3 Microalgal Culture Environment

Beyond the traits of MPs and microalgae, the culture environment adds complexity. Operating conditions and co-pollutants significantly influence microalgae–MP interactions as well as the stability and fate of aggregates. The following sections talk about these influences in detail.

17.5.3.1 Operating Conditions

Operating conditions such as light availability, pH, and ionic strength play a critical role in shaping the interactions between microalgae and MPs and, ultimately, the overall efficiency of bioremediation processes. For instance, MPs in water raise turbidity and form heteroaggregates that block light, limiting chloroplast exposure needed for photosynthesis. This shading effect slows down microalgal growth, reduces pigment and photosynthetic activity, and thereby indirectly limits their ability to aggregate and settle MPs through bioflocculation (Rani-Borges et al. 2021).

In addition to light availability, chemical conditions in the surrounding medium are equally influential. Wang et al. demonstrated that the retention and transport of PS MPs in water-saturated porous media are strongly influenced by environmental factors such as ionic strength and pH, which affect the attachment efficiency of MPs (Wang et al. 2022). Specifically, higher ionic strength compresses the electrical double layer, promoting greater attachment and reduced migration of MP particles. This mechanism suggests that similar physicochemical interactions may govern the MP–microalgae heteroaggregation, where double-layer compression can enhance flocculation and improve pollutant removal.

17.5.3.2 Presence of Co-pollutants

The coexistence of other pollutants alongside MPs can significantly modulate the toxicity and behavior of these particles in the aquatic environment. For instance, Yang et al. demonstrated that exposure to MPs and nonylphenol together caused significant oxidative stress and growth inhibition in *Chlorella pyrenoidosa*, confirming that MPs can intensify the toxicity of coexisting pollutants (Yang et al. 2020). Su et al. also stated a similar observation that when MPs are attached to the microalgae, they can adsorb more coexisting pollutants, such as heavy metals like Cu^{2+}, and this MP–microalgae heteroaggregate showed greater toxicity to microalgal growth than either pollutant alone (Su et al. 2025).

An additional layer of complexity arises from the interaction of MPs with naturally occurring dissolved organic matter like humic acid. Shen et al. reported that the presence of humic acids (natural organic compounds rich in functional groups) can significantly alter how MPs behave as carriers for coexisting pollutants (Shen et al. 2018). When humic acids adsorb onto the surface of MPs, they form a coating that changes the surface properties of the particles, making them less hydrophobic and therefore less likely to attract hydrophobic contaminants such as tetracycline. Moreover, this humic acid layer can introduce electrostatic repulsion between coated MPs and certain pollutant molecules, further limiting the MPs' ability to adsorb certain pollutants.

17.6 Current Challenges in the Microalgal-Based Remediation Process and Feasibility

Compared to conventional methods, microalgae-based bioremediation offers distinct advantages for the removal of MPs from wastewater. However, several key challenges continue to limit its large-scale implementation. These include the long retention times required for effective treatment, the potential toxicity of MPs to microalgal cultures (which can inhibit or slow down the microalgal growth), and the environmental risks associated with the discharge of residual algal biomass or enhanced co-pollutant adsorption by microalgae–MP heteroaggregates (which can be even more toxic). These aspects are discussed in detail in the following sections.

17.6.1 Long Retention Time

One of the primary limitations is the inherently long retention time required to achieve effective removal since stable MP remediation depends on time-intensive biological processes involved: microalgal growth, EPS secretion, MP aggregation, and sedimentation of these aggregates. In practice, this challenge is further amplified by the heterogeneity of wastewater-derived MPs in terms of shape, size, density, and surface charge, which critically influence their interaction with microalgal cells and affect treatment predictability.

17.6.2 Environmental Concerns Linked to Inefficient Microalgal Biomass Recovery

A second concern is the environmental risk posed by residual algal biomass left in the treatment system after remediation. Currently, the most commonly used harvesting methods include flotation, sedimentation, centrifugation, and filtration, with recovery efficiencies ranging approximately from 10%–90%, 50%–90%, slightly over 90%, and 70%–90%, respectively. Although centrifugation offers the highest recovery, it is important to note that it is energy-intensive and economically unviable for large-scale applications. In contrast, less efficient techniques often fail to completely separate biomass, leaving residual algal cells in the effluent. If released, these cells can disrupt native biodiversity and destabilize aquatic ecosystems (Parashar and Hait 2023). Such ecological risks underscore the need for more effective strategies to manage and remove residual biomass after treatment.

17.6.3 Toxicity of MPs to Microalgae

Lastly, another bottleneck in this microalgal remediation process is the inherent toxicity that MPs pose to the microalgae themselves. The presence of MPs in the culture environment can inhibit algal growth and metabolic activity, primarily due to physical stress and the leaching of harmful by-products. For instance, Hadiyanto et al. observed that when *Dunaliella salina* was cultivated in the presence of oxidized HDPE MPs, the microalgae exhibited 40%–80% growth inhibition, indicating substantial stress caused by leached substances from the MPs (Hadiyanto et al. 2022). Such physiological stress responses in microalgae can be understood in terms of their relative sensitivity to different toxicological endpoints. According to the framework proposed by Rani-Borges et al., growth inhibition typically occurs first, followed by disruptions in photosynthesis, enzymatic imbalances, oxidative stress, and finally changes in pigment composition (Rani-Borges et al. 2021). This progression indicates that even when algal viability seems unaffected, sublethal stress may already compromise essential functions required for effective remediation.

This challenge further intensifies when MPs coexist with other co-pollutants, as their combined toxicity can amplify adverse effects on microalgal growth (Su et al. 2025; Yang et al. 2020). Importantly, as mentioned earlier, environmental weathering or aging of MPs also enhances their ability to bind these co-contaminants, thereby increasing their toxicity and undermining treatment performance. Ultimately, pollutant-loaded MP-microalgae aggregates pose a dual risk: they directly threaten algal viability and also pose a broader environmental risk by potentially re-releasing concentrated pollutants into aquatic ecosystems.

Taken together, these challenges highlight the need for continued research to optimize retention time, improve harvesting efficiency, and mitigate the combined toxicity risks of MPs and co-contaminants in future microalgae-based remediation systems.

17.7 Future Direction and Possible Ways

To effectively overcome the challenges associated with MP pollution, future research should focus on several key areas (as highlighted in Fig. 17.4). First, it is essential to continue investigating the occurrence, distribution, properties, and fate of MPs in wastewater and other environmental compartments to establish a comprehensive baseline. Such foundational data are crucial for designing effective mitigation strategies.

Reducing MP emissions to the environment requires identifying and managing the diverse pathways through which these particles infiltrate terrestrial and aquatic ecosystems. MPs are introduced via point sources, such as effluent from wastewater treatment plants, leachate from landfills, and direct disposals at open dumping sites,

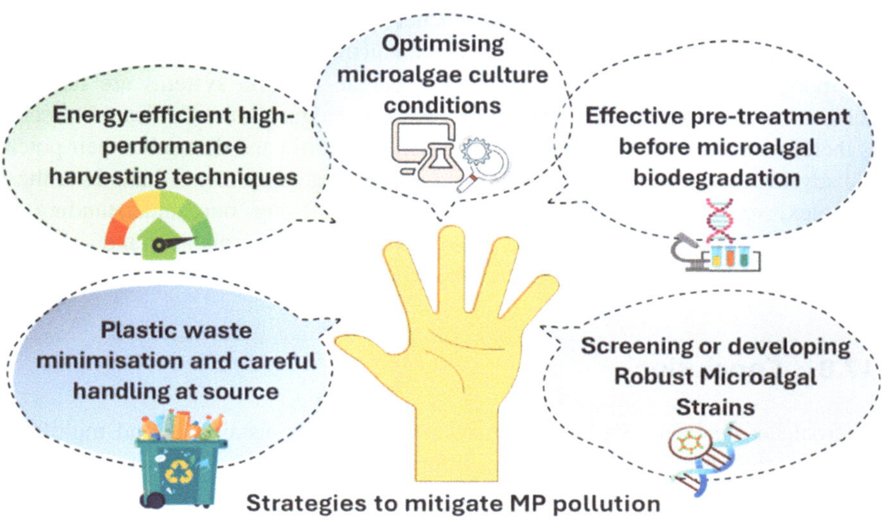

Fig. 17.4 Possible ways to reduce MP pollution through microalgal-based bioremediation

as well as through nonpoint sources, including stormwater runoff and atmospheric deposition. Among these, controlling nonpoint pollution, such as stormwater runoff, a major pathway from land to water, is challenging due to its widespread discharge points. For example, Cho et al. reported that road dust transported by stormwater runoff contributed up to 99% of MPs entering the Han Cheon River in Korea, highlighting the diffuse and hard-to-regulate nature of this pathway (Cho et al. 2023). In contrast, point sources like wastewater plants, though significant, are easier to monitor and regulate due to their centralized infrastructure (Salvador Cesa et al. 2017). Besides managing these pathways, tracing the origins of MPs, the parent plastic products, is also critical to addressing pollution at its source (Yamahara et al. 2024).

Secondly, for microalgae-based removal systems, further research should aim to reduce retention time by selecting the most effective microalgal species, optimizing operational conditions, developing innovative treatment systems, and integrating these techniques with conventional methods to enhance overall performance.

Thirdly, as highlighted earlier, when implementing microalgae-based systems for MP remediation, it is essential to ensure that the residual microalgal biomass is efficiently harvested to prevent any potential secondary environmental risks. To achieve this, highly effective harvesting methods, such as fungal-induced flocculation (Pei et al. 2021) or magnetic flocculation (Liu et al. 2019; Wang et al. 2014), should be further developed and optimized. This will help prevent unintended algal release, which could otherwise lead to secondary pollution or eutrophication.

Finally, future investigations should aim to replicate more realistic contamination scenarios, recognizing that aquatic environments are rarely exposed to single pollutants in isolation. Instead, these systems typically encounter complex MP mixtures of diverse sizes along with other co-pollutants. As environmental stressors have cumulative and interacting effects, the physiological responses of microalgae will inevitably reflect this combined influence. Along with that, it is equally important to recognize that in natural environments, microalgae coexist with other microorganisms such as bacteria, and microalgae–bacteria systems are likely to achieve more effective MP remediation than microalgae alone. Dedicated research is therefore needed to explore these synergistic consortia and determine their potential advantages for large-scale applications. Designing experiments that mimic these complex, real-world conditions is vital for deepening our understanding of microalgae–MP interactions and developing robust, scalable, and efficient microalgae-based MP remediation technologies.

17.8 Conclusion

Microalgae are increasingly recognized as potentially sustainable and multifunctional agents for MP remediation, offering both passive and active removal mechanisms such as EPS-mediated flocculation, heteroaggregation, enzymatic degradation, and biofilm formation. Notably, in this microalgae-based MP remediation system, microalgal EPS plays a dual role—shielding algal cells from stress and creating a bio-supportive matrix for microbial degradation. Microalgal interactions with MPs are influenced by species-specific traits, plastic properties, and environmental conditions, making strain selection and culture optimization critical for effective application. While promising, large-scale deployment faces challenges, including long retention times, performance variability across strains, and MP-induced toxicity that can impair algal function. However, strategies such as plastic pretreatment, co-cultivation with microbial consortia, and genetic enhancement of microalgae show potential to overcome these barriers. In summary, microalgae offer a promising green technology and a hopeful path forward for MP remediation. Future interdisciplinary efforts are needed to optimize these systems and enable their transition from lab-scale success to field-scale application.

Acknowledgments The authors acknowledge the Department of Biotechnology and Medical Engineering of the National Institute of Technology Rourkela for providing the research facility.

References

Amparán MAA, Palacios A, Flores GM, Olivera PMC (2025) Review and future outlook for the removal of microplastics by physical, biological and chemical methods in water bodies and wastewaters. Environ Monit Assess 197:429. https://doi.org/10.1007/s10661-025-13883-0

An L, Liu Q, Deng Y, Wu W, Gao Y, Ling W (2020) Sources of microplastic in the environment. In: He D, Luo Y (eds) Microplastics in terrestrial environments. The handbook of environmental chemistry, vol 95. Springer, Cham, pp 143–159. https://doi.org/10.1007/698_2020_449

Arienzo M, Donadio C (2023) Microplastic–pharmaceuticals interaction in water systems. J Mar Sci Eng 11:1437. https://doi.org/10.3390/jmse11071437

Boucher J, Friot D (2017) Primary microplastics in the oceans: a global evaluation of sources. IUCN, Gland. https://doi.org/10.2305/IUCN.CH.2017.01.en

Chaudhuri R, Balasubramanian P (2023) Evaluating the potential of exopolysaccharide extracted from the spent cultivation media of Spirulina sp. as plant biostimulant. Biomass Convers Biorefinery. https://doi.org/10.1007/s13399-023-04865-8

Chaudhuri R, Balasubramanian P (2025) Enhancing salt tolerance in leafy vegetables using microalgal exopolysaccharides. J Crop Health 77:70. https://doi.org/10.1007/s10343-025-01124-w

Chaudhuri R, Athiyappan KD, Balasubramanian P (2025) From algae to medicine: unveiling the therapeutic potential of microalgal exopolysaccharides. In: Microbial biotechnology: integrated microbial engineering for B3—bioenergy, bioremediation, and bioproducts. Elsevier, pp 271–305. https://doi.org/10.1016/b978-0-443-23796-6.00004-3

Cheng Y-R, Wang H-Y (2022) Highly effective removal of microplastics by microalgae *Scenedesmus abundans*. Chem Eng J 435:135079. https://doi.org/10.1016/j.cej.2022.135079

Cho Y, Shim WJ, Ha SY, Han GM, Jang M, Hong SH (2023) Microplastic emission characteristics of stormwater runoff in an urban area: intra-event variability and influencing factors. Sci Total Environ 866:161318. https://doi.org/10.1016/j.scitotenv.2022.161318

Coyle R, Hardiman G, Driscoll KO (2020) Microplastics in the marine environment: a review of their sources, distribution processes, uptake and exchange in ecosystems. Case Stud Chem Environ Eng 2:100010. https://doi.org/10.1016/j.cscee.2020.100010

Cunha C, Faria M, Nogueira N, Ferreira A, Cordeiro N (2019) Marine vs freshwater microalgae exopolymers as biosolutions to microplastics pollution. Environ Pollut 249:372–380. https://doi.org/10.1016/j.envpol.2019.03.046

Cunha C, Silva L, Paulo J, Faria M, Nogueira N, Cordeiro N (2020) Microalgal-based biopolymer for nano- and microplastic removal: a possible biosolution for wastewater treatment. Environ Pollut 263:114385. https://doi.org/10.1016/j.envpol.2020.114385

Danopoulos E, Jenner LC, Twiddy M, Rotchell JM (2020) Microplastic contamination of seafood intended for human consumption: a systematic review and meta-analysis. Environ Health Perspect 128:126002-1–126002-32. https://doi.org/10.1289/EHP7171

de Souza Machado AA, Lau CW, Kloas W, Bergmann J, Bachelier JB, Faltin E et al (2019) Microplastics can change soil properties and affect plant performance. Environ Sci Technol 53: 6044–6052. https://doi.org/10.1021/acs.est.9b01339

Du Z, Li G, Ding S, Song W, Zhang M, Jia R et al (2023) Effects of UV-based oxidation processes on the degradation of microplastic: fragmentation, organic matter release, toxicity and disinfection byproduct formation. Water Res 237:119983. https://doi.org/10.1016/j.watres.2023.119983

Fan X, Kong L, Wang J, Tan Y, Xu X, Li M et al (2024) Surface-programmed microbiome assembly in phycosphere to microplastics contamination. Water Res 262. https://doi.org/10.1016/j.watres.2024.122064

FAO (2012) FAO Report. The state of world fisheries and aquaculture. Food and Agriculture Organization of the United Nations/Eurospan [distributor]

Ghosh T (2025) Microplastics bioaccumulation in fish: its potential toxic effects on hematology, immune response, neurotoxicity, oxidative stress, growth, and reproductive dysfunction. Toxicol Rep 14:101854. https://doi.org/10.1016/j.toxrep.2024.101854

Gong X, Ge Z, Ma Z, Li Y, Huang D, Zhang J (2023) Effect of different size microplastic particles on the construction of algal-bacterial biofilms and microbial communities. J Environ Manage 343:118246. https://doi.org/10.1016/j.jenvman.2023.118246

Gowthami A, Syed Marjuk M, Santhanam P, Thirumurugan R, Muralisankar T, Perumal P (2025) Marine microalgae—mediated biodegradation of polystyrene microplastics: insights from enzymatic and molecular docking studies. Chemosphere 370:144024. https://doi.org/10.1016/j.chemosphere.2024.144024

Gupta A, Ghosh A, Yadav A, Kirti A, Lenka SS, Jena S et al (2024) Microplastics: the imperative influencer in blueprint of blue economy. J Environ Manag 372:123300. https://doi.org/10.1016/j.jenvman.2024.123300

Hadiyanto H, Khoironi A, Dianratri I, Huda K, Suherman S, Muhammad F (2022) Biodegradation of oxidized high-density polyethylene and oxo-degradable plastic using microalgae *Dunaliella salina*. Environ Pollut Bioavailab 34:469–481. https://doi.org/10.1080/26395940.2022.2128884

Hadiyanto H, Joelyna FA, Khoironi A, Sudarno S, Safaat JA, Pratama WD et al (2025) Harnessing *Chlorella vulgaris–Aspergilus niger* interactions for effective microplastic removal in aquatic ecosystems. Waste Biomass Valoriz. https://doi.org/10.1007/s12649-025-03062-0

Hanif MN, Aijaz N, Azam K, Akhtar M, Laftah WA, Babur M et al (2024) Impact of microplastics on soil (physical and chemical) properties, soil biological properties/soil biota, and response of plants to it: a review. Int J Environ Sci Technol 21:10277–10318. https://doi.org/10.1007/s13762-024-05656-y

Hoang TC (2022) Plastic pollution: where are we regarding research and risk assessment in support of management and regulation? Integr Environ Assess Manag 18:851–852. https://doi.org/10.1002/ieam.4627

Hu J, Lim FY, Hu J (2023) Ozonation facilitates the aging and mineralization of polyethylene microplastics from water: behavior, mechanisms, and pathways. Sci Total Environ 866:161290. https://doi.org/10.1016/j.scitotenv.2022.161290

Iqbal R, Khan MT, Bilal H, Aslam MM, Khan IA, Raja S et al (2022) Microplastics as vectors of environmental contaminants: interactions in the natural ecosystems. Hum Ecol Risk Assess Int J 28:1022–1042. https://doi.org/10.1080/10807039.2022.2120846

Joelyna FA, Hadiyanto H, Khoironi A, Pratama WD (2024) The utilization of exopolysaccharide (EPS) from microalgae *Chlorella vulgaris* in microplastic removal. In: IOP conference series:

earth and environmental science, vol 1414. Institute of Physics. https://doi.org/10.1088/1755-1315/1414/1/012017

Kausar RA, Buhani S, Setiawan A (2021) Synthesis and characterization exopolysaccharide from algae Spirulina sp using technique sol-gel as adsorbent Pb(II) ion. IOP Conf Ser Mater Sci Eng 1173:012014. https://doi.org/10.1088/1757-899X/1173/1/012014

Liu Y, Jin W, Zhou X, Han S-F, Tu R, Feng X et al (2019) Efficient harvesting of *Chlorella pyrenoidosa* and *Scenedesmus obliquus* cultivated in urban sewage by magnetic flocculation using nano-Fe3O4 coated with polyethyleneimine. Bioresour Technol 290:121771. https://doi.org/10.1016/j.biortech.2019.121771

Manosuthi J (2024) Microplastic pollution: the real potential threats to human being. Sci Insights 44:1263–1272. https://doi.org/10.15354/si.24.re925

Mecha AC, Chollom MN (2020) Photocatalytic ozonation of wastewater: a review. Environ Chem Lett 18:1491–1507. https://doi.org/10.1007/s10311-020-01020-x

Molino JVD, Saucedo B, Kang K, Walsh C, Diaz CJ, Tessman M et al (2024) Efficient secretion of a plastic degrading enzyme from the green algae *Chlamydomonas reinhardtii*. Sci Rep. https://doi.org/10.1101/2024.10.23.619606

Moreira JB, Kuntzler SG, Bezerra PQM, Cassuriaga APA, Zaparoli M, da Silva JLV et al (2022) Recent advances of microalgae exopolysaccharides for application as bioflocculants. Polysaccharides 3:264–276. https://doi.org/10.3390/polysaccharides3010015

Ofoegbu O, Onuwa PO, Kelle HI, Onogwu SU, Ajakaye OJ, Lullah-Deh JA et al (2024) The environmental and health implications of microplastics on human and aquatic life. Int J Res Innov Appl Sci IX:367–377. https://doi.org/10.51584/IJRIAS.2024.909031

Parashar N, Hait S (2023) Recent advances on microplastics pollution and removal from wastewater systems: a critical review. J Environ Manag 340:118014. https://doi.org/10.1016/j.jenvman.2023.118014

Park JW, Lee SJ, Hwang DY, Seo S (2021) Removal of microplastics *via* tannic acid-mediated coagulation and *in vitro* impact assessment. RSC Adv 11:3556–3566. https://doi.org/10.1039/D0RA09645H

Park S, Kim I, Jeon W-H, Moon HS (2023) Exploring the vertical transport of microplastics in subsurface environments: lab-scale experiments and field evidence. J Contam Hydrol 257:104215. https://doi.org/10.1016/j.jconhyd.2023.104215

Pei X-Y, Ren H-Y, Liu B-F (2021) Flocculation performance and mechanism of fungal pellets on harvesting of microalgal biomass. Bioresour Technol 321:124463. https://doi.org/10.1016/j.biortech.2020.124463

Pellegrino D, La Russa D, Barberio L (2025) Pollution has no borders: microplastics in Antarctica. Environments 12:77. https://doi.org/10.3390/environments12030077

Ragossnig AM, Agamuthu P (2021) Plastic waste: challenges and opportunities. Waste Manag Res 39:629–630. https://doi.org/10.1177/0734242X211013428

Rani-Borges B, Moschini-Carlos V, Pompêo M (2021) Microplastics and freshwater microalgae: what do we know so far? Aquat Ecol 55:363–377. https://doi.org/10.1007/s10452-021-09834-9

Sacco NA, Zoppas FM, Devard A, González Muñoz MDP, García G, Marchesini FA (2023) Recent advances in microplastics removal from water with special attention given to photocatalytic degradation. Rev Sci Res Microplast 2:278–303. https://doi.org/10.3390/microplastics2030023

Sakali A, Egea-Corbacho A, Coello D, Albendín G, Arellano J, Rodríguez-Barroso R (2024) Analysis of microplastics in the reuse of compost in three agricultural sites (Cádiz, Spain) as a circular economy strategy: detection of micropollutants and incidence of plastic ingestion levels by annelids. Environ Sci Pollut Res 31:51747–51759. https://doi.org/10.1007/s11356-024-34615-w

Salvador Cesa F, Turra A, Baruque-Ramos J (2017) Synthetic fibers as microplastics in the marine environment: a review from textile perspective with a focus on domestic washings. Sci Total Environ 598:1116–1129. https://doi.org/10.1016/j.scitotenv.2017.04.172

Sharma N, Vuppu S (2023) In silico study of enzymatic degradation of bioplastic by microalgae: an outlook on microplastic environmental impact assessment, challenges, and opportunities. Mol Biotechnol. https://doi.org/10.1007/s12033-023-00886-w

Shen X-C, Li D-C, Sima X-F, Cheng H-Y, Jiang H (2018) The effects of environmental conditions on the enrichment of antibiotics on microplastics in simulated natural water column. Environ Res 166:377–383. https://doi.org/10.1016/j.envres.2018.06.034

Shen M, Song B, Zhou C, Hu T, Zeng G, Zhang Y (2022) Advanced oxidation processes for the elimination of microplastics from aqueous systems: assessment of efficiency, perspectives and limitations. Sci Total Environ 842:156723. https://doi.org/10.1016/j.scitotenv.2022.156723

Smith M, Love DC, Rochman CM, Neff RA (2018) Microplastics in seafood and the implications for human health. Curr Environ Health Rep 5:375–386. https://doi.org/10.1007/s40572-018-0206-z

Song C, Liu Z, Wang C, Li S, Kitamura Y (2020) Different interaction performance between microplastics and microalgae: the bio-elimination potential of *Chlorella* sp. L38 and *Phaeodactylum tricornutum* MASCC-0025. Sci Total Environ 723:138146. https://doi.org/10.1016/j.scitotenv.2020.138146

Song Y, Zhang B, Si M, Chen Z, Geng J, Liang F et al (2023) Roles of extracellular polymeric substances on *Microcystis aeruginosa* exposed to different sizes of polystyrene microplastics. Chemosphere 312:137225. https://doi.org/10.1016/j.chemosphere.2022.137225

Sousa PMS, Kinney KA, Sousa CA, Simões M (2025) Microalgae for microplastic removal from water and wastewater: a review. Environ Chem Lett 23:611–648. https://doi.org/10.1007/s10311-024-01809-0

Stride B, Abolfathi S, Bending GD, Pearson J (2024) Quantifying microplastic dispersion due to density effects. J Hazard Mater 466:133440. https://doi.org/10.1016/j.jhazmat.2024.133440

Su Y, Gao L, Peng L, Diao X, Lin S, Bao R et al (2023) Heterogeneous aggregation between microplastics and microalgae: may provide new insights for microplastics removal. Aquat Toxicol 261:106638. https://doi.org/10.1016/j.aquatox.2023.106638

Su Y, Gao L, Xu EG, Peng L, Diao X, Zhang Y et al (2025) When microplastics meet microalgae: unveiling the dynamic formation of aggregates and their impact on toxicity and environmental health. Water Res 273:123008. https://doi.org/10.1016/j.watres.2024.123008

Sui Q, Yang X, Sun X, Zhu L, Zhao X, Feng Z et al (2024) Bioaccumulation of polycyclic aromatic hydrocarbons and their human health risks depend on the characteristics of microplastics in marine organisms of Sanggou Bay, China. J Hazard Mater 473:134622. https://doi.org/10.1016/j.jhazmat.2024.134622

Swain PR, Parida PK, Majhi PJ, Behera BK, Das BK (2025) Microplastics as emerging contaminants: challenges in inland aquatic food web. Water (Basel) 17:201. https://doi.org/10.3390/w17020201

Teran FJC, Mosque VAA, De Oliveira LR, Cuba RMF (2023) Analysis of the efficiency of the electrocoagulation process in the removal of microplastics. Rev Gest Soc Ambient 17:e03280. https://doi.org/10.24857/rgsa.v17n3-027

Thompson RC, Courtene-Jones W, Boucher J, Pahl S, Raubenheimer K, Koelmans AA (2024) Twenty years of microplastic pollution research—what have we learned? Science 386:6720. https://doi.org/10.1126/science.adl2746

Vercauteren M, Lambert S, Hoogerwerf E, Janssen CR, Asselman J (2024) Microplastic-specific biofilm growth determines the vertical transport of plastics in freshwater. Sci Total Environ 910:168399. https://doi.org/10.1016/j.scitotenv.2023.168399

Wang S-K, Wang F, Hu Y-R, Stiles AR, Guo C, Liu C-Z (2014) Magnetic flocculant for high efficiency harvesting of microalgal cells. ACS Appl Mater Interfaces 6:109–115. https://doi.org/10.1021/am404764n

Wang T, Li B, Zou X, Wang Y, Li Y, Xu Y et al (2019) Emission of primary microplastics in mainland China: invisible but not negligible. Water Res 162:214–224. https://doi.org/10.1016/j.watres.2019.06.042

Wang Y, Xu L, Chen H, Zhang M (2022) Retention and transport behavior of microplastic particles in water-saturated porous media. Sci Total Environ 808:152154. https://doi.org/10.1016/j.scitotenv.2021.152154

Yamahara S, Kubota R, Tun TZ, Nakata H (2024) Source traceability of microplastics in road dust using organic/inorganic plastic additives as chemical indicators. Sci Total Environ 932:172808. https://doi.org/10.1016/j.scitotenv.2024.172808

Yang W, Gao X, Wu Y, Wan L, Tan L, Yuan S et al (2020) The combined toxicity influence of microplastics and nonylphenol on microalgae Chlorella pyrenoidosa. Ecotoxicol Environ Saf 195:110484. https://doi.org/10.1016/j.ecoenv.2020.110484

Yang H, Chen G, Wang J (2021) Microplastics in the marine environment: sources, fates, impacts and microbial degradation. Toxics 9:1–19. https://doi.org/10.3390/toxics9020041

Yu RS, Singh S (2023) Microplastic pollution: threats and impacts on global marine ecosystems. Sustainability 15:13252. https://doi.org/10.3390/su151713252

Yu X, Wang B, Han C, Liu L, Han X, Zheng B et al (2023) Physicochemical and biological changes on naturally aged microplastic surfaces in real environments over 10 months. Environ Pollut 337:122522. https://doi.org/10.1016/j.envpol.2023.122522

Zhang Y, Diehl A, Lewandowski A, Gopalakrishnan K, Baker T (2020) Removal efficiency of micro- and nanoplastics (180 nm–125 μm) during drinking water treatment. Sci Total Environ 720:137383. https://doi.org/10.1016/j.scitotenv.2020.137383

Algae-Powered Bioelectronics: Emerging Interfaces Between Living Systems and Green Technology

18

Adeline Juanita, S. Dhanush, and Satya Sundar Mohanty

Abstract

Algae-powered bioelectronics is a combination of biological metabolism and engineered interfaces that deliver renewable electric energy while remediating environmental pollutants at the same time. In this particular field, bio-photoelectrochemical cells (BPECs) exploit whole or subunits of micro- and macroalgae to capture sunlight and enhance the extracellular electron transfer. Cyanobacteria like *Spirulina*, *Synechocystis*, and other photosynthetic algae such as *Chlorella* use their natural photosynthetic systems that are in-built with light-harvesting systems in order to convert sunlight into electrical energy by transmitting electrons to electrodes. During this process, the algae utilize environmental carbon dioxide that indirectly helps in intrinsic carbon sequestration and also reduces the cost of CO2 supply. This concept is extended by photosynthetic microalgae–microbial fuel cells (PMMFCs) that pairs an anodic consortium of bacteria or similar systems with an algal cathode, thereby combining wastewater treatment with electricity generation. Selected targets and microalgae cultivation mitigate the characteristic sluggish oxygen-reduction reaction at the cathode, elevating power outputs and treatment efficiencies. Recent research studies have led to advancements in creating eco-friendly and highly conductive electrodes using biofilms. Using graphene and similar materials arranged in thin, organized layers using Langmuir–Blodgett method provides a large surface and good electrical flow, which helps produce more electricity. Complementary coating strategies such as dip, spin and spray coating, electrophoretic deposition and layer-by-layer assembly provide scalable routes to tailor film thickness, porosity,

A. Juanita · S. Dhanush · S. S. Mohanty (✉)
Division of Biotechnology, Karunya Institute of Technology and Sciences, Coimbatore, Tamil Nadu, India
e-mail: satyasundar@karunya.edu

K. Mohanty et al. (eds.), *Advances in Algal Research*,
https://doi.org/10.1007/978-981-95-7234-2_18

and functionality for diverse device architectures. Together, these innovations position algae-driven BPECs and PMMFCs at the nexus of green technology and biological innovation. By integrating carbon capture, pollutant removal, and clean electricity generation into a single platform. Algae-powered bioelectronics offer a viable, low-cost pathway toward climate-positive energy and water-treatment solutions and emphasize the potential of living systems when combined with advanced materials science.

Keywords

Bio-photoelectrochemical cell · Microalgae · Biofilms · Carbon capture · Wastewater treatment · Sustainable energy

18.1 Introduction

The use of conventional resources for the emerging societal concerns such as wastewater treatments, electricity crisis, increasing pollutants with heavy metal accumulation, and reduced fossil fuel resources has led to the search for renewable, less toxic, and feasible resources, and bioresources have been identified as the best alternatives. Entire or parts of living organisms and their products are the most sustainable and easily renewable resources that could be found in this generation. The role of algae, especially microalgae in the above mentioned crisis, has led to the exploration of the algal world for resources, mainly to deal with the demand in the energy supply. The incorporation and the interaction between algae and electronic designs in energy production, along with efficient storage and transmission, are referred to as bioelectronics, and this system paves a greener and more sustainable solution for reducing the demand in electricity. A bio-photovoltaic (BPV) cell is an upcoming renewable energy technology that is designed to capture and convert the solar energy produced by photosynthesis into usable electrical power. Unlike conventional photovoltaic devices that rely on inorganic semiconductors, BPV devices takes advantage of the intrinsic photosynthetic activity of microalgae, cyanobacteria, or higher plants. These organisms absorb light and drive electron transfer processes during photosynthesis and, in some cases, can be redirected toward an external circuit to generate electricity. Conventional photovoltaic (PV) technology, based on the photoelectric effect in semiconductors, has already achieved high efficiency (over 20%) and is most commonly used in industries. However, bio-photovoltaics (BPV) is a relatively new approach. Certain photovoltaic cells and algal films have already been designed with algal infusions and have sparked its ideology onto upcoming research. The major need for these biointegrated systems came up as the limitations of conventional methods reached its peak. The lack of resources, increasing pollution, higher maintenance, time-consuming and tedious processes, and a lot more factors are the reasons for the design of algae-based electronics as their maintenance cost is lower and the process is made simple. Along with the reduced to zero carbon dioxide production during the process, the ability of algae to accumulate and use environmental Co2 leading to the reduction of existing carbon levels in the environment is seen as an accomplishment in the field of bioremediation that surpasses the core principle in a carbon-neutral economy, where the amount of Co2 produced is equal to the amount of Co2 removed. In this chapter, we discuss the impacts of algae in the field of energy production along with its additional aid in carbon sequestration, its advantages and limitations, and the future scope of bioelectronics (Choi n.d.).

18.2 Algae-Based Integration in Photovoltaic Cells

18.2.1 Principles of Algal Bio-photovoltaics (BPVs)

In bio-photovoltaic (BPV) devices, electricity generation relies on the transfer of electrons from photosynthetic organisms to external electrodes. Unlike conventional solar panels, which are built from rigid materials, these biological systems have the unique capacity to self-replicate and repair, offering environmentally sustainable and durable solutions. Such features make BPVs particularly promising for remote or off-grid applications where traditional photovoltaic technologies may be less effective. Among various photosynthetic microorganisms, cyanobacteria are frequently investigated due to their relatively high photosynthetic efficiency (Yuan et al. 2025). The fundamental principle of algal BPVs is based on the redirection of electrons that is formed during photosynthesis to an external electrical circuit. In microalgae and cyanobacteria, photolysis of water is induced by the light absorbed via photosystem II (PSII). As a result, protons, oxygen, and electrons are produced. The produced electrons subsequently shuttle to photosystem I (PSI) through the photosynthetic electron transport chain (PETC). At this stage, the electrons can either participate in reducing $NADP^+$ to NADPH or be transferred toward the device's anode. The transfer may proceed directly through cell–surface redox proteins [direct electron transfer (DET)] or indirectly through naturally secreted or synthetic electron mediators. Both pathways ultimately allow current generation in BPV systems. Figure 18.1 provides a schematic overview of this process, showing light absorption at PSII, the splitting of water molecules, electron flow through PSI, and the diversion of electrons toward external electrodes for power generation.

18.2.2 Algae in Dye-Sensitized Solar Cells and Biohybrid Photovoltaics

Algae contribute significantly to innovative photovoltaic approaches, especially in dye-sensitized solar cells (DSSCs) and biohybrid systems. In DSSCs, pigments from algae such as chlorophylls, carotenoids, phycobiliproteins, and other light-harvesting molecules acts as a cost efficient, renewable sensitizers that can replace synthetic dyes by generating electricity through photoelectrochemical processes. The conversion efficiency of DSSC is below 5%; however, it can be increased by

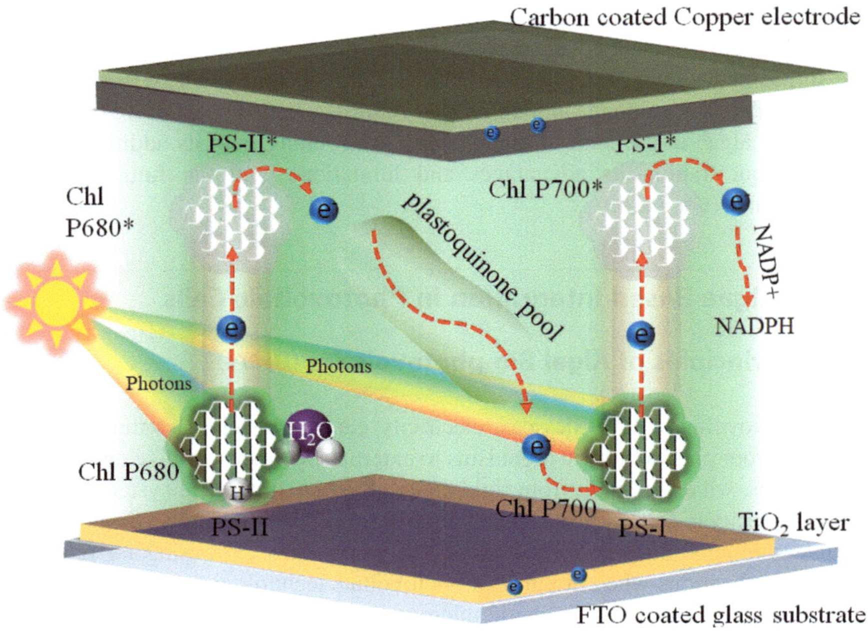

Fig. 18.1 Schematic overview of the principle of algal BPVs (Chatterjee et al. 2024)

integrating diatom frustules containing biogenic silica shells with nanoporous structure into TiO_2 photoanodes and this improves light scattering and photon trapping, which in turn boosts the efficiency by ~30%. Algae also contribute to biohybrid photovoltaics through their photosynthetic proteins. For example, complexes that helps in light harvesting such as the LHC4 from *Chlamydomonas reinhardtii,* immobilized on graphene, gives measurable photocurrents (~40 μA/cm^2) (arXiv). These approaches prove that algae are a natural dye source and functional biocomponents that plays a vital role in advancing sustainable and bio-integrated solar energy systems, particularly in DSSCs and biohybrid photovoltaics. DSSC produces electricity via a photoelectrochemical principle. These natural pigments anchor to the semiconductor surfaces and have broad absorption spectra. Algae-derived components also play a major role in biohybrid photovoltaics through the incorporation photosynthetic proteins. Light-harvesting complexes have been immobilized on conductive substrates such as graphene, which delivers measurable photocurrents (e.g., ~40 μA/cm^2) in prototype devices (Armendáriz-Mireles et al. 2023).

A dye-sensitized solar cell (DSSC) is a type of thin-film solar device that converts sunlight into electricity using a combination of a photoelectrode, a photosensitizing dye, an electrolyte, and a counter electrode. The photoelectrode consists of a transparent conducting glass substrate, usually coated with fluorine-doped tin oxide (FTO) or indium tin oxide (ITO), over which a mesoporous semiconductor layer, typically titanium dioxide (TiO_2), is applied. This nanostructured layer provides larger surface area for dye adsorption, improving light absorption and electron injection, while other semiconductors such as ZnO, SiO_2, and Nb_2O_5 can also be used. The dye forms a monolayer on the semiconductor surface and absorbs sunlight, generating electrons that are injected into the semiconductor. Proper alignment of the dye's molecular orbitals with the semiconductor conduction band is critical for efficient electron transfer, and anchoring groups or co-adsorbents help stabilize the dye and prevent recombination. The electrolyte commonly contains a redox couple like I^-/I_3^- that regenerates the dye by transporting electrons while it should also be conductive, thermally stable, and noncorrosive. This can be replaced by polymer gels or ionic liquids that enhance the stability significantly. The counter electrode is usually coated with platinum or carbon-based catalysts, which completes the circuit by reducing the redox species in the electrolyte and reflects the unabsorbed light back to the photoelectrode. The outline of the mechanism is that when light passes through the glass layer and hits the dye, the light is absorbed by the dye on the mesoporous semiconductor layer, and this excites the electrons from a lower to a higher energy level. Excited electrons are injected into the conduction band of the semiconductor, thus oxidizing the dye. The electrons travel through the semiconductor to the transparent conducting oxide (TCO) and flows through the external circuit to the counter electrode. In the counter electrode, they reduce the electrolyte and regenerate the dye, thereby completing the cycle. The carefully engineered components are put together that allows DSSCs to achieve efficient sunlight-to-electricity conversion while maintaining cost, flexibility and environmental safety. Thus, the main processes involved includes injecting electrons into the semiconductor by the dye, the redox couple transferring electrons to regenerate the dye, recombination of dye cations with electrons from the semiconductor, recombination of the redox couple with electrons, and the decay of the dye from the excited state back to its ground state. Figure 18.2 shows the basic elements of a DSSC and the chemical reactions involved during its working (Chauhan 2022).

18.2.3 Comparative Analysis: Traditional Versus Algae-Integrated PV Cells

Among renewable energy sources, solar energy is perhaps the most well-known. It is generated using photovoltaic panels, which convert sunlight into electricity. Solar energy is a clean, renewable source of energy that is widely available and can be used in a variety of applications, including electricity generation, heating, and lighting. One of the main benefits of solar energy is that it is relatively easy to install and maintain, and it can be used in a variety of locations, including urban, suburban, and rural areas.

Traditional photovoltaic (PV) cells are dominated by crystalline silicon-based technologies, and these have been the major energy production systems since the 1950s. The working principle is based on the photovoltaic effect, as semiconductors absorb photons and release electrons that results in the generation of electric current.

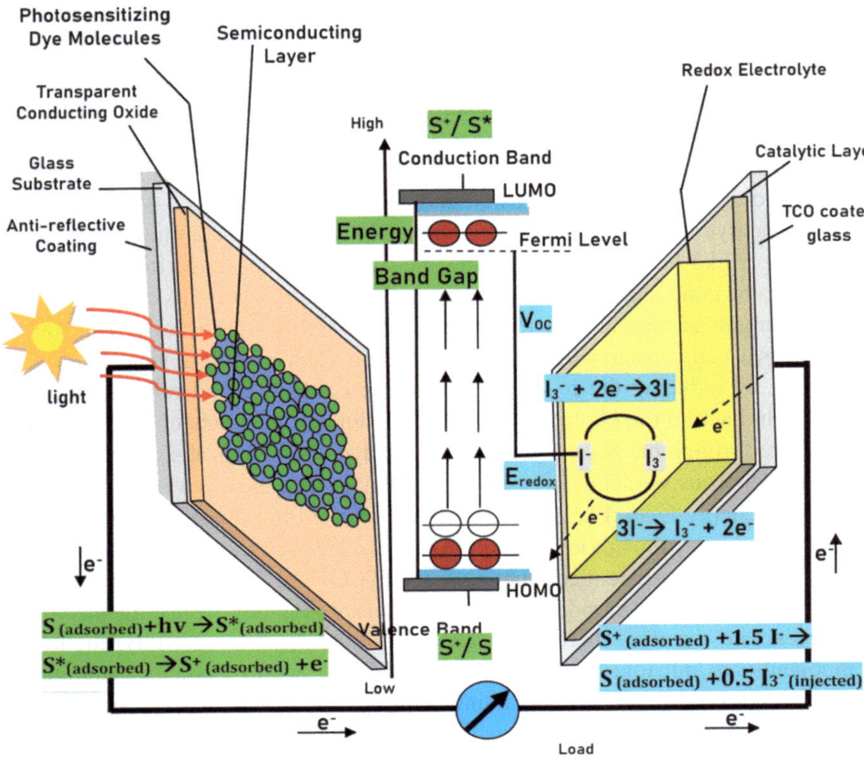

Fig. 18.2 Elements and basic operation of a DSCC (Chauhan 2022)

Traditional PV technology has evolved into three main generations and they are: first-generation silicon wafers, second-generation thin-film cells (CdTe, CIGS, amorphous Si), and third-generation cells such as perovskites, organic photovoltaics (OPVs), and multijunction designs. The conversion efficiencies of commercial silicon PVs are 25%, whereas the laboratory-based tandem and multijunction cells have achieved >40%. Durability, high efficiency, and scalability are their major advantages. However, the limitations such as high energy input for silicon purification, dependence on toxic metalloids (Cd, In, Te), ecological impacts during manufacturing, and challenges in recycling still exists (Solak and Irmak 2023).

Unlike traditional PV cells, algae-integrated PV cells, or bio-photovoltaics (BPVs), refer to a biological alternative that uses photosynthetic processes for electricity generation. Instead of inorganic semiconductors, these systems utilize photosynthetic algae such as *Nannochloropsis oculata* and are grown on conductive substrates (e.g., copper mesh). The released electrons during photosynthesis by algae are captured at the electrode, which acts as a source for producing current. Apart from electricity generation, BPVs add to the ecological benefits, mostly by CO_2 sequestration up to an extent of 89.1% in closed systems. Algae can utilize wastewater or saline water for growth and so cheaper cultivation, and it also produces biomass that could be valorized for biofuels or bioproducts. Moreover, BPVs avoid the usage of toxic materials that are used in traditional PV cells, thus making them more eco-friendly (Al Abdulla et al. 2024).

Parameter	Traditional PV cells (Solak and Irmak 2023)	Algae-integrated PV cells (BPVs) (Al Abdulla et al. 2024)
Power density	Crystalline–Si modules: 150–300 W/m^2	Microbial–algal systems—~1700 mW/m^2 Microfluidic BPVs ~500 mW/m^2
Conversion efficiency	Si PVs: 18%–25% Advanced lab tandem/multi-junction—>40%	BPVs 0.1%–1%
Operational lifetime	20–30 years	Few hours to days
CO_2 sequestration	No CO_2 removal	Up to 89.1% CO_2 removal
Environmental impact	Energy-intensive Uses Cd, In, Te—toxic	Uses algae and consortia—nontoxic
Material sustainability	Relies on mined minerals High-temperature manufacturing	Relies on sunlight and nutrients Can use all types of water for growth
Cost per watt	Less than $0.10 per kWh	$0.20–0.40 per kWh Improved BPV systems: $0.10–0.20 per kWh
Applications	Grid power, rooftops, utility-scale solar farms	Low-power devices, biosensors, self-sustaining microelectronics, environmental monitoring
Material cost	High	Low

18.2.4 Recent Advancements and Prototypes

The major aspect of advancements is focused on improving the transport of electrons to the electrode efficiently, design of the electrode, and overall performance of the cell to increase power output and durability. Modifying cyanobacteria such as *Synechocystis* using genetic engineering to block electron dissipation pathways such as respiratory oxidases, resulted in higher gains, with certain mutant algal cells reaching power densities of 540 mW/m^2. Advanced electrode designs such as the Inverse opal indium tin oxide (IO-ITO) electrodes have optimized pore sizes that allows cyanobacteria to penetrate the structure that boosts photocurrent nearly 300-fold. IO-ITO has been further modified into branched micropillar ITO (BP-ITO) electrodes that have reached photocurrent densities up to 245 mA/cm^2, which is the theoretically ideal maximum limit. Electrode surface alterations using nanomaterials such as carbon nanotubes (CNTs) and redox polymers based on osmium have enhanced electron transfer where certain systems show 35-fold improvements in photocurrent production. The resistance inside the cell has been reduced with the decrease in cell size, and this size reduction also enhances mass transfer. One of the best examples of miniaturized PV cell is the microfluidic BPV devices, which is as small as 0.4 μL and has delivered power outputs of 500 mW/m^2. The most recent trend is the use of synthetic microbial consortia, where algae are combined with heterotrophic, electroactive bacteria like *Shewanella* or *Geobacter*. The continuous electricity generation by these combined systems is sustained by the use of organic metabolites such as sucrose and lactate. A recent consortium containing both photosynthetic and nonphotosynthetic organisms achieved 1700 mW/m^2, and the energy conversion efficiency was up to 0.32%, representing one of the highest outputs to be reported (Zhu et al. 2023).

By combining the ideas of all the above-discussed technologies such as the BPVs, PMMFCs, DSSCs hydrogels, and biofilms, the advances in the algal bioelectronics demonstrate how algae are used as a versatile component for the emerging next-gen energy production systems. Even though each system has different mechanisms, their core principle depends on algae and their properties of sunlight-based metabolism, pigment systems, self-repairing ability, and very less environmental impact. BPVs and PMMFCs show the direct biological energy production, whereas DSSCs produce energy using algal pigments as a natural sensitizer. Biofilms and hydrogels improve the stability and the electron transfer. As a whole, these various approaches prove that algal bioelectronics are not separate but a unified and sustainable pathway toward environmentally friendly energy production that can be used as bio-batteries and carbon-neutral power generators.

18.3 Role of Algal Biofilms

18.3.1 Formation and Architecture of Algal Biofilms

Algal biofilms are surface-associated communities of algal cells enclosed within a matrix of extracellular polymeric substances (EPS), which consist mainly of polysaccharides, proteins, nucleic acids, and lipids. The algal cells secrete these compounds to protect, and adapt themselves to environmental stress such as desiccation, light variation, or nutrient limitation. A structured biofilm is established as the EPS binds the cells together and anchors them to a solid surface (Moreno Osorio et al. 2021). Algal biofilms are formed by going through a series of stages. Initially, the algal cells move toward a suitable solid surface and attach themselves through weak interactions such as van der Waals forces or electrostatic attraction. Once attached, the cells start secreting EPS, which in turn strengthens the adhesion and traps other microorganisms along with debris. This process continues till a dense and layered biofilm structure is formed. The so-formed biofilm commonly consists of a basal layer of firmly attached cells and an upper layer containing loosely attached or motile cells, and both are interconnected through EPS networks (Cheah and Chan 2021).

Biofilm architecture is not uniform, that is, based on the environmental conditions such as light intensity, pH, or nutrient concentration, the biofilm structure also changes. Nutrient exchange and waste removal are done by water channels that are often formed within the EPS matrix. These internal structures help in maintaining the metabolic activity of the biofilm and enhance the cell survival under fluctuating conditions. Different layers perform distinct physiological functions, and this distinct function is supported by the spatial organization of cells within the biofilm. Most of the naturally found algal biofilms are not pure as they have bacterial and fungal symbiosis. However, when the algal component is higher when compared to the other microbial consortia, the structure and metabolism of the community, the term "algal biofilm" is used (Moreno Osorio et al. 2021). However, these mixed consortia often provide synergistic benefits such as enhanced nutrient recycling and stability. Overall, under challenging conditions, algal cells adapt and colonize to form biofilms that help them sustain long-term growth.

18.3.2 Functional Role in Enhancing Electron Transfer

In algae-based bioelectronic systems, an important role of biofilms is to improve electron transfer between the algal cells and the electrode surface. Due to the limited contact between the cells and the conductive surface, algal cells that are suspended in a solution show weaker electrical conductivity. However, when these cells form a biofilm, they form a compact and stable layer that stays in close contact with the electrode, thereby reducing the distance between the cells and electrodes and so, creates a more efficient pathway for electron movement (Conners et al. 2022). The electron transfer mainly occurs through two mechanisms: direct electron transfer (DET) and mediated electron transfer (MET). In DET, electrons are transferred directly from the algal photosynthetic systems, especially from photosystem II and their associated redox carriers, to the electrodes through conductive molecules that are present in the EPS or on the cell surface. In MET, soluble redox mediators such as quinones, phenazines, or other electron shuttles carry electrons between the algal cells and the electrode surface (Cheah and Chan 2021). Both mechanisms benefit from the dense cell-to-surface contact and high cell density found in biofilms.

The EPS matrix in algal biofilms has a dual role. One is to bind cells together and the other is to act as a conductive medium that facilitates charge transfer. Some studies suggest that electron transport within the matrix is enhanced with the help of certain components of EPS such as redox-active proteins or polysaccharides with metal ions (Moreno Osorio et al. 2021). Higher current generation in bio-photoelectrochemical systems is attained as the conductive property of EPS allows electrons generated during photosynthesis to be efficiently transferred to the electrode. Consistent photosynthetic activity is maintained by the biofilm structure as it protects the algal cells from external fluctuations such as light intensity and temperature. Continuous electron flow is ensured by the stability and longevity of biofilms, and this in turn improves the performance of systems like bio-photovoltaic cells and microbial fuel cells. Algae act as bridges between biological electron production and electrochemical conversion. By combining biological adhesion, EPS-mediated conductivity, and protection against environmental stress, these biofilms make electron transfer more stable and efficient when compared to nonbiofilm systems (Conners et al. 2022; Moreno Osorio et al. 2021).

18.3.3 Biofilms as Self-Healing, Living Conductive Matrices

One of the most remarkable properties of algal biofilms is their ability to self-heal and act as living conductive matrices. Unlike artificial conductive materials that degrade over time, algal biofilms continuously grow, repair, and replace the damaged areas. This natural self-repair ability makes them extremely durable and ideal for long-term bioelectronic applications (Moreno Osorio et al. 2021). Additional extracellular polymeric substances (EPS) are produced by algal biofilms to reinforce the structure when they feel any external stress. The EPS functions as a flexible and protective layer that holds the cells together. This enables the community to recover from physical damage or partial detachment. The production of EPS also helps to seal the cracks or voids that are formed within the biofilm, thus maintaining continuous conductivity and electron flow (Cheah and Chan 2021). As long as the environmental conditions allow the growth of cells, biofilms can regenerate naturally as they have metabolically active cells. The damaged regions will be colonized by new cells that divide, which restores the biofilm's physical and electrical integrity. This living property gives biofilms a significant advantage over static conductive materials that are used in artificial systems as they cannot self-repair once damaged (Conners et al. 2022).

Apart from self-healing ability, algal biofilms can adapt to their environment by regulating gene expression and metabolic activity in response to external stress. For example, when light conditions vary, photosynthetic cells adjust their pigment composition to optimize energy capture, thereby ensuring stable electron generation. Cells can also interchange their growth conditions from autotrophic to mixotrophic under fluctuating light conditions. Due to these properties, algal biofilms are referred as "living wires" as they combine biological growth with conductive behavior. Algae are used as components in bio-photovoltaic systems, biosensors, and microbial fuel cells due to their ability to heal, regenerate, and sustain electron transfer over long durations. The continuous renewal and conductivity of these biofilms provide a sustainable and resilient interface between biological and electronic systems (Moreno Osorio et al. 2021; Conners et al. 2022).

18.3.4 Influence on System Stability and Performance

The formation of algal biofilms poses a significant change in the stability and overall performance of bioelectronic systems. Algal biofilms play a central role in achieving long-term operational efficiency and in maintaining consistent electron transfer (Moreno Osorio et al. 2021). A major factor that enables the stability and performance of the biofilms is the EPS matrix that surrounds the algal cells. It acts as both protective and stabilizing medium. It prevents cell damage and maintains a balance against fluctuations in pH or temperature and also offers resistance to mechanical and shear stress. These protective properties ensure that the cells remain metabolically active and produce electrons under various environmental conditions continually. Moreover, EPS slows down nutrient loss and supports the retention of enzymes and pigments that are essential for photosynthesis, enhancing the long-term productivity of the biofilm (Cheah and Chan 2021). Another important factor influencing the performance is the thickness and density of the biofilm. A moderately thick biofilm allows efficient light penetration, nutrient exchange, and gas diffusion. However, if the biofilm turns too dense, it can limit light availability and create internal oxygen gradients that may reduce photosynthetic activity. Therefore, it is necessary to maintain an optimal biofilm thickness to balance stability and performance. Periodic renewal or controlled detachment of older biofilm layers can help sustain the system's efficiency (Moreno Osorio et al. 2021).

The strong adhesion of algal biofilms to electrodes also minimizes operational losses. In planktonic cultures, these are easily washed away, leading to reduced cell density and unstable output. However, biofilm-based systems exhibit higher current density and longer lifespan as the cell–electrode contact and continuous metabolic activity are consistent. Studies have shown that biofilm-based bio-photoelectrochemical systems generate significantly higher power output when compared to systems using free-living algal cells (Conners et al. 2022).

Overall, the presence of algal biofilms enhances both the durability, longevity, and reliability of bioelectronic systems. By providing structural integrity, self-protection, and steady metabolic performance, algal biofilms ensure stable long-term energy generation and efficient electron transfer, making them a vital component of next-generation green energy technologies (Moreno Osorio et al. 2021; Conners et al. 2022).

18.4 Existing Techniques for Algal Biofilm Formation

18.4.1 Natural Versus Engineered Biofilm Development

Algal biofilms can be formed by two ways. This includes naturally occurring biofilms in aquatic and terrestrial environments and engineered biofilms under controlled laboratory conditions for research and industrial applications. Natural biofilm development occurs spontaneously. When algal cells find a suitable surface such as rocks, glass, or other submerged materials, this process takes place. The cells attach with the help of weak physical forces and gradually begin to secrete EPS, which then helps to anchor themselves firmly to the surface and protect them from environmental changes (Moreno Osorio et al. 2021). Over time, these attached cells grow rapidly, multiply, and form dense multilayered communities that can withstand variations in temperature, pH, and nutrient availability.

In contrast, engineered biofilm is formed as it is manipulated by introducing environmental and physical conditions to promote and control biofilm growth. Laboratory systems are designed to optimize parameters such as light intensity, carbon dioxide concentration, nutrient supply, and flow conditions. Uniform biofilm thickness and enhanced electron transfer efficiency can be attained by adjusting these parameters, which are essential for bio-photovoltaic and biofuel cell applications (Cheah and Chan 2021). Engineered biofilms allow researchers to conduct a detailed study on the growth kinetics, optimization for productivity, and improve the mechanical stability of the biofilm compared to naturally formed ones.

Fixed-bed reactors, rotating biological contactors, and photobioreactors equipped with suitable carrier materials are a few cultivation setups that are used to engineer biofilm formation. Carrier materials include glass, stainless steel, indium tin oxide (ITO), carbon cloth, or biocompatible polymers, and these materials provide surfaces for algal attachment. The type and texture of the substrate play an important role. Rough and hydrophilic surfaces are the most preferred carriers as they generally promote better adhesion than smooth or hydrophobic ones (Moreno Osorio et al. 2021). Another engineered approach for biofilm development is the co-cultivation with bacteria. Bacteria assist algae by producing additional signaling molecules and providing essential nutrients that stimulate EPS secretion and biofilm stabilization. This symbiotic relationship results in a thicker and more stable biofilms with enhanced photosynthetic efficiency.

In summary, while natural algal biofilms form without human intervention, engineered biofilms are developed under conditions that are intentionally tampered with to achieve better control over structure, productivity, and stability. Thus, the foundation for designing efficient biofilm-based systems for energy, environmental, and industrial applications is to understand both natural and engineered processes (Cheah and Chan 2021; Moreno Osorio et al. 2021).

18.4.2 Surface Functionalization and Biocompatible Substrates

The surface on which algal biofilms grow also plays a major role in determining the rate of attachment, stability, and productivity of the biofilm. Algal cells prefer to attach to surfaces that are compatible with their physiological properties, and this interaction can be enhanced through a process known as surface functionalization. Surface functionalization refers to the chemical or physical alterations that are done to a substrate to improve its ability to support cell adhesion and biofilm growth (Moreno Osorio et al. 2021). The selection of the carriers depends on the intended application. For example, conductive materials such as carbon cloth and ITO are used in bioelectronic systems as they allow efficient electron transfer between the algal cells and electrodes. Nonconductive materials like glass or plastic are often used for experimental studies that focus on biofilm structure and growth kinetics (Cheah and Chan 2021).

Surface properties such as roughness, hydrophobicity, and surface charge significantly affect algal attachment. Hydrophilic surfaces attract algal cells more effectively than hydrophobic materials and rougher or micro-textured surfaces provide more adhesion sites when compared to smooth ones. To enhance the surface compatibility, carriers are treated with chemical coatings, plasma treatment, or polymer modification. These treatments alter the surface energy and charge, thereby improving the adhesion of algal cells and the retention of EPS layers (Moreno Osorio et al. 2021). In addition to physical and chemical modifications of the carrier surface, biocompatible coatings such as chitosan, poly-L-lysine, and hydrogels have also been used successfully to support cell attachment. These materials mimic natural biological environments and so reduce cell damage, promoting uniform and stable biofilm formation. Some studies have also shown that coating electrodes with biopolymers can enhance electron transfer by maintaining close contact between the biofilm and the conductive surface (Cheah and Chan 2021).

Surface functionalization is therefore an essential part of engineered biofilm systems, as it allows precise control over the initial cell attachment and the biofilm architecture. By choosing suitable carriers and optimizing their surface properties, researchers can enhance both the mechanical stability and electrical performance of algal biofilms, leading to more efficient and reliable bio-electronic systems (Moreno Osorio et al. 2021; Cheah and Chan 2021).

18.4.3 Genetic Engineering Approaches to Promote Biofilm Formation

Recent advancements in the field of bioelectronics using algae have taken a turn as genetic engineering offers innovative ways to enhance algal biofilm formation by modifying the genetic and metabolic characteristics of algal cells. Naturally, algal biofilms depend on environmental factors such as light, nutrient levels, and the presence of other microorganisms to develop. However, through genetic modifications, researchers can strengthen the natural ability of algae to produce EPS, improve surface adhesion, and enhance electron transfer within biofilm-based systems (Roy 2023). The most significant approach involves increasing the EPS production, as it is responsible for binding algal cells together and anchoring them to a rigid surface. This increased EPS production is done by overexpressing genes that are responsible for polysaccharide synthesis, protein secretion, or lipid modification. Enhanced EPS not only stabilizes the biofilm but also improves tolerance toward stress. In addition, the electron conductivity within the biofilm can be enhanced by increasing the ratio of charged or redox-active molecules of the EPS (Moreno Osorio et al. 2021).

Another strategy focuses on enhancing cell-surface adhesion. Adhesion-related proteins such as glycoproteins and hydrophobic surface layer proteins can be genetically modified or overexpressed to improve the attachment of cells to electrodes or carrier materials. These engineered cells adhere more strongly that in turn leads to a faster and more uniform biofilm formation. Some studies have also used coexpression systems where bacteria or other microorganisms are genetically modified to secrete adhesion-promoting molecules that help algal partners in biofilm consortia (Roy 2023). The precise genetic modification of algae is made possible with the help of synthetic biology and one of the most recent advances in genetic engineering itself, that is, CRISPR/Cas9 genome-editing tools. These tools enable the targeted insertion or deletion of genes related to biofilm formation, photosynthetic efficiency, and stress resistance. Through such methods, scientists have been able to create algal strains that form biofilms more rapidly with improved structural stability and generate higher electrical outputs in bioelectronic applications (Roy 2023).

As a result, genetic engineering provides a promising route for altering algae at a genomic level that can form stronger, more conductive, and more stable biofilms. By combining molecular biology with bioengineering, researchers can develop algal systems optimized for bioenergy production, wastewater treatment, and bioelectronic device integration (Roy 2023; Moreno Osorio et al. 2021).

18.4.4 Microfluidics and 3D Scaffolds for Structured Biofilms

More unique ways to control and improve algal biofilm formation were made with the help of advancements in microfluidics and three-dimensional (3D) scaffold technology. Uniform and functional biofilms were made by managing environmental factors, including fluid dynamics using these modern techniques. Unlike traditional biofilm cultivation systems, microfluidics and 3D scaffold-based approaches can replicate the natural microscale environments that influence biofilm development, resulting in more efficient and predictable outcomes (Pousti et al. 2019).

Microfluidic systems, often referred to as "lab-on-a-chip" devices, use microchannels to guide the flow of liquid around small culture chambers containing algal cells. Healthy and active biofilms are maintained by the continuous flow of nutrients and removal of waste products. Real-time monitoring of biofilm growth can be done using this system through optical or electrochemical sensors that are integrated into the microchip. The shape, density, and composition of algal biofilms can be precisely controlled by adjusting flow rates, temperature, and light exposure (Pousti et al. 2019). Such precision is particularly useful in optimizing biofilm-based energy systems and biosensors, where uniform biofilm formation directly influences electron transfer and device performance.

In addition to microfluidics, 3D scaffolds have also become an effective tool for cultivating engineered biofilms. These scaffolds are made of biocompatible materials such as hydrogels, polymers, or nanocomposites that provide high surface area for algal attachment. Nutrients and gases diffuse easily through the porous nature of these materials, and this ensures that all layers of the biofilm receive adequate light and nourishment. Structured 3D biofilms show enhanced electron conductivity and stability as the scaffold supports even distribution of cells and prevents detachment during fluid movement (Moreno Osorio et al. 2021). By combining these systems, advanced platforms for studying and utilizing algal biofilms can be created. These integrated systems mimic natural conditions while maintaining control over biofilm architecture and function. It also offers scalability, making them suitable for applications in bioenergy, environmental sensing, and wastewater treatment.

Overall, microfluidics and 3D scaffold-based techniques represent the next generation of biofilm engineering, thus enabling the development of structured, efficient, and durable algal biofilms for sustainable biotechnological applications (Pousti et al. 2019; Moreno Osorio et al. 2021).

Aspect	Description/focus	Examples/techniques	Impact on performance
Biofilm formation	Attachment of algal cells through secretion of EPS forming a stable surface-associated layer	Natural formation on rocks, glass, or electrodes; engineered systems using optimized light and nutrients	Improves surface adhesion and enables stable electron flow
EPS matrix composition	Composed of polysaccharides, proteins, nucleic acids, and lipids; provides structure and protection	Enhanced through nutrient optimization or genetic modification	Increases stress tolerance, conductivity, and cell–cell cohesion
Electron transfer mechanism	Direct (DET) and mediated (MET) electron transfer via redox-active components	Conductive EPS, close electrode contact, redox mediators	Enhances power density and electrical output
Self-healing and adaptation	Continuous regeneration and EPS production repair physical damage	Living "self-repairing" biofilms on electrodes	Ensures long-term stability and resilience under stress
Surface functionalization	Modification of substrates to improve algal adhesion	Plasma treatment, polymer coatings, biocompatible hydrogels	Promotes uniform and dense biofilm formation
Genetic engineering approaches	Altering genes to enhance EPS secretion or surface protein expression	CRISPR/Cas9, overexpression of adhesion proteins	Produces stronger and more conductive biofilms
Microfluidic and 3D scaffold systems	Controlled environment for biofilm structure and nutrient flow	Lab-on-a-chip devices, hydrogel, or polymer scaffolds	Provides precise architecture, high surface area, and efficient energy conversion
System stability	Structural integrity and consistent performance under varying conditions	Optimized thickness, EPS protection, and biofilm renewal	Sustains steady electron transfer and prolonged device lifespan

18.5 Applications and Potential Benefits

18.5.1 Bioenergy: Microbial Fuel Cells and BPVs

Microalgae have been determined as an efficient source for bioenergy production due to their rapid growth and adaptability and to simultaneously perform photosynthesis while capturing atmospheric CO_2. When it comes to the production of bioelectricity, algae are integrated into microbial fuel cells (MFCs), photosynthetic microbial fuel cells (PMFCs), and bio-photovoltaic (BPV) cells. In an MFC, algae act as a direct electron donor or as a biomass feedstock. Algal biomass is rich in carbohydrates, proteins, and lipids that serve as a substrate for electrochemically active microorganisms, which oxidize the substrates and generate electrons. Conventional electron donors are replaced by live algal species, such as *Chlorella* and *Nitzschia*, contributing directly to current production. Integrating algae with wastewater treatment in MFCs offers dual benefits (i.e., nutrient removal (N and P) and bioelectricity generation). In BPV cells, microalgae act as bio-anodic catalysts that produces electrons as water-splitting reaction takes place during photosynthesis. Oxygen is released as a by-product and the electrons are harvested at the anode. Recent advances in electrode design such as indium tin oxide (ITO) reduce the usage of graphene oxide (rGO), and the electron transfer and stability have been increased drastically due to the biofilm-based immobilization. Marine strains such as *Tetraselmis gracilis* and *Dunaliella* sp. are potentially strong candidates for use in BPV systems; however, limitations such as light intensity optimization, spacing between the electrodes, and biofilm formation still remain. Even though both MFCs and BPVs have limitations in the areas of power density and scalability, they demonstrate greater potential for sustainable and small-scale energy generation. The proof that algae could be the cornerstone of future bioenergy innovation is their ability to combine CO_2 sequestration, wastewater treatment, and renewable electricity production (Tay et al. 2022).

18.5.2 Environmental Sensing and Biosensors

Environmental monitoring has been made simpler with the help of algae-based biosensors as they are naturally sensitive to pollutants and respond rapidly to environmental stressors that makes them the ideal biological components for detecting heavy metals, pesticides, and organic pollutants. Algal biosensing is based on how the pollutants disrupt normal photosynthetic and metabolic functions. Stress responses are indicated by parameters such as chlorophyll fluorescence, photosystem II activity, and oxygen evolution. For example, presence of herbicides or toxic metals is identified by the inhibition of photosynthetic efficiency. Real-time monitoring can be done by assessing these changes that provides an eco-friendly and cost-effective alternative to conventional chemical assays. The sensor stability, along with reusability, is enhanced with the immobilization of algal cells in polymer matrices. Repeated measurements are done with the help of portable biosensors, and biofilm-based sensors use the natural adhesion of algae for continuous water quality monitoring. These make a practical tool for field applications, especially in aquatic ecosystems. A major advantage of algae-based biosensors is their role as early warning systems. As algae are placed at the bottom of the food chain, their sensitivity to toxicants sends a signal to determine the potential harm to the entire ecosystem. Algae also respond to the changes in pH, light intensities, and nutrient levels that extend their application in environmental sensing beyond contaminant detection. Integration of algae with electrochemical transducers and microelectronic systems has increased their use as biosensors due to their multianalyte detection and continuous monitoring of the environment by enhancing signal detection, size reduction, and digital data output (Arunkumar et al. 2023).

18.5.3 Wearable, Portable, and Implantable Electronics

Wearable, portable, and implantable electronics refers to the next-gen biomedical technologies that offer a new perspective and possibility in monitoring health conditions continuously, personalizing the medicine to be administered and therapeutic interventions. The use of traditional electronic systems that are made of hard materials like silicon, metals, and rigid polymers is often limited due to their poor flexibility, mechanical mismatch with soft tissues, and limited long-term biocompatibility as they restrict their integration with the human body. Natural biomaterial-based hydrogels that are derived from proteins like gelatin, collagen, and silk fibroin and polysaccharides such as chitosan, cellulose, alginate, and starch have emerged as promising alternatives. These hydrogels possess good biocompatibility, tissue-like softness, and tunable electrical and mechanical properties, making them ideal materials for bioelectronic interfaces. Wearable bioelectronics are trusted completely when it comes to instant health monitoring, human–machine interfaces, and energy harvesting areas. Continuous and noninvasive monitoring of physiological indicators is done with the help of hydrogel-based wearable devices that has the unique ability to conform closely to the skin. This allows a person to monitor heart activity, neural signals, glucose concentration, and electrolyte levels in sweat. These materials can maintain reliable electrochemical performance even during bodily motion or perspiration due to their inherent softness, elasticity, and stretchability. Practical applications in current lifestyle include hydrogel platforms for sweat-ion detection and patches for glucose responses designed for diabetes care. The design requirements are made stricter for implantable or invasive electronics. They must prove biocompatibility, predictable degradation, and sustained stability within the body over long periods. Immune rejection is reduced when hydrogels are used as they structurally resemble the extracellular matrix. It also provides a favorable condition for tissue integration. When carefully designed and altering the ionic conductivity, mechanical compliance, and viscoelastic behavior, these systems can reliably be used in applications such as neural interfacing, cardiac rhythm monitoring, and muscle stimulation. Hybrid designs have also been introduced. It includes hydrogel–liquid metal composites and conductive bioadhesives that significantly improve device adhesion to the tissues and also enhance the durability under physiological conditions. However, limitations arise as many hydrogel systems still suffer from limited long-term stability, insufficient mechanical strength, and difficulties in large-scale production. In certain cases, the performance has been enhanced significantly with conductive nanomaterials, but this may negatively impact biological compatibility. Therefore, current research focuses on developing a fully bio-derived hydrogel that combines conductivity, self-healing behavior, and multifunctionality without compromising safety (Chen et al. 2025).

In summary, wearable and implantable hydrogel bioelectronics paves a way toward the next-generation healthcare technologies. By closing the gap between the mechanical mismatch between rigid devices and soft tissues, they open new opportunities for continuous health monitoring, accurate diagnostics, and personalized therapeutic interventions.

18.6 Challenges

Despite the growing interest in algae-powered bioelectronics, especially in power generation systems such as bio-photovoltaics (BPVs), photosynthetic microbial fuel cells (PMFCs), and dye-sensitized solar cells (DSSCs), several basic and technical limitations currently affect their scaling up from laboratory prototypes to practical applications.

18.6.1 Low Power Output and Conversion Efficiency

The most widely cited limitation is the exceptionally low power density and overall energy conversion efficiency of these systems. Reported maximum power outputs are typically in the range of 58–540 mW/m^2 for BPVs and MFCs, which is lower compared to conventional fuel cells or silicon-based photovoltaics. For algae-based DSSCs, efficiencies are often below 1%, with some reports as low as 0.01%–0.5% compared to over 14% for synthetic ruthenium-based dyes. These values indicate that current algae-powered devices are only suitable for very low-power applications such as biosensors. In order to overcome limitations regarding the low power output, the design of electrodes can be improved in a way that enhances the electron capture and the contact between algae and the electrodes. Various algal-microbial consortia can also be studied and wider strains can be used to check if the power output is increasing depending on the strain.

18.6.2 Electron Transfer Issues

A central reason for this inefficiency is, the poorly understood and inefficient electron transfer from the photosynthetic electron transport chains within algal cells to the extracellular electrodes. The insulating nature of the cell wall and multiple membranes, combined with the inherently weak exoelectrogenic activity of most photosynthetic microorganisms, severely restrict electron escape. While electron mediators can improve transfer, they often introduce new problems such as cytotoxicity, photochemical instability, and environmental risks. Even mediator-less strategies that rely on biofilm formation have not been able to fully overcome this bottleneck. These barriers can be potentially crossed by the incorporation of nanoparticles such as graphene to improve the conductivity. By immobilizing the algae, the diffusion barriers can be greatly reduced and this also increases the contact between the algae to the electrode surface. The use of genetically altered microbes can enhance the extracellular electron transfer pathways.

18.6.3 Stability and Longevity Issues

Another major limitation is the reduced stability of biological components under operational conditions. Natural pigments that are highly prone to photodegradation and oxidative damage such as chlorophylls and carotenoids lead to a rapid decline in performance over time. Their activity is also highly sensitive to environmental factors such as pH, temperature, and light intensity that make long-term stability a major challenge. In PMFCs, algal photosynthesis leads to fluctuating oxygen levels and respiration can cause unstable power generation, which eventually leads to a phenomenon called "power overshoot," where sudden resistance increases the disruption in current flow. To avoid these technical issues, electrodes can be designed in a way that can withstand any external environmental fluctuations. To prolong the life of biological components, the materials that are used in the devices can be replaced with biological materials or biocomponents that impart lesser damage to the natural compounds, including the pigments. Issues with stress conditions can also be adjusted using algal strains that have enhanced self-repairing ability. Hydrogel matrices are to be prepared with higher concentrations so that they can withstand harsh conditions so that they do not break down due to external forces.

18.6.4 Material and Design Challenges

When it comes to limitations due to materials, algal dyes often suffer, as the adsorption to semiconductor surfaces like TiO_2 is poor. This reduces electron injection efficiency. Similarly, hydrogel-based materials such as the alginate matrices that are derived from algae are mechanically fragile, have low intrinsic conductivity, and can swell or degrade under physiological conditions. This complicates their use in robust bioelectronic devices. Issues of cytotoxicity and long-term compatibility may arise when carbon nanotubes or conductive polymers are used to enhance the conductivity. The use of advanced electrode architectures such as inverse opal or micropillar ITO has reported to produce better laboratory results; however, the use of expensive substrates and catalysts such as platinum and indium tin oxide limits their scalability. The shift from costly to cheaper and affordable materials is necessary. Conductive materials such as carbon and nanoparticle-based materials should be studied further in case of toxicity and how they negatively impact the environment.

18.6.5 Scalability and Standardization

The difficulty in commercialization lies in scaling up these systems while maintaining consistent performance. Mass cultivation of algae requires high operational costs, strict environmental conditions, and problems such as self-shading at high densities or photo-inhibition under strong light may occur. Additionally, there is a lack of standardized protocols in designing the systems, biomass quantification, and performance monitoring that makes it difficult to compare various results that are studied and establish best practices. The need for ionic membranes, buffer solutions, and nutrient inputs adds to the production costs and increases environmental concerns. Standardized protocols should be developed and scalable reactors should be explored more. Computational modeling should be done before the start of pilot and industrial-scale development that greatly helps in the optimization of the design parameters.

Overall Outlook

In summary, even though algae-powered bioelectronics hold promise as eco-friendly and dual-purpose systems capable of both electricity generation and CO_2 capture, their limitations in the areas of power density, electron transfer, stability, material integration, and scalability currently restrict them only to experiments and not commercial applications. Advances in genetic engineering alterations of algal strains, hybridization with electroactive bacteria, improved electrode designs, and standardized reporting frameworks will help overcome these limitations. Until then, algae-based bioelectronics are viewed as an exciting research area rather than a commercially viable alternative to established renewable energy technologies.

18.7 Conclusion

In the pursuit of sustainable technologies, algae have emerged as one of the most promising biological resources. Their ability for rapid growth, adaptability to harsh environments, and potential to produce a wide range of valuable bio-compounds make them essential in modern biotechnology. From bioenergy and environmental management to material production and health applications, algae play a major role in bridging natural biological processes with technological innovation.

This chapter explored the structural, functional, and technological aspects of algae-derived bio-electronic systems that emphasize how biological principles can be applied to create eco-friendly solutions. Modern research has focused on harnessing and enhancing the natural capabilities of algae through scientific advancements. Techniques involving optimization of growth parameters, surface engineering, and genetic modification have transformed algae from simple microorganisms into versatile tools that could make a major impact in sustainable development. The integration of algal systems into emerging fields such as bioelectronics and renewable energy has proven that living organisms can effectively work with or even replace synthetic materials in several industrial processes. In order to address global environmental and energy challenges, these innovations reflect a broader shift toward the usage of renewable, biodegradable, and self-sustaining biological systems.

In conclusion, the detailed research on algae and their applications represent a vital step for a more sustainable and technologically advanced future. A cornerstone of green biotechnology is the incorporation of algae due to their unique abilities that allow us to merge biological efficiency with engineering innovation. Further research and interdisciplinary collaboration will further unlock their potential, and this paves the way for cleaner energy sources, sustainable materials, and environmentally responsible technological solutions.

18.8 Future Perspective

The future of algae-powered electronics move toward resolving the limitations and enhancing the efficiency of the existing systems. Challenges regarding the power density, electron transfer, stability, and scalability are focused. One of the major upcoming fields is the improvement of photosynthetic electron generation and transfer. The design of a safe, biodegradable conductive medium incorporating polymers or multiple consortia to facilitate direct electron transfer is needed. Genetic engineering in the field of algal bioelectronics is also researched. The modification of thylakoids, reduction of cell wall barriers, and the algal–microbial consortia, improve the rate of electron movement to the external electrode. The other part of the future scope for algal bioelectronics is changes in the electrodes that are being used. A mixture of engineered strains, integrated with nanostructured electrodes including graphene foams, nanoporous metals, and micropillar ITO could pave the way for the increased surface contact and electron capture efficiency. The operational stability of the bioelectronics is also an important section that has to be looked into. The pigments of the algae may degrade quickly, and so, methods such as encapsulation or matrices that are rich in antioxidants should be studied in order to extend the lifespan of the device. This use of matrices may also help to maintain the environmental factors such as light, temperature, and pH. Apart from the design and functioning of the bioelectronics, scaling up is another huge task with greater barriers. Making standardized protocols, low-cost algal cultivations, and LCA studies are essential for scaling up to practical applications from prototypes.

Overall, the future progress will focus mainly in the fields of genetic engineering, synthetic biology, materials engineering, and system design to strengthen the performance and improve the advancements of algae-powered bioelectronics toward real-world and sustainable applications.

References

Al Abdulla SA, Al Hammadi K, Al-Ali H, Alami AH, Abdelkareem MA, Olabi AG (2024) Experimental development of a biological photovoltaic cell (BPV) for energy conversion and simultaneous CO2 capture by utilizing marine microalgae on copper mesh. Renew Energy 223 (January):120096. https://doi.org/10.1016/j.renene.2024.120096

Armendáriz-Mireles EN, Calles-Arriaga CA, Pech-Rodríguez W, Castillo-Robles A, Rocha-Rangel E (2023) Alternative sources of natural photosensitizers: role of algae in dye-sensitized solar cell. Colorants 2(1):137–150. https://doi.org/10.3390/colorants2010010

Arunkumar K, Arun A, Raja R, Palaniappan R (2023) Algae materials: applications benefitting health. Elsevier. https://doi.org/10.1016/C2022-0-00271-3

Chatterjee A, Kathirvel A, Manivasagam TG, Batabyal SK (2024) Sustainable power generation from live freshwater photosynthetic filamentous macroalgae Pithophora. J Sci Adv Mater Devices 9(2):100674. https://doi.org/10.1016/j.jsamd.2024.100674

Chauhan R (2022) Scanning prevalent technologies to promote scalable devising of DSSCs: an emphasis on dye component precisely with a shift to ambient algal dyes. Inorg Chem Commun 139(February):109368. https://doi.org/10.1016/j.inoche.2022.109368

Cheah YT, Chan DJC (2021) Physiology of microalgal biofilm: a review on prediction of adhesion on substrates. Bioengineered 12(1):7577–7599. https://doi.org/10.1080/21655979.2021.1980671

Chen B, Zhu Y, Yu R, Feng Y, Han Z, Liu C, Zhu P, Lu L, Mao Y (2025) Recent Progress of biomaterial-based hydrogels for wearable and implantable bioelectronics. Gels 11(6):442. https://doi.org/10.3390/gels11060442

Choi D (n.d.) Developing sustainable water and electricity for impoverished regions using algae-based bio-photovoltaics, pp 1–12

Conners EM, Rengasamy K, Bose A (2022) Electroactive biofilms: how microbial electron transfer enables bioelectrochemical applications Abstract J Ind Microbiol Biotechnol 49(4). https://doi.org/10.1093/jimb/kuac012

Moreno Osorio JH, Pollio A, Frunzo L, Lens PNL, Esposito G (2021) A review of microalgal biofilm technologies: definition applications settings and analysis. Front Chem Eng 3. https://doi.org/10.3389/fceng.2021.737710

Pousti M, Zarabadi MP, Abbaszadeh M, Paquet-Mercier AF, Greener J (2019) Microfluidic bioanalytical flow cells for biofilm studies: a review. The Analyst 144(1):68–86. https://doi.org/10.1039/C8AN01526K

Roy A (2023) Bioelectronic devices for targeted drug delivery and monitoring of microbial electrogenesis. Linköping University Electronic Press. Linköping. https://doi.org/10.3384/9789180751551

Solak EK, Irmak E (2023) Advances in organic photovoltaic cells: a comprehensive review of materials, technologies, and performance. RSC Adv 13(18):12244–12269. https://doi.org/10.1039/d3ra01454a

Tay ZHY, Ng FL, Ling TC, Iwamoto M, Phang SM (2022) The use of marine microalgae in microbial fuel cells, photosynthetic microbial fuel cells and biophotovoltaic platforms for bioelectricity generation. 3 Biotech 12(7):1–22. https://doi.org/10.1007/s13205-022-03214-2

Yuan X, Xu X, Gao X, Liu X, Liang B, Luan G, Lu X (2025) Enhancing the cellular robustness of cyanobacteria to improve the stability and efficiency of bio-photovoltaics. Life 15(2):1–12. https://doi.org/10.3390/life15020299

Zhu H, Wang H, Zhang Y, Li Y (2023) Biophotovoltaics: recent advances and perspectives. Biotechnol Adv 64(October):108101. https://doi.org/10.1016/j.biotechadv.2023.108101

Microalgae: A Sustainable Feedstock for the Production of Biodiesel and Bioplastic

19

B Razia Sultana and Srividya Shivakumar

Abstract

The depletion of fossil fuel reserves driven by rapid urbanization and the escalating impacts of global climate change has intensified the search for renewable and environmentally friendly alternatives. Petrochemical products such as diesel and synthetic plastics are nonrenewable and nonbiodegradable. They contribute significantly to environmental hazards. In this context, photosynthetic microalgae have emerged as a suitable feedstock for biobased products, owing to their global distribution, high biomass productivity, and capacity to sequester atmospheric carbon dioxide. Through photosynthesis, microalgae convert sunlight and carbon dioxide into a range of biomolecules, including carbohydrates, proteins, and lipids. Notably, microalgal lipids can be converted into biodiesel, which is composed of monoalkyl fatty acid esters, while polyhydroxyalkanoates (PHAs) are naturally occurring polymers in microalgae that can be transformed into bioplastics. These microalgal-derived products are sustainable and biodegradable alternatives to the petrochemical-based diesel and plastics, which are major contributors to greenhouse gas emissions, environmental pollution, and long-term ecological degradation. Numerous microalgal species are rich in lipids and PHAs, rendering them highly suitable for large-scale bioproducts production. The effective utilization of microalgal biomass facilitates its integration into circular bioeconomy frameworks and thereby supports the advancement of

B. Razia Sultana
Department of Microbiology and Botany, School of Sciences, JAIN (Deemed-to-be University), Bengaluru, Karnataka, India

S. Shivakumar (✉)
School of Allied Healthcare and Sciences (SAHS), JAIN (Deemed-to-be University), Bengaluru, Karnataka, India
e-mail: sk.srividya@jainuniversity.ac.in

K. Mohanty et al. (eds.), *Advances in Algal Research*,
https://doi.org/10.1007/978-981-95-7234-2_19

green energy technologies aimed at reducing emissions and enhancing resource efficiency. This chapter provides a comprehensive review of lipid and PHA biosynthesis in microalgae, detailing their metabolic pathways, molecular composition, and physicochemical characteristics, while emphasizing their potential in the sustainable production of biodiesel and bioplastics.

Keywords

Microalgae · PHA · Bioplastics · Lipids · Biodiesel

19.1 Introduction

Fossil resources have long served as the backbone of industrial development, providing a wide array of products that have significantly improved human life. However, this convenience has come at a considerable environmental cost. With the rise in the human population, rapid urbanization, and industrial expansion, the demand for fossil-based resources has surged, leading to their accelerated depletion. Compounding this issue is the fact that many fossil-derived products, such as conventional fuels and synthetic plastics, are nonbiodegradable, contributing to long-term pollution, ecological imbalance, and adverse health effects (Yin et al. 2020). The persistent release of exhaust gases—particularly those rich in greenhouse gases—has further intensified environmental degradation, driving climate change and threatening global ecosystems (Tan et al. 2024). These challenges underscore the urgent need to explore sustainable alternatives that can meet industrial demands while mitigating environmental impact. One promising approach involves the replacement of fossil-based products with biomass-derived substitutes, which are renewable, potentially carbon-neutral, and biodegradable (Yin et al. 2020). Biomass-derived products encompass a wide range of materials and chemicals, including biofuels, bioplastics, key intermediates, and high-value compounds, all of which can be derived from plants, algae, or microbial sources. By shifting to such renewable feedstocks, it is possible to reduce dependence on fossil resources, limit greenhouse gas emissions, and promote circularity in production and consumption systems (Irfan et al. 2024).

Microalgae offer a unique advantage in producing value-added products that are directly relevant to current industrial and environmental needs (Kumar et al. 2021a). Lipids accumulated by microalgae serve as precursors for biodiesel, a renewable and biodegradable fuel that can partially or fully replace conventional diesel (Jacob et al. 2021). Additionally, certain microalgae and bacteria synthesize biodegradable polyesters, such as polyhydroxyalkanoates (PHAs), which offer an eco- friendly alternative to conventional petroleum-based plastics (Costa et al. 2019). The simultaneous or sequential production of lipids and PHAs from the same biomass underscores the potential of microalgae in integrated biorefinery systems, thereby maximizing resource utilization and enhancing economic viability (Rahman and Miller 2017).

This chapter highlights the potential of microalgae as a sustainable feedstock for the production of biodiesel and PHA. It explores the metabolic pathways involved in lipid and PHA biosynthesis, along with the molecular composition and physico-chemical properties that define their performance in energy and material applications. By integrating these aspects, the chapter offers a comprehensive perspective on how microalgae can facilitate the transition from fossil-based systems to a circular, bio-based economy.

19.2 Microalgae

Microalgae are photoautotrophic microorganisms that play a vital role in global carbon cycling by converting sunlight and atmospheric carbon dioxide into essential macromolecules, including carbohydrates (4–57%), proteins (6–71%), and lipids (7–23%), which serve as valuable precursors for food, feed, and biofuel production. The biochemical composition of various microalgae is presented in Table 19.1. The macromolecular composition of microalgae varies depending on the species and growth conditions (Ananthi et al. 2021a). This remarkable metabolic flexibility

Table 19.1 Biochemical composition of microalgae: carbohydrates, proteins, and lipids

Microalgal isolate	Carbohydrate content (%)	Protein content (%)	Lipid content (%)	References
Chlorella vulgaris SAG 211-11b	19.3	40.2	16.5	Selvarajan et al. (2015)
Chlorella sorokiniana CCAP 211/8 K	10.0	43.6	10.1	Selvarajan et al. (2015)
Micractinium sp. 99.89	41.5	14.3	28.1	Selvarajan et al. (2015)
Asterarcys quadricellulare	31.0	31.4	24.5	Morsi et al. (2022)
Ankistrodesmus falcatus	14.39	37.60	30.79	Abdelkhalek et al. (2016)
Scenedesmus sp. NC1	21	30	28.3	Kumar et al. (2021b)
Asterarcys quadricellularis AQYS21	21.75	44.34	24.74	Kang et al. (2024)
Monoraphidium griffithii	15.36	32.97	15.89	Vinitha et al. (2022)
Dunaliella sp. ABRIINW-B1	8	19	42	Gharajeh et al. (2020)
Dunaliella sp. ABRIINW-G2/1	4	41	36	Gharajeh et al. (2020)
Monoraphidium sp. NCIM 5585	46.77	14.17	15.55	Pandey et al. (2025)

makes them highly valuable for a range of industrial and environmental applications, including the generation of high-value bioproducts, biofuel production, and bioremediation. (Parmar et al. 2011). Microalgae are predominantly unicellular light-driven organisms, although some species form colonies or filaments with limited multicellularity. Cell size exhibits substantial variation, with diameters spanning 1–50 μm (Rahman and Miller 2017). For instance, the diameter of *Haematococcus pluvialis* cells typically varies between 5 and 25 μm, and that of *Nannochloropsis salina* is 2.5 μm (Dickinson et al. 2017). Taxonomically, microalgae encompass both prokaryotic and eukaryotic lineages. Prokaryotic microalgae comprise the *Cyanophyta* division (cyanobacteria, e.g., *Nostoc* sp.). Whereas eukaryotic microalgae include *Pyrrophyta* (dinoflagellates, e.g., *Noctiluca scintillans*), *Phaeophyta* (brown algae, e.g., *Sargassum natans*), *Chrysophyta* (diatoms, e.g., *Navicula* sp.), and *Chlorophyta* (green algae, e.g., *Chlorella* sp.). These divisions differ significantly in their morphological characteristics, pigment composition, and ecological roles (Sajjadi et al. 2018). It is currently anticipated that over 50,000 microalgal species exist worldwide; however, only around 30,000 species have been formally studied and documented. This highlights the vast untapped biodiversity within the microalgal domain and presents significant opportunities for future research and exploration (Mata et al. 2010).

19.2.1 Methods of Cultivating Microalgae

Cultivation conditions play a key role in determining microalgal growth characteristics and cell composition (Ananthi et al. 2021a). Microalgae are generally cultivated using two primary approaches: open-culture and closed-culture systems. Open cultivation systems, such as raceway ponds and circular ponds, are cost-effective in terms of installation and maintenance and can support large-scale biomass production. However, these systems are prone to contamination and allow only minimal regulation of key parameters, including temperature, light intensity, and water evaporation, leading to poor light utilization and inconsistent productivity (Mehariya et al. 2021). In contrast, closed-culture systems, commonly referred to as photobioreactors, provide a controlled environment in which growth conditions, including light, temperature, aeration, and nutrient supply, can be precisely regulated (Siddiki et al. 2022). A brief comparison between open and closed culture systems is shown in Table 19.2.

19.2.2 Microalgae as a Sustainable Feedstock

Microalgae represent a highly efficient and sustainable biological feedstock for the production of biodiesel and bioplastic due to their rapid growth rates and superior photosynthetic efficiency, resulting in high biomass productivity as compared to plants (Mehariya et al. 2021; Adetunji and Erasmus 2024). Unlike traditional crops, microalgae can be cultivated on nonarable land using minimal amounts of nutrients

Table 19.2 Differences between open and closed culture systems (Ananthi et al. 2021a; Siddiki et al. 2022; Adetunji and Erasmus 2024; Das and Maiti 2021)

Open culture system	Closed culture system
Contamination risk is high	Low contamination
Difficult to control environmental/ growth factors	Controlled environment
Sterile condition is not achievable	Sterile condition is achievable
Energy demand is low	Energy demand is high
Easy maintenance	Difficult maintenance
Low productivity	High productivity
Water loss due to evaporation	Low chances of evaporation
Low capital and operational costs	High capital and operational costs
Raceway ponds, circular ponds	Tubular photobioreactor, vertical photobioreactor
Chlorella, Dunaliella, Spirulina, and *Haematococcus*	*Synechocystis salina*

and water, making their cultivation economically and environmentally advantageous. They are capable of growing in a range of water sources, like freshwater, saline, and brackish water. They are particularly effective in utilizing wastewater by absorbing surplus nutrients such as nitrogen and phosphate (Parmar et al. 2011). Microalgae cultivation is feasible year-round using both open raceway ponds and closed bioreactors, thereby providing operational flexibility (Siddiki et al. 2022). Their oil productivity significantly surpasses that of conventional oilseed crops. Unlike terrestrial crops, microalgae avoid the food-versus-fuel conflict, and their ability to capture CO_2 contributes to climate change mitigation (Mehariya et al. 2021). Additionally, microalgae require low maintenance, with no requirement for herbicides or pesticides, further reducing their environmental footprint (Behera et al. 2015). The growth of microalgae is affected by multiple ecological and biological factors, including light intensity, temperature, pH, salinity, and nutrient availability. These parameters must be carefully optimized to maximize biomass productivity and improve the yields of lipids and PHAs, which serve as precursors to biodiesel and bioplastic, respectively (Siddiki et al. 2022).

19.3 Bioplastics

Bioplastics can be defined as polymers of biological origin that are renewable and may be either biodegradable or nonbiodegradable (Moshood et al. 2022). They are synthesized from feedstocks originating from flora, fauna, or microbes (Adetunji and Erasmus 2024). Microalgae can be employed as a raw material for bioplastic synthesis either through direct utilization of their biomass or by extracting intracellular polymers such as starch and PHAs (Sharma et al. 2024).

19.3.1 Polyhydroxyalkanoate (PHA)

Polyhydroxyalkanoates represent a promising class of bioplastics due to their strong resemblance to polypropylene in both structural and chemical characteristics. PHAs are linear polyesters biosynthesized intracellularly by various microorganisms during the stationary phase and contribute to more than 80% of the cell's weight. These are stored as insoluble cytoplasmic inclusion bodies measuring 0.2–0.7 μm in diameter (Costa et al. 2019). The PHA granule is mainly composed of PHA (97.5%), a low protein content (2%), and phospholipids (0.5%). The granule has a central PHA polyester core and is surrounded by a lipid monolayer (2 nm). The layer is embedded with structural proteins (phasins), regulator proteins, PHA polyesters, PHA synthase, and PHA polymerase. They are synthesized when carbon is in excess and vital nutrients such as phosphorus or nitrogen are limited (Costa et al. 2019; Behera et al. 2022). Depending on the organism or species, the size and number of PHA granules differ per cell (Ojumu et al. 2004).

19.3.2 PHA Accumulation in Microorganisms

Many microorganisms, like bacteria, haloarchaea, and microalgae, under unfavorable conditions, synthesize and accumulate PHA as a source of cellular carbon and energy (Kaur et al. 2017). Several cyanobacterial isolates and a few microalgal species have been reported to produce PHA, although their PHA content is generally lower than that of heterotrophic PHA-producing bacteria (Koch et al. 2020). Nonetheless, microalgae can accumulate a significant amount of PHA under optimized conditions, including sufficient carbon, light, ideal temperature, pH, and appropriate nutrient and metal ion levels in the culture medium (Phalanisong et al. 2021). Under normal growth conditions, PHA accumulation was found to be lower than 3.5% (w/w dwt.) in 133 cyanobacterial strains (Kaewbai-Ngam et al. 2016). The cultivation conditions and corresponding PHA content of selected microalgal isolates under nutrient-replete and nutrient-limited conditions are summarized in Table 19.3.

19.3.3 Classification of PHA

These biopolymers are formed by the polymerization of hydroxy fatty acid monomers bound by ester linkages (Irfan et al. 2024). Based on the number of carbons in the side chain, PHAs are grouped into three categories: (i) short-chain length PHA (SCL-PHA), composed of three to five carbon units, e.g., polyhydroxybutyrate (PHB); (ii) medium-chain length PHA (MCL-PHA), consisting of six to fourteen carbon monomers, e.g., poly (3-hydroxybutyrate-co − 3-hydroxyvalerate) (PHBV); and (iii) long-chain length PHA (LCL-PHA) made of more than fourteen carbon units (Irfan et al. 2024; Behera et al. 2022; Carpine et al. 2020). Out of 155 distinct PHAs, PHB is the most studied polyester (Chong et al. 2021). The SCL-PHAs resemble conventional thermoplastics, while mcl-PHAs are elastomeric, making them promising alternatives to petroleum-based plastics (Phalanisong et al. 2021).

Table 19.3 PHA content of selected microalgae isolates under nutrient-replete and nutrient-limited conditions

Microalgal Isolates	Cultivation conditions	PHA/PHB Content (%)	References
	Under nutrient-replete conditions		
Scenedesmus sp. *UTEX 1589*	BG-11 medium, 25 °C under continuous illumination, pH 8.2	8.61	García et al. (2020)
Arthrospira platensis NLHT3	Zarrouk medium, pH 8.0, 25 °C, 14:10 h light/dark cycle, light intensity of 75 μmolm^{-2} s^{-1}	4.18	Pham et al. (2024)
Microcystis aeruginosa DTB1	Z8 medium pH 7.0, 25 °C, 14:10 h light / dark, photoperiod, light intensity of 75 μmolm^{-2} s^{-1}	3.46	Pham et al. (2024)
Synechocystis sp. PCC 6803	BG-11 medium, pH 7.5, 28 °C, continuous illumination of 50 μmol photon m^{-2} s^{-1}	2.4	Monshupanee and Incharoensakdi (2014)
Nostoc muscorum NCCU- 442	$NaNO_3$ free BG-11 medium, 28 °C, 14: 10 h light/dark cycle at 25 μmol photon m^{-2} s^{-1}	6.44	Ansari and Fatma (2016)
Synechococcus subsalsus	BG-11 medium, 28 °C, 12 h light / dark photoperiod	2.16	Costa et al. (2018a)
Spirulina sp. LEB-18	Zarrouk medium, 28 °C, 12 h light / dark photoperiod	2.02	Costa et al. (2018a)
	Under nutrient-limited (optimized) conditions		
Nostoc muscorum NCCU-442	Phosphorous deficiency, pH 7.5, 30 °C, 10:14 light/dark cycle with the addition of 0.4% glucose and 1.0 g/L of sodium chloride	26.4	Ansari and Fatma (2016)
Scenedesmus sp.	17.6 mM nitrogen, phosphorus omission, 0.5 g/L NaCl, 0.021 mM iron	29.92	García et al. (2020)
Calothrix scytonemicola TISTR 8095	Nitrogen deprivation	25.4	Kaewbai-Ngam et al. (2016)
Chlorella fusca LEB 111	D-xylose supplementation	17.4	Cassuriaga et al. (2018)

PHB: polyhydroxybutyrate

19.3.4 Biosynthesis of PHA

PHA is naturally synthesized through four distinct metabolic pathways (Costa et al. 2019).

19.3.4.1 PHA Synthesis via Pathway I

The most common pathway for PHA biosynthesis is pathway I, which proceeds via a three-step enzymatic pathway. The *phaCAB* operon consists of three genes, *phaA*, *phaB*, and *phaC*, which encode β-ketothiolase, NADPH-dependent acetoacetyl-CoA reductase, and PHA synthase, respectively (Behera et al. 2022).

I. Initially, in the biosynthetic pathway, two molecules of acetyl-CoA condense in a reversible reaction catalyzed by the *phaA*-encoded β-ketothiolase enzyme to form acetoacetyl-CoA.
II. This intermediate is subsequently reduced by *phaB*-encoded NADPH-dependent acetoacetyl-CoA reductase to yield (R)-3-hydroxybutyryl-CoA, which serves as the substrate for PHA synthase.
III. Finally, *phaC*-encoded PHA synthase facilitates the polymerization of (R)-3-hydroxybutyryl-CoA to form polyhydroxybutyrate (PHB) (Singh et al. 2017).

While *phaA* and *phaB* supply the precursor metabolites required for polymer formation, the *phaC* gene product, PHA synthase, is the defining enzyme of the pathway. This enzyme is part of Class I PHA synthases, which are specific for short-chain-length hydroxyalkanoates (C3–C5). PHA synthases are categorized into four classes: class I, II, III, and IV—based on their subunit composition and substrate specificity. Class I PHA synthases utilize short-chain fatty acids (C3–C5), class II enzymes act on medium-chain substrates (C6–C14). In contrast, classes III and IV are associated with the synthesis of longer-chain PHAs. Since each class exhibits distinct substrate preferences, the synthase determines which hydroxyacyl-CoA monomers are incorporated. Consequently, the type of PHA synthesized is governed by the catalytic properties of the synthase (Sharma et al. 2024). Although the *phaCAB* operon functions as a coordinated unit for PHA biosynthesis, classification is determined solely by the catalytic properties of *phaC* (Costa et al. 2019).

19.3.4.2 PHA Synthesis via Pathway II

It involves β-oxidation of fatty acids, resulting in the production of acyl-CoA, which is then converted to 3-hydroxyacyl-CoA monomers. PHA synthase catalyzes the monomers to generate PHA. This pathway synthesizes MCL-PHA, and some of the enzymes involved in this pathway are 3-ketoacyl-CoA reductase, epimerase, (R)-enoyl-CoA hydratase/enoyl-CoA hydratase I, and acyl-CoA oxidase (Sharma et al. 2024).

19.3.4.3 PHA Synthesis via Pathway III

The third pathway involves the conversion of the substrate to 3-hydroxyacyl-ACP, which can then form 3-hydroxy-acyl-CoA and polymerization of PHA by the enzyme PHA synthase. This pathway utilizes two enzymes: 3-hydroxyacyl-ACP-CoA transferase (PhaG) and malonyl-CoA-ACP transacylase (FabD) (Behera et al. 2022). The three metabolic pathways involved in PHA biosynthesis have been illustrated in Fig. 19.1.

19.3.4.4 PHA Synthesis via Pathway IV

In pathway IV, (R)-3-hydroxybutyryl-CoA is oxidized to acetoacetyl-CoA by NADPH-dependent acetoacetyl-CoA reductase, culminating in the synthesis of PHA (Costa et al. 2019).

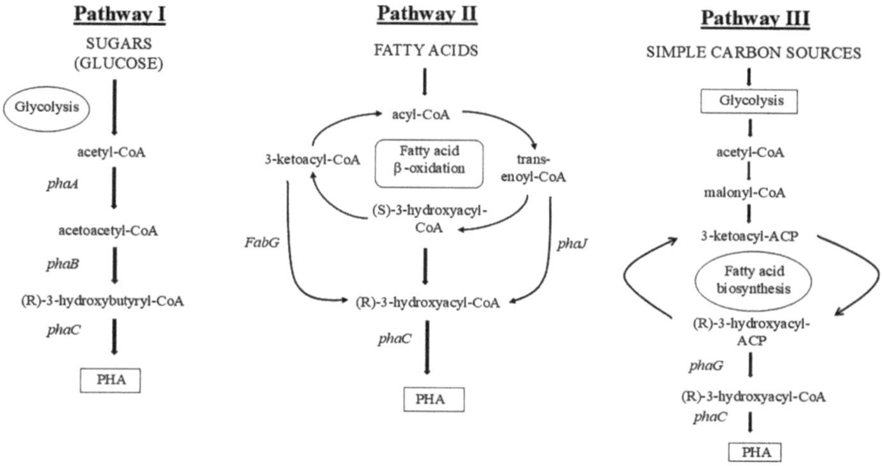

Fig. 19.1 Schematic representation of the three key metabolic pathways for PHA biosynthesis

19.3.5 Biodegradation of Polyhydroxyalkanoate

Under environmental stress or nutrient limitation, the accumulated PHAs are depolymerized by the action of intracellular PHA depolymerases, which catalyze the hydrolytic cleavage of ester linkages within the polymer backbone, thereby releasing soluble hydroxyalkanoate (HA) monomers or oligomers, which are later utilized by cells as a carbon source (Ojumu et al. 2004). Under aerobic conditions, this process yields carbon dioxide and water as terminal products, whereas methane is also generated as part of the degradation pathway under anaerobic conditions (Ojumu et al. 2004; Chauhan et al. 2024). Various factors like pH, temperature, moisture, and nutrient availability influence the rate of PHA degradation (Ojumu et al. 2004).

19.3.6 PHA Extraction, Purification, and Characterization

Under nutrient-limited conditions—particularly when essential elements such as nitrogen, phosphorus, or oxygen are depleted—microalgae divert excess carbon toward the intracellular accumulation of PHA. Following PHA biosynthesis, the algal biomass is harvested through techniques such as centrifugation, flocculation, gravity sedimentation, or filtration (Yin et al. 2020). The recovered biomass is then processed to extract PHA. PHA extraction techniques include chemical, physical, and biological methods, each contributing to the disruption of cells and recovery of intracellular products. The chemical method involves the use of solvents such as sodium hypochlorite, chloroform, and dichloromethane, whereas physical techniques include microwave-assisted or ultrasound-assisted extraction, bead milling, and subcritical water extraction (Jacob et al. 2021; Kartik et al. 2021). Biological methods utilize enzymes such as lysozymes, cellulases, glucanases, nucleases, and proteases to degrade non-PHA components selectively (Chong et al. 2021). Enzymatic action, along with mechanical and chemical methods, boosts PHA recovery (Sharma et al. 2024). Biological and physical processes are preferred for PHA extraction due to the absence of harmful chemicals, although they result in low polymer recovery. However, combining physical, chemical, and biological approaches enhances efficiency by improving cell disruption, shortening processing time, and increasing polymer recovery (Costa et al. 2019). Following extraction, the polymer undergoes purification to remove impurities. Ultrafiltration, electromembrane filtration, aqueous two-phase separation, and three-phase partitioning are some of the purification techniques employed to achieve high-purity PHA (Sharma et al. 2024). The characterization of purified PHA involves various analytical methods such as FTIR, NMR, TGA, DSC, XRD, and GPC to determine its chemical structure, thermal properties, and molecular weight, ensuring its suitability for specific applications (Behera et al. 2022).

19.3.7 PHAs as Functional Bioplastics: From Material Properties to Real-World Uses

Microalgae-derived plastics offer a cost-effective and energy-efficient alternative to conventional plastics. PHAs exhibit characteristics comparable to those of synthetic plastics (Shayesteh et al. 2023). They are biodegradable, biocompatible, and exhibit excellent flexibility. Additionally, they have a reduced carbon footprint and do not produce toxic by-products during degradation, contributing significantly to the advancement of a sustainable circular bioeconomy (Adetunji and Erasmus 2024). The mechanical, physical, and chemical properties of PHAs—particularly short-chain-length types like PHB—are comparable to those of conventional, nonbiodegradable polymers such as polyethylene and polypropylene (Cassuriaga et al. 2018). The mechanical performance of PHA-based materials is typically evaluated through parameters such as tensile strength, elongation at break, hardness, and Young's modulus (Onen Cinar et al. 2020). The physical properties of PHAs, such as melting point (T_m), glass transition temperature (T_g), and crystallinity, are determined by several factors, including the microbial strain, the monomeric composition of the polymer, the extraction and purification techniques, and the specific growth conditions during fermentation (Kaur et al. 2017). PHAs are optically active, highly crystalline in nature, and exhibit piezoelectric properties (Singh et al. 2017).

Microalgal PHAs exhibit superior physicochemical and mechanical properties compared to those produced by heterotrophic bacteria. *Nostoc muscorum* and *Aulosira fertilissima* exhibited crystallinity of 56.8% and 60.7%, and elongation at break of 8.6% and 4.9%, respectively, along with melting temperatures around 174–176 °C and tensile strengths ranging from 31.1 to 37.6 MPa (Ansari and Fatma 2016; Samantaray and Mallick 2012). In contrast, PHAs from heterotrophic bacteria had a melting temperature of 174.2 °C, crystallinity of 40.7%, tensile

Table 19.4 Thermophysical properties of PHA extracted from microalgae

Microalgal Isolates	*Mw* (kDa)	T_{onset} (°C)	T_{decom} (°C)	T_m (°C)	ΔH_m (J/g)	X_c (%)	T_g (°C)	References
Synechococcus subsalus	179.66	286.88	312.19	173.53	54.15	37.09	NA	Costa et al. (2018a)
Spirulina sp. LEB-18	163.26	249.79	269.76	171.69	65.93	45.15	NA	Costa et al. (2018a)
Synechocystis salina	887.5	259.6	272.1	181.6	72.5	49.7	7.5	Meixner et al. (2018)
Nostoc muscorum NCCU- 442	NA	256	284	171	83.25	78.8	6	Ansari and Fatma (2016)

Mw molecular weight, T_{onset} initial decomposition temperature, T_{decom} maximum decomposition temperature, T_m melting temperature, ΔH_m melting enthalpy, X_c degree of crystallinity, T_g glass transition temperature, *NA* not available

strength of 32 MPa, and elongation at break of only 2.5% (Arrieta et al. 2014). Moreover, cyanobacteria utilize atmospheric CO_2 as their sole carbon source, offering a cost-effective alternative to heterotrophic bacteria, which require constant organic carbon input—a key limitation in commercial PHB production (Das and Maiti 2021).

The physicochemical properties of PHAs directly influence their applicability in different sectors (Costa et al. 2018b). PHA, being biocompatible, can be used as a scaffold in tissue engineering, medical implants for biomedical purposes, and as a carrier for therapeutic encapsulation in pharmacology (drug delivery) (Sharma et al. 2024). They are used in agricultural fields for controlled release of fertilizers, and as a biodegradable material for consumer packaging (Shayesteh et al. 2023). They are also gaining attention in 3D printing (Kartik et al. 2021). Thermophysical properties of selected microalgal isolates are listed in Table 19.4.

19.4 Biofuels

Biofuels obtained from renewable sources serve as a viable alternative to petrochemical fuels. Based on their physical state, they are grouped into solid (biochar), liquid (bioethanol and biodiesel), and gaseous (biogas and biohydrogen) (Chen et al. 2018). Biofuels are classified into two categories: primary biofuels and secondary biofuels. Primary biofuels consist of raw organic materials, such as firewood, plants, crop residue, and animal waste, whereas secondary biofuels are derived from processed biomass (Rodionova et al. 2017). Depending on the feedstock and processing methods, secondary biofuel sources are further classified into first, second, and third-generation fuels (Leong et al. 2018). First-generation feedstocks, derived from edible sources such as sunflower, palm oil, soybean, and corn, have triggered food-versus-fuel competition and intensified concerns over land scarcity. These limitations highlight the need for alternative resources, leading to the development of second-generation feedstocks (Jacob et al. 2021). The second-generation feedstocks utilize lignocellulosic material, agricultural waste, and nonedible crops such as jatropha and jojoba. Although second-generation feedstock mitigates the food security issues posed by first-generation fuels, their large-scale cultivation requires adequate irrigation, nutrient management, and careful agronomic practices. Moreover, their economic and practical viability was constrained by low conversion efficiencies and the limited availability of raw materials (Leong et al. 2018). The shortcomings of first and second-generation biofuels have generated growing interest in third-generation biofuels, specifically those produced from oleaginous microorganisms such as microalgae, bacteria, and fungi. Among these, microalgae stand out due to their fast proliferation, high biomass productivity, and remarkable lipid accumulation capacity (70%), at least 15–20 times higher than second-generation biofuels (Alishah Aratboni et al. 2019).

19.4.1 Microalgal Lipids

Microalgal lipids can be broadly classified into polar and nonpolar lipids based on their structural characteristics. Polar lipids, also referred to as structural lipids, are crucial for the formation of cellular membranes. This group includes phospholipids, glycolipids, and sphingolipids. Phospholipids and glycolipids serve as the primary structural components of biological membranes, while sphingolipids play a key role in cell signaling pathways (Chen et al. 2018). In contrast, nonpolar lipids, also known as neutral lipids, primarily function as energy storage molecules. This category comprises triacylglycerols (TAG), steryl esters, and wax esters. Sterols play a role in modulating membrane fluidity and permeability. Neutral lipids are typically stored within lipid bodies that possess a hydrophobic core encased by a phospholipid monolayer (Tang et al. 2020). TAGs are storage lipids that act as energy reserves and electron sinks. They are present in the cytoplasm as lipid granules and are primarily composed of saturated and monounsaturated fatty acids. Each TAG molecule is linked to three fatty acids esterified to a glycerol backbone. These storage lipids serve as a source of carbon and energy reservoir, while supplying essential fatty acids for the production of biodiesel (Morales et al. 2021). The lipid composition of microalgae varies among species and is influenced by growth conditions. Microalgal lipid accumulation varies between 20% to 50% of dry biomass (Chen et al. 2018). The lipid content in *Chlorella* (BIOTECH 4026) and *Chlorolobion* (BIOTECH 4031) was reported to be 10.36% and 8.73%, respectively, when grown in BG-11 media (Arguelles and Martinez-Goss 2021).

19.4.2 TAG Biosynthesis

TAG synthesis occurs in two stages: the first stage involves fatty acid biosynthesis in the chloroplast, and the second stage involves triacylglycerol synthesis in the endoplasmic reticulum via the Kennedy pathway.

19.4.2.1 Fatty Acid Synthesis (De Novo Synthesis)

In photosynthetic microalgae, TAG biosynthesis begins with the capturing of light energy by chlorophyll, initiating the light-dependent reactions of photosynthesis (Ananthi et al. 2021a). These reactions generate ATP and NADPH, which are then utilized in the Calvin cycle within the chloroplast stroma to fix atmospheric CO_2 into glyceraldehyde-3-phosphate (G3P). G3P is subsequently used to synthesize glucose and other sugars, which enter the glycolytic pathway in the cytosol, producing pyruvate along with ATP and NADH. Pyruvate is then transported into the mitochondria (or plastids in microalgae) and decarboxylated by the pyruvate dehydrogenase complex to form acetyl-CoA. Acetyl-CoA is the precursor molecule for lipid production (Pandey et al. 2024). Acetyl-CoA is exported to the cytosol, where it is carboxylated by acetyl-CoA carboxylase (ACC) to form malonyl-CoA. Malonyl-CoA is further converted to malonyl ACP (acyl carrier protein) by the action of malonyl–CoA–ACP–transferase (Chhandama et al. 2021). Malonyl ACP, along with ATP and NADPH, drives the synthesis of 16–18 carbon fatty acids through successive additions of two-carbon units catalyzed by fatty acid synthase (FAS). Fatty acyl-ACP thioesterase (FAT) terminates chain elongation by releasing free fatty acids, which are subsequently exported from the plastid and activated in the cytosol by long-chain acyl-CoA synthase (LACS). In the presence of CoA, these free fatty acids are esterified to form fatty acyl–CoA (Ananthi et al. 2021a).

19.4.2.2 Kennedy Pathway

In the endoplasmic reticulum, fatty acyl-CoA, along with G3P, forms lysophosphatidic acid, which is catalyzed by glycerol 3 phosphate acyltransferase (GPAT). Lysophosphatidic acid acyl-transferase (LPAT) catalyzes the synthesis of phosphatidic acid from lysophosphatidic acid. Phosphatidic acid phosphatase (PAP) removes the phosphate group from phosphatidic acid to form 1,2-diacylglycerol (DAG). In the final step, diacylglycerol acyltransferase (DGAT) catalyzes the acylation of diacylglycerol (DAG), resulting in the formation of triacylglycerols (TAGs) (Pandey et al. 2024). Figure 19.2 shows the two primary lipid biosynthesis pathways in microalgae: de novo synthesis and Kennedy pathway.

19.4.3 Encapsulation of Lipid Bodies

In the final stage of TAG biosynthesis, the accumulated triacylglycerols in the endoplasmic reticulum membrane eventually bud off to form lipid bodies, also known as lipid droplets. These lipid bodies may undergo further processing and trafficking through the Golgi apparatus, facilitating their storage and distribution within the cell (Ananthi et al. 2021a).

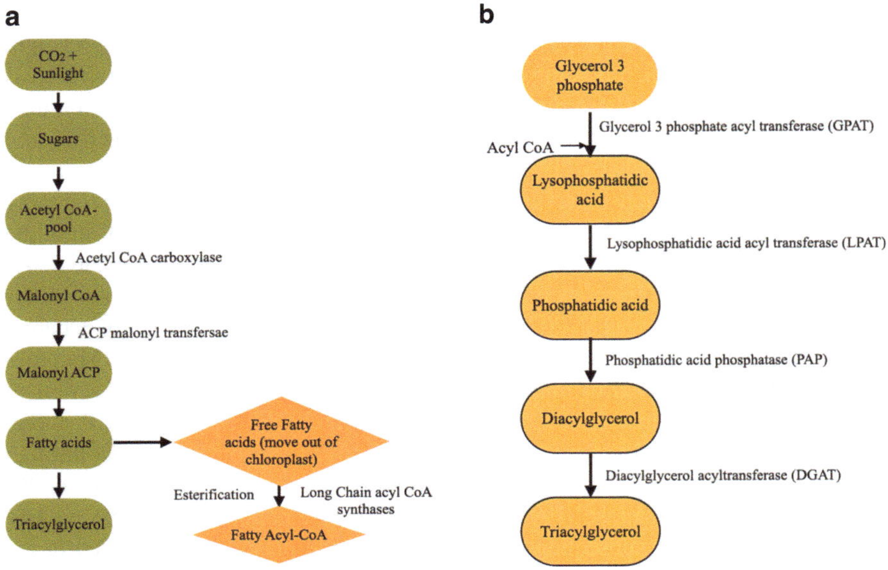

Fig. 19.2 Lipid biosynthesis in microalgae. (**a**) De novo synthesis; (**b**) Kennedy pathway

19.4.4 TAG as an Electron Sink

In phytoplankton, TAG accumulation is induced by disruption in the cellular energy under stress conditions, such as nutrient deprivation, which downregulates anabolic pathways (Calvin cycle). As a result, the surplus energy from the light-dependent reactions causes a hyper-reduction of the electron transport chain, promoting the generation of reactive oxygen species (ROS). TAGs, being highly reduced molecules, function as alternative energy sinks, helping to dissipate excess electrons, limit ROS generation, safeguard photosynthetic function, and stabilize membrane structure and fluidity (Morales et al. 2021).

19.4.5 Biodiesel

Biodiesel is a clean-burning liquid fuel derived from renewable sources such as microalgae. Chemically, biodiesel is a mixture of monoalkyl esters of long-chain fatty acids. The production of biodiesel from microalgae involves four main steps: cultivation, harvesting, lipid extraction, and transesterification. Microalgae are first cultivated under optimized conditions to maximize biomass and lipid accumulation. The biomass is then harvested and processed to extract intracellular lipids. Finally, the extracted lipids undergo transesterification to convert triglycerides into fatty acid methyl esters (biodiesel) (Dickinson et al. 2017; Rodionova et al. 2017). It is anticipated that microalgae can produce 200 times more biodiesel than traditional crops (Rodionova et al. 2017). Biodiesel produced from microalgae serves as an eco-efficient and eco-friendly alternative to conventional diesel. It is biodegradable and is synthesized from nontoxic materials, hence emits lower levels of greenhouse gases, sulfur oxides, and nitrogen oxides (Ananthi et al. 2021a; Ahmad et al. 2011). The closed carbon cycle of microalgae ensures minimal net CO_2 emissions, while its physicochemical properties—such as viscosity, density, and energy content—closely resemble those of conventional diesel, allowing it to be used directly in existing engines, pipelines, and fuel infrastructure without major modifications (Rodionova et al. 2017). Biodiesel exhibits a calorific value comparable to that of fossil fuels and, owing to its renewable origin, constitutes a sustainable bioenergy resource for continuous applications (Leong et al. 2018).

19.4.6 Fatty Acid Composition and Its Impact on Biodiesel Properties

The physicochemical properties of biodiesel are primarily determined by its fatty acids. Fatty acids are the precursor molecules for the synthesis of various lipids in microalgae. Microalgal fatty acids are typical carboxylic acids featuring hydrocarbon chains ranging from 4 to 36 carbon atoms. They are generally classified into saturated fatty acids (SAFA), monounsaturated fatty acids (MUFA), and polyunsaturated fatty acids (PUFA) (Ferreira et al. 2019). The fatty acid composition shows significant variability across algal species, and even within a single species, the profile can shift in response to cultivation conditions and the nutrient composition of the culture medium (Morales et al. 2021). The biodiesel quality is primarily governed by its fatty acid composition, encompassing both the chain length and the degree of saturation of constituent fatty acids. While the overall distribution of SAFAs, MUFAs, and PUFAs influences the fundamental fuel characteristics, the number of double bonds, representing the degree of unsaturation, is a critical determinant of ignition quality, oxidative stability, viscosity, and cold-flow performance. The suitability of the biodiesel produced must be evaluated by determining its key physicochemical properties and comparing them with established international standards, such as ASTM D6751 and EN14214. These standards are designed to ensure that biodiesel can be utilized as a sustainable and economically viable transport fuel with minimal environmental impact (Deshmukh et al. 2019; Anahas and Muralitharan 2015).

The length of the carbon chain and the number of double bonds directly influence viscosity, cetane number, calorific value, ignition quality, oxidative stability, and cold-flow properties. In general, longer-chain SAFAs exhibit higher cetane numbers and better combustion properties, but their elevated viscosity can compromise atomization and result in poor cold flow. Conversely, MUFAs and PUFAs reduce viscosity and improve cold flow properties; however, their cetane number, calorific value, and oxidative stability decline with increasing unsaturation due to the presence of allylic and bis-allylic sites susceptible to autoxidation (Deshmukh et al. 2019). A high fraction of PUFAs (e.g., C18:2, C18:3) is associated with reduced cetane number and poor ignition quality. It leads to increased viscosity and sediment formation, thereby lowering storage and oxidative stability. Nevertheless, a limited proportion of PUFAs can enhance biodiesel flow at low temperatures, particularly in cold climates. Cold flow properties are strongly affected by the proportion of SAFAs; higher SAFA content elevates cloud and pour points, thereby impairing cold flow and increasing the likelihood of solidification at low temperatures (Morales et al. 2021). An optimal biodiesel composition, therefore, requires a balanced mixture of long-chain SAFAs (for higher energy yield, cetane number, and thermal stability) and short-chain MUFAs or limited PUFAs (for improved flow and lower viscosity) (Deshmukh et al. 2019). Palmitic acid (C16:0), stearic acid (C18:0), oleic acid (C18:1), linoleic acid (C18:2), and α-linolenic acid (C18:3) are key fatty acid components that contribute to biodiesel quality. This balance prevents the extremes caused by the dominance of any single fatty acid group and ensures fuel properties that support engine performance, durability, and emission compliance (Morales et al. 2021). Biodiesel properties of some of the microalgae are listed in Table 19.5.

19.5 Factors Affecting the Production of Lipid and PHA

Nutrient availability, particularly nitrogen, phosphorus, and carbon sources, plays a critical role in microalgal metabolism, as the limitation of these essential nutrients often triggers the synthesis of storage compounds such as lipids and PHA. Abiotic parameters such as light intensity, photoperiod, temperature, salinity, and pH modulate metabolic fluxes and enzymatic activity. These factors affect the rate and composition of lipid and PHA accumulation (Ananthi et al. 2021b). Additionally, stress conditions, including oxidative stress or high irradiance, can enhance intracellular carbon storage, while genetic and species-specific traits determine the maximum biosynthetic potential of microalgal cells (Chang et al. 2020; Pancha et al. 2015; Han et al. 2020). The accumulation of lipids and PHAs in microalgae is regulated by three main factors: nutrient availability and assimilation, environmental stress conditions, and the cultivation strategy employed (Sharma et al. 2024; Pandey et al. 2024). Nitrogen is essential for protein synthesis and in biosynthetic pathways. Nitrogen limitation or deficiency negatively affects photosynthetic activity, reducing microalgal growth, but it simultaneously induces the accumulation of lipids and PHA (García et al. 2020; Morales et al. 2021). Phosphorus limitation results in the upregulation of lipid accumulation and protein degradation. At the same time, the synthesis of amino acids, nucleic acids, and photosynthesis is downregulated (Pandey et al. 2024). Light plays a crucial part in the growth and synthesis of secondary metabolites. The wavelength, intensity, and duration of light affect the accumulation of lipids and PHAs (Ansari and Fatma 2016; Zakir Hossain et al. 2021). Polar lipids are synthesized under low light intensities; in contrast, elevated light intensities stimulate the accumulation of storage lipids. Temperature and pH are vital for the enzymatic activity in microalgae. Photoperiod is essential as prolonged exposure to sunlight causes the degradation of chlorophyll, thereby reducing photosynthesis (Ansari and Fatma 2016; Pandey et al. 2024; Zakir Hossain et al. 2021).

Table 19.5 Biodiesel properties of microalgal isolates

Standards	IV (gI_2g/fat)	SV mg KOH/g	CP °C	PP °C	CN	DU Wt. %	LCSF Wt. %	CFPP °C	HHV MJ/kg	OS h	FP °C	ρ gcm^{-3}	ν mm^2s^{-1}	References
EN 14214	≤120	–	–	–	≥51	–	–	≤5/–20	NA	≥ 6	120	0.86–0.90	3.5–5.0	Youssef et al. (2024)
ASTM D6751D-02	NA	–	- 3 to 12	– 15-16.0	≥47	–	–	NA	NA	3	93	0.86–0.90	1.9–6.0	
IS 15607	NA	–	12		≥51	–	–		NA	–	–	0.86–0.90	2.5–6.0	
Biodiesel standard EN 14213	–	–	–	–	–	–	–	–	35	–	–	–	–	
Min/max	Min	Min	Min	Min	Max	Min	Min	Min	Max	Max	Max	Min	Min	
Threshold value	120	–	12	16	51	–	–	18	35	6	120	0.9	6	
Tetradesmus obliquus	38.61	208.0	23.4	18.6	63.9	43.0	5.39	0.46	40.11	6.12	186.3	0.87	4.8	Youssef et al. (2024)
Cyanothece sp.	18.54	228.5	29.7	25.4	66.0	10.3	6.6	4.25	39.41	7.37	204.9	0.88	3.11	Youssef et al. (2024)
Asterarcys sp. RA 100	74	186	6	NA	59	76	NA	7.7	36	NA	NA	0.8	NA	Abed et al. (2025)
Monraphidium griffithii	178.3	195.4	NA	NA	34	NA	NA	NA	46.15	NA	NA	0.89	2.47	Balouch et al. (2023)
Ankistrodesmus falcatus	103.6	199.5	NA	NA	50.3	NA	NA	NA	41.85	NA	NA	0.88	4.01	Balouch et al. (2023)

(continued)

Table 19.5 (continued)

Standards	IV (gI_2g/fat)	SV mg KOH/g	CP °C	PP °C	CN	DU Wt. %	LCSF Wt. %	CFPP °C	HHV MJ/kg	OS h	FP °C	ρ gcm^{-3}	ν mm^2s^{-1}	References
Parachlorella kessleri	83.4	260.8	NA	NA	48.4	NA	NA	NA	42.16	NA	NA	0.88	3.9	Balouch et al. (2023)
Desmodesmus pannonicus	194.9	236.4	NA	NA	25.5	NA	NA	NA	47.44	NA	NA	0.9	2.01	Balouch et al. (2023)

IV iodine value, *SV* saponification value, *CP* cloud point, *PP* pour point, *CN* cetane number, *DU* degree of unsaturation, *LCSF* long-chain saturation factor, *CFPP* cold filter plugging point, *HHV* higher heating value, *OS* oxidative stability, *FP* flash point, ρ density, ν viscosity, *NA* not available

19.6 Genetic Engineering for Enhanced Metabolite Production

Altering growth conditions can significantly influence metabolite production in microalgae; however, achieving high yields requires strategies beyond their inherent metabolic capacity. Genetic engineering approaches, such as gene editing and gene overexpression/silencing, offer a more precise and reliable alternative to conventional culture-based strategies for enhancing metabolite biosynthesis (Sreenikethanam et al. 2022). Advanced tools such as Clustered Regularly Interspaced Short Palindromic Repeats-associated protein 9 (CRISPR-Cas9), Transcription Activator-Like Effector Nucleases (TALENs), RNA interference (RNAi), and Zinc Finger Nucleases (ZFNs) enable targeted modification of key enzymes involved in metabolic bottlenecks, thereby redirecting carbon flux toward desired products. Integrated omics approaches—such as transcriptomics, proteomics, and next-generation sequencing—together with metabolic engineering, can provide deeper insights into lipid and PHA metabolism and their associated genes, leading to significant increases in their production in microalgae. RNA interference effectively enhances lipid accumulation in *Phaeodactylum tricornutum* by silencing the gene coding for nitrate reductase (Tang et al. 2020; Geng et al. 2025). Representative examples of genetic modifications applied in microalgae to enhance lipid and PHA accumulation are summarized in Table 19.6.

19.7 Techno-economic Analysis

A techno-economic analysis (TEA) evaluates the operational efficiency and economic feasibility of a process, providing insights into how the processing of biomass feedstock can be optimized to reduce costs. The economic feasibility is influenced by various factors but is primarily determined by feedstock and processing (Geng et al. 2025). Current bacterial PHA production costs range from \$4–\$6/kg, compared to \$1–\$2/kg for conventional petrochemical plastics, whereas algal biomass can be produced at \$1.13–\$2.04/kg; however, the additional processing required to convert it into PHA results in total production costs that likely exceed those of bacterial PHA (Sarker and Kaparaju 2025). Algal oil is projected to become a major energy feedstock, even though its biodiesel cost (\$1.59–2.12/L) remains higher than that from edible crops (\$0.45–0.55/L), lignocellulosic biomass (\$0.80–1.20/L), and conventional petroleum diesel (~\$0.66/L) (Tang et al. 2020; Geng et al. 2025). Algal biomass production remains more expensive than petrochemical feedstocks, but integrated biorefineries producing co-products such as biodiesel, bioplastics, animal feed, fertilizers, bioenergy, or wastewater remediation can improve overall economics (Laurens et al. 2017). Further cost reductions can be achieved by utilizing low-cost renewable feedstocks, optimizing production methods, adopting greener extraction techniques, and advancing strain engineering (Gundlapalli and Ganesan 2025).

Table 19.6 Genetic modifications in microalgae for enhanced lipid and PHA accumulation

Microalgal strain	Target gene site	Techniques	Results	References
Chlamydomonas reinhardtii CC-124	*Glyceraldehyde-3-phosphate dehydrogenase* (GADPH)	Overexpression under a nitrogen-inducible NIT1 promoter.	2.4-fold increase in lipid content (FAME)	Shin et al. (2024)
Synechocystis sp.	*Acetyl-CoA carboxylase* (ACC)	Overexpression	Lipid content increased by 3.6-fold	Fathy et al. (2021)
Chlamydomonas reinhartdii	*Phospholipase* A_2 (PLA_2)	Knock-out/CRISPR-Cas9	Enhanced lipid content (64.25%)	Shin et al. (2019)
Chlamydomonas reinhardtii	*CrPEPC1 (phosphoenolpyruvate carboxylase)*	94% Down-regulation by CRISPRi-Cas9	Lipid content increased by 74.4% and lipid productivity by 94.2%	Kao and Ng (2017)
Tetraselmis sp.	*ADP-glucose pyrophosphorylase (AGP* gene)	CRISPR-Cas9 RNP	Boosted lipid content in both mutants by 2.7 and 3.1-fold	Chang et al. (2020)
Synechocystis sp. PCC 6803	*sll0944, phaA and phab*	Knockout and overexpression	PHB content increased to 63%	Koch et al. (2020)
Synechocystis sp. PCC 6803	*glutamate decarboxylase*	Knockout/inactivation	2.5-fold increment in the PHB content	Monshupanee et al. (2019)
Synechocystis sp. PCC6803	*phosphotransacetylase* (Pta), *acetyl-CoA hydrolase* (Ach), *phosphoketolase* (XfpK)	Double gene knockout and overexpression	7.3 mg/L/d of increased PHB productivity and PHB content (12%)	Carpine et al. (2017)
Synechocystis Sp. PCC 6714	Phosphate-specific transport system integral membrane protein A (PstA)	UV mutagenesis	2.5-fold increase in PHB productivity	Kamravamanesh et al. (2018)

19.8 Commercial Progress and Industrial Engagement

Recent industrial initiatives demonstrate efforts to overcome these techno-economic barriers. Replacement of fossil fuels with microalgae-based biofuels has driven the focus of several companies on sustainable energy solutions, including Algae.Tec, EcoFuel Laboratories, IBV Biotech, AlgaeBiotech, and Sapphire Energy. Solix Biofuels achieves up to 3000 gallons of biofuel production, and ExxonMobil and Synthetic Genomics have developed genetically modified algae strains with over twice the oil content (Tang et al. 2020). Several case studies across Asia-Pacific, Europe, and North America highlight the advancements and industrial challenges in scaling PHA production. Kaneka Corporation (China) invested $50 million in 2023 to expand PHA production through localized sourcing and process optimization (Kaneka Corporation). TianAn Biologic Materials Co., Ltd. employs bacterial fermentation of agro-industrial waste to lower feedstock costs and improve yield, but the purification steps remain energy-intensive and costly (Ningbo TianAn Biologic Material Co., Ltd.). Meanwhile, Helian Polymers (Netherlands) advances cost reduction via the EU-funded ELLIPSE project by converting low-cost industrial waste streams into valuable resources, thereby aligning with circular economy principles (Gundlapalli and Ganesan 2025). The key players in India are Truegreen, EnviGreen Biotech India Pvt. Ltd., and Pappco Greenware (Nanda and Bharadvaja 2022).

19.9 Challenges and Future Perspective

The large-scale production of lipids and PHAs from microalgae is still constrained by low biomass density, high cultivation costs, and limited metabolite yield. Industrial deployment remains economically challenging due to the expenses associated with land and water requirements for large-scale cultivation, as well as energy-intensive operations such as harvesting and extraction (Tang et al. 2020). These cost barriers persist across both open raceway ponds and closed bioreactors. Furthermore, the intrinsic accumulation of lipids or PHAs in most algal strains is relatively low, which directly impacts overall productivity and commercial feasibility (Das and Maiti 2021; Hoang et al. 2023).

Future progress will depend on a combination of process optimization and advanced biological interventions. Improvements in cultivation strategies, including media optimization and integration with wastewater remediation, can significantly reduce nutrient input costs (Hoang et al. 2023). A biorefinery-based approach is essential, where residual carbohydrates, proteins, and high-value pigments such as chlorophyll or carotenoids are recovered after lipid and PHA extraction to maximize revenue streams (Dickinson et al. 2017; Rodionova et al. 2017). Strain improvement through genetic engineering, metabolic rewiring, omics-based pathway elucidation, and dynamic regulation of key enzymes can further enhance metabolite flux (O'Neill and Cagney 2024; Saini et al. 2020). By combining these strategies, microalgal lipid and PHA production can transition from laboratory-scale feasibility to industrial reality.

19.10 Conclusion

Microalgae represent far more than just another renewable resource—they signify a fundamental shift in how fuels and materials are produced. Their capacity to generate both lipids and biopolymers, coupled with natural CO_2 sequestration, positions them as a truly circular feedstock for biodiesel and bioplastics. Continued advancements in biological optimization, strain engineering, downstream processing, and integrated biorefinery approaches will be key to unlocking their full potential. If scaled strategically, microalgae can ultimately replace fossil-derived fuels and plastics, accelerating the transition toward a carbon-neutral and biodegradable future.

References

Abdelkhalek EA, Mohamed B, Mohammed AM, Lotfi A (2016) Growth performance and biochemical composition of nineteen microalgae collected from different Moroccan reservoirs. Mediterr Mar Sci 17(1):323–332

Abed RM, Al Battashi H, Muthukrishnan T, Al-Bartamani H, Al-Hinai M, Al-Baluchi M (2025) Growth of a newly isolated oleaginous microalgal strain (Asterarcys sp. RA100) in oil produced water and its potential for biodiesel production. PLoS One 20(6):e0325759

Adetunji AI, Erasmus M (2024) Green synthesis of bioplastics from microalgae: a state-of-the-art review. Polymers 16(10):1322

Ahmad AL, Yasin NM, Derek CJ, Lim JK (2011) Microalgae as a sustainable energy source for biodiesel production: a review. Renew Sust Energ Rev 15(1):584–593

Alishah Aratboni H, Rafiei N, Garcia-Granados R, Alemzadeh A, Morones-Ramírez JR (2019) Biomass and lipid induction strategies in microalgae for biofuel production and other applications. Microb Cell Factories 18(1):178

Anahas AM, Muralitharan G (2015) Isolation and screening of heterocystous cyanobacterial strains for biodiesel production by evaluating the fuel properties from fatty acid methyl ester (FAME) profiles. Bioresour Technol 184:9–17

Ananthi V, Raja R, Carvalho IS, Brindhadevi K, Pugazhendhi A, Arun A (2021a) A realistic scenario on microalgae based biodiesel production: third generation biofuel. Fuel 284:118965

Ananthi V, Brindhadevi K, Pugazhendhi A, Arun A (2021b) Impact of abiotic factors on biodiesel production by microalgae. Fuel 284:118962

Ansari S, Fatma T (2016) Cyanobacterial polyhydroxybutyrate (PHB): screening, optimization and characterization. PLoS One 11(6):e0158168

Arguelles ED, Martinez-Goss MR (2021 Feb) Lipid accumulation and profiling in microalgae Chlorolobion sp.(BIOTECH 4031) and Chlorella sp.(BIOTECH 4026) during nitrogen starvation for biodiesel production. J Appl Phycol 33(1):1–1

Arrieta MP, Samper MD, López J, Jiménez A (2014) Combined effect of poly (hydroxybutyrate) and plasticizers on polylactic acid properties for film intended for food packaging. J Polym Environ 22(4):460–470

Balouch H, Zayadan BK, Sadvakasova AK, Kossalbayev BD, Bolatkhan K, Gencer D, Civelek D, Demirbag Z, Alharby HF, Allakhverdiev SI (2023) Prospecting the biofuel potential of new microalgae isolates. Int J Hydrog Energy 48(50):19060–19073

Behera S, Singh R, Arora R, Sharma NK, Shukla M, Kumar S (2015) Scope of algae as third generation biofuels. Front Bioeng Biotechnol 2:90

Behera S, Priyadarshanee M, Das S (2022) Polyhydroxyalkanoates, the bioplastics of microbial origin: properties, biochemical synthesis, and their applications. Chemosphere 294:133723

Carpine R, Du W, Olivieri G, Pollio A, Hellingwerf KJ, Marzocchella A, dos Santos FB (2017) Genetic engineering of Synechocystis sp. PCC6803 for poly-β-hydroxybutyrate overproduction. Algal Res 25:117–127

Carpine R, Olivieri G, Hellingwerf KJ, Pollio A, Marzocchella A (2020) Industrial production of poly-β-hydroxybutyrate from CO2: can cyanobacteria meet this challenge? PRO 8(3):323

Cassuriaga AP, Freitas BC, Morais MG, Costa JA (2018) Innovative polyhydroxybutyrate production by Chlorella fusca grown with pentoses. Bioresour Technol 265:456–463

Chang KS, Kim J, Park H, Hong SJ, Lee CG, Jin E (2020) Enhanced lipid productivity in AGP knockout marine microalga Tetraselmis sp. using a DNA-free CRISPR-Cas9 RNP method. Bioresour Technol 303:122932

Chauhan K, Kaur R, Chauhan I (2024) Sustainable bioplastic: a comprehensive review on sources, methods, advantages, and applications of bioplastics. Polym Plast Technol Mater 63(8): 913–938

Chen Z, Wang L, Qiu S, Ge S (2018) Determination of microalgal lipid content and fatty acid for biofuel production. Biomed Res Int 2018(1):1503126

Chhandama MV, Satyan KB, Changmai B, Vanlalveni C, Rokhum SL (2021) Microalgae as a feedstock for the production of biodiesel: a review. Bioresour Technol Rep 15:100771

Chong JW, Yew GY, Khoo KS, Ho SH, Show PL (2021) Recent advances on food waste pretreatment technology via microalgae for source of polyhydroxyalkanoates. J Environ Manage 293:112782

Costa SS, Miranda AL, Andrade BB, de Jesus AD, Souza CO, de Morais MG, Costa JA, Druzian JI (2018a) Influence of nitrogen on growth, biomass composition, production, and properties of polyhydroxyalkanoates (PHAs) by microalgae. Int J Biol Macromol 116:552–562

Costa JA, Moreira JB, Lucas BF, Braga VD, Cassuriaga AP, Morais MG (2018b) Recent advances and future perspectives of PHB production by cyanobacteria. Ind Biotechnol 14(5):249–256

Costa SS, Miranda AL, de Morais MG, Costa JA, Druzian JI (2019) Microalgae as source of polyhydroxyalkanoates (PHAs)—a review. Int J Biol Macromol 131:536–547

Das M, Maiti SK (2021) Recent progress and challenges in cyanobacterial autotrophic production of polyhydroxybutyrate (PHB), a bioplastic. J Environ Chem Eng 9(4):105379

Deshmukh S, Kumar R, Bala K (2019) Microalgae biodiesel: a review on oil extraction, fatty acid composition, properties and effect on engine performance and emissions. Fuel Process Technol 191:232–247

Dickinson S, Mientus M, Frey D, Amini-Hajibashi A, Ozturk S, Shaikh F, Sengupta D, El-Halwagi MM (2017) A review of biodiesel production from microalgae. Clean Techn Environ Policy 19(3):637–668

Fathy W, Essawy E, Tawfik E, Khedr M, Abdelhameed MS, Hammouda O, Elsayed K (2021) Recombinant overexpression of the Escherichia coli acetyl-CoA carboxylase gene in Synechocystis sp. boosts lipid production. J Basic Microbiol 61(4):330–338

Ferreira GF, Pinto LR, Maciel Filho R, Fregolente LV (2019) A review on lipid production from microalgae: association between cultivation using waste streams and fatty acid profiles. Renew Sust Energ Rev 109:448–466

García G, Sosa-Hernández JE, Rodas-Zuluaga LI, Castillo-Zacarías C, Iqbal H, Parra-Saldívar R (2020) Accumulation of PHA in the microalgae Scenedesmus sp. under nutrient-deficient conditions. Polymers 13(1):131

Geng Y, Shaukat A, Azhar W, Raza QU, Tahir A, Abideen MZ, Zia MA, Bashir MA, Rehim A (2025) Microalgal biorefineries: a systematic review of technological trade-offs and innovation pathways. Biotechnol Biofuels Bioprod 18(1):93

Gharajeh NH, Valizadeh M, Dorani E, Hejazi MA (2020) Biochemical profiling of three indigenous Dunaliella isolates with main focus on fatty acid composition towards potential biotechnological application. Biotechnol Rep (Amst) 26:e00479

Gundlapalli M, Ganesan S (2025) Polyhydroxyalkanoates (PHAs): key challenges in production and sustainable strategies for cost reduction within a circular economy framework. Results Eng 26:105345

Han X, Song X, Li F, Lu Y (2020) Improving lipid productivity by engineering a control-knob gene in the oleaginous microalga Nannochloropsis oceanica. Metab Eng Commun 11:e00142

Hoang AT, Sirohi R, Pandey A, Nižetić S, Lam SS, Chen WH, Luque R, Thomas S, Arıcı M, Pham VV (2023) Biofuel production from microalgae: challenges and chances. Phytochem Rev 22(4): 1089–1126

Irfan F, Tasnim N, Razzak SA, Uddin S (2024) Advances in capturing carbon and producing bioplastics through the microalgal approach towards a sustainable future: a review. J Clean Prod 486:144576

Jacob A, Ashok B, Alagumalai A, Chyuan OH, Le PT (2021) Critical review on third generation micro algae biodiesel production and its feasibility as future bioenergy for IC engine applications. Energy Convers Manag 228:113655

Kaewbai-Ngam A, Incharoensakdi A, Monshupanee T (2016) Increased accumulation of polyhydroxybutyrate in divergent cyanobacteria under nutrient-deprived photoautotrophy: An efficient conversion of solar energy and carbon dioxide to polyhydroxybutyrate by Calothrix scytonemicola TISTR 8095. Bioresour Technol 212:342–347

Kamravamanesh D, Kovacs T, Pflügl S, Druzhinina I, Kroll P, Lackner M, Herwig C (2018) Increased poly-β-hydroxybutyrate production from carbon dioxide in randomly mutated cells of cyanobacterial strain Synechocystis sp. PCC 6714: mutant generation and characterization. Bioresour Technol 266:34–44

Kaneka Corporation. Kaneka Corporation Official Website [Internet]. https://www.kaneka.co.jp/en/

Kang NS, An SM, Jo CR, Ki H, Kim SY, Jeong HG, Choi G, Hong JW, Cho K (2024) Taxonomic, physiological, and biochemical characterization of Asterarcys quadricellularis AQYS21 as a promising sustainable feedstock for biofuels and ω-3 fatty acids. Plants 13(21):3008

Kao PH, Ng IS (2017) CRISPRi mediated phosphoenolpyruvate carboxylase regulation to enhance the production of lipid in Chlamydomonas reinhardtii. Bioresour Technol 245:1527–1537

Kartik A, Akhil D, Lakshmi D, Gopinath KP, Arun J, Sivaramakrishnan R, Pugazhendhi A (2021) A critical review on production of biopolymers from algae biomass and their applications. Bioresour Technol 329:124868

Kaur L, Khajuria R, Parihar L, Singh GD (2017) Polyhydroxyalkanoates: biosynthesis to commercial production-a review. J Microbiol Biotechnol Food Sci 6(4):1098

Koch M, Bruckmoser J, Scholl J, Hauf W, Rieger B, Forchhammer K (2020) Maximizing PHB content in Synechocystis sp. PCC 6803: a new metabolic engineering strategy based on the regulator PirC. Microb Cell Factories 19(1):231

Kumar BR, Mathimani T, Sudhakar MP, Rajendran K, Nizami AS, Brindhadevi K, Pugazhendhi A (2021a) A state of the art review on the cultivation of algae for energy and other valuable products: application, challenges, and opportunities. Renew Sust Energ Rev 138:110649

Kumar N, Banerjee C, Jagadevan S (2021b) Identification, characterization, and lipid profiling of microalgae Scenedesmus sp. NC1, isolated from coal mine effluent with potential for biofuel production. Biotechnol Rep (Amst) 30:e00621

Laurens LM, Markham J, Templeton DW, Christensen ED, Van Wychen S, Vadelius EW, Chen-Glasser M, Dong T, Davis R, Pienkos PT (2017) Development of algae biorefinery concepts for biofuels and bioproducts; a perspective on process-compatible products and their impact on cost-reduction. Energy Environ Sci 10(8):1716–1738

Leong WH, Lim JW, Lam MK, Uemura Y, Ho YC (2018) Third generation biofuels: a nutritional perspective in enhancing microbial lipid production. Renew Sust Energ Rev 91:950–961

Mata TM, Martins AA, Caetano NS (2010) Microalgae for biodiesel production and other applications: a review. Renew Sust Energ Rev 14(1):217–232

Mehariya S, Goswami RK, Verma P, Lavecchia R, Zuorro A (2021) Integrated approach for wastewater treatment and biofuel production in microalgae biorefineries. Energies 14(8):2282

Meixner K, Kovalcik A, Sykacek E, Gruber-Brunhumer M, Zeilinger W, Markl K, Haas C, Fritz I, Mundigler N, Stelzer F, Neureiter M (2018) Cyanobacteria biorefinery—production of poly (3-hydroxybutyrate) with Synechocystis salina and utilisation of residual biomass. J Biotechnol 265:46–53

Monshupanee T, Incharoensakdi A (2014) Enhanced accumulation of glycogen, lipids and polyhydroxybutyrate under optimal nutrients and light intensities in the cyanobacterium Synechocystis sp. PCC 6803. J Appl Microbiol 116(4):830–838

Monshupanee T, Chairattanawat C, Incharoensakdi A (2019) Disruption of cyanobacterial γ-aminobutyric acid shunt pathway reduces metabolites levels in tricarboxylic acid cycle, but enhances pyruvate and poly (3-hydroxybutyrate) accumulation. Sci Rep 9(1):8184

Morales M, Aflalo C, Bernard O (2021) Microalgal lipids: a review of lipids potential and quantification for 95 phytoplankton species. Biomass Bioenergy 150:106108

Morsi H, Eladel H, Maher A (2022) Coupling nutrient removal and biodiesel production by the chlorophyte Asterarcys quadricellulare grown in municipal wastewater. Bioenergy Res 15(1): 193–201

Moshood TD, Nawanir G, Mahmud F (2022) Sustainability of biodegradable plastics: a review on social, economic, and environmental factors. Crit Rev Biotechnol 42(6):892–912

Nanda N, Bharadvaja N (2022) Algal bioplastics: current market trends and technical aspects. Clean Techn Environ Policy 24(9):2659–2679

Ningbo TianAn Biologic Material Co., Ltd. Official Website [Internet]. http://en.tianan-enmat.com

O'Neill E, Cagney M (2024) Strategies for producing high value small molecules in microalgae. Plant Physiol Biochem 214:108942

Ojumu TV, Yu J, Solomon A (2004) Production of polyhydroxyalkanoates, a bacterial biodegradable polymers. Afr J Biotechnol 3(1):18–24

Onen Cinar S, Chong ZK, Kucuker MA, Wieczorek N, Cengiz U, Kuchta K (2020) Bioplastic production from microalgae: a review. Int J Environ Res Public Health 17(11):3842

Pancha I, Chokshi K, Maurya R, Trivedi K, Patidar SK, Ghosh A, Mishra S (2015) Salinity induced oxidative stress enhanced biofuel production potential of microalgae Scenedesmus sp. CCNM 1077. Bioresour Technol 189:341–348

Pandey S, Narayanan I, Selvaraj R, Varadavenkatesan T, Vinayagam R (2024) Biodiesel production from microalgae: a comprehensive review on influential factors, transesterification processes, and challenges. Fuel 367:131547

Pandey S, Varadavenkatesan T, Selvaraj R, Vinayagam R (2025) Biocatalytic conversion of microalgal biomass to biodiesel: optimization of growth conditions and synthesis of CaO bionanocatalyst from Monoraphidium sp. NCIM 5585. Sci Rep 15(1):4309

Parmar A, Singh NK, Pandey A, Gnansounou E, Madamwar D (2011) RETRACTED: cyanobacteria and microalgae: a positive prospect for biofuels. Bioresour Technol 102(22): 10163–10172

Phalanisong P, Plangklang P, Reungsang A (2021) Photoautotrophic and mixotrophic cultivation of polyhydroxyalkanoate-accumulating microalgae consortia selected under nitrogen and phosphate limitation. Molecules 26(24):7613

Pham TT, Nguyen TL, Duong TT, Doan OT, Tran HT, Tran LT (2024) Selection of microalgae and cyanobacteria to produce polyhydroxyalkanoates (PHAs)-a case study in Vietnam. Case Stud Chem Environ Eng 10:100808

Rahman A, Miller CD (2017) Microalgae as a source of bioplastics. In: Algal green chemistry. Elsevier, pp 121–138

Rodionova MV, Poudyal RS, Tiwari I, Voloshin RA, Zharmukhamedov SK, Nam HG, Zayadan BK, Bruce BD, Hou HJ, Allakhverdiev SI (2017) Biofuel production: challenges and opportunities. Int J Hydrog Energy 42(12):8450–8461

Saini DK, Chakdar H, Pabbi S, Shukla P (2020) Enhancing production of microalgal biopigments through metabolic and genetic engineering. Crit Rev Food Sci Nutr 60(3):391–405

Sajjadi B, Chen WY, Raman AA, Ibrahim S (2018) Microalgae lipid and biomass for biofuel production: a comprehensive review on lipid enhancement strategies and their effects on fatty acid composition. Renew Sust Energ Rev 97:200–232

Samantaray S, Mallick N (2012) Production and characterization of poly-β-hydroxybutyrate (PHB) polymer from Aulosira fertilissima. J Appl Phycol 24(4):803–814

Sarker NK, Kaparaju P (2025) Advances in algae-based bioplastics: from strain engineering and fermentation to commercialization and sustainability. Fermentation 11(10):574

Selvarajan R, Felföldi T, Tauber T, Sanniyasi E, Sibanda T, Tekere M (2015) Screening and evaluation of some green algal strains (Chlorophyceae) isolated from freshwater and soda lakes for biofuel production. Energies 8(7):7502–7521

Sharma R, Solanki P, Chaudhary M, Gupta N, Kaur P (2024) Unveiling the potential of microalgae for bioplastic production from wastewater–current trends, innovations, and future prospects. Biotechnol Sustain Mater 1(1):10

Shayesteh H, Laird DW, Hughes LJ, Nematollahi MA, Kakhki AM, Moheimani NR (2023) Co-producing phycocyanin and bioplastic in Arthrospira platensis using carbon-rich wastewater. Biotech 12(3):49

Shin YS, Jeong J, Nguyen TH, Kim JY, Jin E, Sim SJ (2019) Targeted knockout of phospholipase A2 to increase lipid productivity in Chlamydomonas reinhardtii for biodiesel production. Bioresour Technol 271:368–374

Shin SE, Koh HG, Park K, Park SH, Chang YK, Kang NK (2024) Increasing lipid production in Chlamydomonas reinhardtii through genetic introduction for the overexpression of glyceraldehyde-3-phosphate dehydrogenase. Front Bioeng Biotechnol 19(12):1396127

Siddiki SY, Mofijur M, Kumar PS, Ahmed SF, Inayat A, Kusumo F, Badruddin IA, Khan TY, Nghiem LD, Ong HC, Mahlia TM (2022) Microalgae biomass as a sustainable source for biofuel, biochemical and biobased value-added products: an integrated biorefinery concept. Fuel 307:121782

Singh AK, Sharma L, Mallick N, Mala J (2017) Progress and challenges in producing polyhydroxyalkanoate biopolymers from cyanobacteria. J Appl Phycol 29(3):1213–1232

Sreenikethanam A, Raj S, Gugulothu P, Bajhaiya AK (2022) Genetic engineering of microalgae for secondary metabolite production: recent developments, challenges, and future prospects. Front Bioeng Biotechnol 10:836056

Tan CH, Low SS, Cheah WY, Singh J, Chai WS, Tiong SK, Show PL (2024) Futuristic opportunities for pretreatment processes in biofuel production from microalgae. GCB Bioenergy 16(5):e13136

Tang DY, Yew GY, Koyande AK, Chew KW, Vo DV, Show PL (2020) Green technology for the industrial production of biofuels and bioproducts from microalgae: a review. Environ Chem Lett 18(6):1967–1985

Vinitha V, Meignanalakshmi S, Tirumurugaan KG, Sarathchandra G, Sundaram SM (2022) Polyunsaturated fatty acid, biodiesel property and anticancer activity analysis of Monoraphidium griffithii. Curr J Appl Sci Technol 41(41):44–55

Yin Z, Zhu L, Li S, Hu T, Chu R, Mo F, Hu D, Liu C, Li B (2020) A comprehensive review on cultivation and harvesting of microalgae for biodiesel production: environmental pollution control and future directions. Bioresour Technol 301:122804

Youssef AM, Gomaa M, Mohamed AK, El-Shanawany AR (2024) Enhancement of biomass productivity and biochemical composition of alkaliphilic microalgae by mixotrophic cultivation using cheese whey for biofuel production. Environ Sci Pollut Res 31(30):42875–42888

Zakir Hossain SM, Sultana N, Hossain MM, Razzak SA (2021) Optimization of microalgal biomass and lipid productivities for bioenergy production using central composite design with desirability function. Int J Energy Res 45(12):17342–17357

Algal Polyhydroxyalkanoates: Biofabrication of Sustainable Plastics and a SWOT-Based Perspective on Their Commercialization

20

Poonam Kumari and S. Venkata Mohan

Abstract

Amid an escalating plastics crisis, algal polyhydroxyalkanoates (PHAs) produced by microalgae, cyanobacteria, and phototrophic bacteria emerge as a biologically circular alternative that couples CO_2 fixation, wastewater nutrient recovery, and true end-of-life biodegradability. A SWOT frames the field with equal weight on promise and constraints: photosynthetic "green alchemists" that convert sunlight, CO_2, and effluents into polymers, yet still limited by yield, productivity, and solvent-intensive downstream recovery that keep costs above petro-plastics. The chapter follows the trajectory from bench to deployment within multiproduct biorefineries co-valorizing pigments, lipids, and bioactive peptides to amortize CAPEX/OPEX and preserve environmental gains. Through Lenses for the Future, three levers are articulated: technology (redirecting carbon flux, light/gas-transfer-aware photobioreactors, greener recovery), economics (waste-derived feedstocks, scale and learning effects, cost reduction via co-products), and policy (EPR, green procurement, carbon-aligned instruments). Emphasis is placed on co-location with CO_2 point sources and wastewater streams, integration

P. Kumari
Bioengineering and Environmental Sciences Lab, Department of Energy and Environmental Engineering, CSIR-Indian Institute of Chemical Technology (CSIR-IICT), Hyderabad, India

Academy of Scientific and Innovative Research (AcSIR), Ghaziabad, India

S. V. Mohan (✉)
Bioengineering and Environmental Sciences Lab, Department of Energy and Environmental Engineering, CSIR-Indian Institute of Chemical Technology (CSIR-IICT), Hyderabad, India

Academy of Scientific and Innovative Research (AcSIR), Ghaziabad, India

CSIR-National Environmental Engineering Research Institute (CSIR-NEERI), Nagpur, Maharashtra, India
e-mail: mohan.sv@neeri.res.in

K. Mohanty et al. (eds.), *Advances in Algal Research*,
https://doi.org/10.1007/978-981-95-7234-2_20

of data-driven control, and decision-making grounded in TEA/LCA, outlining a credible pathway for photosynthetic PHAs to transition from niche materials to mainstream, economically competitive plastics compatible with planetary boundaries.

Keywords

Bio-based-biodegradable plastics · Circular bioeconomy · Sustainable materials · Wastewater valorization · Metabolic pathway optimization

20.1 Introduction

In the context of an escalating plastic crisis marked by resource depletion, rising greenhouse gas emissions, and persistent environmental pollution, understanding the life cycles of sustainable alternatives versus conventional plastics is critical. Bio-based biodegradable plastics, particularly polyhydroxyalkanoates (PHAs), represent a fundamentally different pathway compared to fossil-derived plastics, offering insights into how material choice shapes environmental outcomes. PHAs are a family of bio-based biodegradable polyesters (Fig. 20.1) first discovered in 1926 by Maurice Lemoigne, who identified intracellular inclusions in *Bacillus megaterium* (Lenz and Marchessault 2005). Initially regarded as microbiological curiosities, PHAs became the subject of intensive research during the latter half of the twentieth century as interest grew in biodegradable alternatives to petrochemical plastics (Pradhan et al. 2024). The earliest studies focused on poly

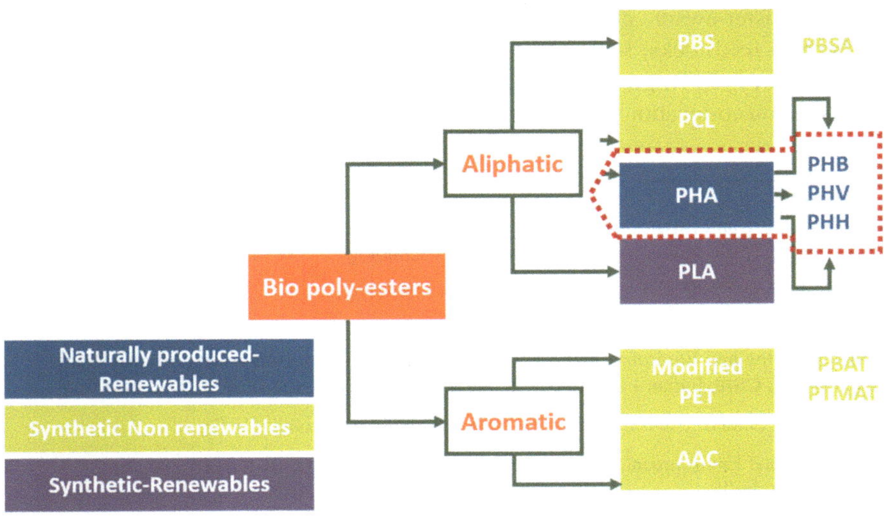

Fig. 20.1 Classification of bio-based polyesters and their derivatives based on origin and structural characteristics

(3-hydroxybutyrate) (PHB), but by the 1980s, advances in fermentation and metabolic engineering enabled the production of PHAs with diverse monomer compositions and improved material properties. Industrial-scale development followed in the 1990s, with companies like Imperial Chemical Industries (ICI) marketing PHAs under the trade name "Biopol" (Chodak 2008). Since then, efforts have expanded into using low-cost renewable substrates, optimizing microbial strains, and tailoring PHA copolymers for applications ranging from packaging to medical devices.

From a biological perspective, PHAs function as natural energy and carbon storage materials for microorganisms, accumulated under nutrient-limiting conditions with excess carbon sources effectively acting as "microbial fat reserves." More than 150 monomer units have been identified, enabling fine-tuning of polymer crystallinity, flexibility, and melting temperature, and allowing PHAs to mimic a wide range of petrochemical plastics from rigid PP-like materials to flexible thermoplastic elastomers (Chodak 2008). The PHA life cycle begins with renewable feedstocks such as sugars and vegetable oils (first generation), agricultural residues and industrial by-products (second generation), and CO_2 fixed by photosynthetic microorganisms like cyanobacteria and algae (third generation) (Kourmentza et al. 2017). PHAs can also be produced from unconventional waste streams such as whey, molasses, waste frying oil, methane, and syngas (Chodak 2008), making them highly versatile in sustainable resource utilization. Under nutrient-limited, carbon-rich conditions, species including *Cupriavidus necator*, *Alcaligenes latus*, and *Halomonas* spp. can accumulate PHAs up to 85% of cell dry weight in optimized fermentations (Pradhan et al. 2018). Extraction methods include solvent-based, enzymatic, and mechanical processes (Kumari et al. 2022). The polymers are then processed into films, molded goods, fibers, and coatings, offering material properties comparable to polypropylene (PP) (Pradhan et al. 2024). Some PHAs also exhibit oxygen barrier properties comparable to PET, making them attractive for food packaging applications (Chodak 2008).

At the end of their functional life, PHAs can undergo complete biodegradation in multiple environments (Fig. 20.2b). Under industrial composting conditions, PHAs mineralize to CO_2 and water within 3–6 months, while in marine environments, significant degradation occurs within 12 months (Mergaert et al. 1992). PHAs also biodegrade in anaerobic sludge digesters, producing methane that can be recovered for energy, adding a renewable energy recovery step to their life cycle. The breakdown process is mediated by microbial depolymerases, which hydrolyze the polymer into monomers, such as 3-hydroxybutyrate, which is subsequently metabolized into benign end products (Chodak 2008). Life cycle assessment studies indicate that PHA production from waste-derived feedstocks can reduce greenhouse gas (GHG) emissions by 50–75% compared to fossil plastics, while enabling waste valorization and CO_2 capture (Kourmentza et al. 2017). Additionally, PHA processing temperatures are lower than those for many petroplastics (160–180 °C), offering potential energy savings during product manufacturing (Chodak 2008).

By contrast, fossil-derived plastics follow a fundamentally linear "take–make–dispose" model, relying entirely on nonrenewable petroleum and natural gas

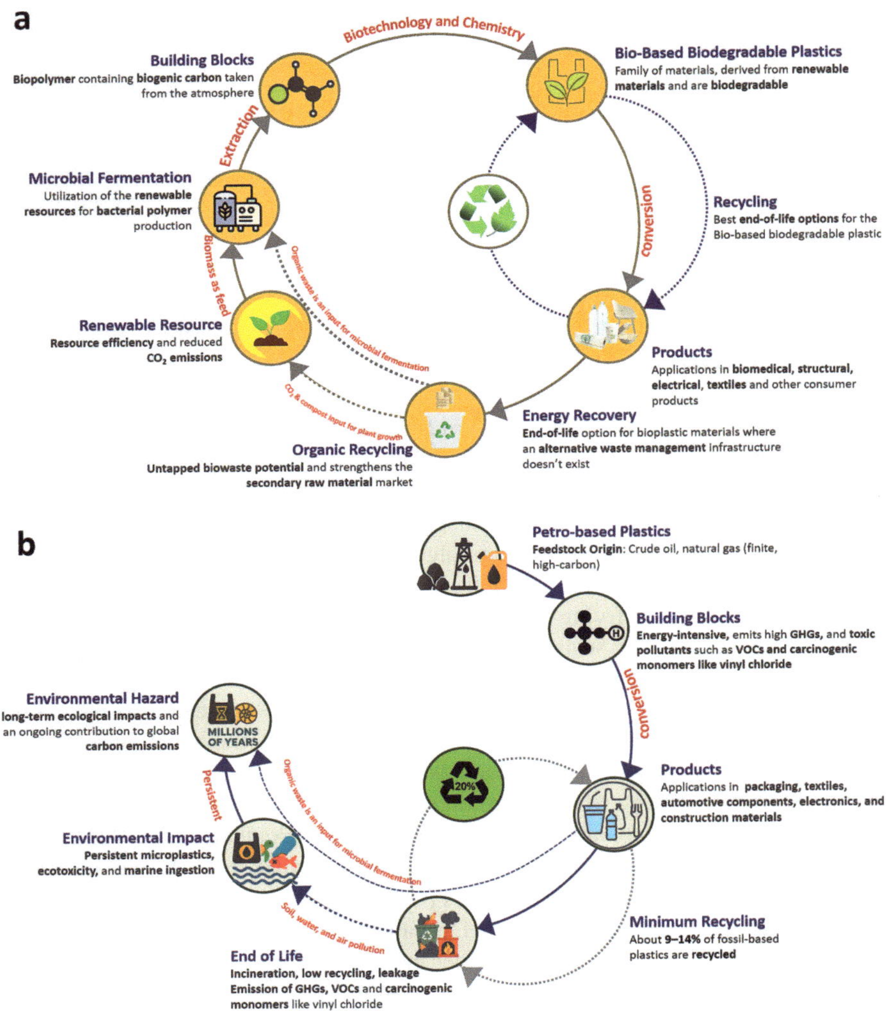

Fig. 20.2 Comparative life cycle of plastics: (**a**) fossil-based (cradle-to-grave); (**b**) PHA bioplastics (cradle-to-circular)

reserves. These raw materials are extracted, transported to refineries, and processed through energy-intensive steam cracking to yield monomers such as ethylene, propylene, styrene, and vinyl chloride, which are polymerized into polyethylene (PE), polypropylene (PP), polyvinyl chloride (PVC), polystyrene (PS), and polyethylene terephthalate (PET). These five categories account for over 90% of global plastic production. In 2019, worldwide production reached 368 million tons, with forecasts predicting more than 1.1 billion tons annually by 2050 (Melchor-Martínez et al. 2022). The manufacturing of fossil plastics is highly carbon-intensive, with single-use plastics alone consuming about 1.6 billion liters of oil annually and emitting substantial quantities of CO_2, CH_4, and N_2O (Chen et al. 2021). Many fossil plastic products, especially packaging, have functional lifetimes of less than 6 months, yet their persistence in the environment spans centuries due to high molecular weight, hydrophobicity, and crystallinity. End-of-life management is dominated by landfilling, incineration for energy recovery (with associated CO_2 and toxin emissions), and limited recycling, only 9–14% globally, hindered by contamination and polymer incompatibility. Furthermore, approximately 11 million tons of plastic waste enter the oceans each year, fragmenting into microplastics that bioaccumulate in marine food webs. Common packaging films are estimated to persist for over 500 years before significant degradation occurs, resulting in long-term ecological impacts and an ongoing contribution to global carbon emissions (Fig. 20.2a). Thus, while PHAs offer a closed-loop carbon cycle with renewable inputs, multiple end-of-life biodegradation pathways, and reduced GHG emissions, fossil-derived plastics lock society into a resource-depleting, carbon-intensive, and environmentally persistent system.

20.2 Photosynthetic Organisms as the Unsung Engineer

Currently, nearly all commercial PHA production is based on heterotrophic bacteria such as *Cupriavidus necator* and *Halomonas bluephagenesis*, cultivated on sugar, plant oil, or other organic carbon sources under nutrient-limited fermentation (Kourmentza et al. 2017; Adetunji and Erasmus 2024). While these systems can achieve intracellular PHA contents of up to 85% of cell dry weight (CDW) and volumetric productivities exceeding 4 g L^{-1} h^{-1}, they rely on expensive feedstocks and large energy inputs for aeration, sterilization, and downstream processing. Furthermore, the carbon footprint of these heterotrophic systems remains tied to the sustainability of the substrate supply chain. By contrast, photosynthetic microorganisms, including cyanobacteria, eukaryotic microalgae, and phototrophic bacteria, offer a largely untapped avenue for PHA production. They directly fix CO_2 into organic carbon using solar energy, potentially eliminating feedstock costs and enabling carbon-negative production (Choi et al. 2020; Sharma et al. 2024a). Cultivation in open ponds or photobioreactors on nonarable land and in saline or wastewater further avoids competition with food production, allowing for the simultaneous recovery of nutrients from waste streams (Bharathiraja et al. 2015; Chaudry et al. 2025). In these organisms, PHA synthesis is integrated with photosynthesis. Light-harvesting pigments within photosystems I and II absorb photons, driving water oxidation and generating reducing equivalents (NADPH) and ATP via the photosynthetic electron transport chain. This energy fuels the Calvin–Benson–Bassham (CBB) cycle, fixing CO_2 into triose phosphates, which are subsequently converted via glycolysis and the pyruvate dehydrogenase complex into acetyl-CoA the key precursor for PHA biosynthesis (Blankenship 2014; Kumari et al. 2022).

While photosynthetic systems avoid organic feedstock costs, their PHA yields are currently lower than heterotrophic systems due to slower growth rates and limited carbon flux toward polymer synthesis. For example: Cyanobacteria (*Synechocystis* sp. PCC 6803): PHB content up to 27% CDW under nitrogen deprivation with acetate supplementation; volumetric productivity ~0.03–0.06 g L^{-1} day^{-1} (Khetkorn et al. 2016). *Spirulina* (Arthrospira) *platensis* is indeed reported as a low PHA accumulator, typically 2–6% of CDW under nutrient stress (mainly nitrogen/phosphate limitation) when cultivated in wastewater or minimal media. (Sudesh et al. 2000). Heterotrophic *Cupriavidus necator* demonstrated PHB content of 80–85% CDW, productivities >4 g L^{-1} h^{-1} (Chen et al. 2021). Even so, photosynthetic PHAs are attractive because their operational costs can be lower when integrated with CO_2 sequestration or wastewater treatment, and life cycle assessment studies show up to 75% lower greenhouse gas emissions compared to fossil plastics (Kourmentza et al. 2017). Algal cultures can fix 1.6–2.0 g CO_2 per g biomass, directly offsetting emissions from industrial sources (Chisti 2007). Cyanobacteria and algae can remove >80% nitrogen and phosphorus from municipal or agro-industrial wastewater, producing PHA while meeting discharge standards (Chaudry et al. 2025). It can be cultivated in deserts, saline lakes, or offshore photobioreactors without competing for agricultural resources.

20.3 SWOT in Storyboards

As the field advances from conceptual promise to technological deployment, the evaluation of PHA production from photosynthetic organisms requires more than a simple presentation of technical facts. A critical and balanced framework is necessary to capture both the transformative potential and the unresolved challenges that define this emerging domain. The SWOT (Strengths, Weaknesses, Opportunities, and Threats) analysis provides such a lens, enabling a systematic exploration of intrinsic advantages, current limitations, future prospects, and external risks shaping the trajectory of photosynthetic PHAs (Fig. 20.3). The Strengths are framed as the "*Green Alchemist*," highlighting how algae and cyanobacteria convert sunlight, CO_2, and wastewater into biodegradable plastics. The Weaknesses emerge as the "*Shy Performer*," reflecting the persistent bottlenecks of low yield, slow growth, and scale-up challenges. The Opportunities appear as the "*Market Challenger*," underlining the potential of integrating PHA production with carbon capture, wastewater valorization, and bioeconomy policies. Finally, the threats are portrayed as the "*Plastic Empire*," where entrenched fossil-derived plastics, with their low production costs and global infrastructure, resist the transition to sustainable alternatives.

20.3.1 Strength: "*The Green Alchemist*"

Photosynthetic organisms embody the role of a "*Green Alchemist*," transforming sunlight, CO_2, and wastewater nutrients into sustainable bioplastics. Their strength lies in simultaneously functioning as carbon sinks and wastewater remediators. Microalgal and cyanobacterial cultures can fix 1.6–2.0 g CO_2 per g biomass, directly offsetting emissions from industrial point sources, while also removing over 80% of

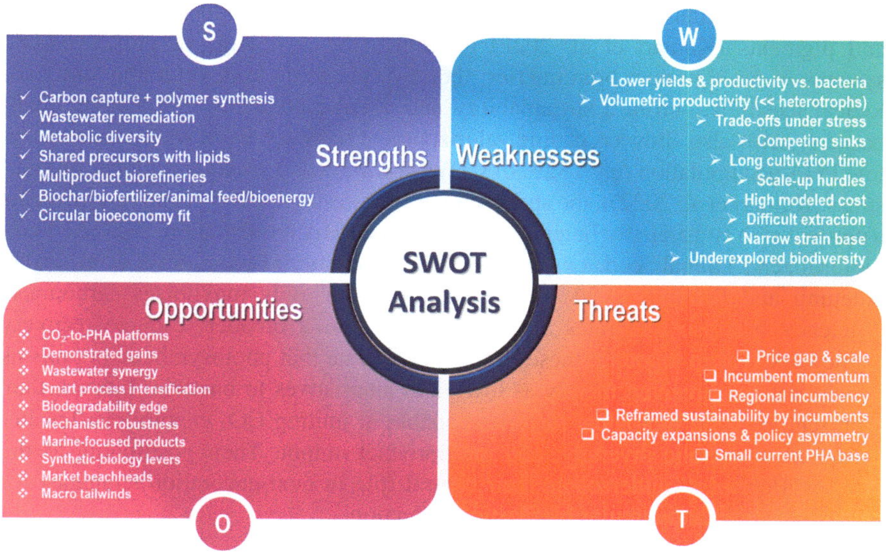

Fig. 20.3 SWOT analysis of photosynthetic PHA production platforms highlighting strengths, weaknesses, opportunities, and threats influencing their industrial scalability and market adoption

nitrogen and phosphorus from municipal and agro-industrial effluents, making them efficient agents of tertiary wastewater treatment (Chisti 2007; Chaudry et al. 2025). This coupling of CO_2 sequestration, nutrient recycling, and PHA biosynthesis positions photosynthetic organisms as uniquely valuable players in the circular bioeconomy.

A critical strength also emerges from their cellular size and metabolic complexity. Algae range from picoplankton such as *Prochlorococcus* (~0.5 μm) to giant multicellular macroalgae like *Macrocystis* that can grow up to 60 m (Raven and Beardall 2020). This variation in size provides diverse intracellular capacities for metabolite storage and opens opportunities for scaling PHA accumulation across different taxa. Unlike bacteria, whose relatively uniform morphology constrains metabolic niches, photosynthetic organisms display a wider spectrum of metabolic pathways. Importantly, while bacterial PHA accumulation is traditionally regarded as a carbon and energy storage response under nutrient limitation, photosynthetic organisms may deploy PHAs in broader roles such as redox balancing, stress tolerance, and photosynthetic carbon flux regulation (Kourmentza et al. 2017; Choi et al. 2020). This underexplored metabolic plasticity suggests that applying a bacterial-only model to algal PHA production risks oversimplification.

Equally compelling is the biochemical overlap between PHAs and fatty acids. Algae are extensively studied for lipid and fatty acid production, and since PHA biosynthesis shares precursors like acetyl-CoA and malonyl-CoA with fatty acid pathways, their metabolic networks are closely linked (Sudesh et al. 2000; Sharma et al. 2024a). This relationship indicates significant potential for flux redirection toward PHA synthesis. Furthermore, the possibility of co-producing PHAs, lipids, and pigments (Kumari et al. 2022) highlights the advantage of photosynthetic systems in multiproduct biorefineries, thereby enhancing both economic and ecological feasibility. From a sustainability and circular economy perspective, algal-based systems also allow cascading valorization of biomass. Even when strains show low PHA accumulation (e.g., 32% of CDW in *Chlorella sorokiniana* under nutrient stress (Kumari et al. 2022)), the spent biomass retains high value. It can be utilized as a nutrient-rich substrate for heterotrophic bacterial PHA fermentation, converted into biofertilizers, animal feed, or processed for bioenergy (Bharathiraja et al. 2015; Adetunji and Erasmus 2024). This dual utility not only maximizes carbon use efficiency but also integrates algal cultivation into a broader resource recovery framework. Taken together, these strengths illustrate that photosynthetic organisms should not be perceived merely as low-yield alternatives to bacteria. Instead, they represent versatile biological factories capable of uniting CO_2 capture, wastewater treatment, metabolic diversity, and multiproduct output. Their positioning as the "*Green Alchemist*" underscores their central role in next-generation strategies for carbon-negative and circular bioplastic production.

20.3.2 Weaknesses: "*The Shy Performer*"

While photosynthetic organisms provide unique ecological and sustainability benefits, their industrial application in PHA production remains limited, earning them the title of "*shy performers.*" The main challenge lies in their relatively low productivity and longer cultivation times compared to heterotrophic bacteria. For example, wild-type *Synechocystis* sp. PCC 6803 generally accumulates ~10–13% PHB of CDW under nutrient limitation, though engineered strains can reach 26–35% CDW with acetate supplementation after 7–9 days of growth (Khetkorn et al. 2016). In comparison, *Cupriavidus necator* can achieve 80–85% PHB of CDW in only 2–3 days, with volumetric productivities exceeding 3–4 g L^{-1} h^{-1} (Madkour et al. 2013; Ilhami et al. 2025).

There are notable exceptions that demonstrate the inherent potential of phototrophic systems. Certain strains, such as *Nostoc muscorum* (69–78% CDW), *Aulosira fertilissima* (77% CDW), and *Chlorella pyrenoidosa* (~79.8% CDW), have been reported to accumulate high levels of PHAs under nutrient stress or mixotrophic conditions. However, these values often require supplementation with organic carbon sources (acetate, glucose, or whey wastewater) and extended cultivation times of 8–12 days, which deviate from cost-effective photoautotrophic operation (Bhati and Mallick 2015; Das et al. 2018; Sharma et al. 2024a). In most practical settings, PHA contents in algae and cyanobacteria range between 5 and 30% CDW, with volumetric productivities around 0.03–0.06 g L^{-1} day^{-1} far below the multigram per liter per hour productivities of bacteria (Costa et al. 2018; Ilhami et al. 2025). Other weaknesses are linked to the metabolic complexity of photosynthetic organisms. Carbon flux is distributed across competing pathways, including fatty acid biosynthesis, pigments, and carbohydrate storage, which might constrain

PHA accumulation. For example, *Spirulina* LEB-18 increased PHB content from 2.0% to 9.6% CDW under nitrogen limitation, but biomass productivity dropped from 0.47 to 0.32 g L^{-1} day^{-1}, illustrating the trade-off between growth and polymer accumulation (Costa et al. 2018). Similarly, *Synechococcus subsalsus* increased PHA to ~7.9% CDW under nitrogen stress but showed impaired protein synthesis and growth (Costa et al. 2018). Furthermore, acetyl-CoA, the precursor for PHAs, is strongly diverted into triacylglycerol (TAG) biosynthesis in many algae under stress, competing for PHAs accumulation as a storage polymer than lipids (Ilhami et al. 2025).

Cultivation time is another critical weakness. Whereas heterotrophic bacteria can reach peak production in 24–48 h, photosynthetic cultures often require 8–12 days to achieve their maximum PHA content (Choi et al. 2020; Madadi et al. 2021). This prolonged growth period increases operational costs, especially for light, CO_2 delivery, and mixing, while also raising risks of contamination. Scalability challenges further limit their competitiveness. Open pond systems are vulnerable to contamination and environmental fluctuations, whereas closed photobioreactors (PBRs), while more controlled, impose high capital and operational costs. Techno-economic analyses reveal that algal PHA production costs are currently estimated at €135–302/kg in closed PBRs, compared to market-competitive fossil plastics at ~€4/kg (Ilhami et al. 2025). Furthermore, robust algal cell walls enriched with polysaccharides and proteins complicate solvent extraction, increasing downstream processing costs and solvent inputs compared to bacterial systems (Adetunji and Erasmus 2024; Ilhami et al. 2025). Strain diversity also remains narrowly explored. Most studies have focused on model cyanobacteria (*Synechocystis*, *Nostoc*, *Spirulina*) and microalgae (*Chlorella*, *Scenedesmus*), leaving the vast biodiversity underutilized. This restricted focus limits the discovery of species with naturally higher PHA accumulation, faster growth kinetics, or novel PHA monomer compositions.

Photosynthetic organisms face multiple weaknesses, including low and inconsistent PHA yields, long cultivation times, strong metabolic competition, narrow strain exploration, and high downstream processing costs (Kumari & Mohan 2025). Yet, these weaknesses do not imply that phototrophic systems are inherently unsuitable. Rather, they highlight the gap between potential and practice. The fact that certain strains can reach PHA levels comparable to bacteria under optimized laboratory conditions suggests that with advances in metabolic engineering, strain discovery, and photobioreactor design, these "*shy performers*" may 1 day evolve into reliable contributors to sustainable, carbon-negative PHA production.

20.3.3 Opportunities: *"The Market Challenger"*

The integration of photosynthetic organisms, including cyanobacteria, microalgae, macroalgae, and photosynthetic bacteria, into PHA production platforms presents a transformative opportunity to achieve sustainable, carbon-negative bioplastics (Fig. 20.4). These systems provide unique ecological and technological advantages,

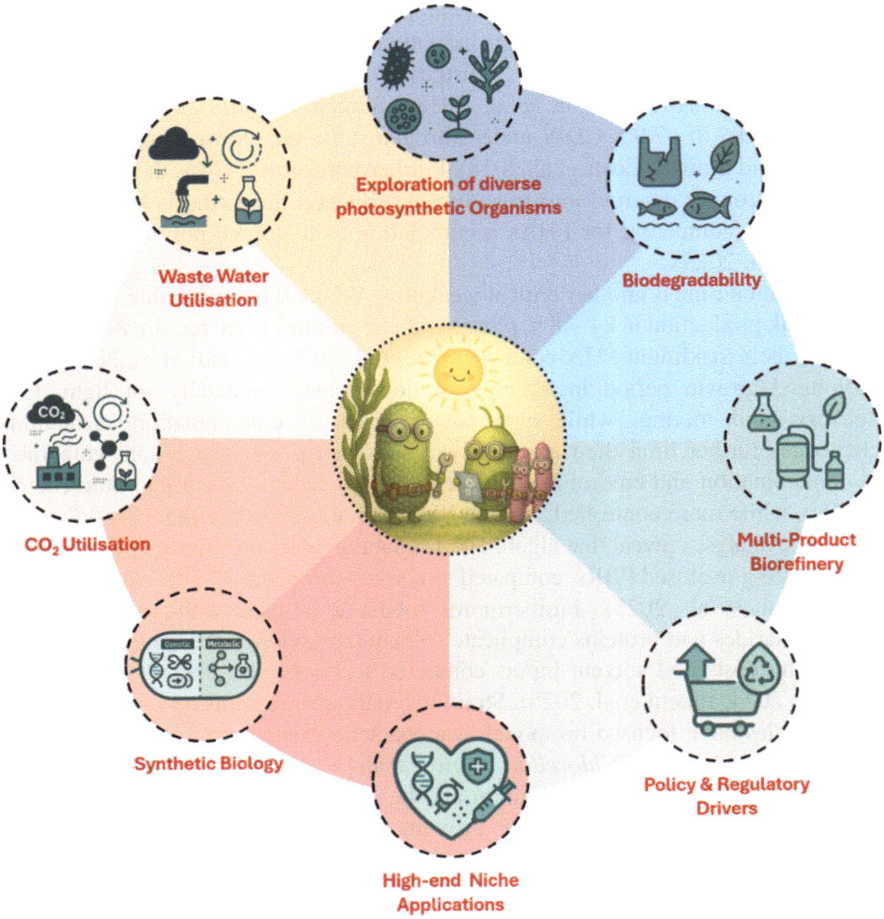

Fig. 20.4 Opportunities and technological enablers for sustainable PHA production using photosynthetic microorganisms

positioning them as credible "*market challengers*" in the bioplastics sector. Unlike conventional heterotrophic platforms that rely on costly sugars or organic substrates, photosynthetic organisms utilize sunlight, CO_2, and minimal inorganic nutrients to synthesize PHA precursors, directly linking carbon capture with bioplastic production while supporting a circular bioeconomy (Nawaz et al. 2024; Ciebiada et al. 2020).

Photosynthetic Carbon Capture Coupled with Wastewater Utilization Photosynthetic systems fix approximately 1.3–1.65 kg CO_2 per kilogram of biomass, a significantly higher rate than terrestrial plants, while simultaneously accumulating PHAs as cellular carbon reserves (Chisti 2007). In contrast to conventional carbon capture and storage (CCS) technologies that permanently sequester CO_2 underground at high energy costs, photosynthetic PHA platforms valorize captured carbon by converting it into high-value products such as PHAs, pigments, biofuels, and biofertilizers, thus integrating carbon sequestration with material production (Koch et al. 2023). Recent advances in metabolic engineering have enabled direct CO_2 to PHA conversion through engineered cyanobacterial cell factories and high-density cultivation strategies. For example, *Synechococcus elongatus* optimized for acetyl-CoA availability and redirected carbon fluxes achieved up to 15 mg PHA per g DCW, and further enhancement through high-density photobioreactors resulted in a 270-fold increase in productivity compared to unoptimized strains (Tan et al. 2022). Similarly, cultivation of *Chlorella vulgaris* in wastewater-fed photobioreactors demonstrated complete removal of free CO_2 within 3 days alongside PHA accumulation (20 mg/L) and substantial nutrient removal, highlighting the feasibility of integrated carbon capture and PHA production systems (Dewangan et al. 2025). Co-locating algal photobioreactors with industrial flue gas streams from cement, steel, or power plants further enhances sustainability by providing on-site CO_2 mitigation while generating renewable bioplastic feedstocks (Koch et al. 2023). Moreover, beyond PHA extraction, residual algal biomass can be valorized into biochar or soil amendments, offering long-term terrestrial carbon sequestration and promoting circular bioeconomy principles. Municipal, agro-industrial, and food-processing wastewater streams offer another opportunity to achieve cost-efficient cultivation of photosynthetic PHA producers. These effluents are enriched with nitrogen (20–150 mg/L), phosphorus (10–50 mg/L), and organic carbon (COD 300–500 mg/L), making them ideal low-cost nutrient sources for cultivating microalgae, cyanobacteria, and photosynthetic bacteria (Pant et al. 2025). Utilizing wastewater not only reduces dependence on synthetic nutrients and lowers operational costs by up to 30–40% but also supports simultaneous wastewater treatment and greenhouse gas mitigation through integrated phyco-remediation (Alcaraz Cercós 2015). For instance, *Chlorella vulgaris* grown in a wastewater-fed photobioreactor achieved a biomass yield of 3.47 g/L and 20 mg/L PHA accumulation, while completely eliminating free CO_2 within 3 days and reducing COD and phosphate levels by ~80% and > 85%, respectively (Sharma et al. 2024b). Further improvements in biomass productivity and resource recovery are achieved using advanced cultivation strategies, such as microalgae-bacteria consortia and membrane photobioreactors (MPBRs) integrated with AI-driven process controls, which enable up to 90% nitrogen and 85% phosphorus recovery while producing high-quality effluents suitable for reuse (Pant et al. 2025). These innovations, combined with residual biomass valorization into biochar or soil enhancers, strengthen the role of photosynthetic systems in fostering a circular bioeconomy.

Biodegradability-Tackling Marine Plastic Pollution The intrinsic biodegradability of PHAs offers a powerful solution to the escalating challenge of plastic pollution, particularly in marine environments. Unlike fossil-based plastics such as PP, PE, and polyethylene PET, which persist for centuries and degrade primarily into microplastics, PHAs comply with the EN 13432 "gold standard" for biodegradability. They achieve ≥90% conversion of carbon to CO_2 in soil within 24 months, degrade in freshwater within 56 days, marine environments within 6 months, home composting systems within <12 months, and industrial composting facilities within <6 months (Lackner et al. 2023; Koller et al. 2025). Meeting all five criteria distinguishes PHAs from alternatives like polylactic acid (PLA), which fails marine biodegradability benchmarks and often requires controlled industrial composting (Koller et al. 2025). Mechanistically, PHA degradation progresses through four coupled stages: biodeterioration (abiotic surface cracking), biofragmentation (extracellular depolymerases cleave chains to oligomers), bioassimilation (uptake of hydroxyacid products), and mineralization (conversion to CO_2 aerobically or CH_4 anaerobically) (Kim et al. 2023; Koller et al. 2025). Material-specific factors such as polymer composition, crystallinity, morphology, and environmental parameters influence degradation kinetics. Importantly, PHAs as natural polyesters are readily metabolized by environmental microbiota, achieving robust end-of-life performance across diverse ecosystems. Given that 8–ten million tons of plastic enter oceans annually and an estimated 50–75 trillion microplastic fragments circulate globally, PHAs represent a critical mitigation pathway (Koller et al. 2025). Recent innovations in PHA copolymer design, such as poly (3-hydroxybutyrate-co-3-hydroxyhexanoate) (P(3HB-co-3HHx)), improve flexibility, processability, and biodegradation rates, broadening applications in marine-sensitive sectors, including biodegradable straws, fishing gear, food packaging, and cutlery. By coupling rapid environmental breakdown with high-performance properties, PHAs present a strategic alternative to conventional plastics and align strongly with global circular bioeconomy goals.

Synthetic Biology and Metabolic Engineering Breakthroughs Parallel to these ecological advantages, synthetic biology and metabolic engineering are reshaping the photosynthetic PHA landscape. In cyanobacteria, strategies such as overexpression of *phaA*, *phaB*, and *phaEC* operons and the suppression of competing glycogen and lipid pathways (*agp*, *ackA*, *accC*, *fabF/H*) significantly enhance carbon flux toward PHA synthesis (Nawaz et al. 2024). Optimization of the CBB cycle through enzymes like phosphoribulokinase, transketolase, and RuBisCO further increases CO_2 fixation efficiency (Ciebiada et al. 2020). The development of light-responsive synthetic promoters allows synchronization of PHA biosynthesis with photosynthetic activity, improving energy utilization. Engineered strains of *Synechococcus elongatus* expressing heterologous *PhaC* and *NphT7* modules have demonstrated enhanced PHB accumulation under light and CO_2-driven cultivation, highlighting compatibility with industrial flue gas integration (Nawaz et al. 2024). In microalgae, the application of synthetic biology has enabled the installation of heterologous PHB/PHA pathways targeting metabolically active organelles like chloroplasts and peroxisomes, improving precursor availability and reducing growth penalties (Zhang et al. 2024). For example, engineered *Chlamydomonas reinhardtii* demonstrated enhanced PHA flux when peroxisomal targeting was used compared to cytosolic expression. Regulatory innovations such as phosphate-limitation–inducible promoters in diatoms like *Phaeodactylum tricornutum* enable conditional expression that maximizes PHB yields without compromising growth (Windhagauer et al. 2022). Episomal vectors and multi-copy expression strategies further enhance pathway stability and photosynthetic efficiency. These advancements are complemented by mixotrophic cultivation approaches, wherein engineered strains simultaneously utilize CO_2 and organic carbon from wastewater, boosting biomass productivity while contributing to resource recovery. Photosynthetic bacteria such as *Rhodopseudomonas palustris*, *Rhodovulum sulfidophilum*, and *Rhodobacter sphaeroides* naturally accumulate PHAs and are increasingly explored as light-driven PHA platforms. Synthetic biology strategies involve granule protein engineering, where phasin (*PhaP*) overexpression and fine-tuning of *phaR* regulatory circuits optimize granule number and size without impairing cellular fitness (Brown et al. 2022). The introduction of class I PHA synthases (*phaC*) from high-performing strains expands monomer incorporation to produce copolymers like PHBV with improved mechanical properties and biodegradability (Higuchi-Takeuchi et al. 2017). Additionally, *phaZ* depolymerase knockouts coupled with biosynthetic pathway overexpression enhance polymer retention (Kobayashi and Kondo 2019), while nanogel scaffold-assisted cultivation boosts PHA yields without extensive genome editing (Srisawat et al. 2022). Leveraging wastewater-derived volatile fatty acids (VFAs) as co-substrates integrates PHA production with wastewater treatment, improving both sustainability and economics.

High-Value Applications and Market Dynamics Despite higher production costs compared to fossil-based plastics, several niche, high-value applications are enabling early market entry for photosynthetically derived PHAs. In the biomedical sector, PHAs are used in biodegradable sutures, drug delivery carriers, orthopedic implants, and tissue engineering scaffolds due to their excellent biocompatibility, tunable degradation, and minimal inflammatory responses (Leong et al. 2017). In the food packaging industry, PHA copolymers such as PHBV and P(3HB-co-4HB) exhibit oxygen and moisture barrier properties comparable to PET, making them suitable for perishable product packaging while ensuring complete biodegradability (Yousefi and Wnek 2025). Similarly, the cosmetics and personal care sectors are adopting PHAs for biodegradable microbeads in scrubs, cleansers, and lotions, replacing banned polyethylene-based alternatives and reducing marine microplastic pollution (Yousefi and Wnek 2025). These applications serve as stepping stones, generating early revenue streams while facilitating technology maturation and supporting industrial-scale cost reductions. Finally, global market dynamics strongly favor the scale-up of PHA production. The global PHA bioplastics market is currently valued at USD 123 million (2024) and is projected to reach ~USD 384 million by 2033, growing at a 15.2% CAGR (PHA Bioplastics Market Research Report 2033). Europe leads the market with a $>$ 35% share, followed by North America, while Asia-Pacific represents the fastest-growing region due to supportive industrial policies and rising demand in China and India. Policy frameworks such as the EU Green Deal and Extended Producer Responsibility (EPR) regulations, along with ongoing UN plastics-treaty negotiations, are accelerating adoption pathways (Gundlapalli and Ganesan 2025). At a broader level, the bioplastics market is expected to surpass USD 30 billion by 2030, with substantial venture capital inflows and industrial investments across Europe, North America, and Asia already driving large-scale fermentation capacity and downstream application development (Bioplastics Market Size, Share, Growth Analysis Report 2030).

By integrating CO_2 capture, wastewater valorization, biodegradability, synthetic biology advances, and global policy support, photosynthetic PHAs are positioned as true "*market challengers*." While cost reduction remains critical, the synergy of ecological benefits, technological innovation, and emerging markets underscores their transformative role in the circular bioeconomy.

20.3.4 Threats: "The Plastic Empire"

The principal headwind for photosynthetic PHAs is the entrenched cost and scale advantage of fossil-based plastics. Mandatory polymers such as PE, PP, PVC, PET, and PS are routinely manufactured at ~USD 1.2–2.0 per kg on mature, fully depreciated assets with optimized logistics, while PHAs still land closer to ~USD 5–7 per kg even with advanced recovery systems, leaving a persistent price delta that constrains penetration in price-sensitive segments like mass packaging. This gap is amplified by the sheer momentum of incumbent volumes: global plastics use stood near 460 Mt. in 2019 and is projected to triple to ~1321 Mt. by 2060, ensuring powerful economies of scale and continuous process improvements across the petrochemical chain (Towards Packaging 2025). Regionally, the Asia-Pacific complex leads fossil plastics, with PE and packaging dominating the 2024 mix, while North America is set for notable growth, particularly polypropylene and electronics reinforcing low benchmark prices and tight, reliable supply for converters (Towards Packaging 2025). These regional dynamics mean PHAs must compete against incumbent polymers that arrive cheap, ubiquitous, and specification-proven in every major downstream market. The incumbent industry is also moving the goalposts on "sustainability." Petrochemical leaders (e.g., ExxonMobil Chemical, Dow, SABIC, LyondellBasell) are scaling mechanical and chemical recycling, optimizing monomaterial designs for better recyclability, and deploying AI/ML in production to cut scrap, energy use, and downtime, thereby lowering unit costs and strengthening a recycled-fossil narrative that slows switching to biodegradable options (Towards Packaging 2025). Capital commitments such as new PP capacity and regional initiatives like India's plastic parks further entrench supply resilience and price competitiveness into the next decade (Towards Packaging 2025). In parallel, policy asymmetry persists: while Europe advances the Green Deal, EPR schemes, and stricter single-use controls, many high-growth markets still subsidize fossil feedstocks or apply uneven enforcement, fragmenting demand signals and making it harder for PHAs to scale on policy pull alone (Towards Packaging 2025).

Finally, even under favorable demand scenarios for bioplastics, absolute market size remains modest in the near term. Recent trackers bracket the PHA market between USD 111.1 M in 2024; USD 169.2 M by 2032 (CAGR ~8.7%) and USD

123 M in 2024; ~USD 384 M by 2033 (CAGR ~15.2%), with broad agreement that Europe leads and Asia-Pacific grows fastest (Data Bridge Market Research 2024; Growth Market Reports 2025). This divergence underscores a strategic risk: if the trajectory follows the conservative curve rather than the accelerated one, unit costs could stay elevated longer, prolonging dependence on niche, high-value applications and delaying parity against low-priced, policy-enabled fossil incumbents. In combination with price pressure, incumbent innovation, capacity expansion, and uneven regulation, the "Plastic Empire" mounts a robust defense that PHAs must overcome through rapid yield gains, low-energy extraction, co-product valorization, and targeted policy instruments tied to end-of-life outcomes and carbon performance.

20.4 Lenses for the Future

The global transition toward sustainable materials is driving unprecedented interest in PHAs produced from photosynthetic microorganisms such as cyanobacteria, microalgae, and photosynthetic bacteria. While these systems offer a unique pathway to carbon-negative bioplastics by directly utilizing CO_2, sunlight, and even wastewater, their current industrial impact remains limited. Existing processes are constrained by low PHA yields, high downstream recovery costs, and the inability to compete economically with fossil-based plastics. Looking forward, the successful transition of photosynthetic PHA production from pilot-scale feasibility to commercial viability will require transformative advancements rather than incremental improvements. This future can be understood through three strategic lenses:

20.4.1 Technology Lens: Advancing Yields, Extraction, and Scalability

Photosynthetic PHA platforms will only become industrially relevant if three technical gaps close together: (i) cellular yield and volumetric productivity, (ii) photobioreactor scale-up under CO_2/O_2 constraints, and (iii) low-impact, low-cost downstream recovery. At the cellular level, engineered cyanobacteria and algae (Chaogang et al. 2010) such as *Synechocystis* have demonstrated promising results, with PHB accumulation reaching 63% of CDW under nitrogen and phosphorus limitation and further increasing to 81% CDW when coupled with acetate co-feeding by deleting the regulatory gene *pirC* and overexpressing the *phaAB* operon (Koch et al. 2020). Similarly, deletion of *adc1* (arginine decarboxylase) in *Synechocystis* channels carbon flux toward PHB synthesis, achieving ~36% CDW under stress conditions (Utharn et al. 2021). These findings highlight the potential of rewiring carbon partitioning pathways to enhance PHA accumulation. In parallel, boosting the CBB cycle by modulating key enzymes such as RuBisCO and phosphoribulokinase has shown measurable improvements in carbon fixation and precursor supply (Tharasirivat and Jantaro 2023), while engineering enhanced acetyl-CoA pathways provides greater substrate availability for PHA biosynthesis

(You et al. 2024). Looking forward, next-generation strategies must integrate light and nutrient-responsive synthetic circuits to dynamically activate PHA synthesis during energy surplus states, minimizing growth penalties while maximizing volumetric productivity, and combine with CRISPRi libraries for rapid pathway balancing (Satta et al. 2023). Introduce bypasses (e.g., *NphT7* thiolase routes) or amplify PDH/PK-PDH shunts; monitor acetyl-CoA/CoA ratios with biosensors to avoid redox/energy choke points (You et al. 2024). Even though this mechanism is identified, this mechanistic need is supported by acetyl-CoA limitation studies. Additionally, combining photoautotrophy with controlled pulses of volatile fatty acids (VFAs) or acetate derived from wastewater valorization represents a promising hybrid cultivation regime, enabling up to a twofold improvement in polymer yields without fully transitioning to heterotrophy.

Beyond cellular engineering, photobioreactor architecture represents a key determinant of process scalability. Optimizing panel depth from ~5–11 mm can double areal productivity (~10 → 21 g m^{-2} day^{-1}) by improving light utilization (FrETR). Thin-layer cascades and flat-panels routinely report 10–20 g m^{-2} day^{-1} long-term, with peaks up to ~38 g m^{-2} day^{-1} on clear days. Long-term productivity rates of 10–20 $g{\cdot}m^{-2}{\cdot}day^{-1}$ are consistently achievable, with peak performances reaching ~38 $g{\cdot}m^{-2}{\cdot}day^{-1}$ under optimized conditions (Wang et al. 2015; Doucha and Lívanský 2009). Among available designs, flat-panel and thin-layer PBRs remain the most efficient geometries for light utilization, while computational fluid -guided engineering has been instrumental in minimizing light-dark cycles and improving culture homogeneity (Chiarini and Quadrio 2021). However, as reactor scales increase, dissolved O_2 removal and CO_2 transfer become major bottlenecks, causing photo-inhibition if unaddressed. Solutions include adopting high surface-area gas exchange, smart sparging, and degassing strategies to maintain balanced gas dynamics (Yahaya et al. 2025).

Scale-up actions can include thin-layer or shallow flat-panel farms (5–15 mm optical depth) with high surface-to-volume ratios; modular stacks allow >100 m^2 illuminated area per skid while keeping mixing energy low (Penloglou et al. 2024). Flue-gas coupling (5–15% CO_2) with online pH/DO control to stabilize carbon supply and strip O_2; combine with intermittent night-time mixing to maintain homogeneity (Shareefdeen et al. 2023). Real-time optimization of irradiance, CO_2, and circulation to maintain maximal photosynthetic efficiency, building on documented improvements in carbon fixation rate under controlled PBRs (Faulkner et al. 2024).

Downstream recovery remains one of the most cost-intensive steps; traditional halogenated-solvent extraction is effective but environmentally and economically limiting. Downstream processing can be 30–50% of total cost in PHA processes. Recent innovations have introduced sustainable alternatives such as thermos-separating aqueous two-phase extraction (ATPE), which has demonstrated 99.9% polymer purity with an estimated recovery cost of ~USD 5.77 per kg at scales of 9000 $t{\cdot}y^{-1}$, achieving a projected return on investment (ROI) exceeding 25% and a payback period of less than 4 years (Leong et al. 2017; Leong 2018). Enzyme-assisted lysis and supercritical CO_2 extraction are being investigated to reduce solvent dependence and preserve polymer molecular weight. Integrating closed-loop purification systems that recycle water and solvents while minimizing energy consumption is expected to further reduce downstream processing costs to below USD 2.0 per kg at commercial scales, making photosynthetic PHAs more competitive with fossil-derived plastics. Meeting these milestones will enable photosynthetic PHA production to transition from laboratory feasibility to industrial-scale sustainability, establishing it as a competitive, environmentally resilient alternative to fossil-derived plastics.

20.4.2 Economic Lens: Achieving Cost Parity with Fossil Plastics

The global PHA market is undergoing rapid transformation, yet achieving cost parity with fossil-based plastics remains one of its biggest challenges. Despite technological advancements, PHAs currently cost 30–50% more than conventional plastics, primarily due to high substrate prices, energy-intensive downstream recovery, and limited production scale. However, the economic landscape is shifting as global PHA production surpassed 35,000 metric tons in 2024 and is projected to exceed 50,000 metric tons by 2025, supported by regulatory frameworks, corporate sustainability goals, and technological breakthroughs (global growth insights). Innovations in fermentation technology and process automation have already led to an average 20% reduction in production costs, while a strategic shift toward low-cost, carbon-rich feedstocks such as industrial CO_2, agro-residues, and wastewater promises further savings, addressing the fact that feedstocks contribute up to 50% of operating expenses. In parallel, multiproduct biorefinery models are emerging, where PHAs are co-produced with pigments, lipids, and bioactive peptides, improving overall process economics by distributing capital and operational expenditures across multiple revenue streams (Fig. 20.5).

Globally, over 100 companies are actively engaged in PHA production and R&D, including Danimer Scientific, Newlight Technologies, Kaneka, Biomer, Honeywell, Siemens, and Emerson Electric, with heavy investments in scaling up production capacities, automation, and energy-efficient processes. Regional dynamics further shape the market: North America and Europe collectively hold 60% of global share, driven by strict regulations like the EU Single-Use Plastics Directive (SUPD) and the U.S. Biodegradable Products Institute (BPI) certification, while Asia-Pacific led by China and Japan is the fastest-growing region, expected to surpass 20,000 metric tons by 2025. Additionally, venture capital funding exceeding $500 million between 2023 and 2024 and over $15 million in US government grants are accelerating large-scale adoption, particularly in packaging, medical, agriculture, and consumer goods sectors (global growth insights).

Recent techno-economic projections suggest that integrating waste-derived feedstocks, automated large-scale photobioreactors, and policy-driven incentives, including carbon pricing, plastic bans, and green procurement, mandates could reduce the minimum selling price (MSP) of PHAs from the current \$4–6 kg^{-1} to around \$1.5–2.0 kg^{-1} by 2035. With growing commitments from global brands like

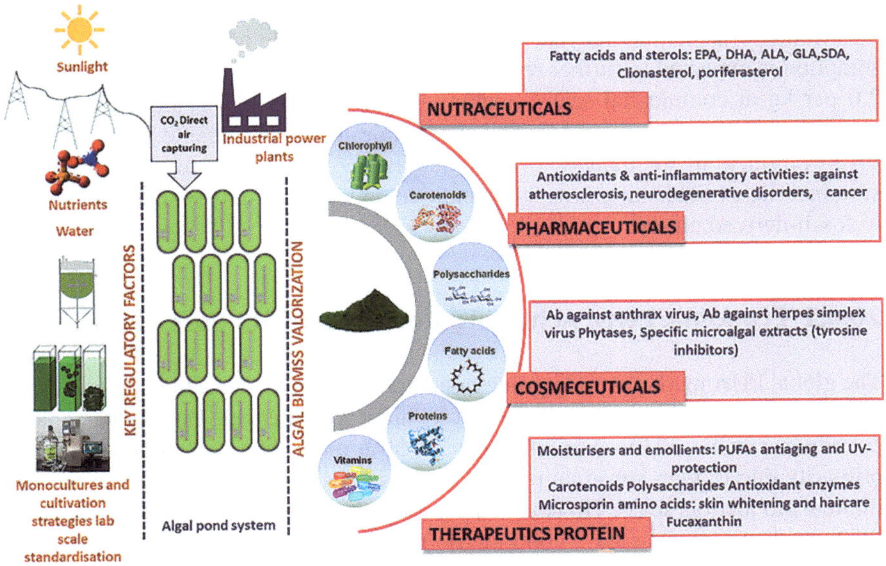

Fig. 20.5 Schematic representation of algal biomass valorization pathways showing diverse applications in nutraceuticals, pharmaceuticals, cosmeceuticals, therapeutic proteins, and biopolymers

Nestlé, PepsiCo, Coca-Cola, and Unilever to adopt PHA-based packaging, the convergence of technological, economic, and regulatory drivers is paving the way for PHAs to transition from niche bioplastics to commercially competitive, scalable solutions at the heart of the circular bioeconomy (global growth insights).

20.5 Conclusion and Future Perspective

The arc of this chapter traces a clear path from problem to possibility: today's fossil-centric plastics economy is linear, carbon-intensive, and pollution-locking, whereas PHAs especially from photosynthetic microorganisms offer a biologically circular route that couples carbon capture, nutrient recovery, and true end-of-life biodegradation. We framed this transition through a storyboard SWOT to stay honest about current limits (slow growth, competing carbon sinks, costly extraction) while keeping sight of differentiating strengths (CO_2 fixation, wastewater valorization, and multi-product biorefinery potential) (Fig. 20.5). The forward lenses then translate that realism into action: technologically, by redirecting flux to PHAs, refining thin-layer and flat-panel PBR architectures, and replacing solvent-heavy recovery with greener, lower-energy routes; economically, by shifting to waste-derived carbon, stacking co-products to amortize costs, scaling intelligently, and aligning with policy instruments that price externalities and reward circularity.

What emerges is not a single silver bullet but a coordinated playbook. Photosynthetic PHAs are most competitive where process integration is strongest co-located with CO_2 point sources and wastewater streams, embedded in biorefineries that valorize pigments, lipids, and peptides, and guided by TEA/LCA so improvements land where they matter. Near-term markets in marine-sensitive packaging and biomedical devices can underwrite maturation, while standards, procurement, and EPR policies accelerate demand signals. If the field pursues these levers in concert engineering for yield and purity, designing reactors for light and gas transfer, and financing for scale and learning effects, PHAs can move from niche promise to mainstream materials, advancing a plastics economy that finally fits within planetary boundaries.

Acknowledgments The authors gratefully acknowledge the Council of Scientific and Industrial Research (CSIR) for supporting this research through project FBR-030308. PK acknowledges the Ministry of Social Justice & Empowerment for providing UGC-NFSC fellowship. The authors wish to thank the Director, CSIR-IICT, for support and encouragement to carry out this work *(IICT/ Pubs./2025/324)*.

Author Contribution
Poonam Kumari performed conceptualization, literature review, analysis, draft writing and reviewing, and visualization. *S. Venkata Mohan* contributed to supervision, draft review, validation, and provision of resources.

Declaration
The authors declare that they have no known competing financial interests or personal relationships that could have appeared to influence the work reported in this paper.

References

Adetunji AI, Erasmus M (2024) Green synthesis of bioplastics from microalgae: a state-of-the-art review. Polymers 16:1322

Alcaraz Cercós E (2015) Bioplastic production from wastes and wastewater. University of Barcelona

Bharathiraja B, Chakravarthy M, Kumar RR, Yogendran D, Yuvaraj D, Jayamuthunagai J, Kumar RP, Palani S (2015) Aquatic biomass (algae) as a future feed stock for bio-refineries: a review on cultivation, processing and products. Renew Sust Energ Rev 47:634–653

Bhati R, Mallick N (2015) Poly (3-hydroxybutyrate-co-3-hydroxyvalerate) copolymer production by the diazotrophic cyanobacterium Nostoc muscorum Agardh: process optimization and polymer characterization. Algal Res 7:78–85

Bioplastics Market Size, Share, Growth Analysis Report (2030)

Blankenship RE (2014) Molecular mechanisms of photosynthesis, 2nd edn. Wiley Blackwell

Brown B, Immethun C, Alsiyabi A, Long D, Wilkins M, Saha R (2022) Heterologous phasin expression in Rhodopseudomonas palustris CGA009 for bioplastic production from lignocellulosic biomass. Metab Eng Commun 14:e00191

Chaogang W, Zhangli H, Anping L, Baohui J (2010) Biosynthesis of Poly-3-Hydroxybutyrate (phb) in the Transgenic Green Alga Chlamydomonas reinhardtii 1. J Phycol 46(2):396–402

Chaudry S, Hurtado-McCormick V, Cheng KY, Willis A, Speight R, Kaksonen AH (2025) Microalgae to bioplastics–routes and challenges. Clean Eng Technol 25:100922

Chen Y, Awasthi AK, Wei F, Tan Q, Li J (2021) Single-use plastics: production, usage, disposal, and adverse impacts. Sci Total Environ 752:141772

Chiarini A, Quadrio M (2021) The light/dark cycle of microalgae in a thin-layer photobioreactor. J Appl Phycol 33(1):183–195

Chisti Y (2007) Biodiesel from microalgae. Biotechnol Adv 25(3):294–306

Chodak I (2008) Polyhydroxyalkanoates: origin, properties and applications. In: Monomers, polymers and composites from renewable resources. Elsevier, pp 451–477

Choi SY, Cho IJ, Lee Y, Kim YJ, Kim KJ, Lee SY (2020) Microbial polyhydroxyalkanoates and nonnatural polyesters. Adv Mater 32(35):1907138

Ciebiada M, Kubiak K, Daroch M (2020) Modifying the cyanobacterial metabolism as a key to efficient biopolymer production in photosynthetic microorganisms. Int J Mol Sci 21(19):7204

Costa SS, Miranda AL, Andrade BB, de Jesus Assis D, Souza CO, de Morais MG, Costa JAV, Druzian JI (2018) Influence of nitrogen on growth, biomass composition, production, and properties of polyhydroxyalkanoates (PHAs) by microalgae. Int J Biol Macromol 116:552–562

Das SK, Sathish A, Stanley J (2018) Production of biofuel and bioplastic from Chlorella pyrenoidosa. Mater Today Proc 5(8):16774–16781

Data Bridge Market Research (2024) https://www.databridgemarketresearch.com/reports/global-polyhydroxyalkanoate-pha-market

Dewangan M, Singh VK, Kumar A (2025) Unraveling the potential of Chlorella vulgaris in sustainable bioplastic production, carbon sequestration, and wastewater treatment using integrated approaches. J Environ Sci Health A 60(1):46–53

Doucha J, Lívanský K (2009) Outdoor open thin-layer microalgal photobioreactor: potential productivity. J Appl Phycol 21(1):111–117

Faulkner M, Andrews F, Scrutton N (2024) Improving productivity of citramalate from CO2 by Synechocystis sp. PCC 6803 through design of experiment. Biotechnol Biofuels Bioprod 17(1): 143

Growth Market Reports (2025) https://www.marketgrowthreports.com/market-reports/pha-market-118292

Gundlapalli M, Ganesan S (2025) Polyhydroxyalkanoates (PHAs): key challenges in production and sustainable strategies for cost reduction within a circular economy framework. Results Eng 26:105345

Higuchi-Takeuchi M, Motoda Y, Kigawa T, Numata K (2017) Class I polyhydroxyalkanoate synthase from the purple photosynthetic bacterium Rhodovulum sulfidophilum predominantly exists as a functional dimer in the absence of a substrate. ACS Omega 2(8):5071–5078

https://pmarketresearch.com/chemi/pha-microspheres-and-pha-bioplastics-market/?utm_source=chatgpt.com

Ilhami S, Rahman SNSA, Iqhrammullah M, Hamid Z, Chai YH, Lam MK (2025) Polyhydroxyalkanoates production from microalgae for sustainable bioplastics: a review. Biotechnol Adv 79:108529

Khetkorn W, Incharoensakdi A, Lindblad P, Jantaro S (2016) Enhancement of poly-3-hydroxybutyrate production in Synechocystis sp. PCC 6803 by overexpression of its native biosynthetic genes. Bioresour Technol 214:761–768

Kim J, Gupta NS, Bezek LB, Linn J, Bejagam KK, Banerjee S, Dumont JH, Nam SY, Kang HW, Park CH, Pilania G (2023) Biodegradation studies of polyhydroxybutyrate and polyhydroxybutyrate-co-polyhydroxyvalerate films in soil. Int J Mol Sci 24(8):7638

Kobayashi J, Kondo A (2019) Disruption of poly (3-hydroxyalkanoate) depolymerase gene and overexpression of three poly (3-hydroxybutyrate) biosynthetic genes improve poly (3-hydroxybutyrate) production from nitrogen rich medium by Rhodobacter sphaeroides. Microb Cell Factories 18(1):40

Koch M, Bruckmoser J, Scholl J, Hauf W, Rieger B, Forchhammer K (2020) Maximizing PHB content in Synechocystis sp. PCC 6803: a new metabolic engineering strategy based on the regulator PirC. Microb Cell Factories 19(1):231

Koch M, Spierling S, Venkatachalam V, Endres HJ, Owsianiak M, Vea EB, Daffert C, Neureiter M, Fritz I (2023) Comparative assessment of environmental impacts of 1st generation (corn feedstock) and 3rd generation (carbon dioxide feedstock) PHA production pathways using life cycle assessment. Sci Total Environ 863:160991

Koller M, Heeney D, Mukherjee A (2025) Biodegradability of polyhydroxyalkanoate (PHA) biopolyesters in nature: a review. Biodegradation 36(4):76

Kourmentza C, Plácido J, Venetsaneas N, Burniol-Figols A, Varrone C, Gavala HN, Reis MA (2017) Recent advances and challenges towards sustainable polyhydroxyalkanoate (PHA) production. Bioengineering 4(2):55

Kumari P, Mohan SV (2025) Metabolic channelling in Rhodopseudomonas palustris for diversification and sustainability assessment of polyhydroxyalkanoates. Green Chem 27(22):6464–6481

Kumari P, Kiran BR, Mohan SV (2022) Polyhydroxybutyrate production by Chlorella sorokiniana SVMIICT8 under nutrient-deprived mixotrophy. Bioresour Technol 354:127135

Lackner M, Mukherjee A, Koller M (2023) What are "bioplastics"? Defining renewability, biosynthesis, biodegradability, and biocompatibility. Polymers 15(24):4695

Lenz RW, Marchessault RH (2005) Bacterial polyesters: biosynthesis, biodegradable plastics and biotechnology. Biomacromolecules 6(1):1–8

Leong YK (2018) Extraction and purification of green polymers using aqueous two-phase extraction (ATPE). (Doctoral dissertation,. University of Nottingham

Leong YK, Show PL, Lan JCW, Loh HS, Lam HL, Ling TC (2017) Economic and environmental analysis of PHAs production process. Clean Techn Environ Policy 19(7):1941–1953

Madadi R, Maljaee H, Serafim LS, Ventura SP (2021) Microalgae as contributors to produce biopolymers. Mar Drugs 19(8):466

Madkour MH, Heinrich D, Alghamdi MA, Shabbaj II, Steinbüchel A (2013) PHA recovery from biomass. Biomacromolecules 14(9):2963–2972

Melchor-Martínez EM, Macías-Garbett R, Alvarado-Ramírez L, Araújo RG, Sosa-Hernández JE, Ramírez-Gamboa D, Parra-Arroyo L, Alvarez AG, Monteverde RPB, Cazares KAS, Reyes-Mayer A (2022) Towards a circular economy of plastics: an evaluation of the systematic transition to a new generation of bioplastics. Polymers 14(6):1203

Mergaert J, Anderson C, Wouters A, Swings J, Kersters KFEMS (1992) Biodegradation of polyhydroxyalkanoates. FEMS Microbiol Rev 9(2–4):317–321

Nawaz T, Gu L, Hu Z, Fahad S, Saud S, Zhou R (2024) Advancements in synthetic biology for enhancing cyanobacterial capabilities in sustainable plastic production: a green horizon perspective. Fuels 5(3):394–438

Pant S, Goel M, Sahoo NK, Yuan Q, Rout PR (2025) Microalgae-based treatment for wastewater management and valorisation. Curr Opin Environ Sci Health 48:100662

Penloglou G, Pavlou A, Kiparissides C (2024) Recent advancements in photo-bioreactors for microalgae cultivation: a brief overview. Processes 12(6):1104

PHA Bioplastics Market Research Report (2033). https://growthmarketreports.com/report/pha-bioplastics-market#:~:text=According%20to%20our%20latest%20research%2C%20the%20global%20PHA,to%20reach%20approximately%20USD%20384%20million%20by%202033

Pradhan S, Dikshit PK, Moholkar VS (2018) Production, ultrasonic extraction, and characterization of poly (3-hydroxybutyrate) (PHB) using bacillus megaterium and Cupriavidus necator. Polym Adv Technol 29(8):2392–2400

Pradhan S, Khan MT, Moholkar VS (2024) Polyhydroxyalkanoates (PHAs): mechanistic insights and contributions to sustainable practices. Encyclopedia 5(4):1933–1947

Raven JA, Beardall J (2020) Energizing the plasmalemma of marine photosynthetic organisms: the role of primary active transport. J Mar Biol Assoc U K 100(3):333–346

Satta A, Esquirol L, Ebert BE (2023) Current metabolic engineering strategies for photosynthetic bioproduction in cyanobacteria. Microorganisms 11(2):455

Shareefdeen Z, Elkamel A, Babar ZB (2023) Recent developments on the performance of algal bioreactors for CO2 removal: focusing on the light intensity and photoperiods. Biotech 12(1):10

Sharma M, Tellili N, Kacem I, Rouissi T (2024a) Microbial biopolymers: from production to environmental applications—a review. Appl Sci 14(12):5081. (a)

Sharma R, Solanki P, Chaudhary M, Gupta N, Kaur P (2024b) Unveiling the potential of microalgae for bioplastic production from wastewater–current trends, innovations, and future prospects. Biotechnol Sustain Mater 1(1):10. (b)

Srisawat P, Higuchi-Takeuchi M, Honda R, Shirai T, Kondo A, Hoshino Y, Numata K (2022) Engineered nanogel particles enhance the photoautotrophic biosynthesis of polyhydroxyalkanoate in marine photosynthetic bacteria. ACS Sustain Chem Eng 10(13): 4133–4142

Sudesh K, Abe H, Doi Y (2000) Synthesis, structure and properties of polyhydroxyalkanoates: biological polyesters. Prog Polym Sci 25(10):1503–1555

Tan C, Tao F, Xu P (2022) Direct carbon capture for the production of high-performance biodegradable plastics by cyanobacterial cell factories. Green Chem 24(11):4470–4483

Tharasirivat V, Jantaro S (2023) Increased biomass and polyhydroxybutyrate production by Synechocystis sp. PCC 6803 overexpressing RuBisCO genes. Int J Mol Sci 24(7):6415

Towards Packaging (2025) https://www.towardspackaging.com/insights/fossil-based-plastics-market-sizing#:~:text=The%20fossil-based%20plastics%20market%20is%20accelerating%2C%20with%20forecasts,plastics%20market%2C%20with%20polyethylene%20and%20packaging%20segments%20dominating

Utharn S, Yodsang P, Incharoensakdi A, Jantaro S (2021) Cyanobacterium Synechocystis sp. PCC 6803 lacking adc1 gene produces higher polyhydroxybutyrate accumulation under modified nutrients of acetate supplementation and nitrogen-phosphorus starvation. Biotechnol Rep 31: e00661

Wang J, Liu J, Liu T (2015) The difference in effective light penetration may explain the superiority in photosynthetic efficiency of attached cultivation over the conventional open pond for microalgae. Biotechnol Biofuels 8(1):49

Windhagauer M, Abbriano RM, Pittrich DA, Doblin MA (2022) Phosphate-inducible poly-hydroxy butyrate production dynamics in CO2 supplemented upscaled cultivation of engineered Phaeodactylum tricornutum. J Appl Phycol 34(5):2259–2270

Yahaya E, Yeo WS, Nandong J, Ngu JCY (2025) CO2 fed microalgae cultivation in photobioreactor: review on challenges and possible solutions. Environ Technol Rev 14(1): 540–564

You D, Rasul F, Wang T, Daroch M (2024) Insufficient acetyl-CoA pool restricts the phototrophic production of organic acids in model cyanobacteria. Int J Mol Sci 25(21):11769

Yousefi AM, Wnek GE (2025) Poly (hydroxyalkanoates): emerging biopolymers in biomedical fields and packaging industries for a circular economy. Biomed Mater Devices 3(1):19–44

Zhang H, Ma S, Dou X, Chen R, Lu H, Chi Z, Xue S, Li-Beisson Y, Kong F (2024) Harnessing algal peroxisomes for efficient poly Hydroxybutyrate production. ACS Sustain Chem Eng 12(8):3312–3321

Non-catalytic and Catalytic Microalgae Pyrolysis: Product Characterization and Economic Perspective

21

Pooja Singh and Kaustubha Mohanty

Abstract

The high lipid and protein content of algal biomass offers a promising pathway for sustainable biofuel production through the pyrolysis of microalgae. This chapter presents a comprehensive distinction between catalytic and non-catalytic pyrolysis processes, emphasizing their effects on the quality and yield of the final product. The physicochemical properties, elemental composition, and energy content of the primary pyrolysis products, that is, bio-oil, biochar, and syngas, were discussed. In the context of bio-oil, the role of several catalysts, such as zeolites and metal oxides, was demonstrated in terms of their ability to improve hydrocarbon selectivity, reduced nitrogenous and acidic compounds, and enhanced deoxygenation. Furthermore, the chapter discussed the role of process parameters, including temperature, heating rate, and residence time, which have a substantial impact on product distribution. An economic perspective was presented in this chapter to analyze the capital and operational costs of catalytic and non-catalytic systems. The chapter concludes by discussing current challenges, research gaps, and prospective directions for expanding microalgae-based pyrolysis technologies within a circular bioeconomy framework.

Keywords

Pyrolysis · Catalytic · Non-catalytic · Catalyst · Microalgae

P. Singh · K. Mohanty (✉)
Department of Chemical Engineering, Indian Institute of Technology Guwahati, Guwahati, Assam, India
e-mail: kmohanty@iitg.ac.in

K. Mohanty et al. (eds.), *Advances in Algal Research*,
https://doi.org/10.1007/978-981-95-7234-2_21

21.1 Introduction

The expanding global need for bioenergy and chemicals, coupled with mounting concerns over climate change, energy security, and the depletion of fossil resources, has intensified the search for sustainable alternatives. Traditional fossil fuels, while prevalent, are progressively acknowledged as unsustainable due to their limited availability and significant greenhouse gas (GHG) emissions. Biofuels and biochemicals sourced from renewable resources present a viable solution to address these difficulties by aiding in decarbonization, diversifying the energy portfolio, and promoting a circular bioeconomy. The selection of feedstock is crucial: although first-generation (food-based) and second-generation (lignocellulosic biomass) (Fig. 21.1) resources have propelled the biofuel industry, their application is limited by food versus fuel controversies, land-use competition, seasonal availability, and intricate pretreatment demands (Deshavath et al. 2021; Ho et al. 2014). Fourth-generation biofuels are formed from developed and genetically engineered microbial systems. While utilizing microalgae, several benefits occur for biofuel production, like short cultivation duration, carbon dioxide sequestration, and carbon capture, and they can survive in polluted wastewater compared to traditional crops. In view of this topic, microalgae have come out as a highly promising third-generation feedstock, with distinct advantages in productivity, resource efficiency, and environmental benefits (Sivaramakrishnan et al. 2022).

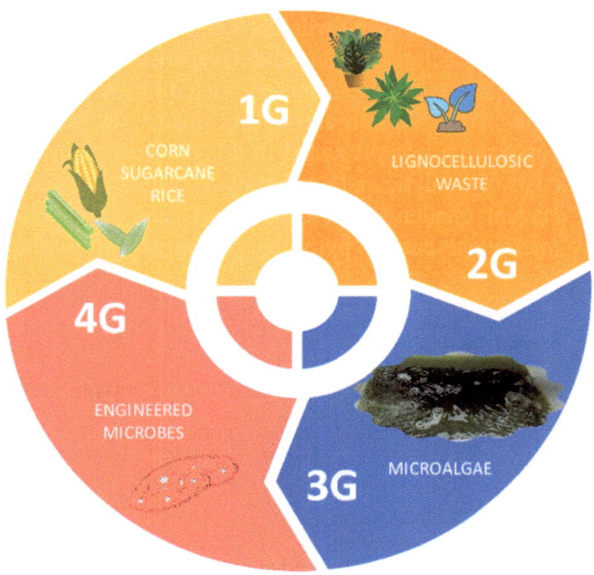

Fig. 21.1 Biofuel production feedstocks categories

Microalgae are photosynthetic microorganisms that may proliferate quickly in diverse aquatic habitats, including nonarable land and wastewater systems (Xia et al. 2022). They demonstrate significantly high biomass and lipid productivities than terrestrial crops, while concurrently sequestering CO_2 from industrial emissions, thereby directly aiding in carbon reduction. Significantly, they do not compete with food crops for arable land or freshwater resources, rendering them an ecologically advantageous alternative. In addition to lipids, microalgal biomass is abundant in proteins, carbohydrates, and precious pigments, presenting prospects for multiproduct biorefineries that encompass fuel, feed, nutraceutical, and pharmaceutical sectors. This biochemical diversity establishes microalgae as a renewable carbon source and a flexible platform for generating fuels and high-value co-products (Sivaramakrishnan et al. 2022), crucial for enhancing the economics of bio-based companies.

Among the numerous conversion pathways for microalgal valorization, such as transesterification for biodiesel (Kothari et al. 2013), anaerobic digestion for biogas (Kusmayadi et al. 2023), and hydrothermal liquefaction for bio-oil (Singh and Mohanty 2025), pyrolysis is distinguished as a versatile thermochemical technology that can transform the entire feedstock into liquid, solid, and gaseous products (Mustapha and Isa 2024). Fast pyrolysis has been extensively researched for its ability to optimize bio-oil yield, whereas slow pyrolysis promotes biochar synthesis, which is utilized for soil enhancement and carbon sequestration (Yu et al. 2017). Pyrolysis provides microalgae with the additional advantage of transforming protein and carbohydrate fractions, in addition to lipids, into viable energy carriers, thus improving overall carbon usage efficiency. Notwithstanding the obstacles posed by the elevated nitrogen content of microalgae, which may result in unwanted nitrogenous compounds in bio-oil (Singh and Mohanty 2025), progress in process optimization and catalytic interventions persistently enhances product quality and process feasibility.

This chapter presents a comparative evaluation of catalytic and non-catalytic pyrolysis of microalgae, with particular attention to their influence on product yield, composition, and upgrading potential. Non-catalytic pyrolysis serves as an essential baseline for understanding thermal decomposition pathways and product distribution, though it typically produces bio-oils with poor stability and elevated oxygen and nitrogen contents. In contrast, catalytic pyrolysis employs solid catalysts such as zeolites and metal oxides to facilitate deoxygenation, denitrogenation, and cracking reactions, thereby enhancing the calorific value, stability, and chemical selectivity of the resulting bio-oils. This chapter illustrates how catalytic processes can shift the product profile towards higher-value fuels and chemicals. In addition, it incorporates an economic perspective, examining the factors that affect the commercial viability of microalgal pyrolysis. By integrating insights from both reaction chemistry and economic considerations, the chapter seeks to provide a basic understanding of how catalytic and non-catalytic pathways can support the advancement of microalgal biofuels and biochemicals toward scalable deployment.

21.2 Fundamentals of Microalgal Pyrolysis

Microalgae are becoming acknowledged as a viable fuel for renewable energy due to their varied biochemical constitution and high growth rates. Microalgae growing in water are usually present in very low cell concentrations when outdoor conditions are maintained. The lower biomass productivity presents a major obstacle in microalgal technology, creating a bottleneck for upstream and downstream algal processes. Figure 21.2 shows a process flow for bio-oil production from microalgae pyrolysis, which includes grinding, sieving, size reduction, pyrolysis, and product separation.

Several literatures have been conducted on pyrolysis studies using microalgae such as *Scenedesmus almeriensis* (Niu et al. 2024), *Scenedesmus obliquus* (Mustapha and Isa 2024; Mustapha et al. 2021), *Chlorella* sp. (Shirazi et al. 2020), and *Isochrysis* sp. (Wang et al. 2016). Microalgal cellular composition (Table 21.1) typically comprises lipids (20–50%), proteins (30–60%), carbohydrates (10–40%), pigments including chlorophylls and carotenoids, and minerals such as phosphorus, potassium, calcium, and trace elements (Yang et al. 2019). The nitrogen proportion in microalgal cells is observed due to major protein components. Also, inorganic nitrogen (10–15%) is present in microalgae as nitrites, nitrates, and ammonia. When these species undergo pyrolysis, the ash is formed as metal oxides. Some minor nitrogen comes from nucleic acids, which constitute of nitrogenous bases. Glucose, cellulose, and hemicellulose are the major carbohydrates in microalgae (Debiagi et al. 2017). TGA and analytical pyrolyser revealed the different degradation behavior of distinct species of carbohydrates in microalgae (Anastasakis et al. 2011). Lipids consist of a fatty acid chain of 14–20 carbon atoms with variations in unsaturation degree. Other inorganic and ash minerals present in microalgae as carbonates, sulphates, phosphates, and chlorides other than N-form. These inorganics converted to oxides, releasing a volatile fraction (Debiagi et al. 2017).

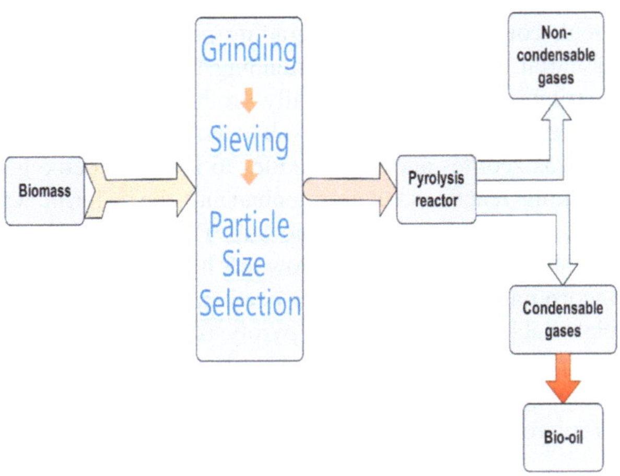

Fig. 21.2 Process flow of microalgae pyrolysis

Table 21.1 Composition of microalgae reported in the previous literature

Unit (wt%)	*Chlorella* sp. (Borges et al. 2014)	*Chlorella vulgaris* (Thangalazhy-Gopakumar et al. 2012)	*Nannochloropsis* sp. (Borges et al. 2014)	*Scenedesmus obliquus* (Mustapha et al. 2021)	*Isochrysis* sp. (Wang et al. 2016)
Ultimate					
C	41.21	44.5 ± 0.3	49.07	45.03	49.26
H	6.21	6.2 ± 0.0	7.59	7.50	7.50
N	7.04	9.6 ± 0.2	6.29	3.59	6.24
S	–	0	1.42	0.98	0.96
0	34.06	29.3 ± 0.5	35.63	42.90	31.74
Proximate					
Moisture	7.28	8.4 ± 0.2	5	–	1.69
Volatile matter	81.15	–	79.69	81.77	79.79
Fixed carbon	–	–	10.64	–	11.63
Ash	11.57	10.3 ± 0.2	5.03	10.17	6.89
Biochemical					
Protein	–	58.13	44	22.08	35.9
Carbohydrate	–	–	21	42.55	15.2
Lipid	–	–	30	21.62	42.9

The main purpose of pyrolysis studies of microalgae is

1. To investigate the viability and practicality of pyrolysis process for processing microalgae.
2. To produce maximum yield of one of the three process products: bio-oil, biochar, or syngas.
3. To explore the influence of catalyst and pretreatment on increasing any product from this process.

As of now, the microalgae-based pyrolysis process has not been commercialized and performed at laboratory and pilot level studies. However, previously, thermogravimetry analysis (TGA) was implemented to replicate slow pyrolysis process and show the effect of temperature on biomass degradation kinetics. Two different modes of pyrolysis process can be performed (Table 21.2). Slow pyrolysis is achieved by slow heating rates, while fast pyrolysis process happens at high heating rate in a fixed bed reactor. Only some publications are reported on fast pyrolysis of microalgae; however, pyrolysis-gas chromatography/mass spectrometry (Py-GC/MS) can be used to analyse volatile products of fast pyrolysis without any secondary reaction products. During slow pyrolysis, a sweeping gas at a constant flow rate was incorporated in the fixed bed reactor to keep an inert atmosphere and remove volatiles from the reactor. The process was heated at 10 °C/s to a required temperature (Yang et al. 2019). Briefly, with slow pyrolysis, the highest bio-oil obtained from *Chlorella Vulgaris*, *Isochrysis* sp., *Isochrysis* sp., and *Scenedesmus almeriensis* were 45.62% at 520 °C (Khodaparasti et al. 2023), 49.36% at 450 °C (Wang et al. 2016), 37% at 500 °C (Rahman et al. 2018), and 42.4% at 500 °C (Niu et al. 2024), respectively. Due to inexpensive and easy operation advantages, slow pyrolysis is the most investigated conversion process. However, this process forms high solid biochar and gas product (Yang et al. 2019). Pyrolytic bio-oil from *Scenedesmus obliquus* was 29.2 wt% at 500 °C with a major proportion of aliphatic and acids (Mustapha and Isa 2024). Also, in one of the study the temperature difference of 25 °C is not enough to make significant change in the bio-oil yield. *Isochrysis* sp. when subjected to 450 °C, 49.36% of bio-oil recovered while almost similar yield was noticed at 475 °C. A major change in decrement of bio-oil was noticed when temperature was increased to 500 °C (Wang et al. 2016). The pyrolysis tests at 500 °C with *Isochrysis* sp. in another study with volatile matter content of 76.9% and high N content (4%) formed bio-oil with major proportion of aliphatic, alcohols, and nitrogen-based compounds. The microalgae and acid pretreated microalgae constitute of ash with 16.8% and 4.5%, respectively, were pyrolyzed. The bio-oil yield with a difference of 42.4% and 43.3% was reported from microalgae and acid-pretreated microalgae, respectively (Niu et al. 2024).

The preferred mode of pyrolysis is fast pyrolysis due to minimization of intermediates secondary cracking, condensation, and polymerization (Yang et al. 2019). Another type of reactor, falling solid reactor composed of an auger screw is an another form of fast pyrolysis reactor system, which showed better liquid yield than slow pyrolysis (Campanella and Harold 2012). Previous work has investigated

Table 21.2 Previous studies on the slow and fast pyrolysis process

Microalgae	Catalyst	Experimental details	References
Slow pyrolysis			
Chlorella sp.	HZSM-5	450 °C, 500 °C, 500 °C in two modes: ex situ and in situ catalytic and catalyst-free pyrolysis	Shirazi et al. (2020)
Chlorella vulgaris	None	500 °C with a heating rate of 48.0 ± 0.3 °C/min	Gong et al. (2014)
D. salina	None	500 °C with a heating rate of 48.0 ± 0.3 °C/min	Gong et al. (2014)
Chlorella sp.	None	450 °C and 550 °C	Borges et al. (2014)
Nannochloropsis sp.	None	450 °C and 550 °C	Borges et al. (2014)
Chlorella sp.	HZSM-5	450 °C and 550 °C with catalyst to biomass ratio of 1:1 and 1:2	Borges et al. (2014)
Nannochloropsis sp.	HZSM-5	450 °C and 550 °C with catalyst to biomass ratio of 1:1 and 1:2	Borges et al. (2014)
Chlorella vulgaris	None	30–800 °C at different heating rates of 10, 20, 30, and 40 °C/min	Thangalazhy-Gopakumar et al. (2012)
Scenedesmus obliquus	None	1 g microalgae sample at heating rate of 10 °C/min and performed at 500 °C	Mustapha et al. (2021)
Scenedesmus obliquus	M/Fe_3O_4-HZSM-5	Microalgae:catalyst =1:1 g/g at temperature between 400 °C and 700 °C	Mustapha et al. (2021)
Isochrysis sp.	Li-LSX-zeolite	500 °C and catalyst to microalgae ratio 1:1	Wang et al. (2016)
Fast pyrolysis			
Chlorella vulgaris	HZSM-5	0.5 mg were loaded at biomass-to-catalyst ratios of 5, 10, and 20 at temperatures of 400 °C, 500 °C, 600 °C, 700 °C, and 800 °C	Wang and Brown (2013)
Chlorella vulgaris	None	650 °C, with the filament ramping rate of 2000 °C/s and holding time of 1.5 min	Thangalazhy-Gopakumar et al. (2012)
Botryococcus braunii	None	Pyrolyzed at 350, 450, and 550 °C for 0.1 min	Subagyono et al. (2023)
Botryococcus braunii	Ni/SBA-15	Pyrolyzed at 350, 450, and 550 °C for 0.1 min	Subagyono et al. (2023)
Laminaria japonica	Hierarchical meso-MFI zeolite	Biomass (2 mg) and catalyst (1 mg) were incorporated in a furnace at 500 °C under a He atmosphere	Lee et al. (2011)

Chlorella (Campanella and Harold 2012) and *Scenedesmus almeriensis* (CCAP-276/24) (Niu et al. 2024) in a fast pyrolysis. Maximum bio-oil from *Chlorella* and *Scenedesmus almeriensis* was found to be 35.5% (500 °C) and 42.2% (500 °C), respectively. It was also observed in one of the studies that steam was used as a carrier gas in pyrolysis reaction, showed a milder effect on product yield but observed a major influence on the bio-oil composition. The steam reduced oxygenated species and increased hydrocarbons in the bio-oil due to steam reforming and deoxygenation (Campanella and Harold 2012). Pyrolysis with *Chlorella sorokiniana* str. SLA-04 produced 62% bio-oil at 550 °C compared to 450 °C and 500 °C. Bio-oil consists of oxygen-based compounds like acetic acid, ketones, aldehydes, and furans generated from carbohydrate degradation; nitrogen-based compounds, namely, pyrroles, indole, pyrazoles, oleoamides, and fatty nitriles from protein fragmentation; fatty ester, aliphatic, and fatty acids from lipid decomposition. The increase in temperature increased the carboxylic acid and aliphatic content and decreased the glyceride formation because of glyceride degradation at higher temperature (Shirazi et al. 2020). It was also reviewed that bio-oil yield was affected if cells were manipulated during the growth stage in terms of the autotrophic or heterotrophic mode of cultivation. The heterotrophic *C. protothecoides* showed three times more bio-oil by fast pyrolysis. Also, the HHV and carbon content of bio-oil from heterotrophic *C. protothecoides* were comparatively higher than autotrophic process-grown microalgae bio-oil (Miao and Wu 2004).

21.2.1 Pyrolysis Products from Microalgae

The bio-oil formed by this process is comprised of complex organic classes, including nitrogenated and oxygenated compounds and hydrocarbons. The nitrogen-based compounds major comprises amides, amines, pyrroles, indoles, pyridines, pyrazines, imidazoles, and their derivatives. While oxygen-based compounds are carboxylic acids, ketones, aldehydes, cyclic aromatics, and phenols. Also, esters and ethers are present in pyrolytic bio-oil (Yang et al. 2019; Razzak et al. 2024; Gong et al. 2014). The pyrolytic bio-oil has a major proportion of carbon in the range between 46 and 77 wt%. The H-content can vary from 7 to 12 wt%. While the N-content creates a significant difference in bio-oil quality, found up to 12.4 wt% (Table 21.3). The molar atomic ratios of H/C, O/C, and N/C give an accurate assessment of microalgal bio-oil quality. The higher HHV of bio-oil ranges from 26.8 to 42 MJ kg^{-1}, which exceeds that of cellulosic biomass bio-oil (15–21 MJ/kg). Its density is approximately 0.98–1.2 kg L^{-1}, while the viscosity varies depending on the water content and chemical composition (Yang et al. 2019).

Another important product produced from the pyrolysis process is solid biochar. The biochar generated from the pyrolysis process is receiving attention toward several applications due to its inexpensive and abundant advantages (Yang et al. 2019). Slow pyrolysis is the preferred route for biochar generation because long reaction time helps in more condensation and polymerization of unstable intermediates and enhances solid product yield (Ronsse et al. 2013; Song and Guo 2012). The process can generate more biochar between 15 and 43 wt% from microalgal biomass. The carbon content of biochar is comparable to lignocellulosic biomass. The nitrogen in microalgal biochar is higher than that present in lignocellulosic material-based biochar. Also, the metals and nonmetals amount are more in microalgal-based biochar than in lignocellulosic material biochar (Borges et al. 2014). Additionally, the surface area of microalgal biochar is comparatively less than lignocellulosic material biochar (Yang et al. 2019; Gong et al. 2014).

Table 21.3 Elemental analysis of bio-oil recovered from the non-catalytic and catalytic pyrolysis

	Microalgae	Catalyst	C (wt %)	H (wt %)	O (wt %)	N (wt%)	HHV (MJ/kg)	Reference
Bio-oil	*C. protothecoides* (autotrophic cultivation)	None	62.07	8.76	19.43	9.74	30	Miao and Wu (2004)
	C. protothecoides (heterotrophic cultivation)	None	76.22	11.61	11.24	0.93	41	Miao and Wu (2004)
	Spirulina	None	74.66	10.57	6.81	7.13	33.62	Jena and Das (2011)
	Spirulina sp.	None	46.05	7.97	36.28	9.7	21.68	Chaiwong et al. (2013)
	Chlorella	None	54.52	9.06	25.69	11.51	25.5	Campanella and Harold (2012)
	Chlorella	H-ZSM5	63.64	8.49	19.41	9.18	29.1	Campanella and Harold (2012)
	Chlorella	Fe-ZSM5	59.67	8.44	23.58	10.96	26.8	Campanella and Harold (2012)
	Chlorella	Cu-ZSM5	60.54	8.60	23.07	10.06	27.5	Campanella and Harold (2012)
	Chlorella	Ni-ZSM-5	58.54	8.99	21.39	10.57	28.22	Campanella and Harold (2012)
	C. vulgaris	ZSM-5 ($SiO_2/Al_2O_3 = 50$)	51.4	10.4	24.8	12.4	18.6	Thangalazhy-Gopakumar et al. (2012)
	C. vulgaris	None	55.1	7.5	27.7	9.3	–	Gong et al. (2014)
	D. salina	None	56.3	7.8	24.8	10.8	–	Gong et al. (2014)
	Scenedesmus obliquus	None	70.77	11.50	11.84	5.46	36.99	Mustapha et al. (2021)
	Scenedesmus obliquus	W/Fe_3O_4-HZSM-5	75.98	10.89	6.57	5.89	38.65	Mustapha et al. (2021)
	Scenedesmus obliquus	Zr/Fe_3O_4-HZSM-5	71.99	12.03	10.37	5.19	38.20	Mustapha et al. (2021)
	Scenedesmus obliquus	Co/Fe_3O_4-HZSM-5	73.00	12.42	9.38	5.25	39.12	Mustapha et al. (2021)
	Scenedesmus obliquus	Mo/Fe_3O_4-HZSM-5	71.47	12.61	10.33	5.25	38.69	Mustapha et al. (2021)
Commercial crude	–	–	85–90	10–14	1–1.5	<0.1–2.0	42	Chaudhuri (2016)

During high-temperature reaction in the pyrolysis of microalgae, gases are also formed. Water-gas, water-gas-shift, and the Boudouard reactions play an important role in gas formation. The major gases formed are hydrogen, methane, carbon dioxide, carbon monoxide, acetylene, ethylene, ethane, propylene, and butane (Dong et al. 2013). Microalgal species *Lyngbya* sp. produced 46.5 wt% of gases with a major proportion of H_2 and CO_2 after pyrolysis at 600 °C. However, gases produced from *Cladophora* sp. formed the highest amount of carbon dioxide, followed by H_2 and CO at 600 °C (Maddi et al. 2011). Microalgae *C. vulgaris* and *D. salina,* when pyrolyzed at 700 °C, formed C_2H_4 and C_2H_6 in major proportions, along with hydrogen, methane, carbon dioxide, and carbon monoxide (Gong et al. 2014).

21.3 Co-pyrolysis with Another Feedstock

Nowadays, researchers are investigating the co-pyrolysis process of microalgae with other waste to enhance production and economic efficiency (Khodaparasti et al. 2023). Also, researchers are focusing on understanding the synergistic interaction between two feedstocks that enhance and weaken the quality and yield of the required product (Abnisa and Wan Daud 2014). A major factor that affects the synergistic effect is the mixing of two feedstocks. The synergistic effect relies on feed type, contact, reaction time, temperature, heating rate, removal or equilibrium of volatiles formed, catalyst, and hydrogen donors (Abnisa and Wan Daud 2014). Co-pyrolysis of two feedstocks has favourably enhanced the oil quantity and quality in previous work. An important advantage of using the co-pyrolysis process is a reduction of waste in terms of consumption as a feedstock, reducing the environmental problems and enhancing energy security (Kumar et al. 2024; Garforth et al. 2004). Additionally, the co-pyrolysis is considered a favourable route for a thermal conversion technology to generate pyrolytic bio-oil (Abnisa and Wan Daud 2014). It is reported that the economic consequences of the synergistic effects are found to be more profitable compared to the single feedstock conversion process (Kuppens et al. 2010). According to their findings given by Khodaparasti et al. (2023), a higher gas yield was observed due to more deoxygenation reactions (Khodaparasti et al. 2023). Presence of metals in feed might have enhanced secondary reactions due to catalytic activity, resulted in enhanced gas formation and reduction in solid residue (Wu et al. 2018). Co-pyrolysis of microalgae and sewage sludge synergistically improved the hydrocarbon content in bio-oil and reduced the nitrogenous and oxygenous compounds (Kumar et al. 2024). In another investigation, the co-pyrolysis process of bio-oil of microalgae and sewage sludge increased the C4 and C7 proportion but decreased C9 in the carbon distribution. While N-heterocyclic compounds was highest in bio-oil (Wang et al. 2016).

This chapter also highlights that numerous studies have explored the potential of co-pyrolysis using different types of biomass wastes, and the outcomes have been highly promising. Various investigations have aimed at producing oil with both high yield and superior quality in accordance with established standards. The technique offers several advantages, including reduced reliance on fossil fuels, mitigation of environmental issues, enhancement of energy security, and improvements in waste management. Additionally, co-pyrolysis is considered relatively economically feasible. A few critical aspects must be considered in the feed system for co-pyrolysis. Achieving high-grade liquid products requires careful selection and optimization of feedstock types and mixing ratios. Effective combinations include industrial sludge (Khodaparasti et al. 2023; Kumar et al. 2024) and coal (Wu et al. 2018), which have shown synergistic effects that enhance pyrolysis oil quality. However, not all feedstock is suitable for this process, like plastic material and coal, which requires higher temperatures for the pyrolysis process (Abnisa and Wan Daud 2014).

21.3.1 Pyrolysis Mechanism of Algae

It is revealed by Debiagi et al. (2017) that the thermal stability of different components of algae biomass differs. The first mass loss relates to the carbohydrate decomposition while the second one initiates at ~300 °C, which demonstrates protein fragmentation. Protein pyrolysis is a multiphase and multistep process (Fig. 21.3). First, depolymerization and degradation led to low molecular weight intermediate product species and gas (Debiagi et al. 2017). Successive pyrolysis of these intermediate species forms nitrogen-containing amides, pyrroles, pyridines, and diketopiperazines (Shen et al. 2024), together with other gas and tar species (Debiagi et al. 2017). C–C/C–N bond breakage and ring-condensation are the major paths toward imines and pyrroline-based compounds during the pyrolysis process, forming NO_x gases (Liu et al. 2021). Amino acid pyrolysis follows decarboxylation, removal of R-group, dehydration, and deamination reaction (Shen et al. 2024). It was demonstrated that different protein feed in TGA showed similar degradation profiles, confirmed the fact that the pyrolysis of these protein species was similar. Tar and gas

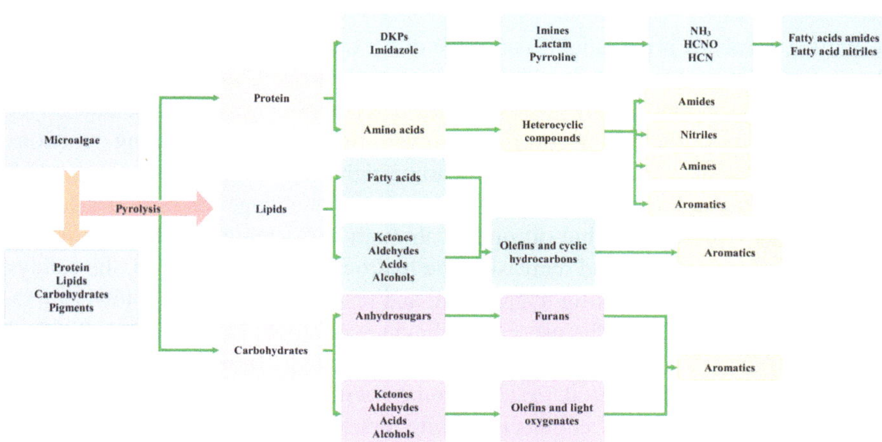

Fig. 21.3 Mechanism during the microalgal pyrolysis. (Data adapted with permission from (Shen et al. 2024), Copyright 2024 Elsevier)

products can have different compositions (Debiagi et al. 2017). Ammonia and carbon dioxide are formed in the first degradation, whereas further process forms tars, carbon monoxide, methane, and ethane (Hansson et al. 2004; Smith et al. 1974). The degradation mechanism of algae carbohydrates is similar to lignocellulosic biomass sugars pyrolysis. Sugars present in algae have higher structural similarities with hemicellulose than cellulose (Debiagi et al. 2017). The carbohydrates are initially decomposed to anhydrosugars and furfurals. Further successive reactions like decarbonylation, deoxygenation, dehydration, cracking, and rearrangement are followed to form oxygenated compounds (aldehydes, ketones, acids, phenols, and alcohols). Some light oxygenates and olefins are formed and cyclized to form O-based aromatic hydrocarbons (Shen et al. 2024). The pyrolysis of lipids occurs with the degradation of fatty acids with a series of intricate reactions, including cracking, decarbonylation, aromatization, isomerization, and other transformations. This complex series of reactions forms acids, alcohols, ketones, and aldehydes (Shen et al. 2024). Cracking reaction at elevated temperatures of triglycerides forms acrolein, together with two fatty acids and one aldehyde molecule according to the lumped reaction of extractive species (Shen et al. 2024; Meier et al. 2015). These intermediates are further converted into a variety of hydrocarbons, carbon monoxide via decarboxylation, cracking, decarbonylation, β-scission, and elimination reactions (Shen et al. 2024). Thermal analysis of lauric acid showed two distinct DTG peaks. Two endotherms signified one peak at low temperature due to melting and another peak at higher temperature because of vaporization (Shen and Alexander 1999). Overall, during algal pyrolysis, carbohydrates are degraded initially below 300 °C, while proteins and lipids decompose at higher temperatures. Decomposition of carbohydrates, proteins, and lipids is predicted by established kinetic models in temperature ranges of 164–497, 209–309, and 200–635 °C (Chen et al. 2018).

21.4 Catalytic Pyrolysis

Catalytic pyrolysis is essential because the direct thermal degradation of microalgal biomass often yields bio-oil with high oxygen, nitrogen, and moisture content, resulting in poor stability, low heating value, and corrosiveness. The use of catalysts helps to upgrade the vapors during pyrolysis by promoting cracking reactions, thereby producing higher-quality hydrocarbon-rich bio-oil that is more compatible with conventional fuels (Li et al. 2019). The major challenge arises in designing a catalyst that enhances the bio-oil quality through performing denitrogenation and deoxygenation reactions and increasing the hydrocarbon amount. Also, the catalyst helps in reducing the cracking temperature and reaction time and thus lowering the energy investment during the process (Rahman et al. 2018; Li et al. 2017, 2019). It was also observed that catalyst addition increases the gas formation and lowers the bio-oil generation (Tang et al. 2021). Zeolites, with a highly porous structure, are widely used in chemical engineering with cracking and dehydration as major reaction routes. H-ZSM-5 is the most widely used zeolite, as its acidic sites facilitate catalytic activity through the carbonium ion mechanism. Also, zeolite helps in enhancing decarboxylation, decarbonylation, and deoxygenation reactions. The catalytic performance of H-ZSM-5 is closely linked to its acidity, which is governed by the Si/Al ratio. A lower Si/Al ratio corresponds to higher acidity, thereby promoting greater carbon incorporation into bio-oil in the form of aromatic hydrocarbons (Thangalazhy-Gopakumar et al. 2012; Wang and Brown 2013; Pan et al. 2010). The strong acidity of H-ZSM-5 facilitates the breakdown of high-molecular-weight compounds into smaller intermediates, which subsequently diffuse through the zeolite pores and undergo further catalytic reactions, ultimately forming light hydrocarbons. This catalytic property makes zeolites particularly effective in microalgae pyrolysis, as they significantly enhance the production of aromatic hydrocarbons. Zeolites are also highly advantageous due to their deoxygenation capability, which reduces oxygenated compounds like phenols, aldehydes, and furans, thereby improving bio-oil quality. Zeolites are relatively cheap and can easily be regenerated to extract the deposited coke. Zeolites operate without the need for hydrogen or other reactive agents and are effective under atmospheric pressure. This makes them highly suitable for both in situ and ex situ catalytic conversion of pyrolysis vapors into partially or fully deoxygenated products, which can serve as blendstocks for upgrading into hydrocarbon transportation fuels (Wang et al. 2015). Among three different zeolites, like HY, H-ZSM-5, and Hβ, different effects were observed in the product distribution during catalytic pyrolysis. The stimulative effect followed the order: HY > Hβ > HZSM-5. This catalytic activity trend was consistent with surface area and micropore volume, suggesting that gas formation is closely linked to the pore structure of the zeolite. The abundant pore network likely enhanced secondary reactions of volatiles, thereby facilitating demethylation, decarbonization, and hydrogen condensation, which in turn promoted the release of methane, hydrogen, and carbon dioxide (Tang et al. 2021). Many investigations have reported that catalytic upgrading of microalgal bio-oil not only increased hydrocarbon yield but also decreased oxygen and nitrogen-containing compounds. For instance, *Chlorella vulgaris* with H-ZSM-5 improved the aromatic hydrocarbon from 0.9% to 25.8% (Thangalazhy-Gopakumar et al. 2012). Du et al. further demonstrated that increasing the catalyst-to-biomass ratio from 1:1 to 5:1 significantly enhanced aromatic hydrocarbon production. In a comparative study involving three zeolites (H-Y, H-ZSM-5, and H-B), H-ZSM-5 exhibited the highest aromatic compounds (18.13%). Moreover, metal-modified catalysts showed additional benefits; for example, copper-impregnated H-ZSM5 enhanced the aromatic content to 21.16%. These catalysts were also effective in reducing oxygen- and nitrogen-containing compounds (Du et al. 2013). However, nitriles were more difficult to hydrodenitrogenate even if zeolite catalyst used, due to the strong stability of C≡N bonds (Tang et al. 2021).

Two main catalytic approaches are employed: *in situ* and *ex situ* catalytic pyrolysis. In in situ catalytic pyrolysis, the catalyst is blended directly with the microalgae, allowing intimate contact between the feedstock and catalyst during thermal decomposition. This often improves reaction efficiency and enhances deoxygenation at the early stage (Li et al. 2019), but it can lead to catalyst deactivation due to coke deposition (Wang et al. 2015) and difficulties in catalyst recovery due to solid microalgae–biochar–catalyst mixing. On the other hand, in an ex situ catalytic pyrolysis, the pyrolysis vapors are first generated in a reactor and then passed through a separate catalyst bed (Li et al. 2019). This configuration provides better control over vapor–catalyst contact, reduces catalyst contamination from ash and minerals, and facilitates easier catalyst regeneration and separation. However, ex situ processes may require additional reactor design for a catalytic process and bring higher operational costs (Wang et al. 2015). Overall, in situ catalytic pyrolysis is relatively simpler and more cost-effective but suffers from faster catalyst deactivation and biomass–catalyst–biochar separation bottleneck, while ex situ offers higher product quality and longer catalyst life due to separate catalyst placement onto wool at the expense of added complexity (Li et al. 2019). Previous studies observed that an ex situ process gives more carbon yield compared to in situ. The ex situ process with *C. pyrenoidosa* and modified ZSM-5 zeolite resulted in 31.9% of light olefins at 650 °C in the presence of steam (Dong et al. 2013). The incorporation of an integrated vapor phase upgrading catalyst can be inexpensive if it removes the multireactor setups and bio-oil handling issues. As bio-oil from microalgae constitutes a complex class of mixtures that distinctly react with the catalyst, two catalytic reactors will be essential to acquire the required product composition (Li et al. 2019). An increment in gaseous product (major carbon monoxide) was observed together with the highest olefins after catalyst addition to microalgae in the 1:1 ratio at 700 °C. The rise in temperature transformed the aliphatic and nitrogeneous compound in bio-oil into carbon monoxide, carbon dioxide, methane, hydrogen, and small molecular weight olefins. Olefins (C_2H_4, C_2H_3, C_3H_6, C_4H_8) were increased from 29.17 to 38.23 wt% when temperature raised from 400 to 700 ° C. These outcomes suggested that Li-LSX-zeolite can be used as a catalyst for better microalgae bio-oil quality with enhanced aliphatics and aromatics that can be blended in gasoline and diesel. However, this catalyst positively increases gas formation at the cost of the bio-oil yield (Rahman et al. 2018). In another study, the Co/Fe_3O_4-H-ZSM-5 resulted in the lowest biochar formation, indicating the highest conversion efficiency compared to other supported metal catalysts. This represented approximately a 30% reduction in biochar formation compared to without catalyst reactions. In addition, Co/Fe_3O_4-HZSM5 achieved the highest bio-oil yield, whereas W/Fe_3O_4-H-ZSM-5 produced the lowest. With respect to gaseous products, the maximum gas yield was obtained using Zr/Fe_3O_4-H-ZSM-5, while Mo/Fe_3O_4-H-ZSM-5 yielded the minimum. Based on bio-oil yield, the catalysts can be ranked as Co/Fe_3O_4-H-ZSM-5 > Zr/Fe_3O_4-H-ZSM-5 > Mo/Fe_3O_4-H-ZSM-5 > W/Fe_3O_4-H-ZSM-5. Furthermore, the bio-oils derived from nutrient-stressed *S. obliquus* exhibited substantially lower oxygen contents (11.84 wt% for non-catalytic and 6.57–10.37 wt% for catalytic pyrolysis) (Mustapha et al. 2021).

21.5 Economic Perspective of Pyrolysis Process

Economic assessments have been implemented on microalgae catalytic pyrolysis and its upgradation, with the primary challenge in this route being the high moisture content in microalgae. Also, the economic feasibility of microalgal pyrolysis is largely influenced by both capital and operating costs, which include expenses related to reactor construction, feedstock harvesting and drying, as well as catalyst use and regeneration. While the technology has the potential to generate multiple revenue streams, the market value of its products varies considerably. Bio-oil can serve as a renewable fuel precursor, biochar holds promise for carbon sequestration and soil amendment, and chemicals derived from pyrolysis offer high-value applications. However, the profitability of the process is highly sensitive to several factors, including feedstock costs, efficiency of catalyst recycling, and the extent to which co-products can be valorized. Microalgae catalytic pyrolysis plant with 2000 metric tons per day (MTPD) could produce 21.4 million gallons of biofuel annually, at an estimated cost of \$6.80/gallon using thermal drying (TDCP), or \$5.61/gallon (\$1.49 per liter) with partial mechanical dewatering (MDCP). The total capital investment is projected at \$346 million for the TDCP and \$409 million for the MDCP. These findings indicate that incorporating mechanical dewatering equipment can help lower the overall cost of algal biofuel production. In the TDCP scenario, the reliance on natural gas for heating and drying raises operational expenses, resulting in a higher minimum fuel selling price (MFSP) (Fig. 21.4).

Sensitivity analysis assesses the influence of multiple variables on the MFSP of algal biofuels. Previous study showed that changes in product yield exert the strongest influence on MFSP, with a 20% increase reducing costs to \$1.51/L in the TDCP and \$1.21/L in the MDCP. Biomass and total capital cost are the important critical factors, with ±20% variations in these three parameters collectively shifting MFSP by more than 5%. Other factors, such as electricity price, have a relatively minor effect due to lower uncertainty, particularly in regulated markets. Overall, the analysis suggested that modest improvements in product yield or, to a lesser degree, reductions in feedstock cost could lead to substantial decreases in MFSP. It is important to note that the sensitivity analysis did not account for upstream

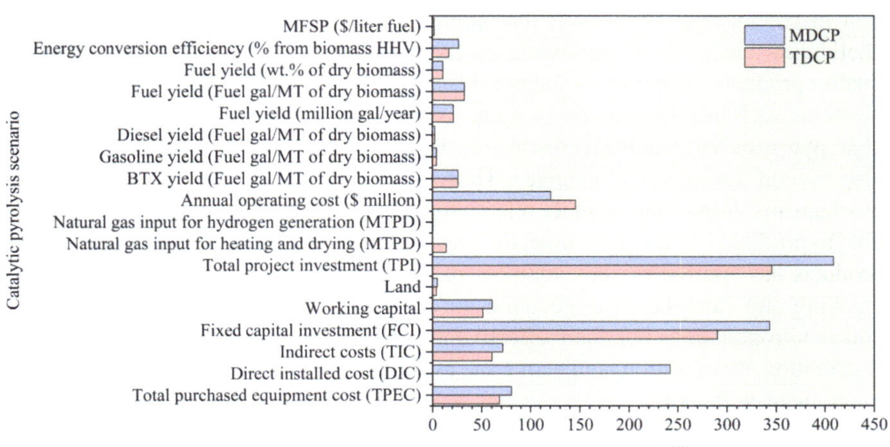

Fig. 21.4 Economic cost comparison of 2000 metric ton per day thermal drying and mechanical dewatering catalytic pyrolysis biorefinery scenarios (in \$ million). (Data adapted with permission from Thilakaratne et al. (2014), Copyright 2014 Elsevier)

uncertainties in algae cultivation and harvesting, owing to limited data availability. While variations in wastewater treatment costs appear unlikely to significantly affect MFSP, fluctuations in algal production costs are expected to translate into feedstock cost variability, which has a notable impact on overall fuel prices. Future research should address these uncertainties as more reliable and publicly available data on algal production systems become accessible (Thilakaratne et al. 2014). It was suggested by Quinn and Davis (2015) that the microalgal bio-fuel system is positively and extremely influenced by enhanced lipid productivity. Economic report analysis reports a selling cost of biofuel between \$1.64 and over \$30 gal^{-1} (Quinn and Davis 2015). Another investigation demonstrated the nonsuitability of microalgae toward biofuel generation in the considered cultivation, harvesting, and conversion process. However, co-production of biofuel with high-value products is more feasible (Jonker and Faaij 2013). Shemfe et al. (2017) performed an economic analysis of biofuel production via bio-oil zeolite upgrading. The integration of a cracking reactor with a single-stage regenerator (fluid cracking unit) has been presented for bio-oil cracking. The researchers also specified that the energy efficiency of two-stage regenerator was 54%, related to MFSP of \$7.48/gallon (gasoline equivalent) (Shemfe et al. 2017). Currently, the commercial status of microalgal pyrolysis is still at a developmental stage, with most efforts limited to laboratory and pilot-scale demonstrations. Overcoming these technical and economic barriers will be essential for transitioning the technology toward large-scale commercialization.

21.6 Conclusions

The pyrolysis process has emerged as a highly promising technology for converting microalgae into valuable chemicals and biofuels, drawing significant research attention in recent years. However, the application of algal biomass presents notable challenges due to the complex composition of algae, which often leads to poor-quality products. Moreover, algal pyrolysis carries the risk of releasing non-required elements such as nitrogen and oxygen. In the broader context of biomass research, algal pyrolysis has become a critical area of focus, with significant progress made but also several unresolved challenges. These include clarifying the detailed pyrolysis mechanisms, improving product selectivity, and reducing the generation of undesirable byproducts. At the same time, the presence of unfavorable elements in pyrolysis products has spurred the development of sustainable catalytic strategies. Catalytic cracking and catalytic hydrogenation are commonly employed for deoxygenation and denitrogenation, but their effectiveness is limited by catalyst structure, pore distribution, deactivation, and active site availability. In recent years, co-pyrolysis of microalgae with other feedstocks has gained considerable research interest, as it offers the dual benefits of enhancing bio-oil quality while facilitating waste management. Studies indicated that incorporating carbon or hydrogen-rich materials can lower the nitrogen content in bio-oil, thereby improving its suitability as a fuel. However, research in this area remains at an early stage. Future investigations should focus on elucidating the chemical interactions between feedstocks, uncovering the underlying mechanisms, and developing appropriate catalytic systems to further optimize the process. Additionally, advanced analytical methods such as TG-FTIR-MS are required to support or unlock the insight for the mechanism and role of catalysts toward the upgradation of bio-oil.

References

Abnisa F, Wan Daud WMA (2014) A review on co-pyrolysis of biomass: an optional technique to obtain a high-grade pyrolysis oil. Energy Convers Manage [Internet] 87:71–85. Available from: https://doi.org/10.1016/j.enconman.2014.07.007

Anastasakis K, Ross AB, Jones JM (2011) Pyrolysis behaviour of the main carbohydrates of brown macro-algae. Fuel [Internet] 90:598–607. Available from: https://doi.org/10.1016/j.fuel.2010.09.023

Borges FC, Xie Q, Min M, Muniz LAÔR, Farenzena M, Trierweiler JO et al (2014) Fast microwave-assisted pyrolysis of microalgae using microwave absorbent and HZSM-5 catalyst. Bioresour Technol [Internet] 166:518–526. Available from: https://doi.org/10.1016/j.biortech.2014.05.100

Campanella A, Harold MP (2012) Fast pyrolysis of microalgae in a falling solids reactor: effects of process variables and zeolite catalysts. Biomass Bioenerg [Internet] 46:218–232. Available from: https://doi.org/10.1016/j.biombioe.2012.08.023

Chaiwong K, Kiatsiriroat T, Vorayos N, Thararax C (2013) Study of bio-oil and bio-char production from algae by slow pyrolysis. Biomass Bioenerg [Internet] 56:600–606. Available from: https://doi.org/10.1016/j.biombioe.2013.05.035

Chaudhuri UR (2016) Crude Petroleum Oil, Fundamentals of petroleum and petrochemical engineering. Boca Raton: CRC Press Pg. 1.

Chen WH, Chu YS, Liu JL, Chang JS (2018) Thermal degradation of carbohydrates, proteins and lipids in microalgae analyzed by evolutionary computation. Energy Convers Manag [Internet] 160:209–219. Available from: https://doi.org/10.1016/j.enconman.2018.01.036

Debiagi PEA, Trinchera M, Frassoldati A, Faravelli T, Vinu R, Ranzi E (2017) Algae characterization and multistep pyrolysis mechanism. J Anal Appl Pyrolysis [Internet]. 128:423–436. Available from: https://doi.org/10.1016/j.jaap.2017.08.007

Deshavath NN, Goud VV, Veeranki VD (2021) Liquefaction of lignocellulosic biomass through biochemical conversion pathway: a strategic approach to achieve an industrial titer of bioethanol. Fuel. 287:119545

Dong X, Chen Z, Xue S, Zhang J, Zhou J, Liu Y et al (2013) Catalytic pyrolysis of microalga Chlorella pyrenoidosa for production of ethylene, propylene and butene. RSC Adv 3:25780–25787

Du Z, Ma X, Li Y, Chen P, Liu Y, Lin X et al (2013) Production of aromatic hydrocarbons by catalytic pyrolysis of microalgae with zeolites: catalyst screening in a pyroprobe. Bioresour Technol [Internet] 139:397–401. Available from: https://doi.org/10.1016/j.biortech.2013.04.053

Garforth AA, Ali S, Hernández-Martínez J, Akah A (2004) Feedstock recycling of polymer wastes. Curr Opin Solid State Mater Sci 8:419–425

Gong X, Zhang B, Zhang Y, Huang Y, Xu M (2014) Investigation on pyrolysis of low lipid microalgae chlorella vulgaris and dunaliella Salina. Energy Fuels. 28:95–103

Hansson KM, Samuelsson J, Tullin C, Åmand LE (2004) Formation of HNCO, HCN, and NH3 from the pyrolysis of bark and nitrogen-containing model compounds. Combust Flame 137:265–277

Ho DP, Ngo HH, Guo W (2014) A mini review on renewable sources for biofuel. Bioresour Technol 169:742–749

Jena U, Das KC (2011) Comparative evaluation of thermochemical liquefaction and pyrolysis for bio-oil production from microalgae. Energy Fuels. 25:5472–5482

Jonker JGG, Faaij APC (2013) Techno-economic assessment of micro-algae as feedstock for renewable bio-energy production. Appl Energy [Internet] 102:461–475. Available from: https://doi.org/10.1016/j.apenergy.2012.07.053

Khodaparasti MS, Khorasani R, Tavakoli O, Khodadadi AA (2023) Optimal co-pyrolysis of municipal sewage sludge and microalgae Chlorella Vulgaris: products characterization, synergistic effects, mechanism, and reaction pathways. J Clean Prod [Internet] 390:135991. Available from: https://doi.org/10.1016/j.jclepro.2023.135991

Kothari R, Prasad R, Kumar V, Singh DP (2013) Production of biodiesel from microalgae Chlamydomonas polypyrenoideum grown on dairy industry wastewater. Bioresour Technol 144:499–503

Kumar A, Jamro IA, Wang J, Ullah A, Kumari L, Cui B et al (2024) Co-pyrolysis of microalgae residue and sewage sludge: an in-depth characterization of kinetics, drivers, and gas-oil-char behaviors. J Anal Appl Pyrolysis [Internet] 179:106438. Available from: https://doi.org/10.1016/j.jaap.2024.106438

Kuppens T, Cornelissen T, Carleer R, Yperman J, Schreurs S, Jans M et al (2010) Economic assessment of flash co-pyrolysis of short rotation coppice and biopolymer waste streams. J Environ Manage [Internet] 91:2736–2747. Available from: https://doi.org/10.1016/j.jenvman.2010.07.022

Kusmayadi A, Huang CY, Kit Leong Y, Lu PH, Yen HW, Lee DJ et al (2023) Integration of microalgae cultivation and anaerobic co-digestion with dairy wastewater to enhance bioenergy and biochemicals production. Bioresour Technol [Internet] 376:128858. Available from: https://doi.org/10.1016/j.biortech.2023.128858

Lee HW, Jeon JK, Park SH, Jeong KE, Chae HJ, Park YK (2011) Catalytic pyrolysis of Laminaria japonica over nanoporous catalysts using Py-GC/MS. Nanoscale Res Lett 6:1–7

Li P, Chen X, Wang X, Shao J, Lin G, Yang H et al (2017) Catalytic upgrading of fast pyrolysis products with Fe-, Zr-, and co-modified zeolites based on Pyrolyzer-GC/MS analysis. Energy Fuels 31:3979–3986

Li F, Srivatsa SC, Bhattacharya S (2019) A review on catalytic pyrolysis of microalgae to high-quality bio-oil with low oxygeneous and nitrogenous compounds. Renew Sust Energ Rev 108: 481–497

Liu J, Zhao W, Yang S, Hu B, Xu M, Ma S et al (2021) Formation mechanism of NO x precursors during the pyrolysis of 2, 5-diketopiperazine based on experimental and theoretical study. Sci Total Environ [Internet] 801:149663. Available from: https://doi.org/10.1016/j.scitotenv.2021.149663

Maddi B, Viamajala S, Varanasi S (2011) Comparative study of pyrolysis of algal biomass from natural lake blooms with lignocellulosic biomass. Bioresour Technol [Internet] 102:11018–11026. Available from: https://doi.org/10.1016/j.biortech.2011.09.055

Meier HF, Wiggers VR, Zonta GR, Scharf DR, Simionatto EL, Ender L (2015) A kinetic model for thermal cracking of waste cooking oil based on chemical lumps. Fuel [Internet] 144:50–59. Available from: https://doi.org/10.1016/j.fuel.2014.12.020

Miao X, Wu Q (2004) High yield bio-oil production from fast pyrolysis by metabolic controlling of Chlorella protothecoides. J Biotechnol 110:85–93

Mustapha SI, Isa YM (2024) Co-pyrolysis of microalgae and sewage sludge over ZnO/MgO/CeO2/HZSM-5 catalyst for energy and water treatment application. J Environ Chem Eng [Internet] 12: 114955. Available from: https://doi.org/10.1016/j.jece.2024.114955

Mustapha SI, Rawat I, Bux F, Isa YM (2021) Catalytic pyrolysis of nutrient-stressed Scenedesmus obliquus microalgae for high-quality bio-oil production. Renew Energy 179:2036–2047

Niu Q, Du X, Li K, Ghysels S, Lu Q, Prins W et al (2024) Coke formation and mineral accumulation on HZSM-5/Al2O3 catalysts during in-situ catalytic fast pyrolysis of microalgae over multiple regeneration cycles. J Anal Appl Pyrolysis [Internet] 182:106701. Available from: https://doi.org/10.1016/j.jaap.2024.106701

Pan P, Hu C, Yang W, Li Y, Dong L, Zhu L et al (2010) The direct pyrolysis and catalytic pyrolysis of Nannochloropsis sp. residue for renewable bio-oils. Bioresour Technol [Internet] 101:4593–4599. Available from: https://doi.org/10.1016/j.biortech.2010.01.070

Quinn JC, Davis R (2015) The potentials and challenges of algae based biofuels: a review of the techno-economic, life cycle, and resource assessment modeling. Bioresour Technol [Internet] 184:444–452. Available from: https://doi.org/10.1016/j.biortech.2014.10.075

Rahman NAA, Fermoso J, Sanna A (2018) Effect of Li-LSX-zeolite on the in-situ catalytic deoxygenation and denitrogenation of Isochrysis sp. microalgae pyrolysis vapours. Fuel Process Technol [Internet] 173:253–261. Available from: https://doi.org/10.1016/j.fuproc.2018.01.020

Razzak SA, Khan M, Irfan F, Shah MA, Nawaz A, Hossain MM (2024) Catalytic co-pyrolysis and kinetic study of microalgae biomass with solid waste feedstock for sustainable biofuel production. J Anal Appl Pyrolysis [Internet] 183:106755. Available from: https://doi.org/10.1016/j.jaap.2024.106755

Ronsse F, van Hecke S, Dickinson D, Prins W (2013) Production and characterization of slow pyrolysis biochar: influence of feedstock type and pyrolysis conditions. GCB Bioenergy 5:104–115

Shemfe M, Gu S, Fidalgo B (2017) Techno-economic analysis of biofuel production via bio-oil zeolite upgrading: an evaluation of two catalyst regeneration systems. Biomass Bioenerg [Internet] 98:182–193. Available from: https://doi.org/10.1016/j.biombioe.2017.01.020

Shen L, Alexander KS (1999) A thermal analysis study of long chain fatty acids. Thermochim Acta 340–341:271–278

Shen W, He S, Mu M, Cao B, Wang S, Naqvi SR et al (2024) A comprehensive review on the intricate processes involved in algae pyrolysis mechanism and possible migration of undesirable chemical elements. J Anal Appl Pyrolysis [Internet] 177:106365. Available from: https://doi.org/10.1016/j.jaap.2024.106365

Shirazi Y, Viamajala S, Varanasi S (2020) In situ and ex situ catalytic pyrolysis of microalgae and integration with pyrolytic fractionation. Front Chem 8:1–12

Singh P, Mohanty K (2025) Bio-oil production via two-stage and direct hydrothermal liquefaction process from high-protein monoraphidium sp. KMC4: a Comparative Study of both processes and an insight into the reaction pathway. Energy Fuels 39:10109–10720

Sivaramakrishnan R, Suresh S, Kanwal S, Ramadoss G, Ramprakash B, Incharoensakdi A (2022) Microalgal biorefinery concepts' developments for biofuel and bioproducts: current perspective and bottlenecks. Int J Mol Sci MDPI 23(5):2623

Smith WT, Harris TB, Patterson JM (1974) Pyrolysis of soybean protein and an amino acid mixture having the same amino acid composition. J Agric Food Chem 22:480–483

Song W, Guo M (2012) Quality variations of poultry litter biochar generated at different pyrolysis temperatures. J Anal Appl Pyrolysis [Internet] 94:138–145. Available from: https://doi.org/10.1016/j.jaap.2011.11.018

Subagyono RRDJN, Putri SA, Manawan M, Mollah M, Nugroho RA, Gunawan R (2023) Catalytic pyrolysis of the green microalgae Botryococcus braunii over Ni/SBA-15 prepared by the ultrasonic-assisted sol-gel method. ACS Omega 8:8582–8595

Tang Z, Chen W, Chen Y, Hu J, Yang H, Chen H (2021) Preparation of low-nitrogen and high-quality bio-oil from microalgae catalytic pyrolysis with zeolites and activated carbon. J Anal Appl Pyrolysis [Internet] 159:105182. Available from: https://doi.org/10.1016/j.jaap.2021.105182

Thangalazhy-Gopakumar S, Adhikari S, Chattanathan SA, Gupta RB (2012) Catalytic pyrolysis of green algae for hydrocarbon production using H +ZSM-5 catalyst. Bioresour Technol [Internet] 118:150–157. Available from: https://doi.org/10.1016/j.biortech.2012.05.080

Thilakaratne R, Wright MM, Brown RC (2014) A techno-economic analysis of microalgae remnant catalytic pyrolysis and upgrading to fuels. Fuel [Internet] 128:104–112. Available from: https://doi.org/10.1016/j.fuel.2014.02.077

Wang K, Brown RC (2013) Catalytic pyrolysis of microalgae for production of aromatics and ammonia. Green Chem 15:675–681

Wang K, Johnston PA, Brown RC (2015) Comparison of in-situ and ex-situ catalytic pyrolysis in a micro-reactor system. Bioresour Technol [Internet] 173:124–131. Available from: https://doi.org/10.1016/j.biortech.2014.09.097

Wang X, Zhao B, Yang X (2016) Co-pyrolysis of microalgae and sewage sludge: biocrude assessment and char yield prediction. Energy Convers Manag [Internet] 117:326–334. Available from: https://doi.org/10.1016/j.enconman.2016.03.013

Wu Z, Yang W, Li Y, Yang B (2018) Co-pyrolysis behavior of microalgae biomass and low-quality coal: products distributions, char-surface morphology, and synergistic effects. Bioresour Technol [Internet] 255:238–245. Available from: https://doi.org/10.1016/j.biortech.2018.01.141

Xia C, Pathy A, Paramasivan B, Ganeshan P, Dhamodharan K, Juneja A et al (2022) Comparative study of pyrolysis and hydrothermal liquefaction of microalgal species: analysis of product yields with reaction temperature. Fuel [Internet] 311:121932. Available from: https://doi.org/10.1016/j.fuel.2021.121932

Yang C, Li R, Zhang B, Qiu Q, Wang B, Yang H et al (2019) Pyrolysis of microalgae: a critical review. Fuel Process Technol 186:53–72

Yu KL, Lau BF, Show PL, Ong HC, Ling TC, Chen WH et al (2017) Recent developments on algal biochar production and characterization. Bioresour Technol. Elsevier Ltd:2–11

Index

K. Mohanty et al. (eds.), *Advances in Algal Research*, https://doi.org/10.1007/978-981-95-7234-2

S

T

V

W

MIX
Papier aus verantwortungsvollen Quellen
Paper from responsible sources
FSC® C105338

If you have any concerns about our products,
you can contact us on
ProductSafety@springernature.com

In case Publisher is established outside the EU,
the EU authorized representative is:
Springer Nature Customer Service Center GmbH
Europaplatz 3, 69115 Heidelberg, Germany

Printed by Libri Plureos GmbH
in Hamburg, Germany